VOLUME NINETY ONE

VITAMINS AND HORMONES

Obesity

VOLUME NINETY ONE

VITAMINS AND HORMONES

Obesity

Editor-in-Chief

GERALD LITWACK, PhD

Toluca Lake, North Hollywood, California

ELSEVIER

AMSTERDAM • BOSTON • HEIDELBERG • LONDON
NEW YORK • OXFORD • PARIS • SAN DIEGO
SAN FRANCISCO • SINGAPORE • SYDNEY • TOKYO
Academic Press is an imprint of Elsevier

Cover photo credit:
Mitsutake, S., and Igarashi, Y.
Sphingolipids in Lipid Microdomains and Obesity
Vitamins and Hormones (2013) **91**, pp. 273–286

Academic Press is an imprint of Elsevier
32 Jamestown Road, London, NW1 7BY, UK
The Boulevard, Langford Lane, Kidlington, Oxford, OX5 1GB, UK
Linacre House, Jordan Hill, Oxford OX2 8DP, UK
225 Wyman Street, Waltham, MA 02451, USA
525 B Street, Suite 1800, San Diego, CA 92101-4495, USA

First edition 2013

ISBN: 978-0-12-407766-9
ISSN: 0083-6729

For information on all Academic Press publications
visit our website at store.elsevier.com

Printed and bound by CPI Group (UK) Ltd, Croydon, CR0 4YY
Transferred to Digital Printing, 2013\

Former Editors

CONTENTS

CONTRIBUTORS

Jesús Argente
Department of Endocrinology, Hospital Infantil Universitario Niño Jesús, Instituto de Investigación La Princesa; Department of Pediatrics, Universidad Autónoma de Madrid, and CIBER Fisiopatología de la Obesidad y Nutrición (CIBERobn), Instituto Carlos III, Madrid, Spain

Tobias Back
Department of Neurology, Saxon Hospital Arnsdorf, Arnsdorf/Dresden, Germany

Vicente Barrios
Department of Endocrinology, Hospital Infantil Universitario Niño Jesús, Instituto de Investigación La Princesa, and CIBER Fisiopatología de la Obesidad y Nutrición (CIBERobn), Instituto Carlos III, Madrid, Spain

Francesca Bertolini
Department of Agricultural and Food Sciences (DISTAL), University of Bologna, Bologna, Italy

Stephen Bloom
Division of Diabetes, Endocrinology, Metabolism, Hammersmith Hospital, Imperial College London, London, United Kingdom

Dongsheng Cai
Department of Molecular Pharmacology, Institute of Aging, Diabetes Research Center, Albert Einstein College of Medicine, Bronx, New York, USA

Guoxun Chen
Department of Nutrition, University of Tennessee at Knoxville, Knoxville, Tennessee, USA

Hyeongjin Cho
Department of Chemistry, Inha University, Incheon, Korea

Sung-E. Choi
Department of Molecular and Integrative Physiology, University of Illinois at Urbana, Urbana, Illinois, USA

Julie A. Chowen
Department of Endocrinology, Hospital Infantil Universitario Niño Jesús, Instituto de Investigación La Princesa, and CIBER Fisiopatología de la Obesidad y Nutrición (CIBERobn), Instituto Carlos III, Madrid, Spain

John F. Cryan
Food for Health Ireland; Laboratory of Neurogastroenterology, Alimentary Pharmabiotic Centre, and Department of Anatomy and Neuroscience, University College Cork, Cork, Ireland

Jude T. Deeney
Department of Medicine, Boston University School of Medicine, Boston, Massachusetts, USA

Gerald V. Denis
Flow Cytometry Core Facility; Nutrition Obesity Research Center; Cancer Research Center, and Department of Pharmacology and Experimental Therapeutics, Boston University School of Medicine, Boston, Massachusetts, USA

Timothy G. Dinan
Food for Health Ireland; Laboratory of Neurogastroenterology, Alimentary Pharmabiotic Centre, and Department of Psychiatry, University College Cork, Cork, Ireland

Luca Fontanesi
Department of Agricultural and Food Sciences (DISTAL), University of Bologna, Bologna, Italy

Salvatore Fusco
Laboratory of Cell Signaling, Institute of General Pathology, Catholic University Medical School, Rome, Italy

Laura L. Gathercole
Centre for Endocrinology, Diabetes and Metabolism, Institute of Biomedical Research, School of Clinical and Experimental Medicine, University of Birmingham, Queen Elizabeth Hospital, Edgbaston, Birmingham, United Kingdom

John T. Heiker
Department of Medicine, Endocrinology and Diabetes, Universität Leipzig, Leipzig, Germany

Yasuyuki Igarashi
Laboratory of Biomembrane and Biofunctional Chemistry, Faculty of Advanced Life Sciences, Hokkaido University, Sapporo, Japan

Mengxi Jiang
Center for Pharmacogenetics, University of Pittsburgh, Pittsburgh, Pennsylvania, USA

Jongsook Kim Kemper
Department of Molecular and Integrative Physiology, University of Illinois at Urbana, Urbana, Illinois, USA

Dong Hyun Kim
Department of Molecular and Integrative Physiology, University of Illinois at Urbana, Urbana, Illinois, USA

Nora Klöting
IFB Adiposity Diseases, Junior Research Group 2 "Animal models of obesity" Universität Leipzig, Leipzig, Germany

Gabriel Ángel Martos-Moreno
Department of Endocrinology, Hospital Infantil Universitario Niño Jesús, Instituto de Investigación La Princesa; Department of Pediatrics, Universidad Autónoma de Madrid, and CIBER Fisiopatología de la Obesidad y Nutrición (CIBERobn), Instituto Carlos III, Madrid, Spain

Susumu Mitsutake
Laboratory of Biomembrane and Biofunctional Chemistry, Faculty of Advanced Life Sciences, Hokkaido University, Sapporo, Japan

Stuart A. Morgan
Centre for Endocrinology, Diabetes and Metabolism, Institute of Biomedical Research, School of Clinical and Experimental Medicine, University of Birmingham, Queen Elizabeth Hospital, Edgbaston, Birmingham, United Kingdom

Dominic S. Ng
Keenan Research Center, Li Ka Shing Knowledge Institute, St Michael's Hospital, and Department of Medicine, University of Toronto, Toronto, Ontario, Canada

Zhen Pang
Department of Immuno-inflammation, Sanofi US, Bridgewater, New Jersey, USA

Giovambattista Pani
Laboratory of Cell Signaling, Institute of General Pathology, Catholic University Medical School, Rome, Italy

Sofia Chiatamone Ranieri
Clinical Chemistry, Department of Laboratory Medicine, Laboratory and Endocrinology Unit, Azienda Ospedaliera ASMN, Istituto di Ricovero e Cura a Carattere Scientifico, Reggio Emilia, Italy

Liangyou Rui
Department of Molecular & Integrative Physiology, University of Michigan Medical School, Ann Arbor, Michigan, USA

Roman Sankowski
Department of Neurology, Philipps-University, Marburg Germany

Ryuichiro Sato
Department of Applied Biological Chemistry, The University of Tokyo, Bunkyo, Tokyo, Japan

Harriët Schellekens
Food for Health Ireland, and School of Pharmacy, University College Cork, Cork, Ireland

Rebecca Scott
Division of Diabetes, Endocrinology, Metabolism, Hammersmith Hospital, Imperial College London, London, and Queen Elizabeth II Hospital, Howlands, Welwyn Garden City, Hertfordshire, United Kingdom

Tricia Tan
Division of Diabetes, Endocrinology, Metabolism, Hammersmith Hospital, Imperial College London, London, United Kingdom

Jeremy W. Tomlinson
Centre for Endocrinology, Diabetes and Metabolism, Institute of Biomedical Research, School of Clinical and Experimental Medicine, University of Birmingham, Queen Elizabeth Hospital, Edgbaston, Birmingham, United Kingdom

Fangnian Wang
IQuum Inc, Marlborough, MA 01752, USA

Jing Wang
Beijing Key Laboratory of Gene Resource and Molecular Development, College of Life Sciences, Beijing Normal University, Beijing, China

Yaroslav Winter
Department of Neurology, Philipps-University, Marburg Germany

Lei Xiao
Key Laboratory for Cell Proliferation and Regulation Biology, Ministry of Education, Biomedicine Research Institute and College of Life Sciences, Beijing Normal University, Beijing, and Cardiovascular Research Center, Xi'an Jiaotong University School of Medicine, Xi'an, China

Wen Xie
Center for Pharmacogenetics, University of Pittsburgh, Pittsburgh, Pennsylvania, USA

Yonggong Zhai
Key Laboratory for Cell Proliferation and Regulation Biology, Ministry of Education, Biomedicine Research Institute, and Beijing Key Laboratory of Gene Resource and Molecular Development, College of Life Sciences, Beijing Normal University, Beijing, China

Yingjiang Zhou
Department of Molecular & Integrative Physiology, University of Michigan Medical School, Ann Arbor, Michigan, USA

PREFACE

Obesity is a nationwide epidemic in the United States. It is interrelated with Type 2 diabetes and other diseases, such as cardiovascular disease, certain cancers, osteoarthritis, and asthma. The chief causes are excessive caloric intake in the forms of fat and sugar, endocrine disorders, lack of exercise, genetics, certain medications, and psychiatric illness. In view of the fact that fat tissue is now understood to be an endocrine gland in its own right and a great deal of basic biochemistry is involved, this book focuses on fundamental aspects of research into this subject.

Chapter 1 is entitled "Hormonal regulation of lipogenesis" by L.L. Gathercole, S.A. Morgan, and J.W. Tomlinson. Next, Y. Winter, R. Sankowski, and T. Back focus on genetics in "Genetic determinants of obesity and related vascular diseases" (Chapter 2). "*Brd2* gene disruption causes 'metabolically healthy' obesity": epigenetic and chromatin-based mechanisms that uncouple obesity from Type 2 diabetes" is the subject of F. Wang, J.T. Denney, and G.V. Denis (Chapter 3). "The *TBC1D1* gene: structure, function, and association with obesity and related traits" is discussed in Chapter 4 by L. Fontanesi and F. Bertolini. In Chapter 5, J.T. Heiker and N. Klöting report on "Replication initiator 1 (Repin 1) in adipose tissue function and human obesity."

Moving more into the endocrine aspects, "Adipokines in childhood obesity" is the topic of G.Á. Martos-Mareno, V. Barrios, J.A. Chowen, and J. Argente (Chapter 6). "Gut hormones and obesity: physiology and therapies" is reviewed by R. Scott, T. Tan, and S. Bloom (Chapter 7). D. Cai discusses "Neuroinflammation in overnutrition-induced diseases" in Chapter 8. S.C. Ranieri, S. Fusco, and G. Pani focus on "$P66^{shcA}$: linking mammalian longevity with obesity-induced insulin resistance" in Chapter 9. "The emerging role of constitutive androstane receptor (CAR) and its crosstalk with liver X receptors (LXRs) and peroxisome proliferator-activated receptor α (PPARα) in lipid metabolism" is the subject of L. Xiao, J. Wang, M. Jiang, W. Xie, and Y. Zhai (Chapter 10).

Other aspects include "Lecithin cholesterol acyltransferase deficiency protects from diet-induced insulin resistance and obesity—novel insights from mouse models" by D.S. Ng (Chapter 11). Next, S. Mitsutake and Y. Igarashi report on "Sphingolipids in lipid microdomains and obesity" in Chapter 12. "Ghrelin: at the interface of obesity and reward" is reviewed by H. Schellekens, T.G. Dinan, and J.F. Cryan (Chapter 13). G. Chen and Z. Pang introduce "Endocannabinoids and obesity" in Chapter 14. "Lipocalin 13 regulation of glucose and lipid metabolism in obesity" is offered by Y. Zhou and L. Rui (Chapter 15).

With regard to specific enzymes, J.K. Kemper, S.-E. Choi, and D.H. Kim report on "Sirtuin 1 deacetylase: a key regulator of hepatic lipid metabolism" (Chapter 16) and H. Cho contributes "Protein tyrosine phosphatase 1B (PTP1B) and obesity" (Chapter 17).

Finally, in Chapter 18, R. Sato reviews "Nomilin as an anti-obesity and anti-hyperglycemic agent."

The illustration on the cover of this book is reproduced from Fig. 12.4 of Chapter 12.

The process of assembly and preparation of the volume for the press was aided by Sarah Latham and Mary Ann Zimmerman.

Gerald Litwack
North Hollywood, California
18 September, 2012

CHAPTER ONE

Hormonal Regulation of Lipogenesis

Laura L. Gathercole, Stuart A. Morgan, Jeremy W. Tomlinson[1]
Centre for Endocrinology, Diabetes and Metabolism, Institute of Biomedical Research, School of Clinical and Experimental Medicine, University of Birmingham, Queen Elizabeth Hospital, Edgbaston, Birmingham, United Kingdom
[1]Corresponding author: e-mail address: j.w.tomlinson@bham.ac.uk

Contents

Abstract

Obesity has reached epidemic proportions with severe heath consequences including type 2 diabetes, nonalcoholic fatty liver disease, and premature cardiovascular mortality. Understanding the biological processes that govern fat deposition in a tissue-specific manner is therefore crucial if we are to try to design novel and efficacious treatments that might limit fat accumulation and improve metabolic phenotype and clinical prognosis.

Vitamins and Hormones, Volume 91
ISSN 0083-6729
http://dx.doi.org/10.1016/B978-0-12-407766-9.00001-8

Lipid accumulation within a given cell type represents a balance between synthesis, mobilization, and utilization. Common endocrine conditions characterized by hormonal excess and deficiency are often associated with profound abnormalities in body composition and fat deposition. This undoubtedly reflects the complex regulation of lipid metabolism by endocrine factors. In this review, we will outline the current literature that has investigated the hormonal regulation of lipogenesis. This is a complex field, and in many studies, its assessment has been oversimplified with a focus on individual hormones acting in isolation and this bears little relationship to the *in vivo* situation where multiple hormones act in concert. Further, regulation may be different between rodents and humans and this will be explored.

Limitation of lipid accumulation still represents a valid therapeutic target, and it is possible that manipulation of hormonal action has the potential to offer a new therapeutic horizon.

1. INTRODUCTION

The current epidemic of obesity and its associated complications, including type 2 diabetes, nonalcoholic fatty liver disease (NAFLD), and premature cardiovascular mortality has heightened the importance of understanding the processes that regulate adipose tissue mass. Increasing adiposity is a complex process driven by both increases in adipocyte size and number and reflects an imbalance between energy intake, storage, and utilization. Hormonal regulation of these processes is critical as evidenced by abnormalities of lipid metabolism and alterations in fat mass seen in many common endocrine conditions. Lipid accumulation through the process of *de novo* lipogenesis (DNL) is a key regulator of lipid accumulation within key metabolic target tissues including adipose tissue and liver. In this chapter, we will review the current understanding of the hormonal regulation of DNL. Where data are available, we will detail the molecular mechanisms underpinning the hormonal regulation of DNL, relating this to the observed phenotypes of hormonal deficiency or excess.

2. DEFINING LIPOGENESIS

In times of calorific excess, energy is stored as triacylglyceride (TAG) in the adipose tissue. TAGs are composed of three fatty acid chains esterified to a glycerol molecule. These fatty acids can originate from two sources, dietary (exogenous) lipids or generation from glucose by DNL, via the metabolic intermediate acetyl-CoA (Fig. 1.1).

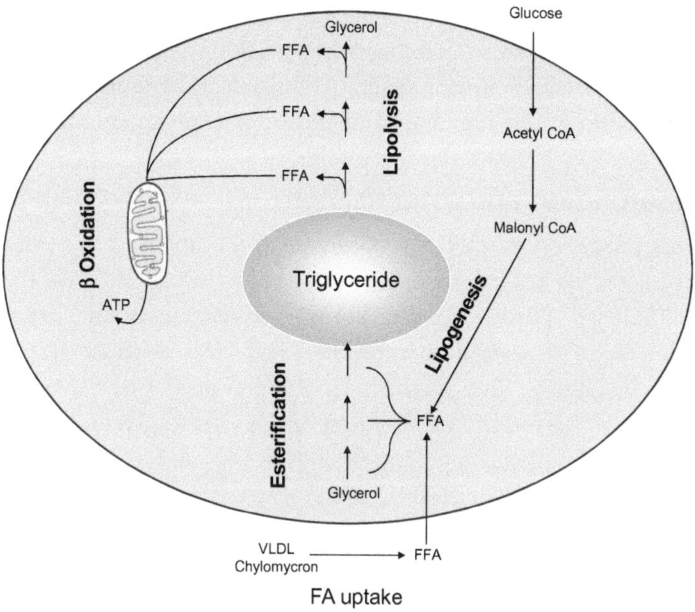

Figure 1.1 Pathways of lipid metabolism. VLDL, very low density lipoproteins; FFA, free fatty acids; ATP, adenosine triphosphate. (See Color Insert.)

2.1. Substrate generation/availability

The substrate for DNL is acetyl-CoA which is generated within the mitochondrion, predominantly from the metabolism of pyruvate. In mitochondria, pyruvate enters the tricarboxylic (TCA) cycle with consequent citrate generation. When mitochondrial adenosine triphosphate (ATP) levels are low, citrate continues through the TCA cycle to generate ATP. However, if ATP levels are high, isocitrate dehydrogenase is inhibited by ATP causing citrate to accumulate which is then able to cross the mitochondrial membrane. Once in the cytosol, ATP citrate lyase (ACL) cleaves citrate releasing oxaloacetate and generating the cytosolic acetyl-CoA that is required for DNL.

2.2. *De novo* lipogenesis

During fatty acid synthesis, carbons from acetyl-CoA are incorporated into a growing fatty acid chain. The first and rate-limiting step is the ATP-dependent carboxylation of acetyl-CoA to malonyl-CoA by acetyl-CoA carboxylase 1 (ACC1) (Lane, Moss, & Polakis, 1974). ACC1 requires the vitamin biotin as a coenzyme, which is covalently bound to a lysyl residue of the carboxylase. The malonyl-CoA produced is subsequently converted by a multistep reaction to the 16-carbon fatty acid palmitate by fatty acid synthase (FAS) (Ruderman, Saha, & Kraegen, 2003). FAS is a multicatalytic, dimeric enzyme, carrying out several reactions and requires the reducing power of nicotinamide adenine dinucleotide phosphate (NADPH).

2.3. Desaturation and esterification

Fatty acids produced by DNL are saturated and converted to monounsaturated fatty acids, the key substrates for TAG generation, by stearoyl-CoA desaturase 1 (SCD1). This is the rate-limiting desaturase in the reaction and activity has been suggested to promote DNL and decrease fat oxidation (Hulver et al., 2005). The biosynthesis of TAG takes place in microsomes and is the successive esterification of fatty acyl-CoAs to glycerol-3-phosphate by glycerol-3-phosphate acyltransferase (GPAT) and diacylglycerol acyltransferase (DGAT) (Leung, 2001). The glycerol-3-phosphate required for TAG synthesis is produced from dihydroxyacetone phosphate (DHAP) by glycerol-3-phosphate dehydrogenase. In the fed state, when glucose is plentiful, DHAP is generated from glucose using the initial reactions in the glycolytic pathway and thus glucose availability, the substrate for both fatty acids and glycerol, can limit the rate of DNL.

3. LIPOGENESIS IN METABOLIC TARGET TISSUES

3.1. Liver

In man, the predominant site of DNL is the liver. Fatty acids that are produced are esterified to glycerol to form TAG, packaged with cholesteryl esters, cholesterol, phospholipid, and protein to form very low density lipoproteins (VLDL) which are then exported to peripheral tissues. At the cell surface, TAG within VLDL is hydrolyzed by lipoprotein lipase to release nonesterified fatty acids (NEFA) and glycerol (Bjorkegren et al., 1996), NEFAs can then enter the cell, where they are either oxidized to generate energy or reesterified and stored as TAG.

Tracer methods have been used to estimate the contribution of DNL to VLDL TAG. In healthy individuals, it has been estimated at approximately 2–10% (Diraison & Beylot, 1998; Diraison, Pachiaudi, & Beylot, 1997; Hellerstein, Neese, & Schwarz, 1993); however, this contribution can greatly increase under certain nutritional or hormonal states, including high carbohydrate diets and hyperinsulinemia (Schwarz, Linfoot, Dare, & Aghajanian, 2003). Hepatic DNL is increased in *ad libitum*-fed obese subjects

(Diraison, Dusserre, Vidal, Sothier, & Beylot, 2002) and in NAFLD, the contribution of DNL to VLDL has been estimated to be as high as 26% and therefore contributes significantly to liver TAG accumulation (Diraison, Moulin, & Beylot, 2003; Donnelly et al., 2005).

3.2. Adipose

Although the key enzymes involved in DNL are present in human white adipose tissue (WAT), it is generally thought that human adipocytes carry out only low levels of DNL, in contrast to rodents. This is thought to be due to low basal levels of ACL, the enzyme that generates cytosolic acetyl-CoA for fatty acid synthesis (Shrago, Spennetta, & Gordon, 1969; Swierczynski et al., 2000). DNL in adipose tissue is estimated to contribute to only about 2% of total daily lipid generation (Guo, Cella, Baum, Ravussin, & Schoeller, 2000). However, studies have shown that in adipose tissue, 20% of newly deposited TAG comes from DNL, although fatty acids synthesized within the liver and delivered via VLDL contribute to this. Importantly, accumulation appears to be above the rates of hepatic delivery alone and this difference may represent adipose DNL (Strawford, Antelo, Christiansen, & Hellerstein, 2004). Although basal rates of DNL in adipose are low they are, at least *in vitro*, regulated and increased by hormones (Gathercole et al., 2011; Wang, Voy, et al., 2004). While few studies have examined hormonal regulation *in vivo* as under high carbohydrate hypercaloric conditions, adipose is an important contributor to whole body DNL (Aarsland, Chinkes, & Wolfe, 1997).

Recently, DNL has been implicated in the regulation of preadipocyte differentiation and may therefore be important not only in hypertrophic but also in hyperplastic adipose tissue expansion (Collins et al., 2011).

3.3. Muscle

The enzymatic pathways required for DNL are also present in skeletal muscle cells although basal levels of DNL are thought to be low. However, in a cell culture model, hyperglycemic conditions increased glucose incorporation into NEFA 88% and TAG 44% (Aas, Kase, Solberg, Jensen, & Rustan, 2004). Peroxisome proliferator-activated receptor gamma coactivator-1alpha (PGC-1α) has a clear role in the catabolism of muscle lipids (Wu, Puigserver, et al., 1999), and recently, it has been suggested to be important in anabolic lipid metabolism and lipid refueling in muscle after exercise (Summermatter, Baum, Santos, Hoppeler, & Handschin, 2010).

4. REGULATION OF LIPOGENESIS

4.1. Transcriptional regulation

4.1.1 Sterol regulatory element-binding protein-1c

Sterol regulatory element-binding protein-1c (SREBP-1c) is a member of the basic helix–loop–helix-leucine zipper family of transcription factors. Synthesized as a membrane-bound precursor, it is located in the endoplasmic reticulum (ER) and when cleaved releases a soluble N terminal fragment that translocates to the nucleus to activate transcription (Wang, Sato, Brown, Hua, & Goldstein, 1994). Two key proteins are required for the regulation of SREBP-1c cleavage, SREBP cleavage-activating protein (SCAP) and insulin-induced gene 1 protein (INSIG1). SCAP is a membrane-bound protein that interacts with the newly synthesized SREBP precursor. When sterols are abundant in the ER, the sterol-bound SCAP is kept in a conformation that permits binding to INSIG1, which anchors the complex within the ER. In the absence of sterol binding, SCAP does not bind INSIG1 and escorts SREBP-1c to the golgi, where resident proteases release the active SREBP-1c from the membrane. In addition, SREBP-1c can be inhibited by phosphorylation (Yellaturu et al., 2009). A recent study has shown that in hepatocytes AMP-activated protein kinase (AMPK) is able to interact with and phosphorylate SREBP-1c. Phosphorylation by AMPK inhibited proteolytic cleavage and nuclear translocation of SREBP-1c, repressing expression of lipogenic genes and reducing DNL. Further, hepatic activation of AMPK, in part through its phosphorylation of SREBP-1c, protected against hepatic steatosis, hyperlipidemia, and accelerated atherosclerosis in rodents fed a high fat, high sucrose diet (Li, Xu, et al., 2011).

SREBP-1c primarily regulates the expression of genes involved in DNL and triglyceride synthesis, including, ACL, ACC, FAS, SCD1, and GPAT (Kim et al., 1998; Magana, Lin, Dooley, & Osborne, 1997; Shimomura, Shimano, Korn, Bashmakov, & Horton, 1998; Tabor, Kim, Spiegelman, & Edwards, 1998). SREBP-1c also increases expression of enzymes that generate NADPH, malic enzyme (ME), glucose 6-phosphate dehydrogenase (G6PD), and phosphoglucaonate dehydrogenase (6PDG). The availability of NADPH is important as it supplies the reducing power for FAS (Shimomura et al., 1998). The adipogenic transcription factor peroxisome proliferator-activated receptor gamma (PPARγ) is a direct target of SREBP-1c (Kim & Spiegelman, 1996), and as such, SREBP-1c has been shown to have an important role in adipocyte differentiation (Klemm et al., 2001).

4.1.2 Liver X receptor

The liver X receptors (LXRs) are important transcriptional regulators of SREBP-1c. They are members of a nuclear receptor superfamily that heterodimerize with retinoid X receptor. Two LXR isoforms have been identified, LXRα and LXRβ, that are activated by oxysterols (metabolites of cholesterol). LXRα null animals have reduced expression of SREBP-1c, FAS, ACC, and SCD1 (Peet et al., 1998; Repa et al., 2000). Additionally, LXRs have been shown to have a direct effect on lipogenic gene expression that does not require SREBP-1c (Joseph et al., 2002).

4.1.3 Carbohydrate response element-binding protein

Carbohydrate response element-binding protein (chREBP) is a glucose responsive transcription factor. Like SREBP-1c, it is a member of the bHLH/ZIP family of transcription factors and forms heterodimers with another member of the bHLHZip family Max-like protein X (Mlx). It binds genes containing carbohydrate response elements (choREs) (Stoeckman, Ma, & Towle, 2004) and chREBP activates a similar pattern of genes to SREBP-1c, but importantly does not affect the expression of SREBP-1c. Activated genes are involved in glycolysis, DNL, and NADPH generation and include pyruvate kinase, pyruvate dehydrogenase, ACL, ACC, FAS, G6PDH, ME (Iizuka, Bruick, Liang, Horton, & Uyeda, 2004; Ishii, Iizuka, Miller, & Uyeda, 2004; Ma, Robinson, & Towle, 2006).

4.1.4 Peroxisome proliferator-activated receptor gamma

The differentiation of preadipocytes to mature adipocytes is a tightly regulated process. A key transcription factor in adipocyte differentiation is PPARγ which acts as a master regulator of adipogenesis (Kubota et al., 1999; Rosen, Walkey, Puigserver, & Spiegelman, 2000; Tontonoz, Hu, & Spiegelman, 1994). PPARy induces the expression of a number of lipogenic genes (Paulauskis & Sul, 1988) across differentiation, and as its levels remain high in the mature adipocyte, it is likely that it is involved in the regulation of DNL as well as adipogenesis (Wu, Rosen, et al., 1999).

4.1.5 Peroxisome proliferator-activated receptor gamma coactivator-1alpha

PGC-1α has been shown to regulate DNL in muscle. Overexpression of PGC-1α increased DNL in cultured myocytes (Espinoza, Boros, Crunkhorn, Gami, & Patti, 2010). In a PGC-1α transgenic mouse model, DNL was enhanced by a number of mechanisms: LXR-dependent activation of FAS expression, increased substrate availability through enhanced glucose uptake, and increased G6PDH activity and cofactor generation (Summermatter et al., 2010).

4.2. Posttranscriptional regulation

The conversion of acetyl-CoA to malonyl-CoA by ACC1 is the rate-limiting step in DNL. As well as being regulated at the transcriptional level, ACC1 protein levels are regulated through proteosomal degradation and activity is regulated by both phosphorylation and allosteric activation.

The E3 ligase, constitutive photomorphogenic protein, working in conjunction with the pseudokinase, Tribbles 3 targets ACC for ubiquitination and is believed to be an important regulatory mechanism to control ACC protein levels in the fasting state (Du, Herzig, Kulkarni, & Montminy, 2003; Qi et al., 2006).

ACC is allosterically activated by the acetyl-CoA precursor citrate. The inactive form of ACC is a protomer (dimer), and citrate binding causes these dimers to polymerize activating ACC. The end product of the pathway, long-chain fatty acyl-CoA, causes depolymerization and inactivation of ACC (Halestrap & Denton, 1974; Munday, Campbell, Carling, & Hardie, 1988).

ACC1 activity is regulated by phosphorylation at a number of sites; phosphorylation at Ser79, 1200, and 1215 are inhibitory to ACC1 activity (Scott, Norman, Hawley, Kontogiannis, & Hardie, 2002). These sites are phosphorylated by AMPK which in turn is activated by a decrease in cyclic adenosine monophosphate (cAMP) levels, reflecting intracellular ATP availability (Hardie, 1992). In addition, ACC1 can be phosphorylated at Ser29; however, the role of phosphorylation at this position remains to be determined (Haystead, Campbell, & Hardie, 1988; Heesom, Moule, & Denton, 1998).

5. HORMONAL REGULATION

5.1. Insulin

Insulin plays a central role in the regulation of carbohydrate and lipid metabolism, and is probably the most important hormonal factor regulating DNL (Fig. 1.2). Insulin has an anabolic effect on metabolism—promoting energy storage and inhibiting energy release. Insulin promotes DNL in both liver and adipose tissue by multiple mechanisms. First, insulin stimulates glucose uptake into key metabolic tissue through a phospatidylinositol-3 kinase-dependent pathway that involves activation of protein kinase B

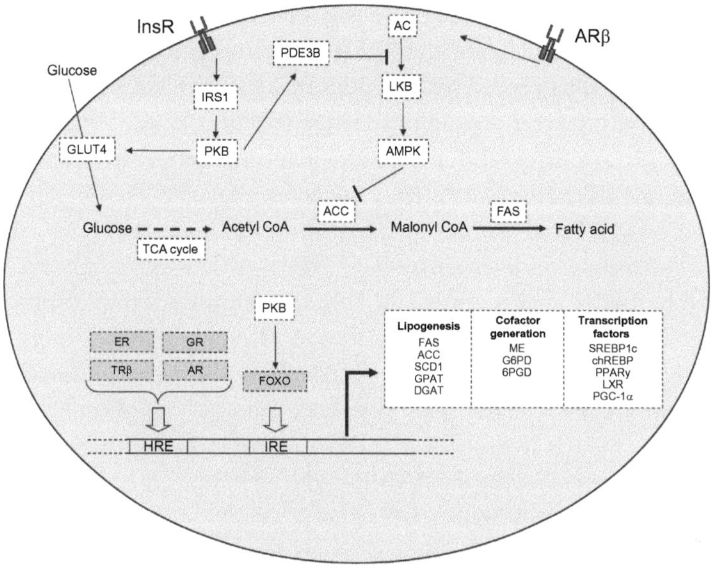

Figure 1.2 Transcriptional and posttranscriptional regulation of *de novo* lipogenesis. InsR, insulin receptor; IRS, insulin receptor substrate; GLUT4, glucose transporter 4; PKB, protein kinase B; PDE3B, phosphodiesterase-3B; AC, adenyl cyclase; LKB, liver kinase B; ARβ, β-adrenergic receptor; AMPK, AMP-activated protein kinase; ACC, acetyl-CoA carboxylase; FAS, fatty acid synthase; ER, estrogen receptor; AR, androgen receptor; TRβ, thyroid receptor β; GR, glucorticoid receptor; SCD1, stearoyl-Coenzyme A desaturase 1; GPAT, glycerol-3-phosphate acyltransferase; DGAT, diacylglycerol acyltransferase; ME, malic enzyme; G6PD, glucose 6 phosphate dehydrogenase; 6PDG, phosphoglucaonate dehydrogenase; SREBP-1c, sterol regulatory element-binding protein-1c; chREBP, carbohydrate response element-binding protein; PPARγ, peroxisome proliferator-activated receptor gamma; LXR, liver X receptor; PGC-1α, peroxisome proliferator-activated receptor gamma coactivator-1alpha.

(PKB) and recruitment of GLUT4 glucose transporters to the plasma membrane, therefore providing substrate for *de novo* fatty acid synthesis. Separately, by increasing SREBP1 expression, insulin increases mRNA levels of FAS and ACC1 (Azzout-Marniche et al., 2000; Foretz et al., 1999; Kim et al., 1998). Insulin also increases transcriptional activation of SREBP1 via MAP-kinase-dependent phosphorylation (Roth et al., 2000). Upstream stimulatory factors (USFs) are ubiquitous bHLH–leucine zipper transcription factors that help initiate transcription by binding to enhancer box (E-box) elements, which are DNA sequences located upstream of a gene. An E-box is present in the promoter of FAS, and mutations within this E-box prevent USF binding and abolish insulin-dependent activation of the FAS promoter (Casado, Vallet, Kahn, & Vaulont, 1999). cAMP response element-binding protein (CREB) positively regulates FAS and ACC1 expression. Insulin stimulates both CREB phosphorylation and transcriptional activation, and thus represents an additional mechanism by which insulin promotes DNL in liver and adipose tissue (Klemm et al., 1998).

Insulin also regulates DNL in a posttranslational manner by modulating the activity of AMPK. In the fasted state, a low intracellular ATP:ADP ratio leads to increased cAMP levels with consequent protein kinase A (PKA) activation, which in turn activates liver kinase B1 (LKB1). AMPK is phosphorylated and activated by LKB1, and increased AMPK activation leads to phosphorylation and the inactivation of ACC1 (Witters & Kemp, 1992; Witters, Watts, Daniels, & Evans, 1988). Insulin blocks AMPK-induced suppression of DNL via activation of PKB. PKB phosphorylates and activates phosphodiesterase-3B (PDE3B), which hydrolyses cAMP to inactive 5′AMP, decreasing PKA activation and thus relieving the AMPK-induced inhibition on ACC, to permit DNL (Kitamura et al., 1999).

5.2. Glucagon

The effects of glucagon are largely reciprocal to those of insulin with levels peaking during fasting, in response to low circulating glucose levels. It is a catabolic hormone, promoting energy release and inhibiting energy storage (Jiang & Zhang, 2003). Glucagon potently inhibits DNL by downregulating SREBP1 expression in hepatocytes, leading to decreased FAS transcription (Foretz et al., 1999), an effect mediated by p38 mitogen-activated protein kinase (Xiong et al., 2007). Glucagon also inhibits DNL by activating AMPK, leading to phosphorylation and inactivation of ACC in both liver and adipose tissue (Hardie, 1992; Holland, Hardie, Clegg, & Zammit, 1985; Kimball, Siegfried, & Jefferson, 2004; Longuet et al., 2008; Swenson & Porter, 1985).

5.3. Glucocorticoids

Glucocorticoids (GCs) are steroid hormones synthesized in the adrenal cortex. In humans, the most potent circulating GC is cortisol (corticosterone in rodents), and its synthesis and secretion are under tight regulation by the hypothalamic-pituitary-adrenal axis. Acutely, GCs are released as part of acute stress response and have many metabolic actions. Chronic GC excess leads to the well-characterized phenotype of Cushing's syndrome, which includes visceral adiposity, insulin resistance, and in some cases, hepatic steatosis.

In human adipocytes, GCs decrease DNL, and this is paralleled by an increased inhibitory phosphorylation of ACC1/2 (Gathercole et al., 2011).

Although AMPK is known to negatively regulate ACC1 activity, several studies have shown GCs to decrease activating phosphorylation of AMPK in human adipose tissue, suggesting that posttranslational regulation of ACC1 by GCs is independent of AMPK in adipocytes (Christ-Crain et al., 2008; Gathercole et al., 2011; Kola et al., 2008). In rodent adipocytes, GC administration reduces FAS expression as well as ACC and FAS activity. Further, rats that have undergone adrenalectomy have elevated ACC and FAS activity in adipose tissue (Volpe & Marasa, 1975). As such, it is clear that GCs have an inhibitory effect upon DNL in adipose tissue, although these observations are likely to most closely reflect the fasting state when insulin levels are low with concomitant high circulating GC levels.

By contrast, GCs alone appear to be without effect upon hepatic DNL, although these studies have largely been performed in rodent models. Basal lipogenic rates in GC-treated primary rat hepatocytes are similar to those of untreated cells (Amatruda, Danahy, & Chang, 1983). Similarly, hepatic ACC and FAS activity and expression are unchanged in rat livers following GC administration or following adrenalectomy (Volpe & Marasa, 1975). However, a single study using cultured human hepatocytes found GCs to increase ACC1 mRNA levels in a dose-dependent manner (Zhao et al., 2010). In contrast to there action in adipose tissue, GCs increase AMPK activation in the liver (Viana et al., 2006), and this maybe important in understanding the tissue-specific regulation of DNL.

Despite the common use of single hormonal treatments in cell culture systems, *in vivo* there is undoubtedly a complex regulation of DNL by the hormonal milieu, and therefore understanding the interactions between hormones acting together is important. Insulin has a stimulatory effect upon DNL, and GCs further augment this effect in adipose tissue (Fig. 1.3) (Gathercole et al., 2011; Wang, Jones, et al., 2004). This may arise as a result of GC-induced insulin sensitization in human adipose tissue with increased glucose uptake and subsequent fatty acid generation (Gathercole, Bujalska, Stewart, & Tomlinson, 2007; Tomlinson et al., 2010). Similarly, GCs have been found to potentiate the actions of insulin in hepatocytes (Amatruda et al., 1983; Zhao et al., 2010). These observations do raise the intriguing possibility of differential nutritional regulation of DNL. In the fasting state, low insulin levels and high GC levels switch off DNL in adipose tissue though serine phosphorylation of ACC1, decreasing fuel storage, and increasing NEFA availability for other more metabolically active tissues. Conversely, in the fed state, insulin levels are high, and insulin and GC may act together to promote lipid storage in both liver

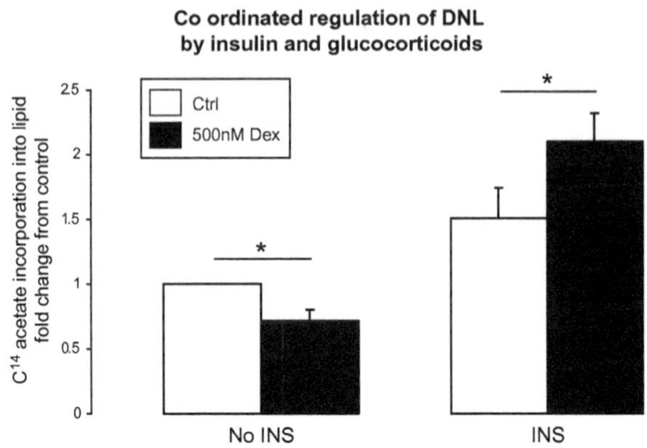

Figure 1.3 In the absence of insulin, dexamethasone (Dex) decreases lipogenesis as measured by 1-[^{14}C]-acetate incorporation into lipid in differentiated Chub-s7 cells. Insulin stimulates lipogenesis, and the actions of insulin are augmented in the presence of 500 nM Dex. $^{*} = p < 0.05$. *Adapted from Gathercole et al. (2011).*

and adipose tissue. However, extrapolation from *in vitro* models to *in vivo* observations is fraught with difficulty and these hypotheses need to be tested in dedicated, well-designed clinical studies.

In addition to GC generated by the adrenal gland, the endoluminal enzyme 11β-hydroxysteroid dehydrogenase type 1 (11β-HSD1) activates GCs at an intracellular level from the inactive precursor cortisone (11-dehydrocorticosterone in rodents). Critically, this enzyme is highly expressed in key metabolic tissues including adipose tissue and liver. In cultured adipocytes, the inhibitory effect of cortisone upon DNL is blocked in the presence of a selective 11β-HSD1 inhibitor (Gathercole et al., 2011). Further, hepatic DNL is reduced in mice injected with a an antisense oligonucleotide to knockdown 11β-HSD1 expression in the liver, parallel by decreased SREBP1 and FAS protein levels (Li, Hernandez-Ono, Crooke, Graham, & Ginsberg, 2011). As such, manipulation of GC availability through selective 11β-HSD1 inhibition has therapeutic potential by altering lipid homeostasis in key metabolic target tissues.

5.4. Thyroid hormone

Thyroid hormones, thyroxine (T4) and triiodothyronine (T3) are produced by the thyroid gland. The predominant circulating hormone is T4 which is converted to the more active T3 intracellularly by deiodinases. Thyroid hormones increase resting energy expenditure (REE) and, as such, patients with thyroid dysfunction exhibit changes in body weight (al Adsani, Hoffer, &

Silva, 1997). Patients with hypothyroidism have a decreased metabolic rate, decreased thermogenesis, and a resultant gain in weight. Patients with hyperthyroidism have an increased metabolic rate, losing weight despite increased appetite (Kim, 2008).

Rates of DNL in the livers from hyperthyroid compared to hypothyroid rats are increased 16-fold. Compared to euthyroid rats, both hyperthyroid and hypothyroid animals had decreased DNL in the retroperitineal fat depot, where as in the subcutaneous depot, it was enhanced under hyperthyroid conditions. Interestingly, in brown adipose tissue (BAT), DNL inversely correlated with thyroid status. Consistent with these observations, levels of ACC and FAS were highest in the livers of hyperthyroid rats, whereas in BAT the highest levels were found in the hypothyroid rats. (Blennemann, Leahy, Kim, & Freake, 1995; Blennemann, Moon, & Freake, 1992; Freake, Schwartz, & Oppenheimer, 1989). While it may seem counterintuitive that there is increased DNL in hyperthyroidism, which is associated with weight loss, this maybe an important contributant to increased REE. It has been proposed that increased REE may arise as a consequence of futile substrate cycling stimulating both gluconeogenesis and glycolysis and lipogenesis and lipolysis (Oppenheimer, Schwartz, Lane, & Thompson, 1991; Shulman, Ladenson, Wolfe, Ridgway, & Wolfe, 1985).

T3 generation is increased after feeding (Davidson & Chopra, 1979) and decreased during fasting (Vagenakis et al., 1977), and it is thought that T3 may be important in regulating dietary effects on DNL. In the liver, the combination of high carbohydrate diet and T3 administration synergistically increased ME, FAS, G6PD, and 6PDG expression while in WAT only, the expression of ME responded synergistically (Mariash, Kaiser, Schwartz, Towle, & Oppenheimer, 1980).

The genomic actions of thyroid hormones are regulated via two nuclear receptors, thyroid receptor α and β. The activation of the lipogenic transcription profile of thyroid hormone is mediated through TRβ. TRβ-null mice fail to regulate expression of liver and adipose FAS, ACC, and ME in TRβ-null mice (Weiss et al., 1998).

Thyroid hormone increases expression of hepatic and adipose chREBP, through TRβ binding to the LXR2 recognition site, but acts independently of LXR. Both FAS or SCD1 contain a choRE, and it is predicted that thyroid hormones could enhance their expression through chREBP (Hashimoto et al., 2009). FAS and ME expression are also regulated by TRβ binding to promoter thyroid response elements (Gauthier et al., 2010; Radenne et al., 2008).

5.5. Catacholamines (adrenaline)

Adrenaline is released from the adrenal medulla as a "fight or flight" response to emotional and physical stresses, including starvation and exercise. Adrenaline drives catabolic metabolism, increasing lipolysis and releasing NEFAs for oxidation (Gordon & Cherkes, 1958).

Once released, adrenaline binds the adrenergic receptors α and β; its impact upon lipid metabolism is mediated via the β-adrenergic receptor. β-Adrenergic receptors are G-protein coupled receptors, and adrenaline binding leads to activation of adenyl cyclase raising intracellular cAMP levels with consequent phosphorylation and activation of AMPK (Omar, Zmuda-Trzebiatowska, Manganiello, Goransson, & Degerman, 2009) (Fig. 1.2).

In vitro, adrenaline increases phosphorylation of AMPK and ACC and increased expression of PPARy in cultured epididymal adipose tissue (Wan et al., 2010). In an animal model with high circulating adrenaline, AMPK activation is enhanced, and there is increased phosphorylation of ACC in WAT. In this model, ACC phosphorylation was blocked by both a β-adrenergic receptor antagonist and AMPK inhibition (Hattori et al., 2010). Through these mechanisms, adrenaline, in addition to increasing lipolysis, may be important in inhibiting anabolic lipid metabolism and preventing futile substrate cycling in times of stress.

5.6. Growth hormone

Growth hormone (GH) is a peptide hormone secreted by the somatotroph cells within the anterior pituitary. GH reduces DNL in adipose tissue, resulting in significant loss of fat mass (Etherton, 2000). These effects are mediated by multiple mechanisms. GH decreases insulin sensitivity, therefore preventing the delivery of the lipogenic substrate into the cell. This effect appears to be driven by GH-induced expression of suppressor of cytokine signaling-1, which interacts with the insulin receptor and prevents downstream activation of PKB (Fasshauer et al., 2004). GH knockout mice have higher basal levels of ACC activity compared to wild-type animals, and administration of GH to both wild-type and GH knockout mice causes a marked reduction of ACC activity (Adamafio & Ng, 1984; Bornstein, Ng, Heng, & Wong, 1983; Ng, Adamafio, & Graystone, 1990). These observations have been endorsed in a variety of animal models where GH reproducibly decreases both FAS and ACC expression and activity in adipose tissue (Beswick & Kennelly, 1998; Borland, Barber, Travers, & Vernon, 1994; Harris et al., 1993; Jose, Alvares, Delgado, Coutinho, & Lanna, 2006; Liu, Grant, Kim, & Mills, 1994).

5.7. Sex steroids

Clinical observations have suggested a role for sex steroid hormones in both men and women in the regulation of adiposity, through either direct or indirect physiological mechanisms. Changes in the hormonal levels at menopause are associated with an increase in total adiposity (Poehlman, 2005). Around the menopause, circulating estrogen levels decline more rapidly than androgen levels (Liu et al., 2001), creating a period of relative androgen excess. This has lead to the suggestion that the relative balance between estrogens and androgens may be a critical regulator of lipid metabolism (Pasquali, 2006).

5.7.1 Testosterone/dihydrotestosterone

The principle androgens in males and females are testosterone and biologically more potent, dihydrotestosterone (DHT). Androgens appear to have a stimulatory effect upon DNL. Mice treated with DHT have increased body weight and visceral fat mass, associated with increased TAG accumulation. Further, DHT treatment increased expression of SREBP2 and FAS, and also decreased inactivating phosphorylation of ACC (McInnes, Corbould, Simpson, & Jones, 2006). Consistent with these findings, both testosterone and DHT inhibit basal AMPK activation in rodent and human adipose cell lines by downregulating LKB expression (McInnes, Brown, Hunger, & Simpson, 2012). In contrast, testosterone administration to men decreases body fat content. The discrepancies between these observations highlight shortfalls in our current understanding of the precise molecular mechanisms by which testosterone and DHT regulate fat mass and lipid accumulation.

5.7.2 Estrogen

Estrogen, in contrast to testosterone and DHT, appears to inhibit DNL. Menopausal woman treated with estrogen have decreased expression of the transcription factor PPARγ, as well as reduced FAS and ACC1 mRNA levels in subcutaneous abdominal adipose. Further, these gene expression changes correlate with plasma triglyceride levels (Lundholm et al., 2008). In rodents, estrogen administration to both male and female rats lowers DNL in adipose tissue (Hansen, Fahmy, & Nielsen, 1980). Further, aromatase knockout mice which are estrogen deficient have increased hepatic triglyceride accumulation, paralleled by elevated hepatic expression of FAS and ACC1; these effects are reversed following estrogen replacement (Hewitt, Pratis, Jones, & Simpson, 2004). Mechanistically (in contrast to testosterone and DHT), estrogen increases LKB expression in both human and rodent adipocyte cell lines, resulting in increased AMPK activation (McInnes et al., 2012).

5.7.3 Dehydroepiandrosterone

Dehydroepiandrosterone (DHEA) is an adrenal-derived androgen that in its sulfated form (DHEAS) the most abundant circulating adrenal steroid in humans. Despite its abundance, its true physiological role is unclear. Although lower mammals secrete almost negligible amounts of DHEA and its metabolites, exogenous administration decreases lipid accumulation in both liver and adipose tissue in rodents (McIntosh & Berdanier, 1991), and this may represent suppression of DNL in these tissues. In support of this, DHEA inhibits glucose incorporation into intracellular lipids in mice liver (Marrero, Prough, Frenkel, & Milewich, 1990) and decreases hepatic SREBP1 and ACC expression in chickens (Tang, Ma, Huang, Miao, & Zou, 2009). However, future detailed analysis is needed to define the role of DHEA in the regulation of DNL.

5.7.4 Progesterone

There is some evidence to suggest that progesterone upregulates DNL in adipose tissue (which may be a physiologically relevant mechanism to preserve energy fat stores required for lactation in adipocytes located in breast tissue). Rats treated with progesterone have increased body weight and adiposity (Mendes, Madon, & Flint, 1985), and progesterone increased SREBP1 mRNA dose-dependently in primary cultured rat preadipocytes. Further, the membrane-bound and mature nuclear forms of SREBP1 protein accumulated in preadipocytes cultured with progesterone (Lacasa, Le, Ferre, & Dugail, 2001).

5.8. Appetite-regulating hormones

Regulation of appetite is a complex process, involving hypothalamic neuropeptides and peripheral signals such as adipokines and gut hormones. Appetite regulation is essential in order to balance energy intake, metabolic requirements, and energy expenditure. Dysregulation of appetite control may contribute to fat accumulation as evidenced by monogenic causes of obesity. However, there is increasing evidence to suggest that a number of appetite-regulating hormones influence DNL in liver and adipose tissue, as discussed below.

5.8.1 Ghrelin

Ghrelin is a peptide hormone secreted by P/D1 cells lining the stomach and the ε-cells of the pancreas and promotes food intake and adiposity (Nakazato et al., 2001). Several *in vivo* studies have shown ghrelin to increase TAG accumulation in both liver and adipose tissue, paralleled by increased mRNA expression of ACC and FAS (Barazzoni et al., 2005). Additionally, ghrelin stimulates DNL by reducing AMPK activating phosphorylation in both liver and adipose tissue (Barazzoni et al., 2005; Kola et al., 2005). In the liver, ghrelin increases total, but not phosphorylated ACC protein levels, indicating a relative reduction of phosphorylated inactive ACC form (Barazzoni et al., 2005). Rats treated with intracerebroventricular (ICV) ghrelin injection have increased expression of ACC1, FAS, and SCD1. Additionally, expression of these enzymes is decreased in ghrelin-deficient mice (Theander-Carrillo et al., 2006). Ghrelin also acts as the endogenous ligand for the GH secretagogue receptor, which when activated drives GH release from the hypothalamus (Kojima et al., 1999), which has the potential to counteract the prolipogenic effects of ghrelin.

5.8.2 Leptin

Leptin is an adipose tissue-derived peptide hormone that has a number of actions including appetite control and regulation of nervous system and reproductive function. Simple obesity is associated with leptin resistance (Munzberg & Myers, 2005), but there is growing evidence to suggest that leptin limits fat storage, not only by inhibiting food intake but also by inhibiting DNL in liver and adipose tissue (Ramsay, 2003). The underpinning mechanism appears to be a leptin-induced downregulation of SREBP1, leading to decreased expression of FAS and ACC (Nogalska, Sucajtys-Szulc, & Swierczynski, 2005; Soukas, Cohen, Socci, & Friedman, 2000). Importantly, leptin also acts centrally in the regulation of DNL, through hypothalamic neural circuits and the autonomic nervous system, in a similar manner to ghrelin (Gallardo et al., 2007).

5.8.3 Glucagon-like peptide-1

Glucagon-like peptide-1 (GLP1) is synthesized in the L-cells of the small intestine, derived from the transcription product of the proglucagon gene. Its major physiological role is the stimulation of glucose-dependent insulin secretion from the pancreas, following oral glucose ingestion. However, it has many other effects including enhancement of pancreatic β-cell proliferation, protection against apoptosis, inhibition of food intake, and decreased gastric emptying. Despite increasing insulin secretion (Ahren & Pacini, 1999), GLP1 appears to inhibit DNL in the liver by increasing hepatic cAMP levels, activating AMPK leading to inactivation of ACC (Ben Shlomo et al., 2011; Ding, Saxena, Lin, Gupta, & Anania, 2006; Svegliati-Baroni et al., 2011). GLP-1 has also been reported to decrease hepatic mRNA expression of SREBP1, SCD1, FAS, and ACC (Ding et al., 2006). Administration of a GLP1 analog to obese humans resulted in significant weight loss (Astrup et al., 2011). Mechanistically, although GLP1 increases DNL in adipose tissue (via modulating AMPK activity) (Vendrell et al., 2011; Villanueva-Penacarrillo, Marquez, Gonzalez, Diaz-Miguel, & Valverde, 2001), it also stimulates lipolysis (Sancho et al., 2006), and this may contribute to a net fat loss in conjunction with the other complex actions of GLP-1.

5.8.4 Neuropeptide Y

Neuropeptide Y (NPY) is an abundant neuropeptide in the central nervous system. The main role of NPY is to increase food intake. ICV or hypothalamus-specific administration of NPY to normal rodents leads to multiple defects including increased DNL in WAT and liver through increased ACC activity (Zarjevski, Cusin, Vettor, Rohner-Jeanrenaud, & Jeanrenaud, 1993). Leptin inhibits *NPY* gene expression, representing an additional mechanism by which leptin suppresses DNL (Morrison, Morton, Niswender, Gelling, & Schwartz, 2005).

In summary, appetite-regulating hormones have complex actions, acting both locally and centrally (via central nervous system neuroendocrine circuits). Many of these hormones work in concert, as exemplified by NPY and ghrelin, both of which stimulate food intake while driving DNL. It may sound counterintuitive to stimulate synthesis of fatty acids for storage when substrate availability is low; however, this may be a priming mechanism that allows more rapid assimilation fatty acids when substrate becomes available in the fed state. The converse may me true for leptin, which inhibits food intake and suppresses DNL.

6. CONCLUSIONS

The hormonal regulation of DNL is complex, with the simultaneous interaction of many hormones acting peripherally and centrally in a time-specific manner. Assessing the impact of individual hormones using cell culture systems is without doubt an over simplification, and the *in vivo* situation is more complex. As has been shown with GCs, regulation of lipid homeostasis may differ in the fed and fasted state. The importance of understanding the hormonal regulation of DNL is clear; it not only allows one to understand the metabolic phenotype characterized by hormonal excess or deficiency but also may point to novel target therapies that may regulate hormonal availability and action. Taking into account the clinical context of the global burden of obesity and its associated complications, morbidity and mortality, understanding and therapeutic targeting of the processes that govern lipid accumulation in metabolic target tissues remains a priority.

REFERENCES

Aarsland, A., Chinkes, D., & Wolfe, R. R. (1997). Hepatic and whole-body fat synthesis in humans during carbohydrate overfeeding. *The American Journal of Clinical Nutrition*, *65*(6), 1774–1782.

Aas, V., Kase, E. T., Solberg, R., Jensen, J., & Rustan, A. C. (2004). Chronic hyperglycaemia promotes lipogenesis and triacylglycerol accumulation in human skeletal muscle cells. *Diabetologia*, *47*(8), 1452–1461.

Adamafio, N. A., & Ng, F. M. (1984). Effects of growth hormone on lipogenesis and glucose oxidation in genetically GH-deficient mice. *Molecular and Cellular Endocrinology*, *37*(2), 241–244.

Ahren, B., & Pacini, G. (1999). Dose-related effects of GLP-1 on insulin secretion, insulin sensitivity, and glucose effectiveness in mice. *The American Journal of Physiology*, *277*(6 Pt. 1), E996–E1004.

al Adsani, H., Hoffer, L. J., & Silva, J. E. (1997). Resting energy expenditure is sensitive to small dose changes in patients on chronic thyroid hormone replacement. *The Journal of Clinical Endocrinology and Metabolism*, *82*(4), 1118–1125.

Amatruda, J. M., Danahy, S. A., & Chang, C. L. (1983). The effects of glucocorticoids on insulin-stimulated lipogenesis in primary cultures of rat hepatocytes. *The Biochemical Journal*, *212*(1), 135–141.

Astrup, A., Carraro, R., Finer, N., Harper, A., Kunesova, M., Lean, M. E., et al. (2011). Safety, tolerability and sustained weight loss over 2 years with the once-daily human GLP-1 analog, liraglutide. *International Journal of Obesity (London)*, *36*, 843–854.

Azzout-Marniche, D., Becard, D., Guichard, C., Foretz, M., Ferre, P., & Foufelle, F. (2000). Insulin effects on sterol regulatory-element-binding protein-1c (SREBP-1c) transcriptional activity in rat hepatocytes. *The Biochemical Journal*, *350*, 2389–2393.

Barazzoni, R., Bosutti, A., Stebel, M., Cattin, M. R., Roder, E., Visintin, L., et al. (2005). Ghrelin regulates mitochondrial-lipid metabolism gene expression and tissue fat distribution in liver and skeletal muscle. *American Journal of Physiology. Endocrinology and Metabolism*, *288*(1), E228–E235.

Ben Shlomo, S., Zvibel, I., Shnell, M., Shlomai, A., Chepurko, E., Halpern, Z., et al. (2011). Glucagon-like peptide-1 reduces hepatic lipogenesis via activation of AMP-activated protein kinase. *Journal of Hepatology*, *54*(6), 1214–1223.

Beswick, N. S., & Kennelly, J. J. (1998). The influence of bovine growth hormone and growth hormone releasing factor on acetyl-CoA carboxylase and fatty acid synthase in primiparous Holstein cows. *Comparative Biochemistry and Physiology-Part C: Pharmacology, Toxicology and Endocrinology*, *120*(2), 241–249.

Bjorkegren, J., Packard, C. J., Hamsten, A., Bedford, D., Caslake, M., Foster, L., et al. (1996). Accumulation of large very low density lipoprotein in plasma during intravenous infusion of a chylomicron-like triglyceride emulsion reflects competition for a common lipolytic pathway. *Journal of Lipid Research*, *37*(1), 76–86.

Blennemann, B., Leahy, P., Kim, T. S., & Freake, H. C. (1995). Tissue-specific regulation of lipogenic mRNAs by thyroid hormone. *Molecular and Cellular Endocrinology*, *110*(1–2), 1–8.

Blennemann, B., Moon, Y. K., & Freake, H. C. (1992). Tissue-specific regulation of fatty acid synthesis by thyroid hormone. *Endocrinology*, *130*(2), 637–643.

Borland, C. A., Barber, M. C., Travers, M. T., & Vernon, R. G. (1994). Growth hormone inhibition of lipogenesis in sheep adipose tissue: Requirement for gene transcription and polyamines. *The Journal of Endocrinology*, *142*(2), 235–243.

Bornstein, J., Ng, F. M., Heng, D., & Wong, K. P. (1983). Metabolic actions of pituitary growth hormone. I. Inhibition of acetyl CoA carboxylase by human growth hormone and a carboxyl terminal part sequence acting through a second messenger. *Acta Endocrinologica (Copenh)*, *103*(4), 479–486.

Casado, M., Vallet, V. S., Kahn, A., & Vaulont, S. (1999). Essential role in vivo of upstream stimulatory factors for a normal dietary response of the fatty acid synthase gene in the liver. *The Journal of Biological Chemistry*, *274*(4), 2009–2013.

Christ-Crain, M., Kola, B., Lolli, F., Fekete, C., Seboek, D., Wittmann, G., et al. (2008). AMP-activated protein kinase mediates glucocorticoid-induced metabolic changes: A novel mechanism in Cushing's syndrome. *The FASEB Journal*, *22*(6), 1672–1683.

Collins, J. M., Neville, M. J., Pinnick, K. E., Hodson, L., Ruyter, B., van Dijk, T. H., et al. (2011). De novo lipogenesis in the differentiating human adipocyte can provide all fatty acids necessary for maturation. *Journal of Lipid Research*, *52*(9), 1683–1692.

Davidson, M. B., & Chopra, I. J. (1979). Effect of carbohydrate and noncarbohydrate sources of calories on plasma 3,5,3'-triiodothyronine concentrations in man. *The Journal of Clinical Endocrinology and Metabolism*, *48*(4), 577–581.

Ding, X. K., Saxena, N. K., Lin, S. B., Gupta, N., & Anania, F. A. (2006). Exendin-4, a glucagon-like protein-1 (GLP-1) receptor agonist, reverses hepatic steatosis in ob/ob mice. *Hepatology*, *43*(1), 173–181.

Diraison, F., & Beylot, M. (1998). Role of human liver lipogenesis and reesterification in triglycerides secretion and in FFA reesterification. *The American Journal of Physiology*, *274*(2 Pt. 1), E321–E327.

Diraison, F., Dusserre, E., Vidal, H., Sothier, M., & Beylot, M. (2002). Increased hepatic lipogenesis but decreased expression of lipogenic gene in adipose tissue in human obesity. *American Journal of Physiology Endocrinology* and *Metabolism*, *282*, E46–E51.

Diraison, F., Moulin, P., & Beylot, M. (2003). Contribution of hepatic de novo lipogenesis and reesterification of plasma non esterified fatty acids to plasma triglyceride synthesis during non-alcoholic fatty liver disease. *Diabetes and Metabolism*, *29*(5), 478–485.

Diraison, F., Pachiaudi, C., & Beylot, M. (1997). Measuring lipogenesis and cholesterol synthesis in humans with deuterated water: Use of simple gas chromatographic/mass spectrometric techniques. *Journal of Mass Spectrometry*, *32*(1), 81–86.

Donnelly, K. L., Smith, C. I., Schwarzenberg, S. J., Jessurun, J., Boldt, M. D., & Parks, E. J. (2005). Sources of fatty acids stored in liver and secreted via lipoproteins in patients with nonalcoholic fatty liver disease. *The Journal of Clinical Investigation*, *115*(5), 1343–1351.

Du, K., Herzig, S., Kulkarni, R. N., & Montminy, M. (2003). TRB3: A tribbles homolog that inhibits Akt/PKB activation by insulin in liver. *Science*, *300*(5625), 1574–1577.

Espinoza, D. O., Boros, L. G., Crunkhorn, S., Gami, H., & Patti, M. E. (2010). Dual modulation of both lipid oxidation and synthesis by peroxisome proliferator-activated receptor-gamma coactivator-1alpha and -1beta in cultured myotubes. *The FASEB Journal*, *24*(4), 1003–1014.

Etherton, T. D. (2000). The biology of somatotropin in adipose tissue growth and nutrient partitioning. *The Journal of Nutrition*, *130*(11), 2623–2625.

Fasshauer, M., Kralisch, S., Klier, M., Lossner, U., Bluher, M., Klein, J., et al. (2004). Insulin resistance-inducing cytokines differentially regulate SOCS mRNA expression via growth factor- and Jak/Stat-signaling pathways in 3T3-L1 adipocytes. *The Journal of Endocrinology, 181*(1), 129–138.

Foretz, M., Pacot, C., Dugail, I., Lemarchand, P., Guichard, C., Le, L. X., et al. (1999). ADD1/SREBP-1c is required in the activation of hepatic lipogenic gene expression by glucose. *Molecular and Cellular Biology, 19*(5), 3760–3768.

Freake, H. C., Schwartz, H. L., & Oppenheimer, J. H. (1989). The regulation of lipogenesis by thyroid hormone and its contribution to thermogenesis. *Endocrinology, 125*(6), 2868–2874.

Gallardo, N., Bonzon-Kulichenko, E., Fernandez-Agullo, T., Molto, E., Gomez-Alonso, S., Blanco, P., et al. (2007). Tissue-specific effects of central leptin on the expression of genes involved in lipid metabolism in liver and white adipose tissue. *Endocrinology, 148*(12), 5604–5610.

Gathercole, L. L., Bujalska, I. J., Stewart, P. M., & Tomlinson, J. W. (2007). Glucocorticoid modulation of insulin signaling in human subcutaneous adipose tissue. *The Journal of Clinical Endocrinology and Metabolism, 92*(11), 4332–4339.

Gathercole, L. L., Morgan, S. A., Bujalska, I. J., Hauton, D., Stewart, P. M., & Tomlinson, J. W. (2011). Regulation of lipogenesis by glucocorticoids and insulin in human adipose tissue. *PLoS One, 6*(10), e26223.

Gauthier, K., Billon, C., Bissler, M., Beylot, M., Lobaccaro, J. M., Vanacker, J. M., et al. (2010). Thyroid hormone receptor beta (TRbeta) and liver X receptor (LXR) regulate carbohydrate-response element-binding protein (ChREBP) expression in a tissue-selective manner. *The Journal of Biological Chemistry, 285*(36), 28156–28163.

Gordon, R. S., Jr., & Cherkes, A. (1958). Production of unesterified fatty acids from isolated rat adipose tissue incubated in vitro. *Proceedings of the Society for Experimental Biology and Medicine, 97*(1), 150–151.

Guo, Z. K., Cella, L. K., Baum, C., Ravussin, E., & Schoeller, D. A. (2000). De novo lipogenesis in adipose tissue of lean and obese women: Application of deuterated water and isotope ratio mass spectrometry. *International Journal of Obesity and Related Metabolic Disorders, 24*(7), 932–937.

Halestrap, A. P., & Denton, R. M. (1974). Hormonal regulation of adipose-tissue acetyl-Coenzyme A carboxylase by changes in the polymeric state of the enzyme. The role of long-chain fatty acyl-Coenzyme A thioesters and citrate. *The Biochemical Journal, 142*(2), 365–377.

Hansen, F. M., Fahmy, N., & Nielsen, J. H. (1980). The influence of sexual hormones on lipogenesis and lipolysis in rat fat cells. *Acta Endocrinologica (Copenh), 95*(4), 566–570.

Hardie, D. G. (1992). Regulation of fatty acid and cholesterol metabolism by the AMP-activated protein kinase. *Biochimica et Biophysica Acta, 1123*(3), 231–238.

Harris, D. M., Dunshea, F. R., Bauman, D. E., Boyd, R. D., Wang, S. Y., Johnson, P. A., et al. (1993). Effect of in vivo somatotropin treatment of growing pigs on adipose tissue lipogenesis. *Journal of Animal Science, 71*(12), 3293–3300.

Hashimoto, K., Ishida, E., Matsumoto, S., Okada, S., Yamada, M., Satoh, T., et al. (2009). Carbohydrate response element binding protein gene expression is positively regulated by thyroid hormone. *Endocrinology, 150*(7), 3417–3424.

Hattori, A., Mawatari, K., Tsuzuki, S., Yoshioka, E., Toda, S., Yoshida, M., et al. (2010). Beta-adrenergic-AMPK pathway phosphorylates acetyl-CoA carboxylase in a high-epinephrine rat model, SPORTS. *Obesity (Silver Spring), 18*(1), 48–54.

Haystead, T. A., Campbell, D. G., & Hardie, D. G. (1988). Analysis of sites phosphorylated on acetyl-CoA carboxylase in response to insulin in isolated adipocytes. Comparison with sites phosphorylated by casein kinase-2 and the calmodulin-dependent multiprotein kinase. *European Journal of Biochemistry, 175*(2), 347–354.

Heesom, K. J., Moule, S. K., & Denton, R. M. (1998). Purification and characterisation of an insulin-stimulated protein-serine kinase which phosphorylates acetyl-CoA carboxylase. *FEBS Letters*, *422*(1), 43–46.

Hellerstein, M. K., Neese, R. A., & Schwarz, J. M. (1993). Model for measuring absolute rates of hepatic de novo lipogenesis and reesterification of free fatty acids. *The American Journal of Physiology*, *265*(5 Pt. 1), E814–E820.

Hewitt, K. N., Pratis, K., Jones, M. E. E., & Simpson, E. R. (2004). Estrogen replacement reverses the hepatic steatosis phenotype in the male aromatase knockout mouse. *Endocrinology*, *145*(4), 1842–1848.

Holland, R., Hardie, D. G., Clegg, R. A., & Zammit, V. A. (1985). Evidence that glucagon-mediated inhibition of acetyl-CoA carboxylase in isolated adipocytes involves increased phosphorylation of the enzyme by cyclic AMP-dependent protein kinase. *The Biochemical Journal*, *226*(1), 139–145.

Hulver, M. W., Berggren, J. R., Carper, M. J., Miyazaki, M., Ntambi, J. M., Hoffman, E. P., et al. (2005). Elevated stearoyl-CoA desaturase-1 expression in skeletal muscle contributes to abnormal fatty acid partitioning in obese humans. *Cell Metabolism*, *2*(4), 251–261.

Iizuka, K., Bruick, R. K., Liang, G., Horton, J. D., & Uyeda, K. (2004). Deficiency of carbohydrate response element-binding protein (ChREBP) reduces lipogenesis as well as glycolysis. *Proceedings of the National Academy of Sciences of the United States of America*, *101*(19), 7281–7286.

Ishii, S., Iizuka, K., Miller, B. C., & Uyeda, K. (2004). Carbohydrate response element binding protein directly promotes lipogenic enzyme gene transcription. *Proceedings of the National Academy of Sciences of the United States of America*, *101*(44), 15597–15602.

Jiang, G., & Zhang, B. B. (2003). Glucagon and regulation of glucose metabolism. *American Journal of Physiology. Endocrinology and Metabolism*, *284*(4), E671–E678.

Jose, A. A. F. B., Alvares, L. E., Delgado, E. F., Coutinho, L. L., & Lanna, D. P. D. (2006). Effect of growth hormone on fatty acid synthase gene expression in porcine adipose tissue cultures. *Genetics and Molecular Biology*, *29*(1), 180–186.

Joseph, S. B., Laffitte, B. A., Patel, P. H., Watson, M. A., Matsukuma, K. E., Walczak, R., et al. (2002). Direct and indirect mechanisms for regulation of fatty acid synthase gene expression by liver X receptors. *The Journal of Biological Chemistry*, *277*(13), 11019–11025.

Kim, B. (2008). Thyroid hormone as a determinant of energy expenditure and the basal metabolic rate. *Thyroid*, *18*(2), 141–144.

Kim, J. B., Sarraf, P., Wright, M., Yao, K. M., Mueller, E., Solanes, G., et al. (1998). Nutritional and insulin regulation of fatty acid synthetase and leptin gene expression through ADD1/SREBP1. *The Journal of Clinical Investigation*, *101*(1), 1–9.

Kim, J. B., & Spiegelman, B. M. (1996). ADD1/SREBP1 promotes adipocyte differentiation and gene expression linked to fatty acid metabolism. *Genes and Development*, *10*(9), 1096–1107.

Kimball, S. R., Siegfried, B. A., & Jefferson, L. S. (2004). Glucagon represses signaling through the mammalian target of rapamycin in rat liver by activating AMP-activated protein kinase. *The Journal of Biological Chemistry*, *279*(52), 54103–54109.

Kitamura, T., Kitamura, Y., Kuroda, S., Hino, Y., Ando, M., Kotani, K., et al. (1999). Insulin-induced phosphorylation and activation of cyclic nucleotide phosphodiesterase 3B by the serine-threonine kinase Akt. *Molecular and Cellular Biology*, *19*(9), 6286–6296.

Klemm, D. J., Leitner, J. W., Watson, P., Nesterova, A., Reusch, J. E., Goalstone, M. L., et al. (2001). Insulin-induced adipocyte differentiation. Activation of CREB rescues adipogenesis from the arrest caused by inhibition of prenylation. *The Journal of Biological Chemistry*, *276*(30), 28430–28435.

Klemm, D. J., Roesler, W. J., Boras, T., Colton, L. A., Felder, K., & Reusch, J. E. (1998). Insulin stimulates cAMP-response element binding protein activity in HepG2 and 3T3-L1 cell lines. *The Journal of Biological Chemistry*, *273*(2), 917–923.

Kojima, M., Hosoda, H., Date, Y., Nakazato, M., Matsuo, H., & Kangawa, K. (1999). Ghrelin is a growth-hormone-releasing acylated peptide from stomach. *Nature, 402*(6762), 656–660.
Kola, B., Christ-Crain, M., Lolli, F., Arnaldi, G., Giacchetti, G., Boscaro, M., et al. (2008). Changes in adenosine 5'-monophosphate-activated protein kinase as a mechanism of visceral obesity in Cushing's syndrome. *The Journal of Clinical Endocrinology and Metabolism, 93*(12), 4969–4973.
Kola, B., Hubina, E., Tucci, S. A., Kirkham, T. C., Garcia, E. A., Mitchell, S. E., et al. (2005). Cannabinoids and ghrelin have both central and peripheral metabolic and cardiac effects via AMP-activated protein kinase. *The Journal of Biological Chemistry, 280*(26), 25196–25201.
Kubota, N., Terauchi, Y., Miki, H., Tamemoto, H., Yamauchi, T., Komeda, K., et al. (1999). PPAR gamma mediates high-fat diet-induced adipocyte hypertrophy and insulin resistance. *Molecular Cell, 4*(4), 597–609.
Lacasa, D., Le, L. X., Ferre, P., & Dugail, I. (2001). Progesterone stimulates adipocyte determination and differentiation 1/sterol regulatory element-binding protein 1c gene expression. Potential mechanism for the lipogenic effect of progesterone in adipose tissue. *The Journal of Biological Chemistry, 276*(15), 11512–11516.
Lane, M. D., Moss, J., & Polakis, S. E. (1974). Acetyl coenzyme A carboxylase. *Current Topics in Cellular Regulation, 8*, 139–195.
Leung, D. W. (2001). The structure and functions of human lysophosphatidic acid acyltransferases. *Frontiers in Bioscience, 6*, D944–D953.
Li, G., Hernandez-Ono, A., Crooke, R. M., Graham, M. J., & Ginsberg, H. N. (2011). Effects of antisense-mediated inhibition of 11beta-hydroxysteroid dehydrogenase type 1 on hepatic lipid metabolism. *Journal of Lipid Research, 52*(5), 971–981.
Li, Y., Xu, S., Mihaylova, M. M., Zheng, B., Hou, X., Jiang, B., et al. (2011). AMPK phosphorylates and inhibits SREBP activity to attenuate hepatic steatosis and atherosclerosis in diet-induced insulin-resistant mice. *Cell Metabolism, 13*(4), 376–388.
Liu, Y., Ding, J., Bush, T. L., Longenecker, J. C., Nieto, F. J., Golden, S. H., et al. (2001). Relative androgen excess and increased cardiovascular risk after menopause: A hypothesized relation. *American Journal of Epidemiology, 154*(6), 489–494.
Liu, C. Y., Grant, A. L., Kim, K. H., & Mills, S. E. (1994). Porcine somatotropin decreases acetyl-CoA carboxylase gene expression in porcine adipose tissue. *Domestic Animal Endocrinology, 11*(1), 125–132.
Longuet, C., Sinclair, E. M., Maida, A., Baggio, L. L., Maziarz, M., Charron, M. J., et al. (2008). The glucagon receptor is required for the adaptive metabolic response to fasting. *Cell Metabolism, 8*(5), 359–371.
Lundholm, L., Zang, H., Hirschberg, A. L., Gustafsson, J. A., Arner, P., & Dahlman-Wright, K. (2008). Key lipogenic gene expression can be decreased by estrogen in human adipose tissue. *Fertility and Sterility, 90*(1), 44–48.
Ma, L., Robinson, L. N., & Towle, H. C. (2006). ChREBP*Mlx is the principal mediator of glucose-induced gene expression in the liver. *The Journal of Biological Chemistry, 281*(39), 28721–28730.
Magana, M. M., Lin, S. S., Dooley, K. A., & Osborne, T. F. (1997). Sterol regulation of acetyl coenzyme A carboxylase promoter requires two interdependent binding sites for sterol regulatory element binding proteins. *Journal of Lipid Research, 38*(8), 1630–1638.
Mariash, C. N., Kaiser, F. E., Schwartz, H. L., Towle, H. C., & Oppenheimer, J. H. (1980). Synergism of thyroid hormone and high carbohydrate diet in the induction of lipogenic enzymes in the rat. Mechanisms and implications. *The Journal of Clinical Investigation, 65*(5), 1126–1134.
Marrero, M., Prough, R. A., Frenkel, R. A., & Milewich, L. (1990). Dehydroepiandrosterone feeding and protein phosphorylation, phosphatases, and lipogenic enzymes in mouse liver. *Proceedings of the Society for Experimental Biology and Medicine, 193*(2), 110–117.

McInnes, K. J., Brown, K. A., Hunger, N. I., & Simpson, E. R. (2012). Regulation of LKB1 expression by sex hormones in adipocytes. *International Journal of Obesity, 36*(7), 982–985.
McInnes, K. J., Corbould, A., Simpson, E. R., & Jones, M. E. (2006). Regulation of adenosine 5', monophosphate-activated protein kinase and lipogenesis by androgens contributes to visceral obesity in an estrogen-deficient state. *Endocrinology, 147*(12), 5907–5913.
McIntosh, M. K., & Berdanier, C. D. (1991). Antiobesity effects of dehydroepiandrosterone are mediated by futile substrate cycling in hepatocytes of BHE/cdb rats. *The Journal of Nutrition, 121*(12), 2037–2043.
Mendes, A. M., Madon, R. J., & Flint, D. J. (1985). Effects of cortisol and progesterone on insulin binding and lipogenesis in adipocytes from normal and diabetic rats. *The Journal of Endocrinology, 106*(2), 225–231.
Morrison, C. D., Morton, G. J., Niswender, K. D., Gelling, R. W., & Schwartz, M. W. (2005). Leptin inhibits hypothalamic Npy and Agrp gene expression via a mechanism that requires phosphatidylinositol 3-OH-kinase signaling. *American Journal of Physiology. Endocrinology and Metabolism, 289*(6), E1051–E1057.
Munday, M. R., Campbell, D. G., Carling, D., & Hardie, D. G. (1988). Identification by amino acid sequencing of three major regulatory phosphorylation sites on rat acetyl-CoA carboxylase. *European Journal of Biochemistry, 175*(2), 331–338.
Munzberg, H., & Myers, M. G., Jr. (2005). Molecular and anatomical determinants of central leptin resistance. *Nature Neuroscience, 8*(5), 566–570.
Nakazato, M., Murakami, N., Date, Y., Kojima, M., Matsuo, H., Kangawa, K., et al. (2001). A role for ghrelin in the central regulation of feeding. *Nature, 409*(6817), 194–198.
Ng, F. M., Adamafio, N. A., & Graystone, J. E. (1990). Effects of exogenous growth hormone on lipid metabolism in the isolated epididymal fat pad of the growth hormone-deficient little mouse. *Journal of Molecular Endocrinology, 4*(1), 43–49.
Nogalska, A., Sucajtys-Szulc, E., & Swierczynski, J. (2005). Leptin decreases lipogenic enzyme gene expression through modification of SREBP-1c gene expression in white adipose tissue of aging rats. *Metabolism, 54*(8), 1041–1047.
Omar, B., Zmuda-Trzebiatowska, E., Manganiello, V., Goransson, O., & Degerman, E. (2009). Regulation of AMP-activated protein kinase by cAMP in adipocytes: Roles for phosphodiesterases, protein kinase B, protein kinase A, Epac and lipolysis. *Cellular Signalling, 21*(5), 760–766.
Oppenheimer, J. H., Schwartz, H. L., Lane, J. T., & Thompson, M. P. (1991). Functional relationship of thyroid hormone-induced lipogenesis, lipolysis, and thermogenesis in the rat. *The Journal of Clinical Investigation, 87*(1), 125–132.
Pasquali, R. (2006). Obesity and androgens: Facts and perspectives. *Fertility and Sterility, 85*(5), 1319–1340.
Paulauskis, J. D., & Sul, H. S. (1988). Cloning and expression of mouse fatty acid synthase and other specific mRNAs. Developmental and hormonal regulation in 3T3-L1 cells. *The Journal of Biological Chemistry, 263*(15), 7049–7054.
Peet, D. J., Turley, S. D., Ma, W., Janowski, B. A., Lobaccaro, J. M., Hammer, R. E., et al. (1998). Cholesterol and bile acid metabolism are impaired in mice lacking the nuclear oxysterol receptor LXR alpha. *Cell, 93*(5), 693–704.
Poehlman, E. T. (2005). Transversing the menopause: Change in energy expenditure and body composition. *Coronary Artery Disease, 16*(8), 511.
Qi, L., Heredia, J. E., Altarejos, J. Y., Screaton, R., Goebel, N., Niessen, S., et al. (2006). TRB3 links the E3 ubiquitin ligase COP1 to lipid metabolism. *Science, 312*(5781), 1763–1766.
Radenne, A., Akpa, M., Martel, C., Sawadogo, S., Mauvoisin, D., & Mounier, C. (2008). Hepatic regulation of fatty acid synthase by insulin and T3: Evidence for T3 genomic and nongenomic actions. *American Journal of Physiology. Endocrinology and Metabolism, 295*(4), E884–E894.

Ramsay, T. G. (2003). Porcine leptin inhibits lipogenesis in porcine adipocytes. *Journal of Animal Science, 81*(12), 3008–3017.

Repa, J. J., Liang, G., Ou, J., Bashmakov, Y., Lobaccaro, J. M., Shimomura, I., et al. (2000). Regulation of mouse sterol regulatory element-binding protein-1c gene (SREBP-1c) by oxysterol receptors, LXRalpha and LXRbeta. *Genes and Development, 14*(22), 2819–2830.

Rosen, E. D., Walkey, C. J., Puigserver, P., & Spiegelman, B. M. (2000). Transcriptional regulation of adipogenesis. *Genes and Development, 14*(11), 1293–1307.

Roth, G., Kotzka, J., Kremer, L., Lehr, S., Lohaus, C., Meyer, H. E., et al. (2000). MAP kinases Erk1/2 phosphorylate sterol regulatory element-binding protein (SREBP)-1a at serine 117 in vitro. *The Journal of Biological Chemistry, 275*(43), 33302–33307.

Ruderman, N. B., Saha, A. K., & Kraegen, E. W. (2003). Minireview: Malonyl CoA, AMP-activated protein kinase, and adiposity. *Endocrinology, 144*(12), 5166–5171.

Sancho, V., Trigo, M. V., Martin-Duce, A., Gonz, L. N., Acitores, A., Arnes, L., et al. (2006). Effect of GLP-1 on D-glucose transport, lipolysis and lipogenesis in adipocytes of obese subjects. *International Journal of Molecular Medicine, 17*(6), 1133–1137.

Schwarz, J. M., Linfoot, P., Dare, D., & Aghajanian, K. (2003). Hepatic de novo lipogenesis in normoinsulinemic and hyperinsulinemic subjects consuming high-fat, low-carbohydrate and low-fat, high-carbohydrate isoenergetic diets. *The American Journal of Clinical Nutrition*, 77(1), 43–50.

Scott, J. W., Norman, D. G., Hawley, S. A., Kontogiannis, L., & Hardie, D. G. (2002). Protein kinase substrate recognition studied using the recombinant catalytic domain of AMP-activated protein kinase and a model substrate. *Journal of Molecular Biology, 317*(2), 309–323.

Shimomura, I., Shimano, H., Korn, B. S., Bashmakov, Y., & Horton, J. D. (1998). Nuclear sterol regulatory element-binding proteins activate genes responsible for the entire program of unsaturated fatty acid biosynthesis in transgenic mouse liver. *The Journal of Biological Chemistry, 273*(52), 35299–35306.

Shrago, E., Spennetta, T., & Gordon, E. (1969). Fatty acid synthesis in human adipose tissue. *The Journal of Biological Chemistry, 244*(10), 2761–2766.

Shulman, G. I., Ladenson, P. W., Wolfe, M. H., Ridgway, E. C., & Wolfe, R. R. (1985). Substrate cycling between gluconeogenesis and glycolysis in euthyroid, hypothyroid, and hyperthyroid man. *The Journal of Clinical Investigation, 76*(2), 757–764.

Soukas, A., Cohen, P., Socci, N. D., & Friedman, J. M. (2000). Leptin-specific patterns of gene expression in white adipose tissue. *Genes & Development, 14*(8), 963–980.

Stoeckman, A. K., Ma, L., & Towle, H. C. (2004). Mlx is the functional heteromeric partner of the carbohydrate response element-binding protein in glucose regulation of lipogenic enzyme genes. *The Journal of Biological Chemistry, 279*(15), 15662–15669.

Strawford, A., Antelo, F., Christiansen, M., & Hellerstein, M. K. (2004). Adipose tissue triglyceride turnover, de novo lipogenesis, and cell proliferation in humans measured with (H2O)-H-2. *American Journal of Physiology. Endocrinology and Metabolism, 286*(4), E577–E588.

Summermatter, S., Baum, O., Santos, G., Hoppeler, H., & Handschin, C. (2010). Peroxisome proliferator-activated receptor gamma coactivator 1{alpha} (PGC-1{alpha}) promotes skeletal muscle lipid refueling in vivo by activating de novo lipogenesis and the pentose phosphate pathway. *The Journal of Biological Chemistry, 285*(43), 32793–32800.

Svegliati-Baroni, G., Saccomanno, S., Rychlicki, C., Agostinelli, L., De Minicis, S., Candelaresi, C., et al. (2011). Glucagon-like peptide-1 receptor activation stimulates hepatic lipid oxidation and restores hepatic signalling alteration induced by a high-fat diet in nonalcoholic steatohepatitis. *Liver International, 31*(9), 1285–1297.

Swenson, T. L., & Porter, J. W. (1985). Mechanism of glucagon inhibition of liver acetyl-CoA carboxylase. Interrelationship of the effects of phosphorylation, polymer-protomer transition, and citrate on enzyme activity. *The Journal of Biological Chemistry*, *260*(6), 3791–3797.

Swierczynski, J., Goyke, E., Wach, L., Pankiewicz, A., Kochan, Z., Adamonis, W., et al. (2000). Comparative study of the lipogenic potential of human and rat adipose tissue. *Metabolism*, *49*(5), 594–599.

Tabor, D. E., Kim, J. B., Spiegelman, B. M., & Edwards, P. A. (1998). Transcriptional activation of the stearoyl-CoA desaturase 2 gene by sterol regulatory element-binding protein/adipocyte determination and differentiation factor 1. *The Journal of Biological Chemistry*, *273*(34), 22052–22058.

Tang, X., Ma, H. T., Huang, G. Q., Miao, J. F., & Zou, S. X. (2009). The effect of dehydroepiandrosterone on lipogenic gene mRNA expression in cultured primary chicken hepatocytes. *European Journal of Lipid Science and Technology*, *111*(5), 432–441.

Theander-Carrillo, C., Wiedmer, P., Cettour-Rose, P., Nogueiras, R., Perez-Tilve, D., Pfluger, P., et al. (2006). Ghrelin action in the brain controls adipocyte metabolism. *The Journal of Clinical Investigation*, *116*(7), 1983–1993.

Tomlinson, J. J., Boudreau, A., Wu, D. M., Salem, H. A., Carrigan, A., Gagnon, A., et al. (2010). Insulin sensitization of human preadipocytes through glucocorticoid hormone induction of forkhead transcription factors. *Molecular Endocrinology*, *24*(1), 104–113.

Tontonoz, P., Hu, E., & Spiegelman, B. M. (1994). Stimulation of adipogenesis in fibroblasts by PPAR gamma 2, a lipid-activated transcription factor. *Cell*, *79*(7), 1147–1156.

Vagenakis, A. G., Portnay, G. I., O'Brian, J. T., Rudolph, M., Arky, R. A., Ingbar, S. H., et al. (1977). Effect of starvation on the production and metabolism of thyroxine and triiodothyronine in euthyroid obese patients. *The Journal of Clinical Endocrinology and Metabolism*, *45*(6), 1305–1309.

Vendrell, J., El Bekay, R., Peral, B., Garcia-Fuentes, E., Megia, A., Macias-Gonzalez, M., et al. (2011). Study of the potential association of adipose tissue GLP-1 receptor with obesity and insulin resistance. *Endocrinology*, *152*(11), 4072–4079.

Viana, A. Y., Sakoda, H., Anai, M., Fujishiro, M., Ono, H., Kushiyama, A., et al. (2006). Role of hepatic AMPK activation in glucose metabolism and dexamethasone-induced regulation of AMPK expression. *Diabetes Research and Clinical Practice*, *73*(2), 135–142.

Villanueva-Penacarrillo, M. L., Marquez, L., Gonzalez, N., Diaz-Miguel, M., & Valverde, I. (2001). Effect of GLP-1 on lipid metabolism in human adipocytes. *Hormone and Metabolic Research*, *33*(2), 73–77.

Volpe, J. J., & Marasa, J. C. (1975). Hormonal regulation of fatty acid synthetase, acetyl-CoA carboxylase and fatty acid synthesis in mammalian adipose tissue and liver. *Biochimica et Biophysica Acta*, *380*(3), 454–472.

Wan, Z., Thrush, A. B., Legare, M., Frier, B. C., Sutherland, L. N., Williams, D. B., et al. (2010). Epinephrine-mediated regulation of PDK4 mRNA in rat adipose tissue. *American Journal of Physiology. Cell Physiology*, *299*(5), C1162–C1170.

Wang, Y., Jones, V. B., Urs, S., Kim, S., Soltani-Bejnood, M., Quigley, N., et al. (2004). The human fatty acid synthase gene and de novo lipogenesis are coordinately regulated in human adipose tissue. *The Journal of Nutrition*, *134*(5), 1032–1038.

Wang, X., Sato, R., Brown, M. S., Hua, X., & Goldstein, J. L. (1994). SREBP-1, a membrane-bound transcription factor released by sterol-regulated proteolysis. *Cell*, *77*(1), 53–62.

Wang, Y. X., Voy, B. J., Urs, S., Kim, S., Soltani-Bejnood, M., Quigley, N., et al. (2004). The human fatty acid synthase gene and de novo lipogenesis are coordinately regulated in human adipose tissue. *The Journal of Nutrition*, *134*(5), 1032–1038.

Weiss, R. E., Murata, Y., Cua, K., Hayashi, Y., Seo, H., & Refetoff, S. (1998). Thyroid hormone action on liver, heart, and energy expenditure in thyroid hormone receptor beta-deficient mice. *Endocrinology*, *139*(12), 4945–4952.

Witters, L. A., & Kemp, B. E. (1992). Insulin activation of acetyl-CoA carboxylase accompanied by inhibition of the 5'-AMP-activated protein kinase. *The Journal of Biological Chemistry*, *267*(5), 2864–2867.

Witters, L. A., Watts, T. D., Daniels, D. L., & Evans, J. L. (1988). Insulin stimulates the dephosphorylation and activation of acetyl-CoA carboxylase. *Proceedings of the National Academy of Sciences of the United States of America*, *85*(15), 5473–5477.

Wu, Z., Puigserver, P., Andersson, U., Zhang, C., Adelmant, G., Mootha, V., et al. (1999). Mechanisms controlling mitochondrial biogenesis and respiration through the thermogenic coactivator PGC-1. *Cell*, *98*(1), 115–124.

Wu, Z., Rosen, E. D., Brun, R., Hauser, S., Adelmant, G., Troy, A. E., et al. (1999). Cross-regulation of C/EBP alpha and PPAR gamma controls the transcriptional pathway of adipogenesis and insulin sensitivity. *Molecular Cell*, *3*(2), 151–158.

Xiong, Y., Collins, Q. F., An, J., Lupo, E., Liu, H. Y., Liu, D. L., et al. (2007). p38 mitogen-activated protein kinase plays an inhibitory role in hepatic lipogenesis. *The Journal of Biological Chemistry*, *282*(7), 4975–4982.

Yellaturu, C. R., Deng, X., Cagen, L. M., Wilcox, H. G., Mansbach, C. M., Siddiqi, S. A., et al. (2009). Insulin enhances post-translational processing of nascent SREBP-1c by promoting its phosphorylation and association with COPII vesicles. *The Journal of Biological Chemistry*, *284*(12), 7518–7532.

Zarjevski, N., Cusin, I., Vettor, R., Rohner-Jeanrenaud, F., & Jeanrenaud, B. (1993). Chronic intracerebroventricular neuropeptide-Y administration to normal rats mimics hormonal and metabolic changes of obesity. *Endocrinology*, *133*(4), 1753–1758.

Zhao, L. F., Iwasaki, Y., Zhe, W., Nishiyama, M., Taguchi, T., Tsugita, M., et al. (2010). Hormonal regulation of acetyl-CoA carboxylase isoenzyme gene transcription. *Endocrine Journal*, *57*(4), 317–324.

CHAPTER TWO

Genetic Determinants of Obesity and Related Vascular Diseases

Yaroslav Winter*, Roman Sankowski*, Tobias Back†,1
*Department of Neurology, Philipps-University, Marburg Germany
†Department of Neurology, Saxon Hospital Arnsdorf, Arnsdorf/Dresden, Germany
[1]Corresponding author: e-mail address: tobias.back@gmx.de

Contents

Abstract

Obesity is one of the major risk factors of vascular diseases, and its prevalence is increasing worldwide. In the past decade, progress has been made in the understanding of genetic determinants of obesity and obesity-associated diseases. Genome-wide association studies identified a number of genetic variants associated with obesity. In addition to common variants, FTO and MC4R, new loci, such as TMEM18, KCTD15, GNPDA2, SH2B1, MTCH2, and NEGR1 have been detected. In the past years, abdominal obesity has been shown to be a more important vascular risk factor than the body mass index. In the context of vascular risk assessment, identification of genetic polymorphisms associated with accumulation of visceral fat is of special importance. Some polymorphisms associated with abdominal obesity, such as variants of gene encoding microsomal triglyceride transfer protein, have been already discovered. In this chapter, we provide a review of genetic determinants of obesity and discuss their role in obesity-related vascular diseases.

1. INTRODUCTION

The prevalence of obesity is increasing worldwide and has already reached epidemic proportions. In the United States, the amount of adults with overweight or obesity more than doubled during the past 30 years (Manson &

Vitamins and Hormones, Volume 91
ISSN 0083-6729
http://dx.doi.org/10.1016/B978-0-12-407766-9.00002-X

Bassuk, 2003). Currently, nearly 70% of the U.S. population are obese or overweight (Hedley et al., 2004). In Europe, up to 50% of adults are obese or overweight, and 15% are obese (Rabin et al., 2007). According to the definition of the World Health Organization, obesity is "an excessive fat accumulation that presents risk to health" (World Health Organization, 2011). Obesity is one of the major risk factors of vascular diseases and is associated with a variety of other vascular risk factors, such as arterial hypertension, diabetes mellitus, dyslipidemia, physical inactivity, and a combination of those factors named metabolic syndrome (Nguyen et al., 2008).

In the past decade, progress has been made in understanding of genetic determinants of obesity and obesity-associated diseases. Until recently, the number of candidate genes for obesity was limited and the identification of new genes of obesity proceeded very slowly. Association studies of obesity candidate genes and linkage analyses detected only limited numbers of susceptibility loci, which were later replicated in larger studies. Since the large-scale genome-wide association studies have been introduced, the number of genetic variants with confirmed association with obesity has substantially increased (Willer et al., 2009). In this chapter, we provide a review of genetic determinants of obesity and obesity-related vascular diseases.

2. GENETIC DETERMINANTS OF OBESITY

In the past, genetic studies of obesity concentrated on the rare monogenic syndromes, which usually cause severe forms of obesity. According to the latest report of the Human Obesity Gene Map, 176 cases of obesity due to single-gene mutations in 11 genes have been described (Rankinen et al., 2006). In the past decade, large-scale association studies with candidate genes and genome scans revealed new common variants of genes associated with obesity (Frayling et al., 2007; Loos et al., 2008; Scuteri et al., 2007). In 2007 and 2008, common variants at two loci, the fat mass and obesity (*FTO*)-associated gene and melanocortin 4 receptor (*MC4R*) gene, were identified. The effect of FTO and MC4R on the body mass index (BMI) is modest (OR: 1.1–1.3 for obesity) (Frayling et al., 2007; Loos et al., 2008; Scuteri et al., 2007). Common variants in the *PCSK1* gene, encoding the proprotein convertase 1/3 (PC1/3), also contribute to human obesity (Benzinou et al., 2008). Mutations causing impairment of PC1/3 function have a mild association with increased risk of common obesity and are present in 0.83% of extreme obesity phenotypes (Creemers et al., 2012). Common variants of FTO, MC4R, and PCSK1 and rare variants are explanatory for only a small proportion of genetic contribution to human obesity. In order to promote the identification of genes associated with obesity, the Genetic Association of Anthropometric Traits consortium performed recently a large-scale meta-analysis of data from 15 genome-wide association studies (Willer et al., 2009). Six previously unreported loci were found to be reproducibly associated with obesity (TMEM18, KCTD15, SH2B1, MTCH2, GNPDA2, and NEGR1). For each of the six newly discovered variants, odds ratios for obesity were 1.03–1.25 and for overweight 1.03–1.14 (Willer et al., 2009). This finding is supported by the data of Thorleifson et al., which also showed evidence for associations of TMEM18, KCTD15, SH2B1, MTCH2, and NEGR1 with obesity (Thorleifsson et al., 2009). FTO, MC4R, and five of newly discovered genes (*TMEM18*, *KCTD15*, *SH2B1*, *GNPDA2*, and *NEGR1*) are highly expressed in the brain, suggesting that genetic variation contributes to common human obesity through processes in the central nervous system (CNS). However, effects of pro-obesity genes in other organs and tissues are also possible. In particular, genes involved in topological distribution of fat mass, such as in case of abdominal obesity, were detected outside the CNS (Berthier et al., 2004).

The above-mentioned genetic variants show association with BMI. In the past years, data from studies in vascular diseases provided evidence that the abdominal type of obesity is a more relevant vascular risk factor than BMI (Hu et al., 2007; Matsuzawa, 2008; Winter et al., 2008a; Yusuf et al., 2005). Identification of genetic polymorphisms that are associated with accumulation of visceral fat is of special importance in the context of risk assessment and prevention of vascular diseases. A number of genetic variants associated with visceral adipose tissue have been already discovered. Berthier et al. revealed that some polymorphisms of gene encoding microsomal triglyceride transfer protein (MTP) in the French Canadian population are associated with accumulation of visceral fat mass (Berthier et al., 2004). MTP plays an important role in the assembly of lipoproteins and is highly expressed in entero- and hepatocytes (Nielsen et al., 1998). An association with waist-to-hip ratio (WHR), which is a marker of abdominal obesity, was supposed for the common T60N polymorphism of the *lymphotoxin-α* (*LTA*) gene (Hamid et al., 2005). *LTA* is associated with myocardial infarction, ischemic stroke, type 2 diabetes, and other phenotypes of metabolic syndrome (Hamid et al., 2005; Ozaki et al., 2002). After correction for multiple testing, the apparent effect of *LTA* on WHR became nonsignificant (Hamid et al., 2005). In the Hispanic population, an association between the gene encoding adiponectin (*APM1*) and different measures of obesity, including waist circumference (WC) and WHR, was reported (Sutton et al., 2005). Associations with body fat distribution measured by WHR, WC, and sagittal diameter were also reported for genetic variants of *ACE* (Katzov et al., 2004), *ADRB2* (Lange et al., 2005), *APOA2* (Lara-Castro et al., 2005a), *FABP2* (Lara-Castro et al., 2005b), *PLIN* (Qi et al., 2004), *PPARG* (Kim et al., 2004), and *UCP1* (Kim et al., 2005).

The number of quantitative trait loci (QTLs) of human obesity detected by genome scans is continuously increasing. The latest update of the Human Obesity Gene Map reported 253 QTLs for obesity-related phenotypes, which were obtained from 61 genome scans (Rankinen et al., 2006). Of these, 52 QTLs were replicated by at least one study. There are 127 obesity candidate genes, and for 22 of them association with obesity was replicated in at least five studies (Rankinen et al., 2006). For the detailed information concerning currently available obesity candidate genes, we refer to the latest update of the Human Obesity Gene Map (Rankinen et al., 2006).

3. VASCULAR DISEASES RELATED TO OVERWEIGHT AND OBESITY

3.1. Mechanisms of vascular damage

Taken the major clinical sequelae of arteriosclerosis together, vascular diseases account for more than half of deaths of adults aged over 60 in industrialized nations. Arteriosclerosis, also named atherosclerosis, is a pandemic among physically hypoactive and inappropriately nourished children, adolescents, and adults. The description of mechanisms of vascular damage should take into account the influence of vascular risk factors, the localization of vascular damage, the stage and nature of vascular disease, and pathophysiological processes that contribute to arteriosclerosis as the common (but not obligatory) basis of vascular disease.

(1) Atherosclerosis is an age-dependent stiffening of arteries that is characterized by deposition of cholesterol, fatty acids, and fibrous material; consecutive thickening of the arterial wall; and sometimes calcification. Endothelial lesions (inner vessel wall) serve as origin of thrombus formation which can be dislocated downstream as emboli. Large- and medium-sized arteries show changes with dilatation or narrowing called arterial macroangiopathy (dilatative or stenotizing). But also small arteries, so-called arterioles, may be affected which is described as arterial microangiopathy. In many cases, the arterial wall changes with progressive narrowing of the vessel lumen remain for a long time subclinical until ischemia of end organs

occur (e.g., heart, brain, leg). Ischemic tissue changes, local thrombosis, or embolism manifest in vascular diseases as coronary heart disease, stroke, peripheral artery occlusive disease, or sudden vascular death.

To date, there are two major hypotheses to explain the evolution of arterial disease: first, the "response-to-injury hypothesis" that stresses any kind of damage to the endothelium (the inner layer of the vessel wall) as trigger of processes and cellular interactions that culminate in the lesions of arteriosclerosis (Ross, 1999). Hypertension is a typical factor that may cause endothelial damage. Second, the "lipoprotein-induced atherosclerosis hypothesis" claims that the oxidation of low-density lipoproteins (LDLs) by macrophages triggers a cascade of inflammatory processes that finally result in arteriosclerotic plaques and atheromas of arteries (Ross, 1999). High blood concentrations of LDL cholesterol are a typical condition that enhances this type of inflammation of the arterial wall.

(2) We know a large variety of modifiable or biologically given risk factors that contribute to a variable extent to vascular disease. Aging, male gender, and genetic factors contribute as biological factors to arteriosclerosis. Typical modifiable risk factors are arterial hypertension, smoking, high LDL cholesterol and dyslipidemia, obesity (overweight) and body fat distribution, sedentary inactive lifestyle, diabetes mellitus, low levels of high-density lipoprotein (HDL) cholesterol, chronic inflammatory diseases, infections, and stress among others (Goldstein et al., 2006). There is overwhelming evidence that a "low risk lifestyle" including nonsmoking, normal body weight, regular physical activity, low alcohol intake, and healthy diet reduces the overall mortality by four to five times compared to persons at high risk (Chiuve et al., 2008; Khaw et al., 2008). Interestingly, not only the cardiovascular risk is grossly reduced by modification of "vascular" risk factors but also—to a lesser degree—the risk of cancer.

Risk factors interact in a complex manner thereby driving mechanisms of vascular damage. However, the individual impact of those factors can be addressed by calculating the relative or population attributable risks. Overweight and obesity predispose to or influence in a negative manner like other major vascular risk factors such as arterial hypertension, diabetes, hyperlipidemia, or physical inactivity. For example, obese persons carry a more than threefold increased relative risk to develop diabetes and a two- to threefold increased risk of coronary heart disease. The corresponding population attributable risks are 69% for diabetes and 40–69% for coronary heart disease (Colditz, 1999; Kenchaiah et al., 2002; Yusuf et al., 2005). This means that those proportions of disease could be avoided if patients achieve normal body weight.

Arterial hypertension is the most frequent disease that accompanies obesity. In obese persons, the incidence of hypertension is threefold higher compared to people with normal body weight. In a near linear correlation, an increased BMI is associated with elevated systolic and diastolic blood pressure (Jones et al., 1994). Abdominal adiposity measured by the WC causes an increased prevalence of hypertension in males (29%) and females (44%) with an adiposity attributable risk of 27.3% in males and 36.5% in females (Okosun et al., 2004). A population-based study in the Netherlands showed in males that abdominal adiposity is associated 2.5–4.5 times more frequently with the occurrence of vascular risk factors compared to men with normal WC (Lean et al., 1998). These findings among many others suggest that the visceral part of body fat distribution plays a particular pathophysiological role.

In fact, the abdominal or visceral fat mass (comprising 5–10% of the whole body fat) has particular importance for the risk profile: the veins drain into the portal vein system and the visceral fat is metabolically very active in comparison with subcutaneous fat. Visceral fat tissue produces substrates, enzymes, proteins, hormones, and zytokines. The concentration of fatty acids is increased, adipokines like leptin and interleukin-6 are secreted, and adiponectin concentrations are subnormal. The widespread metabolic changes of visceral fat mass in obese persons reduce the insulin sensitivity and carry proinflammatory and proatherogenic effects. Adiposity associated mechanisms that contribute to the development of hypertension are the increase of fat mass (activating the renin-angiotensin-aldosteron system), elevated sympathicotonus, reduced insulin sensitivity, increased blood volume, and intravascular volume, all of them leading to an increase in cardiac output volume (Fig. 2.1).

Obesity can be regarded as a key risk factor that tends to be combined with other risk profiles. In the preclinical phase, an endothelial dysfunction can be observed far more frequently in persons with adiposity (Arcaro et al., 1999) than with normal weight. Autopsy studies showed even in young adults a positive correlation between BMI or abdominal fat mass and fatty streaks or raised lesions (plaques) in coronary arteries (McGill et al., 2002). The metabolic syndrome (combined abdominal adiposity, diabetes or altered glucose tolerance, hypertension, and disturbed lipid metabolism) represents a combination of factors that are all interrelated, caused by a passive and sedentary lifestyle and result in high morbidity and mortality (Phillips & Prins, 2008). This complex pattern of interaction contributes in a differential and individual way to the development of coronary,

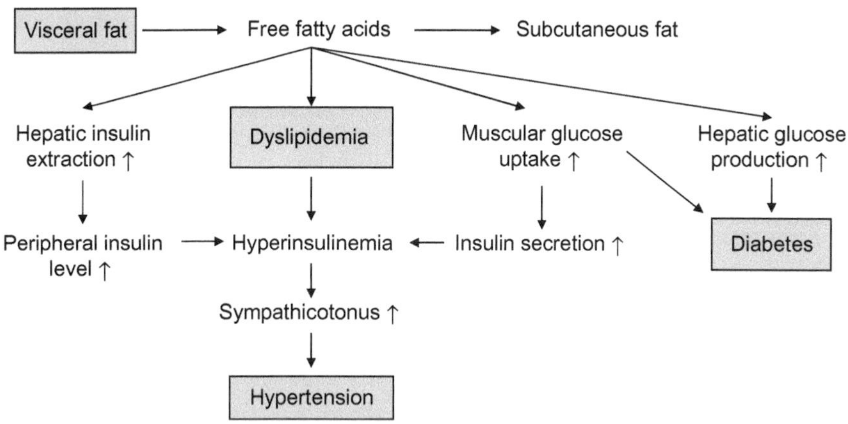

Figure 2.1 Pathophysiology of the metabolic syndrome. *Modified after Wirth (2008).*

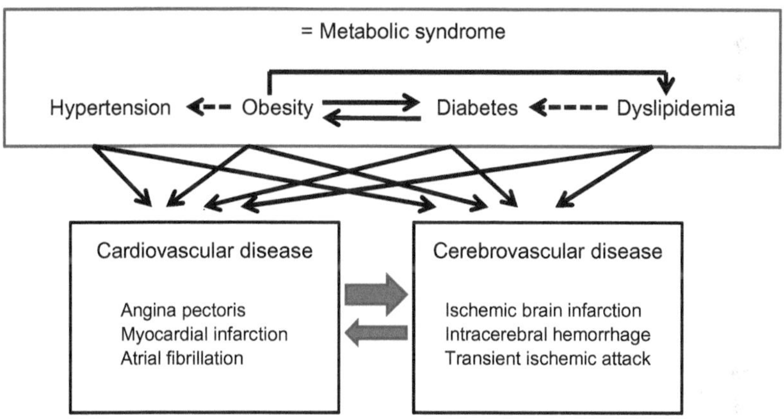

Figure 2.2 Interrelations between obesity, other risk factors, and vascular diseases. (See Color Insert.)

cerebral, and peripheral artery occlusive disease (Fig. 2.2). For example, hypertension is the most important (frequent) risk factor to develop brain infarction, whereas disturbances of the lipid metabolism have a higher impact for the genesis of coronary heart disease. The common pathological process behind all these clinical entities is atherosclerosis.

(3) In individual patients, there is a predilection for the vascular territory affected. This predilection is influenced by genetic determinants, individual constitution, risk profile, and life style. Based on the data of the Framingham Heart Project, a sex-specific multivariable risk factor algorithm can be used to assess general cardiovascular disease (CVD) risk and risk of individual

vascular events (coronary, cerebrovascular, and peripheral arterial diseases and heart failure). The estimated absolute CVD event rates can be used to quantify risk and to guide preventive care (D'Agostino et al., 2008). Coronary heart disease may present as angina pectoris, myocardial infarction, ischemic heart failure, disorders of cardiac rhythm, or sudden cardiac death. Stenotizing macroangiopathic changes (90%) of the major coronary arteries and microvascular pathology (10%), rarely also coronary vasospasms, contribute to myocardial ischemia. The major cause of heart infarction is, however, the rupture of an arteriosclerotic atheroma of coronary vessels that provoke local thrombosis and occlusion.

Cerebrovascular disease is caused either by large vessel disease of major cerebral arteries, small vessel disease, embolic brain infarcts or by hemodynamic infarction due to high-grade stenosis/occlusion of brain supplying arteries (in most cases the ipsilateral internal carotid artery) (Fig. 2.3). Besides, hemorrhagic brain infarcts may occur in which arteriosclerotic changes and arterial aneurysms are involved (Dodel et al., 2010). All those types of vessel pathology cause typical patterns of ischemic or hemorrhagic brain tissue changes that can be readily detected by means of modern CT and MR imaging methods (von Kummer & Back, 2006). Atrial fibrillation, as a frequent manifestation of ischemic heart disease, allows thrombus formation in the left atrium. They may be dislocated and travel as emboli into the cerebral circulation causing embolic territorial brain infarcts (about one-third of all ischemic strokes).

The manifestation of peripheral vascular disease is due in over 90% to stenotizing and occlusive arteriosclerosis of the arteries of the lower extremities. The most prevalent risk factors are smoking and diabetes mellitus. Since

	Infarct type	Vessel pathology	Basic disease
Ischemic stroke (80%)	Atherothrombotic (15–20%)	Cerebral macroangiopathy	Atherosclerosis
	Cardio-embolic (30%)	Not obligatory	Atrial fibrillation, cardiomyopathy, valvular dysfunction
	Hemodynamic (5–10%)	Extra/intracranial stenosisof brain supplying arteries	Atherosclerosis
	Lacunar (25%)	Cerebral microangiopathy	Hypertension, diabetes, obesity
Hemorrhagic stroke (20%)	Intracerebral hemorrhage (15%)	Atherosclerosis (not obligatory)	Hypertension, amyloid angiopathy, AV malformation
	Subarachnoidal hemorrhage (5%)	Cerebral aneurysm	Aneurysm, atherosclerosis

Figure 2.3 Pathogenesis of cerebrovascular disease.

heart and brain infarctions are the most prevalent manifestations of vascular disease, the following chapters will describe how obesity influences and alters the risk of coronary heart disease and cerebrovascular disorders as the main causes for morbidity and mortality in developed populations.

3.2. Obesity, obesity genes, and risk of CVD

Obesity is an important risk factor of CVD. It has been shown to be associated with high morbidity and mortality (Van Gaal et al., 2006) of CVD. In particular, abdominal obesity has been identified as a risk factor for CVD (Yusuf et al., 2005). As shown above, obesity is associated with risk factors of CVD, such as diabetes, high blood pressure, hyperlipoproteinemia, and elevated fibrinogen levels (Din-Dzietham et al., 2007; Ritchie & Connell, 2007). Cardiovascular function is impaired in obese individuals. They usually have an increased total blood volume and cardiac output as compared to individuals with normal weight (Lavie et al., 2009). The cardiac workload is also greater in obesity. The increase in cardiac output in obese individuals is largely caused by an elevated stroke volume (Messerli et al., 1987). In addition, there is a mild increase in heart rate associated with obesity. The enlargement of the left ventricular chamber caused by increased filling volume and pressure is often observed in obese individuals (Alpert, 2001; Lavie et al., 2009). Along with hypertension and age, obesity is an independent risk factor of left ventricular hypertrophy (Lavie et al., 2007). With increased circulating blood volume and abnormal left ventricular diastolic filling, obese individuals develop left atrial dilatation, which is associated with increased risk of atrial fibrillation and cardioembolic stroke. In the meta-analysis of population-based studies with 78,602 individuals, obesity increased the risk of atrial fibrillation by 49% in the general population (Wanahita et al., 2008). This risk was growing with increasing BMI.

A number of studies looked for associations between obesity genes and CVD. In the recently published Finnish Diabetes Prevention Study, the common single nucleotide polymorphism (SNP) rs9939609 (T/A) of the *FTO*-associated gene was investigated for its contribution to the risk of CVD (Lappalainen et al., 2011). The AA genotype was shown to be associated with increased risk of CVD in men (OR: 2.1; 95%CI: 1.17–3.73). The association of rs9939609 with a myocardial infarction was replicated in the subgroup of men with type 2 diabetes in a large cross-sectional population-based random sample of 6229 Finnish men (The METabolic Syndrome In Men Study) (Stancakova et al., 2011). In a prospective cohort study of 21,674 U.S. women, the risk allele (A) of the *FTO* rs8050136 showed a significant association with CVD (OR: 1.14; 95%CI: 1.01–1.28). This association was attenuated after adjustment for BMI (OR: 1.10; 95%CI: 0.97–1.23). In this cohort, the increased risk of CVD mediated by BMI also showed an interaction with physical activity. Among carriers of the risk allele, only individuals with reduced physical activity showed an increased risk of CVD. In a study from Prague with 1092 male patients with acute coronary syndrome (ACS) and 1191 healthy controls, *FTO* rs17817449 variant (G allele) was significantly associated with ACS (OR: 1.32; 95%CI: 1.01–1.72) (Hubacek et al., 2010).

He et al. investigated the role of obesity genotype in 1395 women with type 2 diabetes mellitus (He et al., 2010). After adjustment for BMI and other potential covariates, significant associations with increased risk of CVD were found for rs9939609 (*FTO*), rs11084753 (*KCTD15*), rs10838738 (*MTCH2*), and rs10938397 (*GNPDA2*). However, individual genetic effects of these genetic variants on CVD were moderate. In order to increase the effect power, the authors calculate a genetic risk score (GRS), which was derived by adding up risk alleles for each individual (He et al., 2010). This approach was applied to provide a more precise characterization of the individual risk profile. The risk of developing CVD was increased by 53% in individuals in the highest quartile of GRS compared to individuals in the lowest quartile. Interestingly, a higher GRS was associated with decreased adiponectin levels ($p < 0.01$ after adjustment for BMI and other vascular risk factors). The plasma levels of adiponectin are known to be inversely related to the risk of CVD (Pischon et al., 2004; Qi et al., 2005). Secreted by adipocytes, adiponectin improves lipid metabolism and insulin action (Berg et al., 2001). An inverse association with HDL, which is a protective factor for CVD, was also observed for GRS (He et al., 2010). In the study by Brscic et al., various polymorphisms of pro-obesity genes were investigated for their association with myocardial infarction (Brscic et al., 2000). Among other genetic polymorphisms (angiotensin I converting enzyme, angiotensin II type I receptor, apolipoprotein E (ApoE), endothelial constitutive nitric oxide synthase, and platelet glycoprotein IIIa) included in this study, only ApoE (epsilon4 allele) was an independent predictor of myocardial infarction (OR: 6.8; 95%CI: 2.0–22.0).

3.3. Obesity, obesity genes, and risk of cerebrovascular disease

Cerebrovascular diseases are among the major causes of death and the leading cause of disability in adults worldwide (Wolf et al., 1998). Stroke and its sequelae exert a major burden to public health and patients' families

(Epifanov et al., 2007; Kolominsky-Rabas et al., 2006; Meyer et al., 2010; Winter et al., 2008b). The incidence of cerebrovascular diseases is age dependent. It increases exponentially from 3–4/100,000 in people aged 30–40 years to 3000/100,000 in those aged 80–90 years (Bonita, 1992). The common cerebrovascular risk factors are hypertension, diabetes mellitus, hyperlipidemia, smoking, and physical inactivity (Fig. 2.2). Being an important vascular risk factor (Yusuf et al., 2005), obesity has been intensively investigated for its role in CVD in the past decades. However, studies aimed to estimate the cerebrovascular risk related to obesity are complicated by the fact that cerebrovascular diseases represent a heterogeneous group of diseases of both ischemic and hemorrhagic etiologies (Fig. 2.3), which show different associations to different types of obesity.

The best evidence is available for association between abdominal type of obesity and risk of stroke/TIA. Recently, we reported results of the Mannheim–Heidelberg Stroke Study, which showed an association of markers of abdominal obesity with increased risk of stroke/TIA (Winter et al., 2008a). In this large case–control study with 1137 participants (379 cases and 758 controls), cerebrovascular risk increased in graded manner with increasing WHR (Fig. 2.4). After adjustment for age and sex, patients

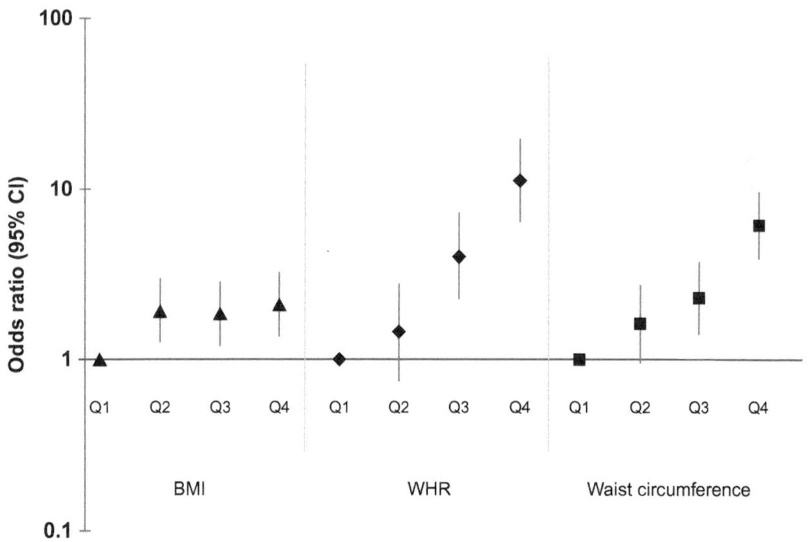

Figure 2.4 Association of the obesity markers BMI, WHR, and waist circumference with risk of stroke. Values are quartiles (Q), adjusted for other risk factors. Odd ratios are shown in quartiles (Q1–Q4). BMI, body mass index; WHR, waist-to-hip ratio. *Modified after Winter et al. (2008a).*

in the highest tertile of WHR had a 12.8-fold (95%CI: 7.8–20.9, $p < 0.001$) risk of stroke/TIA in comparison with the lowest tertile. This association remained significant after adjustment for arterial hypertension, diabetes mellitus, and physical inactivity (OR: 7.7; 95%CI: 4.5–13.0, $p < 0.001$) (Winter et al., 2008a). The association of WHR with risk of stroke/TIA was found both in men and women, showing higher cerebrovascular risk of abdominal obesity in women. Other markers of abdominal obesity, such as WC and waist-to-stature ratio (WSR), also showed a positive association with increased cerebrovascular risk in that study, although these associations were weaker than for WHR (Fig. 2.4). In case of WC, the risk in the highest tertile group was 4.3-fold (95%CI: 2.7–6.8, $p < 0.001$ after adjustment for age, sex, and vascular risk factors) higher than in the lowest tertile group. The comparison of highest and lowest tertile of WSR also revealed significantly increased risk for stroke/TIA (OR: 4.7; 95%CI: 2.8–7.7). The results of the Mannheim–Heidelberg Stroke Study are supported by some previous studies, investigating markers of abdominal obesity for its association with cerebrovascular risk (Hu et al., 2007; Ohira et al., 2006; Suk et al., 2003; Walker et al., 1996). A comparison of highest and lowest quintiles of WHR in a longitudinal study with >28,000 male U.S. professionals revealed an age-adjusted 2.3-fold risk of stroke (95%CI: 1.3–4.4) (Walker et al., 1996). The risk for ischemic stroke was increased in the case–control Northern Manhattan Stroke Study by factor 3.0 (Suk et al., 2003). In a large-scale longitudinal cohort study from Finland, WHR was weakly associated with cerebrovascular risk in men (OR: 1.6; 95%CI: 1.1–2.3, highest vs. lowest tertiles) (Hu et al., 2007). This study did not include patients with transient ischemic attacks and patients with coronary heart disease at baseline. The contribution of obesity to cerebrovascular risk may differ depending on stroke etiology. For example, the longitudinal Atherosclerosis Risk in Communities study found a positive association of abdominal obesity with risk for nonlacunar, but not for lacunar (microangiopathic) stroke. There are inconsistent data on contribution of obesity to risk of hemorrhagic stroke with studies showing positive (Kurth et al., 2002), reverse (Rexrode et al., 1997), and no associations (Jood et al., 2004). Future research should concentrate on the more precise evaluation of obesity-related cerebrovascular risk in different subtypes of stroke.

Unlike markers of abdominal obesity, BMI is apparently not an appropriate marker of cerebrovascular risk. In the Mannheim–Heidelberg Stroke Study, BMI did not show a graded relation to risk of stroke/TIA (Winter et al., 2008a). The association of BMI and cerebrovascular risk became nonsignificant after adjustment for vascular risk factors. Similarly, the Copenhagen City Heart Study and two other studies did not find associations of BMI with risk of stroke (Khaw et al., 1984; Lindenstrom et al., 1993; Lu et al., 2006).

Data on contribution of obesity genes to risk of stroke are limited. In the genetics of stroke one should differentiate between the genetics of well-defined rare causes of monogenic cerebrovascular disease, such as Fabry's disease or cerebral autosomal dominant arteriopathy with subcortical infarcts and leukoencephalopathy, and common stroke, which is believed to have a polygenic mode of inheritance. The majority of genetic association studies evaluating candidate genes of common stroke concentrated on genes involved in regulation of coagulation pathways (Casas et al., 2004) and disregarded genes associated with vascular risk factors, such as arterial hypertension, diabetes, hyperlipidemia, and obesity. Studies investigating the association of genes of diabetes mellitus with risk of stroke are scarce and their results are inconsistent (Lee et al., 2006; Zafarmand et al., 2008). To the best of our knowledge, our recently published study was the first to evaluate the contribution of obesity candidate genes in the risk of common stroke (Winter et al., 2011). In that study, the following SNPs of obesity genes were evaluated: rs9937053 (*FTO*, obesity risk allele A) (Frayling et al., 2007), rs2229616 (*MC4R* V103I, obesity risk allele A) (Geller et al., 2004), and rs17782313 (188 kb downstream of *MC4R*, obesity risk allele C) (Loos et al., 2008). As it is known from large-scale genetic studies, common variants of *FTO* and *MC4R* are associated with BMI (Frayling et al., 2007; Loos et al., 2008). The association of *MC4R* with abdominal obesity and its role in the development of insulin resistance and metabolic syndrome was discussed in previous publications (Chambers et al., 2008; Heid et al., 2008). The association with vascular risk factors (e.g., metabolic syndrome) was also reported for *FTO* (Al-Attar et al., 2008). There are data showing a trend toward increased carotid intima-media thickness, a marker of atherosclerosis, in individuals with the A allele of the *FTO* SNP rs9939609 (Kivimaki et al., 2008). Although previous studies showed an association of *MC4R* and FTO with cerebrovascular risk factors, no association with risk of stroke was found in our above-mentioned study (Winter et al., 2011). However, a weak association of *FTO* with risk of TIA was found (OR: 1.6; 95%CI: 0.9–2.6, $p=0.08$).

The role of gene *TCF7L2*, the association of which with type 2 diabetes mellitus has been replicated in different studies (Cauchi et al., 2007), has also been evaluated for its risk of stroke/TIA (Winter et al., 2011). The *TCF7L2* polymorphism rs7903146 has been reported to have pro-obesity effects and to be associated with increased mortality of vascular disease (Sousa et al., 2009). Similarly to *FTO* and *MC4R*, no association of *TCF7L2* with risk of stroke was found (Winter et al., 2011). Interestingly, a sex-specific influence of *TCF7L2* (T allele) on the functional outcome after stroke has been demonstrated (Winter et al., 2011) and deserves further investigation.

Studies investigating genetic determinants of outcome in cerebrovascular disease are rare. There are reports on association of genotypes of interleukin-6, insulin-like growth factor I, and interleukin 1 receptor antagonist with disability in cerebrovascular diseases (Greisenegger et al., 2003; Gromadzka et al., 2007; van Rijn et al., 2006). A large genetic study investigating genetic determinants of cerebrovascular risk and outcome after stroke is currently underway (Ramoni et al., 2009). In general, future studies investigating candidate genes of common stroke should pay more attention to the evaluation of genes associated with vascular risk factors.

4. CONCLUSIONS

The number of identified genetic polymorphisms associated with obesity is continuously increasing. However, all discovered polymorphisms taken together contribute to less than 2% of common human obesity. In order to facilitate the progress in discovery of new obesity-related polymorphisms, large-scale efforts, such as meta-analysis of multiple genome-wide association studies, are required (Willer et al., 2009).

Despite the fact, that (abdominal) obesity is an independent vascular risk factor, the role of genetic determinants of obesity in vascular diseases is not sufficiently investigated. In particular, there are only few studies on obesity-related genes in cerebrovascular disease (Winter et al., 2011). While searching for candidate genes of common stroke, genetic association studies should concentrate on genes that are associated with vascular risk factors, such as obesity, diabetes mellitus, dyslipidemia, arterial hypertension, and the metabolic syndrome. In addition, the role of abdominal obesity and its genetic determinants in the development of vascular disease should be considered. Obviously, there are differences in the expression of genes in human tissues depending on their role in the morphological distribution of fat mass. The majority of genes associated with BMI are highly expressed in the brain, suggesting genetic contribution to common human obesity through CNS processes. In case of genes associated with abdominal obesity, expression was detected outside the CNS in the endoplasmatic reticulum of enterocytes and hepatocytes (Berthier et al., 2004). Finally, a better understanding of epigenetic factors, which contribute to development of obesity and obesity-related vascular diseases, will be of great importance.

REFERENCES

Al-Attar, S. A., Pollex, R. L., Ban, M. R., Young, T. K., Bjerregaard, P., Anand, S. S., et al. (2008). Association between the FTO rs9939609 polymorphism and the metabolic syndrome in a non-Caucasian multi-ethnic sample. *Cardiovascular Diabetology*, 7, 5.

Alpert, M. A. (2001). Obesity cardiomyopathy: Pathophysiology and evolution of the clinical syndrome. *The American Journal of the Medical Sciences*, *321*, 225–236.

Arcaro, G., Zamboni, M., Rossi, L., Turcato, E., Covi, G., Armellini, F., et al. (1999). Body fat distribution predicts the degree of endothelial dysfunction in uncomplicated obesity. *International Journal of Obesity and Related Metabolic Disorders*, *23*, 936–942.

Benzinou, M., Creemers, J. W., Choquet, H., Lobbens, S., Dina, C., Durand, E., et al. (2008). Common nonsynonymous variants in PCSK1 confer risk of obesity. *Nature Genetics*, *40*, 943–945.

Berg, A. H., Combs, T. P., Du, X., Brownlee, M., & Scherer, P. E. (2001). The adipocyte-secreted protein Acrp30 enhances hepatic insulin action. *Nature Medicine*, 7, 947–953.

Berthier, M. T., Houde, A., Paradis, A. M., Couture, P., Gaudet, D., Després, J. P., et al. (2004). Molecular screening of the microsomal triglyceride transfer protein: Association between polymorphisms and both abdominal obesity and plasma apolipoprotein B concentration. *Journal of Human Genetics*, *49*, 684–690.

Bonita, R. (1992). Epidemiology of stroke. *The Lancet*, *339*, 342–344.

Brscic, E., Bergerone, S., Gagnor, A., Colajanni, E., Matullo, G., Scaglione, L., et al. (2000). Acute myocardial infarction in young adults: Prognostic role of angiotensin-converting enzyme, angiotensin II type I receptor, apolipoprotein E, endothelial constitutive nitric oxide synthase, and glycoprotein IIIa genetic polymorphisms at medium-term follow-up. *American Heart Journal*, *139*, 979–984.

Casas, J. P., Hingorani, A. D., Bautista, L. E., & Sharma, P. (2004). Meta-analysis of genetic studies in ischemic stroke: Thirty-two genes involving approximately 18,000 cases and 58,000 controls. *Archives of Neurology*, *61*, 1652–1661.

Cauchi, S., Meyre, D., Choquet, H., Deghmoun, S., Durand, E., Gaget, S., et al. (2007). TCF7L2 is reproducibly associated with type 2 diabetes in various ethnic groups: A global meta-analysis. *Journal of Molecular Medicine*, *85*, 777–782.

Chambers, J. C., Elliott, P., Zabaneh, D., Zhang, W., Li, Y., Froguel, P., Balding, D., et al. (2008). Common genetic variation near MC4R is associated with waist circumference and insulin resistance. *Nature Genetics*, *40*, 716–718.

Chiuve, S. E., et al. (2008). Primary prevention of stroke by healthy lifestyle. *Circulation*, *118*, 947–954.

Colditz, G. A. (1999). Economic costs of obesity and inactivity. *Medicine and Science in Sports and Exercise*, *31*, S663–S667.

Creemers, J. W., Choquet, H., Stijnen, P., Vatin, V., Pigeyre, M., Beckers, S., et al. (2012). Heterozygous mutations causing partial prohormone convertase 1 deficiency contribute to human obesity. *Diabetes*, *61*, 383–390.

D'Agostino, R. B., Vasan, R. S., Pencina, M. J., Wolf, P. A., Cobain, M., Massaro, J. M., et al. (2008). General cardiovascular risk profile for use in primary care: The Framingham Heart Study. *Circulation*, *117*, 743–753.

Din-Dzietham, R., Liu, Y., Bielo, M. V., & Shamsa, F. (2007). High blood pressure trends in children and adolescents in national surveys, 1963 to 2002. *Circulation*, *116*, 1488–1496.

Dodel, R., Winter, Y., Ringel, F., Spottke, A., Gharevi, N., Müller, I., et al. (2010). Cost of illness in subarachnoid hemorrhage: A german longitudinal study. *Stroke, 41*, 2918–2923.

Epifanov, Y., Dodel, R., Haacke, C., Schaeg, M., Schöffski, O., Hennerici, M., et al. (2007). Costs of acute stroke care on regular neurological wards: A comparison with stroke unit setting. *Health Policy, 81*, 339–349.

Frayling, T. M., Timpson, N. J., Weedon, M. N., Zeggini, E., Freathy, R. M., Lindgren, C. M., et al. (2007). A common variant in the FTO gene is associated with body mass index and predisposes to childhood and adult obesity. *Science, 316*, 889–894.

Geller, F., Reichwald, K., Dempfle, A., Illig, T., Vollmert, C., Herpertz, S., et al. (2004). Melanocortin-4 receptor gene variant I103 is negatively associated with obesity. *The American Journal of Human Genetics, 74*, 572–581.

Goldstein, L. B., Adams, R., Alberts, M. J., Appel, L. J., Brass, L. M., Bushnell, C. D., et al. (2006). Primary prevention of ischemic stroke: A guideline from the American Heart Association/American Stroke Association Stroke Council: Cosponsored by the Atherosclerotic Peripheral Vascular Disease Interdisciplinary Working Group; Cardiovascular Nursing Council; Clinical Cardiology Council; Nutrition, Physical Activity, and Metabolism Council; and the Quality of Care and Outcomes Research Interdisciplinary Working Group. *Circulation, 113*, e873–e923.

Greisenegger, S., Endler, G., Haering, D., Schillinger, M., Lang, W., Lalouschek, W., et al. (2003). The (−174) G/C polymorphism in the interleukin-6 gene is associated with the severity of acute cerebrovascular events. *Thrombosis Research, 110*, 181–186.

Gromadzka, G., Sarzynska-Dlugosz, I., & Czlonkowska, A. (2007). IL1RN intron 2 polymorphism caused by variable number tandem repeats is associated with 1-year outcome in patients with ischaemic stroke. *Journal of Neurology, Neurosurgery, and Psychiatry, 78*, 183–186.

Hamid, Y. H., Urhammer, S. A., Glümer, C., Borch-Johnsen, K., Jørgensen, T., Hansen, T., et al. (2005). The common T60N polymorphism of the lymphotoxin-alpha gene is associated with type 2 diabetes and other phenotypes of the metabolic syndrome. *Diabetologia, 48*, 445–451.

He, M., Cornelis, M. C., Franks, P. W., Zhang, C., Hu, F. B., & Qi, L. (2010). Obesity genotype score and cardiovascular risk in women with type 2 diabetes mellitus. *Arteriosclerosis, Thrombosis, and Vascular Biology, 30*, 327–332.

Hedley, A. A., Ogden, C. L., Johnson, C. L., Carroll, M. D., Curtin, L. R., & Flegal, K. M. (2004). Prevalence of overweight and obesity among US children, adolescents, and adults, 1999–2002. *The Journal of the American Medical Association, 291*, 2847–2850.

Heid, I. M., Vollmert, C., Kronenberg, F., Huth, C., Ankerst, D. P., Luchner, A., et al. (2008). Association of the MC4R V103I polymorphism with the metabolic syndrome: The KORA Study. *Obesity (Silver Spring), 16*, 369–376.

Hu, G., Tuomilehto, J., Silventoinen, K., Sarti, C., Männistö, S., & Jousilahti, P. (2007). Body mass index, waist circumference, and waist-hip ratio on the risk of total and type-specific stroke. *Archives of Internal Medicine, 167*, 1420–1427.

Hubacek, J. A., Stanek, V., Gebauerová, M., Pilipcincová, A., Dlouhá, D., Poledne, R., et al. (2010). A FTO variant and risk of acute coronary syndrome. *Clinica Chimica Acta, 411*, 1069–1072.

Jones, D. W., Kim, J. S., Andrew, M. E., Kim, S. J., & Hong, Y. P. (1994). Body mass index and blood pressure in Korean men and women: The Korean National Blood Pressure Survey. *Journal of Hypertension, 12*, 1433–1437.

Jood, K., Jern, C., Wilhelmsen, L., & Rosengren, A. (2004). Body mass index in mid-life is associated with a first stroke in men: A prospective population study over 28 years. *Stroke, 35*, 2764–2769.

Katzov, H., Bennet, A. M., Kehoe, P., Wiman, B., Gatz, M., Blennow, K., et al. (2004). A cladistic model of ACE sequence variation with implications for myocardial infarction, Alzheimer disease and obesity. *Human Molecular Genetics, 13*, 2647–2657.

Kenchaiah, S., Evans, J. C., Levy, D., Wilson, P. W., Benjamin, E. J., Larson, M. G., et al. (2002). Obesity and the risk of heart failure. *The New England Journal of Medicine, 347*, 305–313.

Khaw, K. T., Barrett-Connor, E., Suarez, L., & Criqui, M. H. (1984). Predictors of stroke-associated mortality in the elderly. *Stroke, 15*, 244–248.

Khaw, K. T., Wareham, N., Bingham, S., Welch, A., Luben, R., & Day, N. (2008). Combined impact of health behaviours and mortality in men and women: The EPIC-Norfolk prospective population study. *PLoS Medicine, 5*, e12.

Kim, K. S., Choi, S. M., Shin, S. U., Yang, H. S., & Yoon, Y. (2004). Effects of peroxisome proliferator-activated receptor-gamma 2 Pro12Ala polymorphism on body fat distribution in female Korean subjects. *Metabolism, 53*, 1538–1543.

Kim, K. S., Cho, D. Y., Kim, Y. J., Choi, S. M., Kim, Y. J., Shin, S. U., et al. (2005). The finding of new genetic polymorphism of UCP-1 A-1766G and its effects on body fat accumulation. *Biochimica et Biophysica Acta, 1741*, 149–155.

Kivimaki, M., Smith, G. D., Timpson, N. J., Lawlor, D. A., Batty, G. D., Kähönen, M., et al. (2008). Lifetime body mass index and later atherosclerosis risk in young adults: Examining causal links using Mendelian randomization in the cardiovascular risk in Young Finns Study. *European Heart Journal, 29*, 2552–2560.

Kolominsky-Rabas, P. L., Heuschmann, P. U., Marschall, D., Emmert, M., Baltzer, N., Neundörfer, B., et al. (2006). Lifetime cost of ischemic stroke in Germany: Results and national projections from a population-based stroke registry: The Erlangen Stroke Project. *Stroke, 37*, 1179–1183.

Kurth, T., Gaziano, J. M., Berger, K., Kase, C. S., Rexrode, K. M., Cook, N. R., et al. (2002). Body mass index and the risk of stroke in men. *Archives of Internal Medicine, 162*, 2557–2562.

Lange, L. A., Norris, J. M., Langefeld, C. D., Nicklas, B. J., Wagenknecht, L. E., Saad, M. F., et al. (2005). Association of adipose tissue deposition and beta-2 adrenergic receptor variants: The IRAS family study. *International Journal of Obesity, 29*, 449–457.

Lappalainen, T., Kolehmainen, M., Schwab, U. S., Tolppanen, A. M., Stančáková, A., Lindström, J., et al. (2011). Association of the FTO gene variant (rs9939609) with cardiovascular disease in men with abnormal glucose metabolism—The Finnish Diabetes Prevention Study. *Nutrition, Metabolism, and Cardiovascular Diseases, 21*, 691–698.

Lara-Castro, C., Hunter, G. R., Lovejoy, J. C., Gower, B. A., & Fernández, J. R. (2005a). Apolipoprotein A-II polymorphism and visceral adiposity in African-American and white women. *Obesity Research, 13*, 507–512.

Lara-Castro, C., Hunter, G. R., Lovejoy, J. C., Gower, B. A., & Fernández, J. R. (2005b). Association of the intestinal fatty acid-binding protein Ala54Thr polymorphism and abdominal adipose tissue in African-American and Caucasian women. *The Journal of Clinical Endocrinology and Metabolism, 90*, 1196–1201.

Lavie, C. J., Milani, R. V., Ventura, H. O., Cardenas, G. A., Mehra, M. R., & Messerli, F. H. (2007). Disparate effects of left ventricular geometry and obesity on mortality in patients with preserved left ventricular ejection fraction. *The American Journal of Cardiology, 100*, 1460–1464.

Lavie, C. J., Milani, R. V., & Ventura, H. O. (2009). Obesity and cardiovascular disease: Risk factor, paradox, and impact of weight loss. *Journal of the American College of Cardiology, 53*, 1925–1932.

Lean, M. E., Han, T. S., & Seidell, J. C. (1998). Impairment of health and quality of life in people with large waist circumference. *The Lancet, 351*, 853–856.

Lee, B. C., Lee, H. J., & Chung, J. H. (2006). Peroxisome proliferator-activated receptor-gamma2 Pro12Ala polymorphism is associated with reduced risk for ischemic stroke with type 2 diabetes. *Neuroscience Letters*, *410*, 141–145.

Lindenstrom, E., Boysen, G., & Nyboe, J. (1993). Lifestyle factors and risk of cerebrovascular disease in women. The Copenhagen City Heart Study. *Stroke*, *24*, 1468–1472.

Loos, R. J., Lindgren, C. M., Li, S., Wheeler, E., Zhao, J. H., Prokopenko, I., Inouye, M., et al. (2008). Common variants near MC4R are associated with fat mass, weight and risk of obesity. *Nature Genetics*, *40*, 768–775.

Lu, M., Ye, W., Adami, H. O., & Weiderpass, E. (2006). Prospective study of body size and risk for stroke amongst women below age 60. *Journal of Internal Medicine*, *260*, 442–450.

Manson, J. E., & Bassuk, S. S. (2003). Obesity in the United States: A fresh look at its high toll. *The Journal of the American Medical Association*, *289*, 229–230.

Matsuzawa, Y. (2008). The role of fat topology in the risk of disease. *International Journal of Obesity*, *32*(Suppl. 7), S83–S92.

McGill, H. C., Jr., McMahan, C. A., Herderick, E. E., Zieske, A. W., Malcom, G. T., Tracy, R. E., et al. (2002). Obesity accelerates the progression of coronary atherosclerosis in young men. *Circulation*, *105*, 2712–2718.

Messerli, F. H., Nunez, B. D., Ventura, H. O., & Snyder, D. W. (1987). Overweight and sudden death. Increased ventricular ectopy in cardiopathy of obesity. *Archives of Internal Medicine*, *147*, 1725–1728.

Meyer, B., Ringel, F., Winter, Y., Spottke, A., Gharevi, N., Dams, J., et al. (2010). Health-related quality of life in patients with subarachnoid haemorrhage. *Cerebrovascular Diseases*, *30*, 423–431.

Nguyen, N. T., Magno, C. P., Lane, K. T., Hinojosa, M. W., & Lane, J. S. (2008). Association of hypertension, diabetes, dyslipidemia, and metabolic syndrome with obesity: Findings from the National Health and Nutrition Examination Survey, 1999 to 2004. *Journal of the American College of Surgeons*, *207*, 928–934.

Nielsen, L. B., Véniant, M., Borén, J., Raabe, M., Wong, J. S., Tam, C., et al. (1998). Genes for apolipoprotein B and microsomal triglyceride transfer protein are expressed in the heart: Evidence that the heart has the capacity to synthesize and secrete lipoproteins. *Circulation*, *98*, 13–16.

Ohira, T., Shahar, E., Chambless, L. E., Rosamond, W. D., Mosley, T. H., Jr., & Folsom, A. R. (2006). Risk factors for ischemic stroke subtypes: The Atherosclerosis Risk in Communities Study. *Stroke*, *37*, 2493–2498.

Okosun, I. S., Boltri, J. M., Anochie, L. K., & Chandra, K. M. (2004). Racial/ethnic differences in prehypertension in American adults: Population and relative attributable risks of abdominal obesity. *Journal of Human Hypertension*, *18*, 849–855.

Ozaki, K., Ohnishi, Y., Iida, A., Sekine, A., Yamada, R., Tsunoda, T., et al. (2002). Functional SNPs in the lymphotoxin-alpha gene that are associated with susceptibility to myocardial infarction. *Nature Genetics*, *32*, 650–654.

Phillips, L. K., & Prins, J. B. (2008). The link between abdominal obesity and the metabolic syndrome. *Current Hypertension Reports*, *10*, 156–164.

Pischon, T., Girman, C. J., Hotamisligil, G. S., Rifai, N., Hu, F. B., & Rimm, E. B. (2004). Plasma adiponectin levels and risk of myocardial infarction in men. *The Journal of the American Medical Association*, *291*, 1730–1737.

Qi, L., Shen, H., Larson, I., Schaefer, E. J., Greenberg, A. S., Tregouet, D. A., et al. (2004). Gender-specific association of a perilipin gene haplotype with obesity risk in a white population. *Obesity Research*, *12*, 1758–1765.

Qi, L., Li, T., Rimm, E., Zhang, C., Rifai, N., Hunter, D., et al. (2005). The +276 polymorphism of the APM1 gene, plasma adiponectin concentration, and cardiovascular risk in diabetic men. *Diabetes*, *54*, 1607–1610.

Rabin, B. A., Boehmer, T. K., & Brownson, R. C. (2007). Cross-national comparison of environmental and policy correlates of obesity in Europe. *European Journal of Public Health, 17*, 53–61.

Ramoni, R. B., Himes, B. E., Sale, M. M., Furie, K. L., & Ramoni, M. F. (2009). Predictive genomics of cardioembolic stroke. *Stroke, 40*, S67–S70.

Rankinen, T., Zuberi, A., Chagnon, Y. C., Weisnagel, S. J., Argyropoulos, G., Walts, B., et al. (2006). The human obesity gene map: The 2005 update. *Obesity (Silver Spring), 14*, 529–644.

Rexrode, K. M., Hennekens, C. H., Willett, W. C., Colditz, G. A., Stampfer, M. J., Rich-Edwards, J. W., et al. (1997). A prospective study of body mass index, weight change, and risk of stroke in women. *The Journal of the American Medical Association, 277*, 1539–1545.

Ritchie, S. A., & Connell, J. M. (2007). The link between abdominal obesity, metabolic syndrome and cardiovascular disease. *Nutrition, Metabolism, and Cardiovascular Diseases, 17*, 319–326.

Ross, R. (1999). Atherosclerosis—An inflammatory disease. *The New England Journal of Medicine, 340*, 115–126.

Scuteri, A., Sanna, S., Chen, W. M., Uda, M., Albai, G., Strait, J., et al. (2007). Genome-wide association scan shows genetic variants in the FTO gene are associated with obesity-related traits. *PLoS Genetics, 3*, e115.

Sousa, A. G., Marquezine, G. F., Lemos, P. A., Martinez, E., Lopes, N., Hueb, W. A., et al. (2009). TCF7L2 polymorphism rs7903146 is associated with coronary artery disease severity and mortality. *PLoS One, 4*, e7697.

Stancakova, A., Paananen, J., Soininen, P., Kangas, A. J., Bonnycastle, L. L., Morken, M. A., et al. (2011). Effects of 34 risk loci for type 2 diabetes or hyperglycemia on lipoprotein subclasses and their composition in 6,580 nondiabetic Finnish men. *Diabetes, 60*, 1608–1616.

Suk, S. H., Sacco, R. L., Boden-Albala, B., Cheun, J. F., Pittman, J. G., Elkind, M. S., et al. (2003). Abdominal obesity and risk of ischemic stroke: The Northern Manhattan Stroke Study. *Stroke, 34*, 1586–1592.

Sutton, B. S., Weinert, S., Langefeld, C. D., Williams, A. H., Campbell, J. K., Saad, M. F., et al. (2005). Genetic analysis of adiponectin and obesity in Hispanic families: The IRAS Family Study. *Human Genetics, 117*, 107–118.

Thorleifsson, G., Walters, G. B., Gudbjartsson, D. F., Steinthorsdottir, V., Sulem, P., Helgadottir, A., et al. (2009). Genome-wide association yields new sequence variants at seven loci that associate with measures of obesity. *Nature Genetics, 41*, 18–24.

Van Gaal, L. F., Mertens, I. L., & De Block, C. E. (2006). Mechanisms linking obesity with cardiovascular disease. *Nature, 444*, 875–880.

van Rijn, M. J., Slooter, A. J., Bos, M. J., Catarino, C. F., Koudstaal, P. J., Hofman, A., et al. (2006). Insulin-like growth factor I promoter polymorphism, risk of stroke, and survival after stroke: The Rotterdam study. *Journal of Neurology, Neurosurgery and Psychiatry, 77*, 24–27.

von Kummer, R., & Back, T. (Eds.), (2006). *MR imaging in ischemic stroke*. Berlin/Heidelberg/New York: Springer-Verlag.

Walker, S. P., Rimm, E. B., Ascherio, A., Kawachi, I., Stampfer, M. J., & Willett, W. C. (1996). Body size and fat distribution as predictors of stroke among US men. *American Journal of Epidemiology, 144*, 1143–1150.

Wanahita, N., Messerli, F. H., Bangalore, S., Gami, A. S., Somers, V. K., Steinberg, J. S., et al. (2008). Atrial fibrillation and obesity—Results of a meta-analysis. *American Heart Journal, 155*, 310–315.

Willer, C. J., Speliotes, E. K., Loos, R. J., Li, S., Lindgren, C. M., et al. (2009). Six new loci associated with body mass index highlight a neuronal influence on body weight regulation. *Nature Genetics, 41*, 25–34.

Winter, Y., Rohrmann, S., Linseisen, J., Lanczik, O., Ringleb, P. A., Hebebrand, J., et al. (2008a). Contribution of obesity and abdominal fat mass to risk of stroke and transient ischemic attacks. *Stroke, 39*, 3145–3151.

Winter, Y., Wolfram, C., Schöffski, O., Dodel, R. C., Back, T., et al. (2008b). Long-term disease-related costs 4 years after stroke or TIA in Germany. *Der Nervenarzt, 79*, 918–920 922–4, 926.

Winter, Y., Back, T., Scherag, A., Linseisen, J., Rohrmann, S., Lanczik, O., et al. (2011). Evaluation of the obesity genes FTO and MC4R and the type 2 diabetes mellitus gene TCF7L2 for contribution to stroke risk: The Mannheim-Heidelberg Stroke Study. *Obesity Facts, 4*, 290–296.

Wirth, A. (2008). *Adipositas*. Heidelberg: Springer.

Wolf, P., et al. (1998). Epidemiology of stroke. In M. Ginsberg & J. Bogousslavsky (Eds.), *Cerebrovascular disease: Pathophysiology, diagnosis and management* (pp. 834–850). Malden, MA: Blackwell Science.

World Health Organization. (2011). Available at: http://www.who.int/topics/obesity/en/.

Yusuf, S., Hawken, S., Ounpuu, S., Bautista, L., Franzosi, M. G., Commerford, P., et al. (2005). Obesity and the risk of myocardial infarction in 27000 participants from 52 countries: A case-control study. *The Lancet, 366*, 1640–1649.

Zafarmand, M. H., van der Schouw, Y. T., Grobbee, D. E., de Leeuw, P. W., & Bots, M. L. (2008). Peroxisome proliferator-activated receptor gamma-2 P12A polymorphism and risk of acute myocardial infarction, coronary heart disease and ischemic stroke: A case-cohort study and meta-analyses. *Vascular Health and Risk Management, 4*, 427–436.

CHAPTER THREE

Brd2 Gene Disruption Causes "Metabolically Healthy" Obesity: Epigenetic and Chromatin-Based Mechanisms that Uncouple Obesity from Type 2 Diabetes

Fangnian Wang[*], Jude T. Deeney[†], Gerald V. Denis[‡,§,¶,||,1]
[*]IQuum Inc, Marlborough, MA 01752, USA
[†]Department of Medicine, Boston University School of Medicine, Boston, Massachusetts, USA
[‡]Flow Cytometry Core Facility, Boston University School of Medicine, Boston, Massachusetts, USA
[§]Nutrition Obesity Research Center, Boston University School of Medicine, Boston, Massachusetts, USA
[¶]Cancer Research Center, Boston University School of Medicine, Boston, Massachusetts, USA
[||]Department of Pharmacology and Experimental Therapeutics, Boston University School of Medicine, Boston, Massachusetts, USA
[1]Corresponding author: e-mail address: gdenis@bu.edu

Contents

Abstract

Disturbed body energy balance can lead to obesity and obesity-driven diseases such as Type 2 diabetes, which have reached an epidemic level. Evidence indicates that obesity-induced inflammation is a major cause of insulin resistance and Type 2 diabetes. Environmental factors, such as nutrients, affect body energy balance through epigenetic or chromatin-based mechanisms. As a bromodomain and external domain family

Vitamins and Hormones, Volume 91
ISSN 0083-6729
http://dx.doi.org/10.1016/B978-0-12-407766-9.00003-1

transcription regulator, Brd2 regulates expression of many genes through interpretation of chromatin codes and participates in the regulation of body energy balance and immune function. In the severely obese state, Brd2 knockdown in mice prevented obesity-induced inflammatory responses, protected animals from insulin resistance, glucose intolerance and pancreatic beta cell dysfunction, and thus uncoupled obesity from diabetes. Brd2 provides an important model for investigation of the function of transcription regulators and the development of obesity and diabetes; it also provides a possible, innovative target to treat obesity and diabetes through modulation of the function of a chromatin code reader.

ABBREVIATIONS

BET bromodomain and extraterminal domain
BMI body mass index
HOMA homeostatic metabolic assessment
MHC major histocompatibility complex
PPAR peroxisome proliferator-activated receptor
SNP single nucleotide polymorphisms
TNF tumor necrosis factor
TWEAK TNF-related weak inducer of apoptosis
TZD thiazolidinedione

1. INTRODUCTION: THE PROBLEM OF OBESITY AND ITS COMPLICATIONS

Over the course of recorded history, many different diseases and risks have threatened human survival on a wide scale, including epidemics of infectious organisms, war, starvation, famine, sudden climate change, and ethnic cleansing. As recently as the end of the Second World War, a chronic lack of calories in Europe and Japan was widely acknowledged as a major contributor to poor health, elevated infant mortality, increased disease susceptibility (particularly to tuberculosis), and its attendant consequences of economic and political instability. The end of that war brought the benefits of modern peace: a dramatic transformation of the international economy, accompanied by huge increases in industrial and agricultural output, vast international trade in manufactured goods, raw materials and foods (both processed and unprocessed), and a pronounced shift from manual to sedentary labor. These tremendous structural and economic changes, combined with increased life expectancy and the rise of agribusiness, advertising, and a hugely successful international food industry, have contributed to the transformation of obesity from a quaint feature of royal families in former times to an alarming, international public health worry that affects all socio-economic classes at present. Thus, of the serious and widespread menaces to public health in human history, obesity is the newest challenge.

Obesity is defined as a body mass index (BMI) of $\geq$30, calculated as body weight in kilograms divided by the square of height in centimeters. Lean individuals are defined as BMI <25 and individuals with intermediate BMI ($\geq$25 to <30) are classified as overweight. Morbid obesity is defined as BMI $\geq$40 and super obesity is defined as BMI $\geq$50. These higher categories of obesity are used to be extremely rare, and they still are relatively rare. However, as the U.S. population distribution shifts to higher mean BMI, these extreme forms of obesity are increasing in prevalence at a greater rate than the mean. According to the most recent statistics from the U.S. Centers for Disease Control (CDC, 2010 data), all U.S. states now report at least 20% prevalence of obesity among adults. Mississippi, West Virginia, and Alabama currently report the highest rates (34.0%, 32.5%, and 32.2%, respectively), but several other states are not far behind. Worldwide, 1.7 billion people are classified as overweight (Haslam & James, 2009). These statistics pose fundamental and novel problems for health maintenance and delivery systems.

The speed with which overweight and obesity have become a public health crisis is breathtaking http://www.cdc.gov/obesity/data/adult.html. For a useful meta-analysis, see Finucane et al. (2011). Obesity in the United States is associated with a dramatic increase in the prevalence of obesity-associated diseases, particularly cardiovascular disease and Type 2 diabetes, as well as hypertension, stroke, metabolic syndrome and insulin resistance, nonalcoholic fatty liver disease, and some forms of cancer, notably breast cancer in postmenopausal women and adult colorectal cancer in both men and women. About 90% of Type 2 diabetes is attributable to excess weight (Hossain, Kawar, & El Nahas, 2007). The incidence of Type 2 diabetes, in particular, has now reached epidemic proportions in the United States and internationally, affecting about 12% of U.S. adults and more than 25% of those over the age of 65. Type 2 diabetes is thought to be responsible for 4.6 million deaths worldwide each year. An oft-cited early estimate of the incidence of Type 2 diabetes was 366 million worldwide by 2030 (Wild, Roglic, Green, Sicree, & King, 2004). However, more recent estimates have pegged the anticipated number of diabetic individuals to be 439 million by 2030 (Shaw, Sicree, & Zimmet, 2010), almost entirely due to obesity-driven metabolic complications. There is reason to worry that this prediction will soon have to be revised upward yet again.

Unless reversed, the rapidly worsening problem of obesity predicts an epidemic of comorbidities that will strain or break many health care delivery systems, both in the United States and internationally. In the absence of national agreement about the proper role of U.S. health care delivery systems with controlled costs, or political will to deliver well-designed solutions, the U.S. economy is uniquely vulnerable to this expected impact. Thus, obesity poses a critical challenge of overarching importance for U.S. public health, particularly pediatric health: many American children are already being swept along with this national wave of progressive metabolic dysfunction in obesity. For the purpose of this chapter, some of the genes and epigenetic forces that drive obesity and metabolic dysfunction will be examined.

2. COMPLEX POLYGENIC INTERACTIONS WITH THE ENVIRONMENT AND EPIGENETICS IN OBESITY

We live in an era of genetic determinism. Major effort has been invested to identify "obesity-associated genes" with the translational goal of isolating the "drivers" from the "passengers," much the same way that oncogenes have been identified, in order to develop novel therapeutic targets. Evidence for a genetic component that influences risk for obesity controlling for environment has been developed through studies of adoption and twins (Maes, Neale, & Eaves, 1997, Silventoinen, Rokholm, Kaprio, & Sørensen, 2010). Genetic factors appear to explain as much as 90% of the variance in BMI among certain monozygotic twins. Furthermore, several recent genome-wide association studies have established convergent validity in the discovery of genes that are strongly associated with obesity throughout the human lifespan. The first robust association was reported for *FTO* (fat mass and obesity associated) (Frayling et al., 2007). This association was joined by insulin-induced gene 2 (*INSIG2*), melanocortin 4 receptor (*MC4R*), transmembrane protein 18 (*TMEM18*), glucosamine-6-phosphate deaminase 2 (*GNPDA2*), neuronal growth regulator 1 (*NEGR1*), brain-derived neurotrophic factor (*BDNF*), and potassium channel tetramerization domain containing 15 (*KCTD15*), as associated with pediatric obesity (den Hoed et al., 2010; Zhao et al., 2009). Intensive research in molecular genetics continues on other well-established loci of obesity-associated genes, including those that encode the uncoupling proteins (Oppert et al., 1994; Ricquier, Casteilla, & Bouillaud, 1991), peroxisome proliferator-activated receptor (PPAR)γ (Lefebvre et al., 1998; Vidal-Puig et al., 1997), low-density lipoprotein receptor (Morris, Zee, & Robinson, 1994), hormone-sensitive lipase (Klannemark et al., 1998; Stich et al., 1997), beta adrenergic receptors (Reynisdottir, Ellerfeldt,

Wahrenberg, Lithell, & Arner, 1994; Walston et al., 1995), and inflammatory cytokine genes such as tumor necrosis factor (TNF)-α (Saghizadeh, Ong, Garvey, Henry, & Kern, 1996) and interleukin-6 (Kern, 1997; Kern, Ranganathan, Li, Wood, & Ranganathan, 2001). The hope of these studies has been that, through identification and characterization of candidate genes for obesity and metabolic syndrome, potent new drugs will be developed that specifically target these proteins or pathways and will have therapeutic value for prediabetic or diabetic obese patients.

Although more than 100 genes have been identified that influence body weight (Leibel, 2008), the usual patterns and rules of Mendelian inheritance have proven inadequate to understand the inheritance of predisposition to metabolic syndrome and Type 2 diabetes. At least 18 genes have been directly associated with Type 2 diabetes (Florez, 2008; Ridderstråle & Groop, 2009). This fact indicates that, unlike inborn errors of metabolism, such as Tay–Sachs disease, for which a straightforward causal connection exists between a mutant allele and a well-characterized dysfunction (Schneck et al., 1970), the genetic mechanisms that promote increased risks in obesity for Type 2 diabetes, cardiometabolic diseases, and obesity-associated cancers are not so straightforward.

It is also clear that the interaction of genes alone is insufficient to explain the contemporary problem of obesity. In view of the new international environment of widespread overnutrition, it also seems important to ask how genes respond to weight gain or weight loss, not just how specific genes make us fat or thin. Dietary caloric excess that is totally misaligned with physical energy expenditure characterizes this environment. Epidemiologists and U.S. public health officials have taken note of critical features of the current American environment, particularly easy access to cheap calories, the profligate use of sugar sweeteners like "high fructose corn syrup" in the food industry (Duffey & Popkin, 2008; Popkin, 2007; Popkin & Nielsen, 2003) and widespread overconsumption of fried, fatty, "fast" foods.

Ten thousand years of human evolution have tended to produce a lean physical phenotype. Indeed, severe pediatric obesity of genetic origin is observed in less than 0.01% of the population (Farooqi & O'Rahilly, 2006). Thus, the dramatically increased prevalence of overweight, obese, and morbidly obese phenotypes in the past decades of the twentieth century and first decade of the twenty-first century must be attributable primarily to environmental and consumption changes, i.e., environmental obesogenic pollutants, reduced energy expenditure and diet-induced obesity. Strong evidence for the role of the environment when genetic factors are relatively well controlled was obtained from studies of the Pima Indians, an aboriginal population that lives in the desert southwest of the United States and in an area of northern Mexico with a similar climate. The genetic variation between the two populations is relatively small. Pima Indians in the United States are on average 25 kg heavier than Pima Indians in Mexico (Ravussin, 1995), with corresponding differences in cardiometabolic risk. These differences have been attributed primarily to diet. Alternatively, population patterns of obesity may be attributed in part to epigenetic shifts, not to mutation in human DNA, because the timescale on which these changes in public health have occurred are far too rapid to be due primarily to evolution.

There is new interest in the maternal–fetal environment and an effort to understand a potential role for epigenetic mechanisms in the transmission of metabolic risk factors. The epigenetic modifications of nucleosomal histones and DNA that underlie these phenomena are poorly understood. These modifications include DNA methylation and histone acetylation or phosphorylation and are critical for proper control of the response of promoters to signal transduction in the differentiated cells of the adult (Bernstein, Meissner, & Lander, 2007; Li, Carey, & Workman, 2007), as well as for embryonic and fetal development (Bernstein et al., 2006). Indeed, some of the most interesting work on epigenetic and developmental mechanisms has been conducted in rat dams fed a high-fat diet. Investigators have identified alterations in homeobox gene expression patterns, such as HoxA10, that likely influence stem cell fate choice in osteogenic and adipogenic cell lineages (Chen et al., 2012). These alterations of transcriptional programming depend critically on patterns of acetylation and methylation of histone lysines in nucleosomal chromatin at key target genes. This concept suggests that patterns of obesity and metabolic dysfunction may indeed run in families, but not in the same way or subject to the same mechanisms as patterns of familial cancers.

One famous and well-studied case that strongly supports an epigenetic mechanism for cardiometabolic risk is the Dutch "Hunger Winter" of 1944–1945. This wartime event, when severe starvation affected the western Netherlands, revealed in detail for the first time that epigenetic mechanisms play a critical role in human BMI and insulin sensitivity. Specifically, maternal hunger created by cold, harsh living conditions, and caloric deprivation of pregnant mothers was linked to insulin resistance, obesity, an atherogenic lipid profile, and elevated cardiovascular risk in the surviving children as they aged (Kyle & Pichard, 2006). Of particular interest, children born to the same mothers in times of less severe deprivation did not show the same risk patterns. More recent work has shown that infants born to obese, overweight, and Type 2 diabetic mothers display increased adiposity and elevated risk for later metabolic disease (Heerwagen, Miller, Barbour, & Friedman, 2010). In addition, a UK study found that males who were most underweight at birth were seven times more likely to develop metabolic dysfunction and Type 2 diabetes later in life than males who were heaviest at birth (Hales et al., 1991). There is strong support for an inverse relationship between birth weight and hypertension in the adult, for both men and women, in which low birth weight predicts risk among the highest weight adults (Barker, Osmond, Golding, Kuh, & Wadsworth, 1989). These observations suggest that maternal uterine environments vary in ways that create stable and lasting consequences for the metabolic patterns of offspring exposed to that environment during gestation (Barker, 1995). Many of these patterns have been ascribed to epigenetic modification of key genes that regulate energy metabolism. Such mechanisms have been and continue to be explored and validated in rodent models (Levin & Govek, 1998; Nathanielsz, Poston, & Taylor, 2007; Samuelsson et al., 2008; Shankar et al., 2008). Thus, new research focusing on the specific modifications of nucleosomal histones in chromatin and DNA is worthy of attention.

Interesting work has shown that manipulation of DNA methylation pathways by metabolic supplementation can alter the intergenerational, epigenetic patterns of obesity (Waterland, Travisano, Tahiliani, Rached, & Mirza, 2008), which directly addresses mechanisms that may be relevant to the Hunger Winter offspring. Such approaches, if validated with animal models and in clinical trials, offer the possibility of a "personalized medicine" approach to obesity therapeutics (Martínez, Cordero, Campión, & Milagro, 2012). Innovative clinical investigations to measure how an altered maternal uterine environment during gestation affects DNA methylation patterns have already been reported (Cooper et al., 2012).

Although chromatin modifications are well studied in embryogenesis and cancer (Sharma, Kelly, & Jones, 2010), much less is known about their role in metabolic diseases. Intense research is now focused on how DNA sequence-specific transcription factors such as PPARγ and transcription coregulators such as PGC-1α (Puigserver et al., 1998) interact with histones and nucleosome remodelers (Pedersen, Kowenz-Leutz, Leutz, & Nerlov, 2001) to alter chromatin and reprogram gene expression networks (Jaenisch & Bird, 2003). These transcriptional coregulator functions are also important in inflammation (Freund, Orjalo, Desprez, & Campisi, 2010; Orjalo, Bhaumik, Gengler, Scott, & Campisi, 2009), which plays a critical role in the development of insulin resistance in obesity (Bastard et al., 2006; Shoelson, Herrero, & Naaz, 2007).

Histone acetylation is also being considered as a relevant target for epigenetic manipulation to affect energy metabolism. For example, genome-wide patterns of histone H3 acetylation at lysine 9 and 18 have been reported in INS-1 cells, a rat model for the pancreatic β-cell, upon exposure to peptide incretin hormones (Kim, Nian, & McIntosh, 2009). These modifications alter transcriptional coregulator function which likely coordinates reprogramming of transcriptional networks that respond to glucose signal transduction in the β-cell. This approach may have translational significance for obese, insulin-resistant patients, because these peptide hormones potentiate glucose-stimulated insulin secretion in the islet, among other effects. On the other hand, increased histone acetylation has been widely appreciated for decades as a mark of transcriptional activation at numerous loci. Absent a more specific, promoter-defined, or β-cell-specific signature of histone acetylation that is unique and resolvable from the general somatic pattern, these insights have relatively modest utility and present no new therapeutic targets for Type 2 diabetes. The emerging practicality of manipulation of chromatin-controlled transcriptional programs for therapeutic benefit is potentially useful in view of the new availability of a class of bromodomain protein-directed, small-molecule inhibitors (Belkina & Denis, 2012; Filippakopoulos et al., 2010; Muller et al., 2011; Nicodeme et al., 2010). If specific, functional targets of histone acetylation could be identified, small-molecule therapeutics for aberrant histone acetylation/transcriptional coactivation may develop an experimental basis for animal model experiments. However, the field has not yet reached this stage of development.

Activators of PPARγ function, such as pioglitazone, also appear to play a role in transcriptional programming controlled by PPARγ in part through alteration of histone modifications. In the β-cell, increased methylation of histone H3 at lysine 4 in the promoter regions of *Ins1*, *Ins2*, and *Glut2* genes is maintained through the dimethyl-lysine regulating complex Set7/9 methyltransferase and is required for proper transcription of these loci (Deering, Ogihara, Trace, Maier, & Mirmira, 2009). Pioglitazone treatment of obese *db/db* (leptin receptor-deficient) mice or high-fat diet-fed wild-type mice, which is well known to improve metabolism, dramatically improves *in vivo* transcription of *Ins1*, *Ins2*, and *Glut2* genes. This improvement is associated with increased histone acetylation of these key target genes (Evans-Molina et al., 2009). Similar enrichment in methylation of histone H3 at lysine 4 has been reported in β-cell lines and islets, and parallel increases transcription of these same genes (Francis, Chakrabarti, Garmey, & Mirmira, 2005).

In normal human pancreatic islets, histone lysine methylation patterns have been identified that define transcriptionally "primed" as well as transcriptionally active promoters (Bhandare et al., 2010), but this vein of research is relatively underdeveloped, compared to the study of histone lysine methylation in cancer. Additional experiments to test mechanistic hypotheses are critically needed to deepen understanding of how manipulation of chromatin "readers" and "writers," and DNA methylation enzymes, might be mobilized therapeutically (Belkina & Denis, 2012) to improve β-cell dysfunction. Most of the relevant work to date has been no more than correlative.

3. THE "*brd2 lo*" MOUSE MODEL OF "METABOLICALLY HEALTHY" OBESITY

How are these engines of histone modification (the "writers" of the epigenome) and the transcriptional machines that respond to signal transduction to interpret these modifications (the "readers" of the epigenome) to be studied and understood in their regulation of diverse transcriptional networks? Animal models that enable manipulation of chromatin and epigenetic mechanisms will obviously become increasingly important as investigators seek to understand the relevant genes and signal transduction pathways at work in obesity. In particular, one model with an epigenetic basis for obesity has received attention recently and offers a novel interpretive tool: the Brd2-deficient model for "metabolically healthy" obesity. In both humans and mice, the *BRD2* gene encodes an unusual transcriptional coregulator that contains double bromodomains (Haynes et al., 1992; Horn & Peterson, 2001; Jeanmougin, Wurtz, Le Douarin, Chambon, & Losson, 1997; Winston & Allis, 1999). Brd2 belongs to the bromodomain and extraterminal domain (BET) family of transcriptional coregulators defined by two tandem, mutually related bromodomains at the amino-terminus of the protein that bind to acetylated lysines in nucleosomal chromatin (Kanno et al., 2004; Nakamura et al., 2007), particularly acetyl histone H4 (Umehara et al., 2010) and transcriptionally couple histone acetylation to gene activation (LeRoy, Rickards, & Flint, 2008). The amino-terminal dual bromodomains in this family are followed by an "extraterminal" domain that is involved in protein–protein interactions (Lin et al., 2008; Rahman et al., 2011), thus the "BET" family name. The bromodomain is the only protein structural motif (Dhalluin et al., 1999) that is capable of "reading" sites of histone acetylation in nucleosomal chromatin (Sanchez & Zhou, 2009). Brd2 studies have been important because they have revealed remarkable and unexpected roles for chromatin regulation in energy metabolism.

3.1. Gene targeting of the *Brd2* locus in mice

Targeted disruption of the *Brd2* gene, which is located in the class II major histocompatibility complex (MHC) near *Tnf*, causes extreme obesity with hyperinsulinemia, but also hypoglycemia, hyperadiponectinemia, and improved glucose tolerance quite distinct from other animal models of obesity (Wang et al., 2009). *Brd2* had no previously known link to obesity, insulin sensitivity, or energy metabolism (Belkina & Denis, 2010; Denis, Nikolajczyk, & Schnitzler, 2010). The gene was disrupted in mouse embryonic stem cells by insertion of a *lacZ* cassette that encodes β-galactosidase. The cells were engineered by BayGenomics at the University of California, Davis, which is an arm of the Mutant Mouse Regional Resource Centers, and which receives support from the National Institutes of Health (http://www.mmrrc.org/catalog/overview_BG.php). Two types of embryonic stem cells were developed, one with a *lacZ* insertion in the coding region of the gene (designated RRE050) and one with a *lacZ* insertion in the promoter region (designated RRT234). These latter cells were used to develop the *brd2 lo* mice (Wang et al., 2009). The phenotype could not have been predicted (Wang et al., 2009), yet holds out the possibility of a more unified, epigenetically-based mechanism that underlies several human comorbidities associated with obesity and inflammation (Denis, 2010).

3.2. Systemic, protective phenotypes in *brd2 lo* mice

Despite severe obesity approaching 100 g, *brd2 lo* animals show improved whole-body insulin sensitivity and do not develop insulin resistance or glucose intolerance. They show better whole-body insulin sensitivity than wild type, control animals on the C57Bl6/J background, despite dramatic obesity (Jornayvaz et al., unpublished data). Several whole-body mechanisms likely contribute to this protection (Wang et al., 2009). For example, although *brd2 lo* animals, both males and females, consume slightly more chow than age- and sex-matched wild-type controls, they burn slightly more calories as heat in isothermal housing. The animals carry more interscapular brown adipose tissue and their white adipose tissue expresses higher levels of uncoupling proteins, which provides a mitochondrial mechanism for increased heat production. This phenotype is known to be protective of metabolism in obesity. In addition, *brd2 lo* mice show a respiratory exchange ratio of oxygen consumption and carbon dioxide production that is suggestive of fat metabolism at all times of day and night, fed state or fasting state, rather than a shift between carbohydrate metabolism expected for the fed state to fat metabolism expected for the fasted state (Wang et al., 2009). Continuous β-oxidation of fatty acid as the primary energy source might be expected to be metabolically protective, although the mitochondrial basis for this preferred mechanism of energy metabolism is not understood in these mice. Finally, production of high-molecular weight adiponectin, an insulin-sensitizing adipokine, was elevated almost to the serum levels seen in adiponectin transgenic mice on the *ob/ob* (leptin deficient) background (Kim et al., 2007). Elevated adiponectin confers yet another layer of metabolic protection in obesity. Yet it is important to point out here that the adiponectin promoter has not been directly genetically manipulated in the source embryonic stem cells; this result suggests that Brd2 coregulator function is ordinarily required for corepression of adiponectin transcription. This hypothesis has not yet been tested experimentally.

3.3. The *brd2 lo* phenotype protects adipose tissue

Remarkably, knocked-down expression of Brd2 in the 3T3-L1 differentiation model also protects cultured adipocytes from TNF-α-induced insulin resistance *in vitro*, probably by uncoupling TNF receptor signaling from transcription (Wang et al., 2009). The effect of reduced Brd2 levels also phenocopies the action of thiazolidinedione (TZD) drugs, such as rosiglitazone and pioglitazone, which are used as insulin sensitizers. Brd2 coactivator function opposes the action of PPARγ; Brd2 protein associates with PPARγ protein complexes and knockdown of Brd2 strongly stimulates PPARγ-dependent transcription and adipogenesis in 3T3-L1 cells (Wang et al., 2009).

3.4. Insulin resistant obesity is an inflammatory disease

Insulin resistance in the context of obesity is associated with a chronic state of subclinical inflammation (Bastard et al., 2006; Shoelson, Lee, & Goldfine, 2006; Weisberg et al., 2003; Xu et al., 2003), including increased serum concentrations of C-reactive protein, interleukin-6, interleukin-8, and TNF-α in patients and different animal models of obesity (Kahn et al., 2006). In insulin-resistant obesity, production of TNF-α in liver, fat, and muscle by infiltrating, proinflammatory (M1) adipose tissue macrophages promotes insulin resistance directly (Hotamisligil, Shargill, & Spiegelman, 1993). Here, M1 macrophages are distinguished from the more remodeling-purposed, alternatively polarized "M2" type of macrophage. Commonly, proinflammatory macrophages infiltrate white adipose tissue in patients and different animal models of obesity (Kahn et al., 2006), forming characteristic, histologically identifiable patterns of leukocytes called "crown-like structures" (Apovian et al., 2008; Cancello et al., 2005, Cinti et al., 2005) that surround stressed, dead, and dying adipocytes, and are closely associated with metabolic risk. These structures have recently been shown also to contain B cells (McDonnell et al., 2012), and both T cells and B cells engage in crosstalk with peripheral blood monocytes to define the proinflammatory and anti-inflammatory balance of cytokines in the obese, insulin-resistant adult (Jagannathan et al., 2009, 2010). To a first approximation, the greater the proinflammatory balance, the greater the metabolic dysfunction and the more advanced the disease. However, much work remains to be done to understand the specific T cell, B cell, and macrophage subtypes that are required for disease progression, the anti-inflammatory and homeostatic mechanisms that oppose aggravated chronic inflammation, and the kinetics of entry onto and departure from the metabolic "theatres of action" of each of these cell types: the central and peripheral adipose depots, the liver, the pancreas, and the blood.

3.5. Metabolically protective phenotypes in the *brd2 lo* immune system

The *brd2 lo* mice exhibit broad-spectrum protection against the inflammatory complications of obesity (Wang et al., 2009). Inflammatory responses are mildly deficient in these animals (Belkina, Blanton, Wang, Liu, & Denis, 2010), which likely contributes to their protection against metabolic dysfunction. These observations lead to novel hypotheses about the interactions and crosstalk between adipocytes, macrophages, and T cells that control the inflammatory state of adipose tissue. White adipose tissue shows reduced inflammation: bone marrow-derived macrophages underproduce proinflammatory cytokines (Belkina et al., 2010, 2013), T cell migration is ablated (G. V. Denis et al., unpublished observations), and regulatory T cells are expanded (G. V. Denis et al., unpublished observations). Furthermore, because polarization of macrophages to the anti-inflammatory, proremodeling (M2) state requires PPARγ (Odegaard et al., 2007), increased PPARγ activity in *brd2 lo* macrophages and elevated serum adiponectin (Ohashi et al., 2009) is likely to promote M2 polarization, also protecting against inflammation-driven insulin resistance. By transcriptionally uncoupling obesity from insulin resistance, the mild hypo-inflammatory phenotype of the *brd2 lo* model focuses attention on T cell and adipose tissue macrophage mechanisms that likely protect "metabolically healthy" obese individuals and might be harnessed to protect the larger population of obese patients with insulin resistance. Consistent with the inflammatory mechanism, a report in *Nature*

Genetics has identified human single nucleotide polymorphisms (SNPs) in the *BRD2* locus that are significantly associated with rheumatoid arthritis, which is driven by autoimmune and inflammatory processes (Mahdi et al., 2009). Furthermore, *BRD2* variants may be associated with BMI in Pima Indians (Muller, Abdussamad, et al., 2011). Upregulation of proinflammatory cytokine genes has long been appreciated to promote insulin resistance and glucose intolerance, which precede serious metabolic dysregulation that leads to Type 2 diabetes (Pickup & Crook, 1998).

Furthermore, reduced inflammation protects against insulin resistance and characterizes human "metabolically healthy" obese populations. These human subjects display attenuation of each of: glucose intolerance, dyslipidemia, hyperuricemia, and hypertension (Bonora et al., 1991; Wildman et al., 2008). Therefore, control of adipose tissue-infiltrating inflammatory cells and their chromatin-regulated gene expression holds promise for insulin-resistance patients. In obesity, macrophages, T cells, and adipocytes produce proinflammatory and anti-inflammatory cytokines, chemokines, and adipokines. Given that these factors synergize with, antagonize, and regulate each other, there are many unanswered questions about what happens during the adipocyte–macrophage–T cell interactions (Kintscher et al., 2008), and how and when insulin resistance develops. We have proposed that these animals offer a novel model for human "metabolically healthy" obesity. The model motivates interesting and informative hypotheses to test the links among leukocyte migration, proinflammatory cytokine production, and insulin resistance in obesity.

4. WHO ARE "METABOLICALLY HEALTHY" OBESE HUMANS?

In most humans, the main effect of increasing BMI is a nonlinear deterioration in metabolic and cardiovascular health. This process has been described as a gradual clustering of traits that include visceral obesity (as distinct from elevated BMI), insulin resistance, dyslipidemia, hypercholesterolemia, and hypertension. In the clinic, insulin resistance status is often determined by the homeostatic metabolic assessment (HOMA), which is assessed using the formula: fasting insulin (μU/mL) $\times$ fasting glucose (mg/dL)/405 (Ditschuneit, Flechtner-Mors, Johnson, & Adler, 1999; Matthews et al., 1985). As a stringent measure of insulin resistance, a value of HOMA >2 is often applied. According to the original Adults Treatment Panel III of 2001, a metabolic syndrome diagnosis is rendered when three or more of these five criteria are fulfilled: fasting plasma glucose concentration of at least 110 mg/dL, waist circumference ≥88 cm for women and 102 cm for men, serum high-density lipoprotein concentration <50 mg/dL for women and <40 for men, blood pressure of ≥130/85 mmHg, and serum triglyceride concentration of ≥150 mg/dL (Grundy et al., 2005). The International Diabetes Federation/American Heart Association/National Heart Lung and Blood Institute unified criteria are similar and require three of the above criteria, including the above cut points for triglycerides and blood pressure, and a cut point for fasting glucose more stringent than ATP III (>100 mg/dL) (Alberti et al., 2009) with the added criteria of central obesity, drug treatment for diabetes, elevated triglycerides, low levels of high-density lipoprotein, or hypertension. Nevertheless, there is no universally accepted definition of metabolic syndrome. There has been disagreement in the literature about the optimal criteria to use for metabolic dysfunction in obesity because risk in certain human populations, such as specific ethnic groups, for certain diseases such as cardiovascular disease, may be better predicted by some criteria, such as glucose intolerance, than by others, such as metabolic syndrome. Oral glucose tolerance tests and criteria for impaired fasting glucose and impaired glucose tolerance have been used to supplement the diagnosis of metabolic syndrome, but discussion of this literature is beyond the scope of this chapter.

However, it is pertinent that, as mentioned above, certain obese individuals are "metabolically healthy" and enjoy reduced risk for cardiovascular disease and Type 2 diabetes (Klöting et al., 2010; Sims, 2001; Succurro et al., 2008; Wildman et al., 2008). "Metabolically healthy" obese adults have been defined as abdominally obese (BMI ≥30) but lacking metabolic syndrome (Meigs et al., 2006). Such individuals comprise about 25% of the adult obese population in the United States, however, this prevalence depends on inclusion criteria, with some reported disagreement in prevalence, from a minimum of 11% of obese subjects in an Italian study (Calori et al., 2011) to 47.9% of obese subjects in a Korean study (Lee, 2009). The "metabolically healthy" obese phenotype is best conceptualized as a continuous distribution of preserved insulin sensitivity as a function of increasing BMI (Blüher, 2010). Some of these individuals show protective, elevated levels of adiponectin (Aguilar-Salinas et al., 2008) and maintain normal glucose tolerance despite startlingly high BMI.

Factors that couple obesity to insulin resistance and metabolic syndrome are of great medical interest, because they underlie the etiology of obesity-driven Type 2 diabetes. Thus, the "metabolically healthy" obese individual is likely to provide a goldmine of information. The study of this population for genes and pathways that couple obesity to insulin resistance has the potential to identify novel, "druggable" targets to help unhealthy obese patients avoid the worst comorbidities of their condition. Significantly, the "metabolically healthy" obese phenotype is associated with a reduced inflammatory profile (Karelis et al., 2005; Romano et al., 2003). Thus, it is likely that inflammatory functions of the innate and adaptive immune system are essential to link obesity to insulin resistance, cardiometabolic risk, and Type 2 diabetes. Greater detail concerning the immune cell subtypes, their cytokine production profiles, and kinetics of mobilization in the insulin-resistant obese subject, and how these differ in critical ways form the "metabolically healthy" obese subject is urgently required.

5. OTHER ANIMAL MODELS OF "METABOLICALLY HEALTHY" OBESITY

Additional mechanistic understanding of how this population of humans is protected from obesity-driven comorbidity will be achieved through hypothesis building and testing in animal models. Beyond the example of *brd2 lo* mice, there are other, fundamentally different types of animal models available, some that are primarily immunological, while others are adipose tissue directed. These different molecular and cellular perturbations or deficiencies will enable more precise mechanistic exploration of the relevant pathways that couple obesity to insulin resistance.

1. Low-inflammatory models include
   a. *Interleukin-1 Receptor 1 knockout*: A reduced inflammatory profile, particularly lower levels of TNF and interleukin-6, appears to protect these animals from high-fat diet-induced insulin resistance and glucose intolerance (McGillicuddy et al., 2011).
   b. *Inducible nitric oxide synthase knockout*: Deficient polarization of macrophages as a result of knockout of *Nos2* protects against obesity-induced skeletal muscle insulin resistance, and this is associated with improved phosphoinositide 3-kinase/Akt activity (Perreault & Marette, 2001).
   c. *Ablation of TNF*: Antibody against TNF improves insulin resistance in obesity (Hotamisligil et al., 1993), and mice deficient in TNF signaling are protected from insulin resistance in obesity (Uysal, Wiesbrock, Marino, & Hotamisligil, 1997).
   d. *TWEAK knockout*: TNF-related weak inducer of apoptosis (TWEAK), a cytokine of the TNF superfamily, is important not only

for tissue remodeling after injury (Burkly, Michaelson, Hahm, Jakubowski, & Zheng, 2007) but also for remodeling of adipose tissue to accommodate increased storage in obesity (Li et al., 2009). The TWEAK pathway is activated in obese Type 2 diabetic patients (Chacón et al., 2006). TWEAK deficiency (Campbell et al., 2006) shifts macrophage polarization to the alternatively-activated "low-inflammatory" phenotype, increases collagen turnover, and decreases JNK activation in gonadal adipose tissue, conferring metabolic protection in obesity (M.S. Obin, Tufts University, School of Medicine, personal communication).

e. *IκBα superrepressor*: Expression in the liver of the repressor of NF-κB signaling protects against high-fat diet-induced and low level NF-κ B-induced insulin resistance (Cai et al., 2005).

2. Adipose tissue models include

a. *Collagen 6 knockout*: Increased capability of adipose depots to remodel and accept increased storage in obesity appears to reduce adipocyte stress and apoptosis (Khan et al., 2009). This increased capacity depends on loss of collagen 6, and improves fasting glucose and glucose tolerance perhaps by relaxing physical, steric constraints on adipose depots.

b. *Adiponectin transgenic*: Severe obesity with insulin resistance on an *ob/ob* (leptin deficient) background can be ameliorated in a dramatic fashion by transgenic expression of adiponectin (Kim et al., 2007), a factor that sensitizes cells to insulin signaling.

Such mouse models will be useful to reveal different kinds of metabolic protection and how crosstalk may also protect organ systems from comorbidities (Denis, 2010).

6. DEREPRESSION OF INSULIN TRANSCRIPTION IN THE "*brd2 lo*" ENVIRONMENT

A deeper understanding of the mechanisms that control maternal–fetal transmission of increased risk for Type 2 diabetes is critical at this stage of the obesity epidemic. There is significant evidence, obtained initially from studies of the Pima Indians, that mothers with Type 2 diabetes can confer elevated diabetes risk to offspring (Dabelea, 2007; Dabelea & Pettitt, 2001; Dabelea et al., 2000, 2008). The development of Type 2 diabetes is dependent on both the gradually decreasing metabolic health of the obese individual as insulin resistance and inflammation increases, as well as strain on β-cell production of insulin that eventually leads to β-cell failure. The question of how epigenetic mechanisms influence the distinct but related risks to declining insulin action and insulin production has not been well studied.

Increased β-cell proliferation and differentiation, as well as increased insulin transcription and release from pancreatic β-cells *in vivo*, undoubtedly protects obese *brd2 lo* animals from progression to glucose intolerance and β-cell failure. Islets show no signs of apoptosis or stress but are expanded from an early age, likely as an early perturbation to islet homeostasis that is directly attributable to Brd2 reduction, not to insulin resistance in the periphery (Wang et al., 2009).

We explored the potential significance of reduced Brd2 expression during development. We wanted to know if reduced Brd2 expression in the embryonic stem cells of RRE050 or RRT234 origin (Wang et al., 2009) had altered biology relevant to energy metabolism. Based on published results of increased *insulin 1* gene transcription in β-cell lines, we hypothesized that insulin transcription would be potentiated in the RRE050 or RRT234 cells. Accordingly, we transfected RRE050 embryonic stem cells with an eGFP reporter construct for the *Ins1* promoter, obtained as a generous gift from Dr. Manami Hara, then selected the cells under hygromycin as shown in Fig. 3.1.

Remarkably, this result shows that reduced levels of Brd2 protein potentiate transcription of the insulin gene extremely early in mouse development, even before the embryonic stem cell has lost totipotent characteristics during the course of *in vitro* differentiation. One potential implication of these data is that targeted inhibition of Brd2 or genetic modification of embryonic stem cells could provide a therapeutic strategy for β-cell failure through regeneration of β-cell mass, or for gene therapy for Type 1 diabetes. Epigenetically based therapeutics for metabolic dysfunction are therefore feasible, although detailed mechanistic studies are obviously now required.

7. TRANSLATIONAL IMPLICATIONS OF EPIGENETIC REPROGRAMMING: CONCLUSIONS

Apart from the maternal–fetal transmission of increased cardiometabolic risk in the Hunger Winter case, other processes with a potential epigenetic component are likely at work in obesity. For example, weight regain after bariatric surgery has emerged as a worrisome problem for clinicians (Magro et al., 2008). Over the long term, a significant fraction of bariatric patients (20.4% for morbidly obese patients and 34.9% for super obese

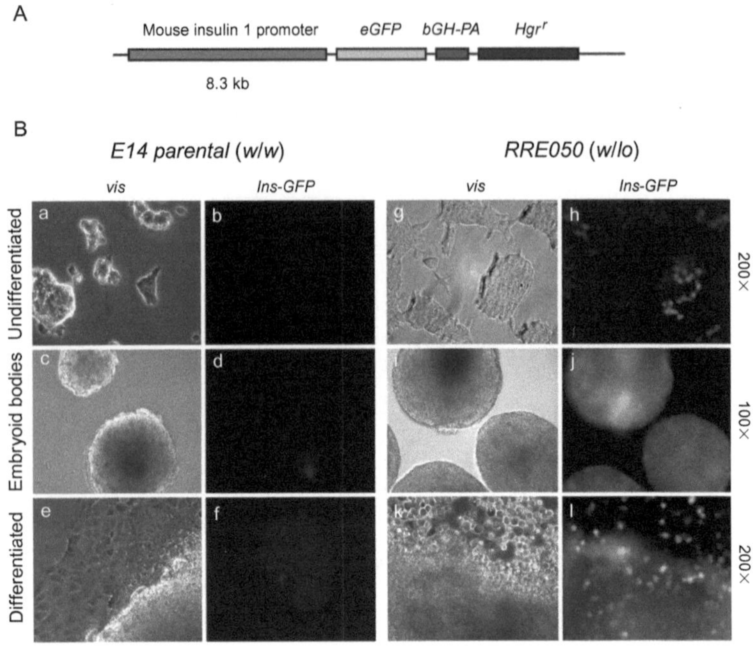

Figure 3.1 Reduced expression of Brd2 potentiates insulin transcription in embryonic stem cells. (A) Schematic of mouse insulin 1 promoter that drives expression of eGFP (*eGFP*). Promoter elements are also shown: bovine growth hormone poly A sequence (*bGH-PA*) and hygromycin resistance (*Hgr^r*). (B) Visual (*vis*) and fluorescence (*Ins-GFP*) micrographs of ES cells transfected with the construct in (A) and permitted to undergo differentiation. Undifferentiated cells were cultured on gelatin-coated tissue culture plastic in the presence of leukemia inhibitory factor (LIF) (undifferentiated), then LIF was withdrawn to permit the formation of embryoid bodies for 2 weeks (embryoid bodies), and culture was continued for an additional 2 weeks to permit additional differentiation (differentiated). Magnification is shown at the right-hand edge of the figure. The parental ES cells (*E14 parental; w/w*) were compared to Brd2 KO ES cells (*RRE050; w/lo*) as described in Wang et al. (2009), at the same stage of differentiation. (See Color Insert.)

patients in one study) (Christou, Look, & MacLean, 2006) regain significant weight. For the fraction of patients for whom regain is unrelated to surgical failure, it is possible that epigenetic factors play a role in the difficulty with maintaining healthy weight in the 10-year-period that follows surgery. Likewise, diet and lifestyle modification for the less-morbidly obese also shows only mixed success.

Important, outstanding questions remain. It is unclear whether there is an epigenetic component to the "metabolically healthy" obese phenotype, either in humans or in mice. Studies to measure a maternal genetic contribution to Brd2-regulated metabolism have yet to be performed in *brd2 lo* mice. There is some evidence for maternally defined DNA methylation patterns in *brd2 lo* mice (F. Wang, unpublished observations), but additional experiments are required to address this issue. Because *Brd2* is homologous to a known "maternal effect" gene in *Drosophila*, called *female sterile (1) homeotic* (Beck, Hanson, Kelly, Pappin, & Trowsdale, 1992; Denis & Green, 1996; Digan et al., 1986), we have proposed that Brd2-regulated metabolism may have a maternal effect in humans and mice (Belkina & Denis, 2012; Denis, 2010). Finally, we speculate that epigenetic modifications that attenuate expression of the *BRD2* locus in humans, nearby *TNF*, or other genes that encode proinflammatory factors, will be present in "metabolically healthy" obese patients, but modifications that exacerbate expression of such genes will be present in at-risk, "pro-inflammatory" obese populations. In support of this idea, recent work identifies increased inflammatory signatures in the uterine environment of obese female rats, which leads to weight gain and metabolic consequences for offspring (Shankar et al., 2011). The details of maternal effect or epigenetic contribution to body composition and fat distribution in humans are not well understood and will require further investigation in clinical studies and animal model systems.

More urgently, based on our published and preliminary data and other published data discussed here, we speculate that surgical, dietary, and exercise interventions may be insufficient to ensure sustained weight loss for obese, morbidly obese, and super obese patients if key genes important for energy metabolism have become epigenetically modified or reprogrammed in a way that resists whole-body return to healthy BMI. Such a mechanism has not been studied in patients or animal models after weight loss, but offers a compelling possible explanation for the persistent failure of these therapeutic interventions among many patients. Epigenetic "correction" might be possible with drugs that directly affect "writers" of histone methylation and acetylation marks such as histone deacetylase inhibitors, or "readers," such as the small-molecule BET protein inhibitors. Re-randomization of certain epigenetic marks of histone acetylation, methylation, or of DNA methylation, might help patients to lose weight by encouraging key tissues of the body to "forget" the unfortunate metabolic history of chronic obesity. Thus, metabolic "set-point" could be reset to a healthy orexigenic state and BMI would be easier to normalize after weight loss. Without this kind of drug intervention to correct the epigenetic status quo of chronic obesity, successful and sustained, long-term weight loss, even after bariatric surgery, may well be impossible for most patients.

ACKNOWLEDGMENTS

We thank Dr. Minami Hara of the University of Chicago for the generous gift of a mouse insulin promoter-eGFP reporter construct. We also thank Caroline Apovian, Barbara Corkey, Barbara Nikolajczyk, and Martin Obin for useful discussions and valuable suggestions. G. V. D. is supported by the National Institutes of Health (R56 DK090455), the Boston University Clinical and Translational Science Institute (UL1-TR000157), and two NIH-supported Centers: the Boston Area Diabetes and Endocrinology Research Center (P30 DK057521; PI: Joseph Avruch) and the Boston Nutrition Obesity Research Center (P30 DK046200; PI: Susan K. Fried). G. V. D. is currently the Chair of Basic Science Section of The Obesity Society, and gratefully acknowledges the Society for its support and promotion of scientific interactions and collaborations.

REFERENCES

Aguilar-Salinas, C. A., García, E. G., Robles, L., Riaño, D., Ruiz-Gomez, D. G., García-Ulloa, A. C., et al. (2008). High adiponectin concentrations are associated with the metabolically healthy obese phenotype. *The Journal of Clinical Endocrinology and Metabolism, 93*, 4075–4079.

Alberti, K. G., Eckel, R. H., Grundy, S. M., Zimmet, P. Z., Cleeman, J. I., Donato, K. A., et al. (2009). Harmonizing the metabolic syndrome: A joint interim statement of the International Diabetes Federation Task Force on Epidemiology and Prevention; National Heart, Lung, and Blood Institute; American Heart Association; World Heart Federation; International Atherosclerosis Society; and International Association for the Study of Obesity. *Circulation, 120*, 1640–1645.

Apovian, C. M., Bigornia, S., Mott, M., Meyers, M. R., Ulloor, J., Gagua, M., et al. (2008). Adipose macrophage infiltration is associated with insulin resistance and vascular endothelial dysfunction in obese subjects. *Arteriosclerosis, Thrombosis, and Vascular Biology, 28*, 1654–1659.

Barker, D. J. (1995). Fetal origins of coronary heart disease. *British Medical Journal, 311*, 171–174.

Barker, D. J., Osmond, C., Golding, J., Kuh, D., & Wadsworth, M. E. (1989). Growth in utero, blood pressure in childhood and adult life, and mortality from cardiovascular disease. *British Medical Journal, 298*, 654–657.

Bastard, J. P., Maachi, M., Lagathu, C., Kim, M. J., Caron, M., Vidal, H., et al. (2006). Recent advances in the relationship between obesity, inflammation, and insulin resistance. *European Cytokine Network, 17*, 4–12.

Beck, S., Hanson, I., Kelly, A., Pappin, D. J. C., & Trowsdale, J. (1992). A homologue of the *Drosophila female sterile homeotic (fsh)* gene in the class II region of the human MHC. *DNA Sequence, 2*, 203–210.

Belkina, A. C., Blanton, W., Wang, F., Liu, H., & Denis, G. V. (2010). Whole body Brd2 deficiency protects obese mice from insulin resistance by creating a low inflammatory environment. *Obesity, 18*, S58.

Belkina, A. C., & Denis, G. V. (2010). Obesity genes and insulin resistance. *Current Opinion in Endocrinology, Diabetes, and Obesity, 17*, 472–477.

Belkina, A. C., & Denis, G. V. (2012). BET domain co-regulators in obesity, inflammation and cancer. *Nature Reviews. Cancer, 12*, 465–477.

Belkina, A. C., Nikolajczyk, B. S., & Denis, G. V. (2013). BET protein function is required for inflammation: Brd2 genetic disruption and BET inhibitor JQ1 impair macrophage inflammatory responses. *Journal of Immunology*, In press.

Bernstein, B. E., Meissner, A., & Lander, E. S. (2007). The mammalian epigenome. *Cell, 128*, 669–681.

Bernstein, B. E., Mikkelsen, T. S., Xie, X., Kamal, M., Huebert, D. J., Cuff, J., et al. (2006). A bivalent chromatin structure marks key developmental genes in embryonic stem cells. *Cell, 125*, 315–326.

Bhandare, R., Schug, J., Le Lay, J., Fox, A., Smirnova, O., Liu, C., et al. (2010). Genome-wide analysis of histone modifications in human pancreatic islets. *Genome Research, 20*, 428–433.

Blüher, M. (2010). The distinction of metabolically 'healthy' from 'unhealthy' obese individuals. *Current Opinion in Lipidology, 21*, 38–43.

Bonora, E., Willeit, J., Kiechl, S., Oberhollenzer, F., Egger, G., Bonadonna, R., et al. (1991). U-shaped and J-shaped relationships between serum insulin and coronary heart disease in the general population. The Bruneck Study. *Diabetes Care, 21*, 221–230.

Burkly, L. C., Michaelson, J. S., Hahm, K., Jakubowski, A., & Zheng, T. S. (2007). TWEAKing tissue remodeling by a multifunctional cytokine: Role of TWEAK/Fn14 pathway in health and disease. *Cytokine, 40*, 1–16.

Cai, D., Yuan, M., Frantz, D. F., Melendez, P. A., Hansen, L., Lee, J., et al. (2005). Local and systemic insulin resistance resulting from hepatic activation of IKK-beta and NF-kappaB. *Nature Medicine, 11*, 183–190.

Calori, G., Lattuada, G., Piemonti, L., Garancini, M. P., Ragogna, F., Villa, M., et al. (2011). Prevalence, metabolic features, and prognosis of metabolically healthy obese Italian individuals: The Cremona Study. *Diabetes Care, 34*, 210–215.

Campbell, S., Burkly, L. C., Gao, H., Berman, J. W., Su, L., Browning, B., et al. (2006). Proinflammatory effects of Tweak/Fn14 interactions in glomerular mesangial cells. *The Journal of Immunology, 176*, 1889–1898.

Cancello, R., Henegar, C., Viguerie, N., Taleb, S., Poitou, C., Rouault, C., et al. (2005). Reduction of macrophage infiltration and chemoattractant gene expression changes in white adipose tissue of morbidly obese subjects after surgery-induced weight loss. *Diabetes, 54*, 2277–2286.

Centers for Disease Control and Prevention (CDC) (2010). Vital signs: state-specific obesity prevalence among adults — United States, 2009. *MMWR Morb Mortal Wkly Rep, 59*(30), 951–955.

Chacón, M. R., Richart, C., Gómez, J. M., Megía, A., Vilarrasa, N., Fernández-Real, J. M., et al. (2006). Expression of TWEAK and its receptor Fn14 in human subcutaneous adipose tissue. Relationship with other inflammatory cytokines in obesity. *Cytokine, 33*, 129–137.

Chen, J. R., Zhang, J., Lazarenko, O. P., Kang, P., Blackburn, M. L., Ronis, M. J., et al. (2012). Inhibition of fetal bone development through epigenetic down-regulation of HoxA10 in obese rats fed high-fat diet. *The FASEB Journal, 26*, 1131–1141.

Christou, N. V., Look, D., & MacLean, L. D. (2006). Weight gain after short- and long-limb gastric bypass in patients followed for longer than 10 years. *Annals of Surgery, 244*, 734–740.

Cinti, S., Mitchell, G., Barbatelli, G., Murano, I., Ceresi, E., Faloia, E., et al. (2005). Adipocyte death defines macrophage localization and function in adipose tissue of obese mice and humans. *Journal of Lipid Research, 46*, 2347–2355.

Cooper, W. N., Khulan, B., Owens, S., Elks, C. E., Seidel, V., Prentice, A. M., et al. (2012). DNA methylation profiling at imprinted loci after periconceptional micronutrient supplementation in humans: Results of a pilot randomized controlled trial. *The FASEB Journal, 26*, 1782–1790.

Dabelea, D. (2007). The predisposition to obesity and diabetes in offspring of diabetic mothers. *Diabetes Care, 30*(Suppl. 2), S169–S174.

Dabelea, D., Hanson, R. L., Lindsay, R. S., Pettitt, D. J., Imperatore, G., Gabir, M. M., et al. (2000). Intrauterine exposure to diabetes conveys risks for type 2 diabetes and obesity: A study of discordant sibships. *Diabetes, 49*, 2208–2211.

Dabelea, D., Mayer-Davis, E. J., Lamichhane, A. P., D'Agostino, R. B., Jr., Liese, A. D., Vehik, K. S., et al. (2008). Association of intrauterine exposure to maternal diabetes and obesity with type 2 diabetes in youth: The SEARCH case-control study. *Diabetes Care, 31*, 1422–1426.

Dabelea, D., & Pettitt, D. J. (2001). Intrauterine diabetic environment confers risks for type 2 diabetes mellitus and obesity in the offspring, in addition to genetic susceptibility. *Journal of Pediatric Endocrinology & Metabolism, 14*, 1085–1091.

Deering, T. G., Ogihara, T., Trace, A. P., Maier, B., & Mirmira, R. G. (2009). Methyltransferase Set7/9 maintains transcription and euchromatin structure at islet-enriched genes. *Diabetes, 58*, 185–193.

den Hoed, M., Ekelund, U., Brage, S., Grontved, A., Zhao, J. H., Sharp, S. J., et al. (2010). Genetic susceptibility to obesity and related traits in childhood and adolescence: Influence of loci identified by genome-wide association studies. *Diabetes, 59*, 2980–2988.

Denis, G. V. (2010). Bromodomain coactivators in cancer, obesity, type 2 diabetes and inflammation. *Discovery Medicine, 10*, 489–499.

Denis, G. V., & Green, M. R. (1996). A novel, mitogen-activated nuclear kinase is related to a *Drosophila* developmental regulator. *Genes and Development, 10*, 261–271.

Denis, G. V., Nikolajczyk, B. N., & Schnitzler, G. R. (2010). An emerging role for bromodomain-containing proteins in chromatin regulation and transcriptional control of adipogenesis. *FEBS Letters, 584*, 3260–3268.

Dhalluin, C., Carlson, J. E., Zeng, L., He, C., Aggarwal, A. K., & Zhou, M.-M. (1999). Structure and ligand of a histone acetyltransferase bromodomain. *Nature, 399*, 491–496.

Digan, M. E., Haynes, S. R., Mozer, B. A., Dawid, I. B., Forquignon, F., & Gans, M. (1986). Genetic and molecular analysis of *fs(1)h*, a maternal effect homeotic gene in *Drosophila*. *Developmental Biology, 114*, 161–169.

Ditschuneit, H. H., Flechtner-Mors, M., Johnson, T. D., & Adler, G. (1999). Metabolic and weight-loss effects of a long-term dietary intervention in obese patients. *The American Journal of Clinical Nutrition, 69*, 198–204.

Duffey, K. J., & Popkin, B. M. (2008). High-fructose corn syrup: Is this what's for dinner? *The American Journal of Clinical Nutrition, 88*, 1722S–1732S.

Evans-Molina, C., Robbins, R. D., Kono, T., Tersey, S. A., Vestermark, G. L., Nunemaker, C. S., et al. (2009). Peroxisome proliferator-activated receptor gamma activation restores islet function in diabetic mice through reduction of endoplasmic reticulum stress and maintenance of euchromatin structure. *Molecular and Cellular Biology, 29*, 2053–2067.

Executive summary of the third report of the National Cholesterol Education Program (NCEP) expert panel on detection, evaluation and treatment of high blood cholesterol in adults (Adult Treatment Panel III). 2001. *Journal of the American Medical Association, 285*, 2486–2497.

Farooqi, S., & O'Rahilly, S. (2006). Genetics of obesity in humans. *Endocrine Reviews, 27*, 710–718.

Filippakopoulos, P., Qi, J., Picaud, S., Shen, Y., Smith, W. B., Fedorov, O., et al. (2010). Selective inhibition of BET bromodomains. *Nature, 468*, 1067–1073.

Finucane, M. M., Stevens, G. A., Cowan, M. J., Danaei, G., Lin, J. K., Paciorek, C. J., et al. (2011). Global Burden of Metabolic Risk Factors of Chronic Diseases Collaborating Group (Body Mass Index). National, regional, and global trends in body-mass index since 1980: Systematic analysis of health examination surveys and epidemiological studies with 960 country-years and 9·1 million participants. *The Lancet, 377*, 557–567.

Florez, J. (2008). The genetics in Type 2 diabetes: A realistic appraisal in 2008. *The Journal of Clinical Endocrinology and Metabolism, 93*, 4633–4642.

Francis, J., Chakrabarti, S. K., Garmey, J. C., & Mirmira, R. G. (2005). Pdx-1 links histone H3-Lys-4 methylation to RNA polymerase II elongation during activation of insulin transcription. *The Journal of Biological Chemistry, 280*, 36244–36253.

Frayling, T. M., Timpson, N. J., Weedon, M. N., Zeggini, E., Freathy, R. M., Lindgren, C. M., et al. (2007). A common variant in the FTO gene is associated with body mass index and predisposes to childhood and adult obesity. *Science, 316*, 889–894.

Freund, A., Orjalo, A. V., Desprez, P. Y., & Campisi, J. (2010). Inflammatory networks during cellular senescence: Causes and consequences. *Trends in Molecular Medicine, 16*, 238–246.

Grundy, S. M., Cleeman, J. I., Daniels, S. R., Donato, K. A., Eckel, R. H., Franklin, B. A., et al. (2005). Diagnosis and management of the metabolic syndrome: An American Heart Association/National Heart, Lung, and Blood Institute Scientific Statement. *Circulation, 112*, 2735–2752 Erratum in: Circulation 112, e297, e298 (2005).

Hales, C. N., Barker, D. J., Clark, P. M., Cox, L. J., Fall, C., Osmond, C., et al. (1991). Fetal and infant growth and impaired glucose tolerance at age 64. *British Medical Journal, 303*, 1019–1022.

Haslam, D. W., & James, W. P. (2009). Obesity. *The Lancet, 366*, 1197–1209.

Haynes, S. R., Dollard, C., Winston, F., Beck, S., Trowsdale, J., & Dawid, I. B. (1992). The bromodomain: A conserved sequence found in human, *Drosophila* and yeast proteins. *Nucleic Acids Research, 20*, 2603.

Heerwagen, M. J., Miller, M. R., Barbour, L. A., & Friedman, J. E. (2010). Maternal obesity and fetal metabolic programming: A fertile epigenetic soil. *American Journal of Physiology. Regulatory, Integrative and Comparative Physiology, 299*, R711–R722.

Horn, P. J., & Peterson, C. L. (2001). The bromodomain: A regulator of ATP-dependent chromatin remodeling? *Frontiers in Bioscience, 6*, D1019–D1023.

Hossain, P., Kawar, B., & El Nahas, M. (2007). Obesity and diabetes in the developing world—A growing challenge. *The New England Journal of Medicine, 356*, 213–215.

Hotamisligil, G. S., Shargill, N. S., & Spiegelman, B. M. (1993). Adipose expression of tumor necrosis factor-α: Direct role in obesity-linked insulin resistance. *Science, 259*, 87–91.

Jaenisch, R., & Bird, A. (2003). Epigenetic regulation of gene expression: How the genome integrates intrinsic and environmental signals. *Nature Genetics, 33*(Suppl.), 245–254.

Jagannathan, M., Hasturk, H., Liang, Y., Shin, H., Hetzel, J. T., Kantarci, A., et al. (2009). TLR cross-talk specifically regulates cytokine production by B cells from chronic inflammatory disease patients. *The Journal of Immunology, 183*, 7461–7470.

Jagannathan, M., McDonnell, M., Liang, Y., Hasturk, H., Hetzel, J., Rubin, D., et al. (2010). Toll-like receptors regulate B cell cytokine production in patients with diabetes. *Diabetologia, 53*, 1461–1471.

Jeanmougin, F., Wurtz, J.-M., Le Douarin, B., Chambon, P., & Losson, R. (1997). The bromodomain revisited. *Trends in Biochemical Sciences, 22*, 151–153.

Kahn, S. E., Zinman, B., Haffner, S. M., O'Neill, M. C., Kravitz, B. G., Yu, D., et al. (2006). Obesity is a major determinant of the association of C-reactive protein levels and the metabolic syndrome in type 2 diabetes. *Diabetes, 55*, 2357–2364.

Kanno, T., Kanno, Y., Siegel, R. M., Jang, M. K., Lenardo, M. J., & Ozato, K. (2004). Selective recognition of acetylated histones by bromodomain proteins visualized in living cells. *Molecular Cell, 13*, 33–43.

Karelis, A. D., Faraj, M., Bastard, J. P., St-Pierre, D. H., Brochu, M., Prud'homme, D., et al. (2005). The metabolically healthy but obese individual presents a favorable inflammation profile. *The Journal of Clinical Endocrinology and Metabolism, 90*, 4145–4150.

Kern, P. A. (1997). Potential role of TNFalpha and lipoprotein lipase as candidate genes for obesity. *The Journal of Nutrition, 127*, 1917S–1922S.

Kern, P. A., Ranganathan, S., Li, C., Wood, L., & Ranganathan, G. (2001). Adipose tissue tumor necrosis factor and interleukin-6 expression in human obesity and insulin resistance. *American Journal of Physiology. Endocrinology and Metabolism, 280*, E745–E751.

Khan, T., Muise, E. S., Iyengar, P., Wang, Z. V., Chandalia, M., Abate, N., et al. (2009). Metabolic dysregulation and adipose tissue fibrosis: Role of collagen VI. *Molecular and Cellular Biology, 29*, 1575–1591.

Kim, S. J., Nian, C., & McIntosh, C. H. (2009). Glucose-dependent insulinotropic polypeptide and glucagon-like peptide-1 modulate beta-cell chromatin structure. *The Journal of Biological Chemistry, 284*, 12896–12904.

Kim, J. Y., van de Wall, E., Laplante, M., Azzara, A., Trujillo, M. E., Hofmann, S. M., et al. (2007). Obesity-associated improvements in metabolic profile through expansion of adipose tissue. *The Journal of Clinical Investigation, 117*, 2621–2637.

Kintscher, U., Hartge, M., Hess, K., Foryst-Ludwig, A., Clemenz, M., Wabitsch, M., et al. (2008). T-lymphocyte infiltration in visceral adipose tissue: A primary event in adipose tissue inflammation and the development of obesity-mediated insulin resistance. *Arteriosclerosis, Thrombosis, and Vascular Biology, 28*, 1304–1310.

Klannemark, M., Orho, M., Langin, D., Laurell, H., Holm, C., Reynisdottir, S., et al. (1998). The putative role of the hormone-sensitive lipase gene in the pathogenesis of Type II diabetes mellitus and abdominal obesity. *Diabetologia, 41*, 1516–1522.

Klöting, N., Fasshauer, M., Dietrich, A., Kovacs, P., Schön, M. R., Kern, M., et al. (2010). Insulin-sensitive obesity. *American Journal of Physiology. Endocrinology and Metabolism, 299*, E506–E515.

Kyle, U. G., & Pichard, C. (2006). The Dutch Famine of 1944–1945: A pathophysiological model of long-term consequences of wasting disease. *Current Opinion in Clinical Nutrition and Metabolic Care, 9*, 388–394.

Lee, K. (2009). Metabolically obese but normal weight (MONW) and metabolically healthy but obese (MHO) phenotypes in Koreans: Characteristics and health behaviors. *Asia Pacific Journal of Clinical Nutrition, 18*, 280–284.

Lefebvre, A. M., Laville, M., Vega, N., Riou, J. P., van Gaal, L., Auwerx, J., et al. (1998). Depot-specific differences in adipose tissue gene expression in lean and obese subjects. *Diabetes, 47*, 98–103.

Leibel, R. L. (2008). Energy in, energy out, and the effects of obesity-related genes. *The New England Journal of Medicine, 359*, 2603–2604.

LeRoy, G., Rickards, B., & Flint, S. J. (2008). The double bromodomain proteins Brd2 and Brd3 couple histone acetylation to transcription. *Molecular Cell, 30*, 51–60.

Levin, B. E., & Govek, E. (1998). Gestational obesity accentuates obesity in obesity-prone progeny. *The American Journal of Physiology, 275*, R1374–R1379.

Li, B., Carey, M., & Workman, J. L. (2007). The role of chromatin during transcription. *Cell, 128*, 707–719.

Li, H., Mittal, A., Paul, P. K., Kumar, M., Srivastava, D. S., Tyagi, S. C., et al. (2009). Tumor necrosis factor-related weak inducer of apoptosis augments matrix metalloproteinase 9 (MMP-9) production in skeletal muscle through the activation of Nuclear Factor-κB-inducing kinase and p38 mitogen-activated protein kinase. *The Journal of Biological Chemistry, 284*, 4439–4450.

Lin, Y. J., Umehara, T., Inoue, M., Saito, K., Kigawa, T., Jang, M. K., et al. (2008). Solution structure of the extraterminal domain of the bromodomain-containing protein BRD4. *Protein Science, 17*, 2174–2179.

Maes, H. H., Neale, M. C., & Eaves, L. J. (1997). Genetic and environmental factors in relative body weight and human adiposity. *Behavior Genetics, 27*, 325–351.

Magro, D. O., Geloneze, B., Delfini, R., Pareja, B. C., Callejas, F., & Pareja, J. C. (2008). Long-term weight regain after gastric bypass: A 5-year prospective study. *Obesity Surgery, 18*, 648–651.

Mahdi, H., Fisher, B. A., Källberg, H., Plant, D., Malmström, V., Rönnelid, J., et al. (2009). Specific interaction between genotype, smoking and autoimmunity to citrullinated alpha-enolase in the etiology of rheumatoid arthritis. *Nature Genetics, 41*, 1319–1324.

Martínez, J. A., Cordero, P., Campión, J., & Milagro, F. I. (2012). Interplay of early-life nutritional programming on obesity, inflammation and epigenetic outcomes. *The Proceedings of the Nutrition Society, 71*, 276–283.

Matthews, D. R., Hosker, J. P., Rudenski, A. S., Naylor, B. A., Treacher, D. F., & Turner, R. C. (1985). Homeostasis model assessment: Insulin resistance and β-cell function from fasting plasma glucose and insulin concentrations in man. *Diabetologia*, *28*, 412–419.

McDonnell, M. E., Ganley-Leal, L. M., Mehta, A., Bigornia, S. J., Mott, M., Rehman, Q., et al. (2012). B lymphocytes in human subcutaneous adipose crown-like structures. *Obesity (Silver Spring)*, *20*, 1372–1378.

McGillicuddy, F. C., Harford, K. A., Reynolds, C. M., Oliver, E., Claessens, M., Mills, K. H. G., et al. (2011). Lack of interleukin-1 receptor I (IL-1RI) protects mice from high-fat diet-induced adipose tissue inflammation coincident with improved glucose homeostasis. *Diabetes*, *60*, 1688–1698.

Meigs, J. B., Wilson, P. W., Fox, C. S., Vasan, R. S., Nathan, D. M., Sullivan, L. M., et al. (2006). Body mass index, metabolic syndrome, and risk of type 2 diabetes or cardiovascular disease. *The Journal of Clinical Endocrinology and Metabolism*, *91*, 2906–2912.

Morris, B. J., Zee, R. Y., & Robinson, B. G. (1994). Significant relationships of plasma lipids and body mass index with polymorphisms at the linked low-density-lipoprotein receptor gene and insulin receptor gene loci (19p13.2) in essential hypertensive patients. *Clinical Science (London, England)*, *86*, 583–592.

Muller, Y., Abdussamad, M., Hanson, R., Knowler, W. C., Bogardus, C., & Baier, L. (2011). Variants in/near *BRD2* are associated with body mass index in Pima Indians. *Diabetes*, *60*(Suppl. 1), A1400.

Muller, S., Filippakopoulos, P., & Knapp, S. (2011). Bromodomains as therapeutic targets. *Expert Reviews in Molecular Medicine*, *13*, e29.

Nakamura, Y., Umehara, T., Nakano, K., Jang, M. K., Shirouzu, M., Morita, S., et al. (2007). Crystal structure of the human BRD2 bromodomain: Insights into dimerization and recognition of acetylated histone H4. *The Journal of Biological Chemistry*, *282*, 4193–4201.

Nathanielsz, P. W., Poston, L., & Taylor, P. D. (2007). In utero exposure to maternal obesity and diabetes: Animal models that identify and characterize implications for future health. *Clinics in Perinatology*, *34*, 515–526.

Nicodeme, E., Jeffrey, K. L., Schaefer, U., Beinke, S., Dewell, S., Chung, C. W., et al. (2010). Suppression of inflammation by a synthetic histone mimic. *Nature*, *468*, 1119–1123.

Odegaard, J. I., Ricardo-Gonzalez, R. R., Goforth, M. H., Morel, C. R., Subramanian, V., Mukundan, L., et al. (2007). Macrophage-specific PPARgamma controls alternative activation and improves insulin resistance. *Nature*, *447*, 1116–1120.

Ohashi, K., Parker, J. L., Ouchi, N., Higuchi, A., Vita, J. A., Gokce, N., et al. (2009). Adiponectin promotes macrophage polarization towards an anti-inflammatory phenotype. *The Journal of Biological Chemistry*, *285*, 6153–6160.

Oppert, J. M., Vohl, M. C., Chagnon, M., Dionne, F. T., Cassard-Doulcier, A. M., Ricquier, D., et al. (1994). DNA polymorphism in the uncoupling protein (UCP) gene and human body fat. *International Journal of Obesity and Related Metabolic Disorders*, *18*, 526–531.

Orjalo, A. V., Bhaumik, D., Gengler, B. K., Scott, G. K., & Campisi, J. (2009). Cell surface-bound IL-1alpha is an upstream regulator of the senescence-associated IL-6/IL-8-cytokine network. *Proceedings of the National Academy of Science of the United States of America*, *106*, 17031–17036.

Pedersen, T. A., Kowenz-Leutz, E., Leutz, A., & Nerlov, C. (2001). Cooperation between C/EBPalpha TBP/TFIIB and SWI/SNF recruiting domains is required for adipocyte differentiation. *Genes and Development*, *15*, 3208–3216.

Perreault, M., & Marette, A. (2001). Targeted disruption of inducible nitric oxide synthase protects against obesity-linked insulin resistance in muscle. *Nature Medicine*, *7*, 1138–1143.

Pickup, J. C., & Crook, M. A. (1998). Is type II diabetes mellitus a disease of the innate immune system? *Diabetologia, 41*, 1241–1248.

Popkin, B. M. (2007). The world is fat. *Scientific American, 297*, 88–95.

Popkin, B. M., & Nielsen, S. J. (2003). The sweetening of the world's diet. *Obesity Research, 11*, 1325–1332.

Puigserver, P., Wu, Z., Park, C. W., Graves, R., Wright, M., & Spiegelman, B. M. (1998). A cold-inducible coactivator of nuclear receptors linked to adaptive thermogenesis. *Cell, 92*, 829–839.

Rahman, S., Sowa, M. E., Ottinger, M., Smith, J. A., Shi, Y., Harper, J. W., et al. (2011). The Brd4 extraterminal domain confers transcription activation independent of pTEFb by recruiting multiple proteins, including NSD3. *Molecular and Cellular Biology, 31*, 2641–2652.

Ravussin, E. (1995). Metabolic differences and the development of obesity. *Metabolism, 9* (Suppl. 3), 12–14.

Reynisdottir, S., Ellerfeldt, K., Wahrenberg, H., Lithell, H., & Arner, P. (1994). Multiple lipolysis defects in the insulin resistance (metabolic) syndrome. *The Journal of Clinical Investigation, 93*, 2590–2599.

Ricquier, D., Casteilla, L., & Bouillaud, F. (1991). Molecular studies of the uncoupling protein. *The FASEB Journal, 5*, 2237–2242.

Ridderstråle, M., & Groop, L. (2009). Genetic dissection of type 2 diabetes. *Molecular and Cellular Endocrinology, 297*, 10–17.

Romano, M., Guagnano, M. T., Pacini, G., Vigneri, S., Falco, A., Marinopiccoli, M., et al. (2003). Association of inflammation markers with impaired insulin sensitivity and coagulative activation in obese healthy women. *The Journal of Clinical Endocrinology and Metabolism, 88*, 5321–5326.

Saghizadeh, M., Ong, J. M., Garvey, W. T., Henry, R. R., & Kern, P. A. (1996). The expression of TNF alpha by human muscle. Relationship to insulin resistance. *The Journal of Clinical Investigation, 97*, 1111–1116.

Samuelsson, A. M., Matthews, P. A., Argenton, M., Christie, M. R., McConnell, J. M., Jansen, E. H., et al. (2008). Diet-induced obesity in female mice leads to offspring hyperphagia, adiposity, hypertension, and insulin resistance: A novel murine model of developmental programming. *Hypertension, 51*, 383–392.

Sanchez, R., & Zhou, M.-M. (2009). The role of human bromodomains in chromatin biology and gene transcription. *Current Opinion in Drug Discovery and Development, 12*, 659–665.

Schneck, L., Valenti, C., Amsterdam, D., Friedland, J., Adachi, M., & Volk, B. W. (1970). Prenatal diagnosis of Tay–Sachs disease. *The Lancet, 1*, 582–584.

Shankar, K., Harrell, A., Liu, X., Gilchrist, J. M., Ronis, M. J., & Badger, T. M. (2008). Maternal obesity at conception programs obesity in the offspring. *American Journal of Physiology. Regulatory, Integrative and Comparative Physiology, 294*, R528–R538.

Shankar, K., Zhong, Y., Kang, P., Lau, F., Blackburn, M. L., Chen, J. R., et al. (2011). Maternal obesity promotes a proinflammatory signature in rat uterus and blastocyst. *Endocrinology, 152*, 4158–4170.

Sharma, S., Kelly, T. K., & Jones, P. A. (2010). Epigenetics in cancer. *Carcinogenesis, 31*, 27–36.

Shaw, J. E., Sicree, R. A., & Zimmet, P. Z. (2010). Global estimates of the prevalence of diabetes for 2010 and 2030. *Diabetes Research and Clinical Practice, 87*, 4–14.

Shoelson, S. E., Herrero, L., & Naaz, A. (2007). Obesity, inflammation and insulin resistance. *Gastroenterology, 132*, 2169–2180.

Shoelson, S. E., Lee, J., & Goldfine, A. B. (2006). Inflammation and insulin resistance. *The Journal of Clinical Investigation, 116*, 1793–1801.

Silventoinen, K., Rokholm, B., Kaprio, J., & Sørensen, T. I. (2010). The genetic and environmental influences on childhood obesity: A systematic review of twin and adoption studies. *International Journal of Obesity*, *34*, 29–40.

Sims, E. A. H. (2001). Are there persons who are obese, but metabolically healthy? *Metabolism*, *50*, 1499–1504.

Stich, V., Harant, I., De Glisezinski, I., Crampes, F., Berlan, M., Kunesova, M., et al. (1997). Adipose tissue lipolysis and hormone-sensitive lipase expression during very-low-calorie diet in obese female identical twins. *The Journal of Clinical Endocrinology and Metabolism*, *82*, 739–744.

Succurro, E., Marini, M. A., Frontoni, S., Hribal, M. L., Andreozzi, F., Lauro, R., et al. (2008). Insulin secretion in metabolically obese, but normal weight, and in metabolically healthy but obese individuals. *Obesity (Silver Spring)*, *16*, 1881–1886.

Umehara, T., Nakamura, Y., Jang, M. K., Nakano, K., Tanaka, A., Ozato, K., et al. (2010). Structural basis for acetylated histone H4 recognition by the human BRD2 bromodomain. *The Journal of Biological Chemistry*, *285*, 7610–7618.

Uysal, K. T., Wiesbrock, S. M., Marino, M. W., & Hotamisligil, G. S. (1997). Protection from obesity-induced insulin resistance in mice lacking TNF-alpha function. *Nature*, *389*, 610–614.

Vidal-Puig, A. J., Considine, R. V., Jimenez-Liñan, M., Werman, A., Pories, W. J., Caro, J. F., et al. (1997). Peroxisome proliferator-activated receptor gene expression in human tissues. Effects of obesity, weight loss, and regulation by insulin and glucocorticoids. *The Journal of Clinical Investigation*, *99*, 2416–2422.

Walston, J., Silver, K., Bogardus, C., Knowler, W. C., Celi, F. S., Austin, S., et al. (1995). Time of onset of non-insulin-dependent diabetes mellitus and genetic variation in the beta 3-adrenergic-receptor gene. *The New England Journal of Medicine*, *333*, 343–347.

Wang, F., Liu, H., Blanton, W. P., Belkina, A., LeBrasseur, N. K., & Denis, G. V. (2009). Brd2 disruption in mice causes severe obesity without type 2 diabetes. *The Biochemical Journal*, *425*, 71–83.

Waterland, R. A., Travisano, M., Tahiliani, K. G., Rached, M. T., & Mirza, S. (2008). Methyl donor supplementation prevents transgenerational amplification of obesity. *International Journal of Obesity*, *32*, 1373–1379.

Weisberg, S. P., McCann, D., Desai, M., Rosenbaum, M., Leibel, R. L., & Ferrante, A. W., Jr. (2003). Obesity is associated with macrophage accumulation in adipose tissue. *The Journal of Clinical Investigation*, *112*, 1796–1808.

Wild, S., Roglic, G., Green, A., Sicree, R., & King, H. (2004). Global prevalence of diabetes: Estimates for the year 2000 and projections for 2030. *Diabetes Care*, *27*, 1047–1053.

Wildman, R. P., Muntner, P., Reynolds, K., McGinn, A. P., Rajpathak, S., Wylie-Rosett, J., et al. (2008). The obese without cardiometabolic risk factor clustering and the normal weight with cardiometabolic risk factor clustering: Prevalence and correlates of 2 phenotypes among the US population (NHANES 1999-2004). *Archives of Internal Medicine*, *168*, 1617–1624.

Winston, F., & Allis, C. D. (1999). The bromodomain: A chromatin-targeting module? *Nature Structural Biology*, *6*, 601–604.

Xu, H., Barnes, G. T., Yang, Q., Tan, G., Yang, D., Chou, C. J., et al. (2003). Chronic inflammation in fat plays a crucial role in the development of obesity-related insulin resistance. *The Journal of Clinical Investigation*, *112*, 1821–1830.

Zhao, J., Bradfield, J. P., Li, M., Wang, K., Zhang, H., Kim, C. E., et al. (2009). The role of obesity-associated loci identified in genome-wide association studies in the determination of pediatric BMI. *Obesity (Silver Spring)*, *17*, 2254–2257.

CHAPTER FOUR

The *TBC1D1* Gene: Structure, Function, and Association with Obesity and Related Traits

Luca Fontanesi[1], Francesca Bertolini
Department of Agricultural and Food Sciences (DISTAL), University of Bologna, Bologna, Italy
[1]Corresponding author: e-mail address: luca.fontanesi@unibo.it

Contents

Abstract

This review summarizes what is currently known on the TBC1 (tre-2/USP6, BUB2, cdc16) domain family, member 1 (*TBC1D1*) gene and gives an overview of its links with human obesity and related traits, taking also information from other animals. Several studies indicated that TBC1D1 contributes to the development of obesity by regulating skeletal muscle insulin sensitivity. A polymorphic site (p.R125W) in the human gene has been associated with female body mass index (BMI) in a few different cohorts. Few other polymorphisms in this gene were associated with BMI and diabetic nephropathy. Following a candidate gene approach, polymorphisms in the pig *TBC1D1* gene were associated with fat deposition traits. In chicken, a few quantitative trait loci for growth and fat deposition were localized in the region containing *TBC1D1*. These complementary results could contribute to further understand the role of TBC1D1 in developing obesity.

Vitamins and Hormones, Volume 91
ISSN 0083-6729
http://dx.doi.org/10.1016/B978-0-12-407766-9.00004-3

1. INTRODUCTION

During the past few years, a quite large number of human genes have been linked to monogenic obesity (usually with rare variants) and/or polygenic obesity (usually with common variants), by exploring different approaches, including candidate gene studies, genome-wide linkage analyses in family-based pedigrees, and genome-wide association (GWA) studies. Among this list of genes, the TBC1 (tre-2/USP6, BUB2, cdc16) domain family, member 1 gene, known as *TBC1D1*, has been recognized to be one of the possible contributors to severe obesity. The full name of this gene reflects the homology of its protein product with other proteins originally studied in yeast that contributed to understand its functions: TRE2, homologous to USP6 (ubiquitin-specific peptidase 6 (Tre-2 oncogene)) in higher organisms; yeast Bub2; and cdc16 (cell division cycle 16). Even if *TBC1D1* has not been repetitively associated with obesity as frequently as other genes, for example, *FTO*, its role, only recently clarified with more precision, may contribute to understand the relationships between metabolic syndrome, obesity, and muscle metabolism. This review summarizes what is currently known on TBC1D1 in humans and other species and gives an overview of its links with obesity and related traits, adding information recently obtained in livestock that could help to refine the biological role of this gene.

2. *TBC1D1* GENE STRUCTURE AND EXPRESSION

The human *TBC1D1* is localized on chromosome 4p14, with genomic coordinates (GRCh37) 4:37,892,708–38,140,796. It is constituted by 20 core exons, 19 of which contain coding sequences, that encode for a 1168 consensus amino acid protein (Fig. 4.1; CCDS Database, http://www.ncbi.nlm.nih.gov/CCDS/CcdsBrowse.cgi). The structure of the mouse *Tbc1d1* gene and that of other mammals is similar to the human gene, with a few additional coding exons predicted *in silico* or derived from available expressed sequence information. For example, in mouse 6, protein isoforms are listed in Ensembl database (GRCm38 genome version). Eighteen human splice variants, deduced from expressed sequence tags or other RNA sequencing data, are listed in Ensembl database (release 68—July 2012; http://www.ensembl.org): seven of which potentially encode for additional protein products. Direct gene expression data and tissue distribution for different splicing variants and/or protein isoforms are available only from a few studies.

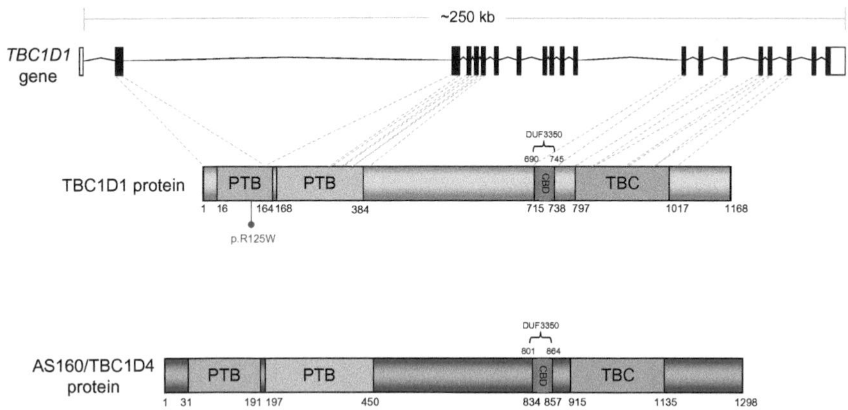

Figure 4.1 Structure of the human *TBC1D1* gene and functional domains in the TBC1D1 and TBC1D4 proteins. Domains have been identified using Simple Modular Architecture Research Tool—SMART (http://smart.embl-heidelberg.de/) analysis on Q86TI0 (TBC1D1) and O60343 (TBC1D4) or by BLAST and PFAM analysis (CBD). The protein domains are PTB: phosphotyrosine-binding domains; TBC: domains in Tre-2, BUB2p, and Cdc16p—Rab-GAP (GTPase-activating protein) domains; CBD: calmodulin-binding domains. DUF3350 regions (domain of unknown function) are annotated in the NP_055988 (TBC1D1) and NP_055647 (TBC1D4) protein sequences. To identify the putative phosphorylation sites, please see these entries in the NCBI database: http://www.ncbi.nlm.nih.gov/guide/proteins/. Vertical bars in the schematic representation of the gene structure representing exons are not on scale. (See Color Insert.)

Expression experiments in mice confirm the existence of at least two *Tbc1d1* transcripts, a long and a short one, with the short lacking a 93-amino acid sequence, amino acids 631–724 (derived from alternative splicing of exons 12 and 13; Chadt et al., 2008; Roach, Chavez, Mîinea, & Lienhard, 2007). The long transcript has been reported to be highly expressed in skeletal muscle, whereas, in adipose tissue, both transcripts have been detected (Taylor et al., 2008). Chadt et al. (2008) reported the expression of this gene in skeletal muscle, pancreas, kidney, and hypothalamus in C57BL/6J and NZO (New Zealand obese) mice, whereas low expression was detected in white and brown adipose tissues, liver, intestine, and colon. In SJL mice (lean), *Tbc1d1* expression was, in general, lower than in the other lines. Western blot data reported predominant expression in skeletal muscle, lower expression in heart, pancreas, colon, kidney, and hypothalamus, and no presence in 3T3-L1 adipocytes (Chadt et al., 2008). However, other data reported for 3T3-L1 adipocytes indicated expression of the short variant that increased during adipocyte differentiation and decreased after cellular differentiation (Chavez, Roach,

Keller, Lane, & Lienhard, 2008). Differences of protein expression were detected also in murine skeletal muscles. *Musculus tibialis* showed a 10 times higher expression than the soleus muscle (Taylor et al., 2008).

The human TBC1D1 protein has been studied mainly in skeletal muscle through biopsies, to evaluate its level in exercise model experiments or in subjects with different disease status (Frøsig, Pehmøller, Birk, Richter, & Wojtaszewski, 2010; Jensen et al., 2012; Jessen et al., 2011; Peck et al., 2009). No differences in TBC1D1 protein expression were evidenced among muscles (soleus, gastrocnemius, and vastus lateralis) and before and after exercise (Frøsig et al., 2010; Jensen et al., 2012). No differences in protein level were evidenced in quadriceps femoris between subjects with type 2 diabetes and health controls (Peck et al., 2009). Three splice variants of *TBC1D1* were detected in human skeletal muscle. Their predicted weight was ~140 kDa (short form), 146–148 kDa (medium form), and ~155 kDa (long form), which was the predominant variant (Jessen et al., 2011). Some phosphorylation sites are lost in the short and medium forms, with potential functional relevance (Roach et al., 2007). The presence of a larger number of different *TBC1D1* transcripts in databases might indicate the existence of additional spliced variants.

A large number of human *TBC1D1* polymorphisms are listed in public databases. In particular, about 70 missense mutations and a few stop gained and frameshift mutations are reported, but their frequency has been validated or deduced just for a few of them (Table 4.1; and data not shown). Therefore, it is not clear if most of them could be relevant or not in case they would be *in silico* predicted to have a functional role. A specific discussion about the p.R125W variant (Stone et al., 2006) is reported below. It is expected that more variants in this gene will be detected as whole genome resequencing of human genomes is becoming a routine analysis. Therefore, considering the position of these variants in the TBC1D1 protein, in addition to *in vitro* and *in vivo* experiments, could be important to understand the possible role this gene might have in affecting phenotypes and disease-related traits.

3. TBC1D1 PROTEIN STRUCTURE AND FUNCTION

Structural domain analysis has shown that TBC1D1 contains two phosphotyrosine-binding (PTB) domains at its NH_2-terminus, a Rab-GAP (GTPase-activating protein) domain (TBC domain) near the COOH-terminus (detected using Simple Modular Architecture Research Tool—SMART; http://smart.embl-heidelberg.de/), and a putative calmodulin-binding domain

Table 4.1 Validated nonsynonymous mutations in the human *TBC1D1* gene available in databases

rs number	Chromosome position[a]	mRNA position[b]	MAF[c]	Alleles (ref/alt)[d]	aa[e]	Validation[f]	SIFT[g]	Database[h]
rs2279027	37903756	398	0.3021	T/C	p.S14P	A B C D F	1	NCBI/Ensembl/UniProt
rs4008480	37903880	522	NA	C/T	p.T55I	A B D	0.08	NCBI/Ensembl/UniProt
rs35859249	37904089	731	0.080–0.10	C/T	p.R125W	A B F G	0.02	Ensembl/UniProt
rs11935412	38016371	1017	NA	C/A	p.A220E	E	0.33	NCBI/Ensembl
rs10501	38016395	1041	0.4229	T/G	p.V228G	A B D F G	0.57	NCBI/Ensembl
rs139404167	38016517	1163	NA	A/G	p.I269V	A F	0.51	NCBI/Ensembl/UniProt
rs145932557	38019979	1245	NA	G/A	p.G296R	A B F	0	NCBI/Ensembl
rs112261209	38022219	1338	0.0036	G/A	p.R327K	A B F	1	NCBI/Ensembl
rs145572665	38022282	1401	NA	T/A	p.V348R	B	0	NCBI/Ensembl
rs187586429	38023221	1450	NA	G/C	p.M364I	F	0.06	NCBI/Ensembl
rs61731610	38023294	1523	0.0350	G/A	p.G389S	A B F	0.72	NCBI/Ensembl
rs146950338	38029435	1595	NA	G/A	p.E413H	B	0.06	NCBI/Ensembl
rs74955942	38029466	1626	0.0018	A/G	p.N423S	F	0.06	NCBI/Ensembl
rs140673717	38037244	1696	NA	A/T	p.D446E	F	0.05	NCBI/Ensembl
rs141092310	38045998	1787	NA	G/A	p.A477T	F	0.33	NCBI/Ensembl
rs148919679	38055845	2291	NA	C/A	P.L646I	B	0.61	Ensembl
rs7677030	38091556	2409	NA	A/C	p.Y685S	A	0.39	Ensembl/UniProt

Continued

Table 4.1 Validated nonsynonymous mutations in the human *TBC1D1* gene available in databases—cont'd

rs number	Chromosome position	mRNA position	MAF	Alleles (ref/alt)	aa	Validation	SIFT	Database
rs148919679	38055845	2576	NA	C/A	p.L740I	B	NC	NCBI
rs149746733	38126619	3354	0.105	A/C	p.K1000T	B	0	Ensembl
rs13110318	38138856	3738	0.0922	G/A	p.R1127Q	B	NC	NCBI
rs13110318	38138856	3762	0.105	G/A	p.R1136Q	A B F	0.23	Ensembl/UniProt
rs145834587	38138916	3798	NA	G/A	p.R1147Q	B	0.12	NCBI/Ensembl
rs145834587	38138916	3822	NA	G/A	p.R1156Q	A B F	0.19	Ensembl

[a]Human chromosome 4 (GRCh37).
[b]The mRNA position is reported according to NCBI database information (NCBI dbSNP Build 137 for Human). When the mutation is not present in this database, reported position is according to Ensembl database (release 68—July 2012).
[c]MAF = minor allele frequency; NA = not available.
[d]ref = reference allele; alt = alternative allele.
[e]aa = amino acid substitution.
[f]A = validated by multiple, independent submissions to the refSNP cluster; B = validated by frequency or genotype data: minor alleles observed in at least two chromosomes; C = validated by submitter confirmation; D = all alleles have been observed in at least two chromosomes apiece; E = genotyped by HapMap project; F = SNP has been sequenced in 1000 Genome project; G = information available in the literature.
[g]SIFT = score of the amino acid substitution obtained using the SIFT program (Kumar, Henikoff, & Ng, 2009). NC = not calculated. Scores below 0.05 indicate putative disrupting substitutions.
[h]NCBI = NCBI dbSNP Build 137 for Human (http://www.ncbi.nlm.nih.gov/SNP/index.html); Ensembl = variation table available for the human *TBC1D1* gene (ENSG00000065882) (http://www.Ensembl.org/Homo_sapiens/); UniProt = information available for the Q86TI0 accession number at the UniProt database, release 2012_07—July 11, 2012 (http://www.uniprot.org/).

(CBD; Roach et al., 2007; see Fig. 4.1). The presence of these domains suggested a potential functional role of this protein, inferred starting from its closely related and first studied paralog, AS160 (Akt substrate of 160 kDa), also known as TBC1D4. There is an overall identity of about 50% between TBC1D1 and TBC1D4 proteins, even if identity between the GAP domains and the putative CBDs increases by 79% (Roach et al., 2007), suggesting indirectly important and similar roles of the two proteins derived from these conserved regions. The crystal structures of the human TBC1D1 and TBC1D4 Rab-GAP domains have been recently obtained by Park, Jin, Woo, and Shoelson (2011) who showed critical functional elements for GLUT4 (insulin-regulated glucose transporter protein type 4) translocation.

Phylogenetic analysis of *TBC1D1* and *TBC1D4* suggested that *TBC1D1* is the most ancient gene and *TBC1D4* is derived by duplication and subsequent divergence in vertebrates (Chen et al., 2008). This hypothesis is supported by the presence of only one *TBC1D1/4* gene in *Drosophila* (*pollux*) and lower organisms.

3.1. Regulation of GLUT4 traffic: The role of TBC1D1 and TBC1D4

Analysis of the TBC1D1 structure and that of its related TBC1D4 suggested several potential interactions of these two proteins with other components already known or supposed to be involved in the regulation of GLUT4 traffic, linking signaling and membrane trafficking pathways (Koumanov & Holman, 2007; Sakamoto & Holman, 2008). A schematic representation of the insulin- and muscle contraction-dependent pathway regulating GLUT4 translocation in skeletal muscle centered on TBC1D1 and TBC1D4 is reported in Fig. 4.2. This is part of a complex homeostatic mechanisms evolved for maintaining a stable glucose concentration in the blood and for providing a constant supply of this primary substrate for energy production (Bogan, 2012; Rowland, Fazakerley, & James, 2011). In skeletal muscle, adipocytes, and liver, insulin signaling is initiated when insulin is bound to the alpha subunit of the insulin receptor on the extracellular surface, which causes the autophosphorylation of the beta-subunit, that in turn causes the phosphorylation of the insulin receptor substrate (IRS) protein. This process is regulated by the mammalian target of rapamycin complex 1 (mTORC1)/S6 kinase 1 pathway (Veilleux, Houde, Bellmann, & Marette, 2010). Phosphorylated IRS proteins recruit and activate class 1A phosphatidylinositol 3-kinase that catalyzes the production of phosphatidylinositol 3,4,5-triphosphate (PIP3) from phosphatidylinositol 4,5-biphosphate. PIP3

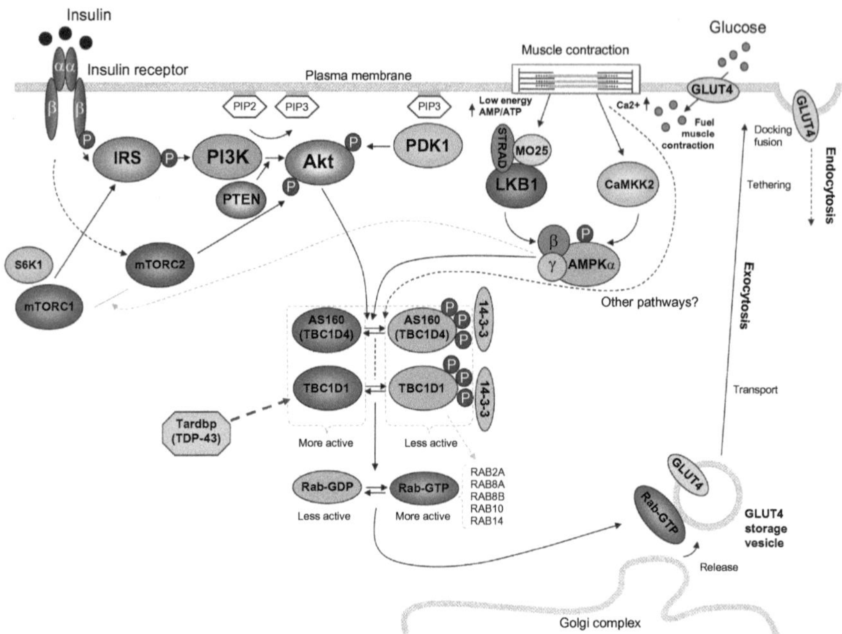

Figure 4.2 Schematic representation of the insulin and muscle contraction-dependent pathways regulating GLUT4 translocation in skeletal muscle showing the role of TBC1D1 and TBC1D4. Explanation of the different steps is given in the text. 14-3-3: 14-3-3 protein; AKT: protein kinase B (PKB); AMP: adenosine monophosphate; AMPK: 5′ adenosine monophosphate-activated protein kinase; AS160: Akt substrate of 160 kDa; ATP: adenosine-5′-triphosphate; CAMKK2: calcium/calmodulin-dependent protein kinase kinase 2; GLUT4: glucose transporter type 4; GMP: guanosine monophosphate; GTP: guanosine-5′-triphosphate; IRS: insulin receptor substrate protein; LKB1: liver kinase B1; MO25: mouse protein-25; mTORc1: mammalian target of rapamycin, in multiprotein complex 1; mTORc2: mammalian target of rapamycin, in multiprotein complex 2; PDK1: pyruvate dehydrogenase lipoamide kinase isozyme 1; PI3K: phosphatidylinositol 3-kinase; PIP2: phosphatidylinositol (3,4,5)-biphosphate; PIP3: phosphatidylinositol (3,4,5)-triphosphate; PTEN: phosphatase and tensin homolog; Rab: rab protein (member of the Ras superfamily of monomeric G proteins); STRAD: STe20-related adaptor; Tardbp: Tat activating regulatory DNA-binding protein; TBC1D1: TBC1 (tre-2/USP6, BUB2, cdc16) domain family, member 1; TBC1D4: TBC1 (tre-2/USP6, BUB2, cdc16) domain family, member 4; TDP-43: TAR DNA-binding protein 43. (See Color Insert.)

level is negatively regulated by a phosphatidylinositol-3,4,5-trisphosphate 3-phosphatase (PTEN: phosphatase and tensin homolog). PIP3 recruits both pyruvate dehydrogenase lipoamide kinase isozyme 1 (known as phosphoinositide-dependent protein kinase 1, PDK1) and protein kinase B (PKB or Akt) to the phospholipid and allows the activation of Akt through phosphorylation of Thr^{308} by PDK1. Akt is also phosphorylated by the mammalian target of rapamycin complex 2 (mTORC2) in a second site (Ser^{473}). This additional phosphorylation is needed for full Akt activity (Zhou et al., 2008). Akt-mediated phosphorylation of TBC1D1 and TBC1D4 represents the intersection between the insulin signaling and vesicle trafficking that determines GLUT4 translocation (Kane et al., 2002). Phosphorylated Akt binds a PTB domain in TBC1D1 and TBC1D4, then downstream transmission of this signal is carried out by the Rab-GAP domain of the two proteins. Rab-GTPases are small GTPases that serve as organizers of almost all membrane trafficking processes (Stenmark, 2009). Phosphorylation of TBC1D1 and TBC1D4 by Akt in response to an upstream insulin signal recruits 14-3-3 which binds to the two target proteins (Ramm, Larance, Guilhaus, & James, 2006). Several phosphomotifs have been predicted *in silico* and *in vitro* on TBC1D1 and TBC1D4, with potential different functional effects on the activity of the two proteins (Sakamoto & Holman, 2008). These phosphorylation and binding events inhibit Rab-GAP activity allowing GTP loading and activation of the Rab proteins (Sano et al., 2007). Several cognate Rab proteins of TBC1D1 have been identified (RAB2A, RAB8A, RAB8B, RAB10, and RAB14; Stenmark, 2009). GTP loading increases the concentration of activated Rab proteins (Rab-GTP) on GLUT4 storage vesicles that promote their translocation and fusion to the plasma membrane (Rowland et al., 2011).

Another route of regulation of TBC1D1 and TBC1D4 activity is through muscle contraction and energy depletion. Exercise stimulates GLUT4 translocation by phosphorylation of the two proteins through 5′ adenosine monophosphate-activated protein kinase (AMPK) and calmodulin pathways (Bruss, Arias, Lienhard, & Cartee, 2005; Deshmukh et al., 2006; Pehmøller et al., 2009). Phosphorylation of AMPK is also regulated by liver kinase B1 (LKB1 or STK11) after the depletion of ATP in energy-stressed cells activated, by forming a complex with STRAD (Ste20-related adaptor) and MO25 (Mouse protein-25; Filippi & Alessi, 2008).

TBC1D1 has a more important role than TBC1D4 on contraction-stimulated glucose uptake. TBC1D1 is mainly expressed in skeletal muscle and TBC1D4 in adipose tissues, suggesting indirectly that these two genes could have a few specific and complementary functions. TBC1D1 includes an important phosphomotif (Ser^{237}) that is not present in TBC1D4. Phosphorylation of this site is increased with AMPK activation but not with an insulin stimulus (An et al., 2010; Chen et al., 2008). The upstream signal for phosphorylation of Ser^{237} derives from muscle contraction (Frøsig et al., 2010).

Another information on the signals affecting TBC1D1 derives from the regulation of its gene expression. Chiang et al. (2010) demonstrated that Tat activating regulatory DNA-binding protein (Tardbp or TDP-43, TAR DNA-binding protein 43), a highly conserved metazoan DNA/RNA binding protein thought to be involved in RNA transcription and splicing, regulates *Tbc1d1* transcription. This study provided an explanation to the altered body fat metabolism derived from *TDP-43* deletion in mice.

Summarizing, based on what briefly illustrated above, it is now possible to understand the general role of TBC1D1. However, several other details on the function of this protein on muscle contraction-induced GLUT4 translocations are still waiting to be fully understood.

4. THE *TBC1D1* GENE IN OBESITY AND RELATED TRAITS

4.1. Of mice and men (women)

The role of TBC1D1 (and TBC1D4) as a link between signaling and GLUT4 trafficking pathways, mainly exploited in skeletal muscle, provides functional clues to understand the potential effects of mutations in this gene. Vice versa, it is also true that genetic analysis of this gene in mice has contributed to understand the function of Tbc1d1.

Quantitative trait loci (QTL) mapping in crossbreeding experiments using the NZO and the lean SJL mouse strains identified a major QTL on chromosome 5 for high-fat diet-induced obesity (*Nob1*; Giesen, Plum, Kluge, Ortlepp, & Joost, 2003; Jurgens et al., 2007; Kluge et al., 2000). Subsequent backcross experiments showed that the $Nob1^{SJL}$ allele had an obesity-suppressed effect in high-fat-fed animals (Chadt et al., 2008). For positional cloning of the *Nob1* gene, fine mapping reduced the QTL interval to a block of ~8 Mb that included 19 protein-coding genes. The coding regions of these 19 genes were sequenced and gene expression analysis was conducted for 302 genes in the entire *Nob1* QTL region (Chadt et al., 2008). *Tbc1d1*, that mapped in the QTL peak region, was downregulated by 70% in skeletal muscle of SJL mice, suggesting its possible involvement in determining the *Nob1* QTL. Sequencing of the *Tbc1d1* gene revealed a 7-bp deletion in exon 18 present only in the SJL line. This deletion represents a loss-of-function mutation as the derived protein lacks part of the fundamental TBC-GAP domain. Using a subcongenic line which harbors ~10 Mb of the *Nob1* region from SJL containing the mutated *Tbc1d1* gene on a C57BL/6J background, Chadt et al. (2008) showed that homozygous carriers of the $Nob1^{SJL}$ allele had lower body weight than did the control normal mice, derived from lower fat mass in the former. Hence, the mutated *Tbc1d1* gene exerted a weight-suppressive effect in mice and protected TBC1D1-deficient animals on high-fat diet against the development of obesity due partly to increased whole body lipid oxidation and increased skeletal muscle lipid oxidation compared to control mice (Chadt et al., 2008). This study provided the first direct and compelling demonstration of the involvement of *Tbc1d1* in fat deposition, energy metabolism, and, in turn, obesity. A more detailed physiological characterization of *Tbc1d1*-deficient mice (homozygous for the $Nob1^{SJL}$ allele) was recently reported by Szekeres et al. (2012). These authors showed that even on a low-fat diet, *Tbc1d1*-deficient mice displayed impaired insulin-stimulated glucose transport in glycolytic extensor digitorum longus muscle, which is accompanied by a reduction in GLUT4 protein abundance. The use of $Nob1^{SJL}$ mice suggested that functional Tbc1d1 is required for

Table 4.2 List of studies that showed association or linkage between chromosome 4 microsatellites and SNPs in the *TBC1D1* gene and human obesity or other related traits

References	Cohorts/type of study	Most significant markers	Associated traits
Stone et al. (2002)	Multigenerational Utah pedigrees with at least two members highly overweight, including patients with gastric bypass and their families (1182 individuals)/QTL mapping	D4S3350	Female BMI
Arya et al. (2004)	Four hundred and thirty individuals from 27 pedigrees from Mexican Americans (San Antonio Family Diabetes Study)/QTL mapping	D4S2912	BMI
Stone et al. (2006)	Multigenerational Utah pedigrees with at least two members highly overweight, including patients with gastric bypass and their families/resequencing, linkage analysis	p.R125W	Female BMI
Fox et al. (2007)	Families from the Framingham Heart Study 100k project/GWA study	rs10517461	Waist circumference
Meyre et al. (2008)	1109 French Caucasian pedigrees with obese children or adults/association study	p.R125W	Female BMI
Savage et al. (2008)	Type 1 diabetic patients with diabetic nephropathy/association study with SNPs in candidate genes	rs10501	Diabetic nephropathy

insulin and energy-sensing signals controlling glucose metabolism in skeletal muscle. Hence, mutant forms of Tbc1d1 may indirectly contribute to the development of obesity by regulating skeletal muscle insulin sensitivity (Szekeres et al., 2012).

In humans, a few studies linked the *TBC1D1* gene to obesity (Table 4.2). First, Stone et al. (2002), conducting a genome-wide search with 628 microsatellite markers in multigenerational Utah selected pedigrees containing 1182 subjects, identified a major predisposition locus for severe obesity (body mass index: BMI) in female patients at 4p15-p14, with the higher multipoint heterogeneity LOD score in the correspondence of marker D4S2632. Analysis with BMI information from males only generated no evidence of linkage to this region, whereas analysis with both female and male data produced only weak evidence of linkage.

Another evidence of a QTL for BMI on chromosome 4p was obtained using data from 27 Mexican–American pedigrees including 430 individuals who participated in the San Antonio Family Diabetes Study (Arya et al., 2004). The most significant evidence for linkage with BMI was near marker D4S2912 at position 4p15.1, confirming the region detected by Stone et al. (2002). The study in Mexican Americans did not report any sex-specific effect.

The first work of Stone et al. (2002) was subsequently refined by the same authors who resequenced the coding region and part of introns in 16 candidate genes in the genomic interval that most likely contained the gene of interest on chromosome 4p15-p14 (Stone et al., 2006). Ten disease-associated haplotypes were defined, and 16 out of 68 detected variants were selected as possibly segregating with high BMI. Subsequent filtering was conducted by comparing their frequency within the 10 disease-associated haplotypes with their frequency in a control population. Only three variants were significantly enriched on disease-associated haplotypes, of which only one variant (the most significantly associated; *P* nominal value = 0.000008) could have functional effects. This variant is a missense mutation (p.R125W, with W associated with higher BMI; frequency in Caucasian population = 0.08–0.10) in the first coding exon of the *TBC1D1* gene, located in a conserved position (R or Q in mammals) of the first PTB domain. These results were further strengthened by adding other individuals in the linked pedigrees and additional control subjects from the Utah population. In addition, excluding from the linkage analysis in the entire pedigrees, females carrying the p.R125W substitution, the heterogeneity LOD score dropped just below 3.0, indicating an important role of this mutation in determining the QTL results. However, the drop was not as high as expected, suggesting that additional disease-associated variants in *TBC1D1* could segregate in the analyzed cohort (Stone et al., 2006). A small case and control confirmation study reported by the same authors did not provide the same evidences as the pedigree-based experimental design supported. This negative result raised the possibility that the p.R125W variant conferred risk for obesity in the context of specific gene/gene interactions. To evaluate this hypothesis, Stone et al. (2006) used families that segregated the p.R125W mutation in an adjusted linkage analysis. Confirming their hypothesis, they identified another putative QTL region for female BMI on 4q34-35.

Meyre et al. (2008) replicated the association results between the p.R125W variant and BMI in 1109 French Caucasian pedigrees with obese children or adults for a total of 5080 subjects. Testing the additive mode of inheritance, using the 97th BMI percentile as threshold for obesity and restricting the association analysis to women only, the genotyped variant was associated with obesity ($P=0.008$). The same analysis restricted to men did not show any association with BMI. None of the additional single nucleotide polymorphisms (SNPs) around the p.R125W mutation, genotyped in the same individuals using the Illumina 370 CNV duo array, were significantly associated with BMI. These data might exclude the possibility that other variants in or very close to the *TBC1D1* gene could affect obesity phenotypes. Moreover, other analyses did not confirm any QTL for BMI in the 4q34-35 region (Meyre et al., 2008). The p.R125W SNP was then genotyped in the French prospective DESIR cohort ($n=4634$) to assess the contribution of this polymorphism to BMI variability in a population not enriched in subjects with severe familial obesity. The results in this cohort did not indicate any association of this variant to obesity or obesity predisposition (Meyre et al., 2008).

The only GWA study carried out using high-density SNP chips that showed association between a *TBC1D1* gene marker and obesity was published by Fox et al. (2007). This investigation was based on 1345 participants from the Framingham Heart Study 100K project. Association analysis indicated that rs10517461 (in the *TBC1D1* gene) is associated with waist circumference. In another genetic association analysis for diabetic nephropathy including a large number of nonsynonymous SNPs in a case and control experimental design, the TBC1D1 p.Val228Gly variant resulted associated with this disease (Savage et al., 2008). However, it is difficult to suppose a functional relevance of this mutation in the associated disease status as *in silico* prediction of its effect did not suggest an important role of this polymorphic site (Table 4.1).

Several other large GWA studies of BMI, waist circumference or correlated diseases (type 2 diabetes), and glycemic or metabolomic traits did not report any significant association with *TBC1D1* markers, confirming the elusive nature of variability in this gene in affecting obesity.

4.2. Association with fat deposition and production traits in livestock

Fat deposition and correlated traits (lean meat deposition, growth rate, feed conversion rate, etc.) are important phenotypes in livestock species that affect production efficiency and profitability. These traits have, in general, medium–high heritability, and breeding strategies are designed to optimize selection plans for their improvement according to the needs of production systems. These aspects are particularly relevant in pigs, chicken (broilers), and beef cattle. Therefore, the identification of DNA markers associated to economic traits is an important and preliminary step to subsequently improve and drive selection efficiency through marker-assisted selection. Different approaches have been used in livestock to identify the association between DNA polymorphisms and production traits. A candidate gene approach takes advantage from information available on the function of the gene derived from its involvement in physiological mechanisms and biological pathways that might be linked directly or indirectly to a phenotype of economic importance (e.g., backfat thickness or lean meat carcass content). Hence, natural polymorphisms in the gene could explain part of the variability of phenotypic traits.

Starting from evidences available in mice and humans and based on its function, *TBC1D1* has been recently investigated in pigs as a candidate gene for fat and lean meat deposition (Fontanesi et al., 2011, 2012). By sequencing parts of this gene, a few SNPs were identified. In the first exploratory study (Fontanesi et al., 2011), one intronic polymorphism (FN677935: g.219G > *A*) was analyzed in Italian Large White and Italian Duroc pigs using a selective genotyping approach, including pigs with extreme and divergent estimated breeding values for backfat thickness, lean cut weight (Italian Large White), and visible intermuscular fat (Italian Duroc) (Fontanesi et al., 2011). Significant differences of allele frequencies between the extreme tails were obtained in all three groups of pigs. Significance was reduced when related animals were excluded from the analysis. Other groups of Italian Large White and Italian Duroc pigs showed only weak signal of association between the analyzed SNP and production traits, probably due to the small number of pigs investigated (Fontanesi et al., 2011). Part of this study was replicated in a larger experiment based again on selective genotyping, including additional Italian Large White pigs with extreme and divergent estimated breeding value for backfat thickness (Fontanesi et al., 2012). Animals were all sows ($n=560$), chosen within a performance-tested population of about 12,000 pigs. These pigs were genotyped for several hundred SNPs chosen in many candidate genes for fat deposition. The *TBC1D1* SNP was the third most significant associated marker after two other polymorphisms in genes already well known to affect fatness in pigs (*IGF2* and *MC4R*). These results confirmed the suggestive evidences obtained in the exploratory study, suggesting that polymorphisms in the *TBC1D1* gene can explain an important fraction of backfat thickness genetic variability in the Italian Large White pig breed. Interestingly, in the second study, all investigated animals were females, matching, to some extent, results obtained in humans. However, direct proof of sex-specific effects of porcine *TBC1D1* mutations is still to be demonstrated. Further analyses of the *TBC1D1* gene in pigs are currently under way.

Indirect evidences of a possible involvement of the *TBC1D1* gene in affecting production traits in chicken derive from QTL experiments. Nassar, Goraga, and Brockmann (2012), using reciprocal F2 intercross populations generated from crosses between the dual-purpose New Hampshire inbred line and the layer inbred line White Leghorn, identified an important QTL affecting visceral, subcutaneous neck and total adipose tissues on chromosome 4. The confidence interval region contains the *TBC1D1* gene. In another GWA study of body weight in a chicken F2 resource population derived from the cross between Silky Fowl and Plymouth Rock (two chicken breeds with extreme phenotypes in body weight and growth rate), Gu et al. (2011) reported a highly significant QTL on chicken chromosome 4, in the same position of that of the study mentioned above. A marker positioned only 92 kb downstream of the *TBC1D1* gene was highly significant for body weight at 12 weeks of age. Three other studies in different chicken resource populations reported QTL for growth rate and body weight in the same region (Ambo et al., 2009; Sewalem et al., 2002; Zhou, Deeb, Evock-Clover, Ashwell, & Lamont, 2006). Interestingly, whole genome resequencing of different chicken lines and breeds revealed that a sweep *TBC1D1* haplotype has been originated during the early development of broiler lines (Rubin et al., 2010). Functional studies of different chicken *TBC1D1* variants might provide additional information on the role of this gene in growth and fat deposition.

5. CONCLUSIONS

Information available on the TBC1D1 function provides a general picture about the potential role of this gene in obesity (or leanness). An et al. (2010) provided a preliminary functional analysis of the p.R125W mutation. Overexpression of the obesity-associated mutant form (p.125W) in mouse skeletal muscle significantly decreased insulin-stimulated glucose uptake, which could be a mechanism for the development of obesity associated with this variant. However, it remains to fully demonstrate how the p. R125W mutation could cause obesity in humans. Another open question is the female-specific effect of this polymorphic site, for which, at present, there is no biological explanation. Only two independent studies have so far confirmed association between the p.R125W and BMI in females, but many other GWA studies failed to identify *TBC1D1* as an important determinant of common obesity. Additional studies are needed to clarify these inconsistencies. Evidences in model animals (rodents) and emerging results from pigs and chicken can complement and complete the general picture we have now, adding useful biological details to understand the role of *TBC1D1* in human obesity. In addition, the results obtained in pigs and chicken further confirm and strengthen the role of these two species as important animal models for human obesity.

ACKNOWLEDGMENT

The authors would like to thank the support of Italian MiPAAF INNOVAGEN project.

REFERENCES

Ambo, M., Moura, A. S., Ledur, M. C., Pinto, L. F., Baron, E. E., Ruy, D. C., et al. (2009). Quantitative trait loci for performance traits in a broiler 3 layer cross. *Animal Genetics*, *40*, 200–208.

An, D., Toyoda, T., Taylor, E. B., Yu, H., Fujii, N., Hirshman, M. F., et al. (2010). TBC1D1 regulates insulin- and contraction-induced glucose transport in mouse skeletal muscle. *Diabetes*, *59*, 1358–1365.

Arya, R., Duggirala, R., Jenkinson, C. P., Almasy, L., Blangero, J., O'Connell, P., et al. (2004). Evidence of a novel quantitative-trait locus for obesity on chromosome 4p in Mexican Americans. *The American Journal of Human Genetics*, *74*, 272–282.

Bogan, J. S. (2012). Regulation of glucose transporter translocation in health and diabetes. *Annual Review of Biochemistry*, *81*, 507–532.

Bruss, M. D., Arias, E. B., Lienhard, G. E., & Cartee, G. D. (2005). Increased phosphorylation of Akt substrate of 160 kDa (AS160) in rat skeletal muscle in response to insulin or contractile activity. *Diabetes*, *54*, 41–50.

Chadt, A., Leicht, K., Deshmukh, A., Jiang, L. Q., Scherneck, S., Bernhardt, U., et al. (2008). *Tbc1d1* mutation in lean mouse strain confers leanness and protects from diet-induced obesity. *Nature Genetics*, *40*, 1354–1359.

Chavez, J. A., Roach, W. G., Keller, S. R., Lane, W. S., & Lienhard, G. E. (2008). Inhibition of GLUT4 translocation by Tbc1d1, a Rab GTPase-activating protein abundant in skeletal muscle, is partially relieved by AMP-activated protein kinase activation. *The Journal of Biological Chemistry*, *283*, 9187–9195.

Chen, S., Murphy, J., Toth, R., Campbell, D. G., Morrice, N. A., & Mackintosh, C. (2008). Complementary regulation of TBC1D1 and AS160 by growth factors, insulin and AMPK activators. *The Biochemical Journal*, *409*, 449–459.

Chiang, P. M., Ling, J., Jeong, Y. H., Price, D. L., Aja, S. M., & Wong, P. C. (2010). Deletion of TDP-43 down-regulates Tbc1d1, a gene linked to obesity, and alters body fat metabolism. *Proceedings of the National Academy of Sciences of the United States of America*, *107*, 16320–16324.

Deshmukh, A., Coffey, V. G., Zhong, Z., Chibalin, A. V., Hawley, J. A., & Zierath, J. R. (2006). Exercise-induced phosphorylation of the novel Akt substrates AS160 and filamin A in human skeletal muscle. *Diabetes*, *55*, 1776–1782.

Filippi, B. M., & Alessi, D. R. (2008). Novel role for the LKB1 pathway in controlling monocarboxylate fuel transporters. *The Journal of Cell Biology*, *183*, 7–9.

Fontanesi, L., Colombo, M., Tognazzi, L., Scotti, E., Buttazzoni, L., Dall'Olio, S., et al. (2011). The porcine TBC1D1 gene: Mapping, SNP identification, and association study with meat, carcass and production traits in Italian heavy pigs. *Molecular Biology Reports*, *38*, 1425–1431.

Fontanesi, L., Galimberti, G., Calò, D. G., Fronza, R., Martelli, P. L., Scotti, E., et al. (2012). Identification and association analysis of several hundred single nucleotide polymorphisms within candidate genes for back fat thickness in Italian Large White pigs using a selective genotyping approach. *Journal of Animal Science*, *90*, 2450–2464.

Fox, C. S., Heard-Costa, N., Cupples, L. A., Dupuis, J., Vasan, R. S., & Atwood, L. D. (2007). Genome-wide association to body mass index and waist circumference: The Framingham Heart Study 100K project. *BMC Medical Genetics*, *8*(Suppl. 1), S18.

Frøsig, C., Pehmøller, C., Birk, J. B., Richter, E. A., & Wojtaszewski, J. F. (2010). Exercise-induced TBC1D1 Ser237 phosphorylation and 14-3-3 protein binding capacity in human skeletal muscle. *The Journal of Physiology*, *588*, 4539–4548.

Giesen, K., Plum, L., Kluge, R., Ortlepp, J., & Joost, H. G. (2003). Diet-dependent obesity and hypercholesterolemia in the New Zealand obese mouse: Identification of a quantitative trait locus for elevated serum cholesterol on the distal mouse chromosome 5. *Biochemical and Biophysical Research Communications*, *304*, 812–817.

Gu, X., Feng, C., Ma, L., Song, C., Wang, Y., Da, Y., et al. (2011). Genome-wide association study of body weight in chicken F2 resource population. *PLoS One*, *6*, e21872.

Jensen, T. E., Leutert, R., Rasmussen, S. T., Mouatt, J. R., Christiansen, M. L. B., Jensen, B. R., et al. (2012). EMG-normalised kinase activation during exercise is higher in human gastrocnemius compared to soleus muscle. *PLoS One*, 7, e31054.

Jessen, N., An, D., Lihn, A. S., Nygren, J., Hirshman, M. F., Thorell, A., et al. (2011). Exercise increases TBC1D1 phosphorylation in human skeletal muscle. *American Journal of Physiology. Endocrinology and Metabolism*, *301*, E164–E171.

Jurgens, H. S., Neschen, S., Ortmann, S., Scherneck, S., Schmolz, K., Schüler, G., et al. (2007). Development of diabetes in obese, insulin-resistant mice: Essential role of dietary carbohydrate in beta cell destruction. *Diabetologia*, *50*, 1481–1489.

Kane, S., Sano, H., Liu, S. C., Asara, J. M., Lane, W. S., Garner, C. C., et al. (2002). A method to identify serine kinase substrates. Akt phosphorylates a novel adipocyte protein with a Rab GTPase-activating protein (GAP) domain. *The Journal of Biological Chemistry, 277*, 22115–22118.

Kluge, R., Giesen, K., Bahrenberg, G., Plum, L., Ortlepp, J. R., & Joost, H. G. (2000). Quantitative trait loci for obesity and insulin resistance (Nob1, Nob2) and their interaction with the leptin receptor allele (LeprA720T/T1044I) in New Zealand obese mice. *Diabetologia, 43*, 1565–1572.

Koumanov, F., & Holman, G. D. (2007). Thrifty Tbc1d1 and Tbc1d4 proteins link signalling and membrane trafficking pathways. *The Biochemical Journal, 403*, e9–e11.

Kumar, P., Henikoff, S., & Ng, P. C. (2009). Predicting the effects of coding non-synonymous variants on protein function using the SIFT algorithm. *Nature Protocols, 4*, 1073–1081.

Meyre, D., Farge, M., Lecoeur, C., Proenca, C., Durand, E., Allegaert, F., et al. (2008). R125W coding variant in TBC1D1 confers risk for familial obesity and contributes to linkage on chromosome 4p14 in the French population. *Human Molecular Genetics, 17*, 1798–1802.

Nassar, M. K., Goraga, Z. S., & Brockmann, G. A. (2012). Quantitative trait loci segregating in crosses between New Hampshire and White Leghorn chicken lines: III. Fat deposition and intramuscular fat content. *Animal Genetics*. http://dx.doi.org/10.1111/j.1365-2052.2012.02365.x, in press.

Park, S. Y., Jin, W., Woo, J. R., & Shoelson, S. E. (2011). Crystal structures of human TBC1D1 and TBC1D4 (AS160) RabGTPase-activating protein (RabGAP) domains reveal critical elements for GLUT4 translocation. *The Journal of Biological Chemistry, 286*, 18130–18138.

Peck, G. R., Chavez, J. A., Roach, W. G., Budnik, B. A., Lane, W. S., Karlsson, H. K., et al. (2009). Insulin-stimulated phosphorylation of the Rab GTPase-activating protein TBC1D1 regulates GLUT4 translocation. *The Journal of Biological Chemistry, 284*, 30016–30023.

Pehmøller, C., Treebak, J. T., Birk, J. B., Chen, S., Mackintosh, C., Hardie, D. G., et al. (2009). Genetic disruption of AMPK signaling abolishes both contraction-and insulin-stimulated TBC1D1 phosphorylation and 14-3-3 binding in mouse skeletal muscle. *American Journal of Physiology. Endocrinology and Metabolism, 297*, E665–E675.

Ramm, G., Larance, M., Guilhaus, M., & James, D. E. (2006). A role for 14-3-3 in insulin-stimulated GLUT4 translocation through its interaction with the RabGAP AS160. *The Journal of Biological Chemistry, 281*, 29174–29180.

Roach, W. G., Chavez, J. A., Mîinea, C. P., & Lienhard, G. E. (2007). Substrate specificity and effect on GLUT4 translocation of the Rab GTPase-activating protein Tbc1d1. *The Biochemical Journal, 403*, 353–358.

Rowland, A. F., Fazakerley, D. J., & James, D. E. (2011). Mapping insulin/GLUT4 circuitry. *Traffic, 12*, 672–681.

Rubin, C. J., Zody, M. C., Eriksson, J., Meadows, J. R., Sherwood, E., Webster, M. T., et al. (2010). Whole-genome resequencing reveals loci under selection during chicken domestication. *Nature, 464*, 587–591.

Sakamoto, K., & Holman, G. D. (2008). Emerging role for AS160/TBC1D4 and TBC1D1 in the regulation of GLUT4 traffic. *American Journal of Physiology. Endocrinology and Metabolism, 295*, E29–E37.

Sano, H., Eguez, L., Teruel, M. N., Fukuda, M., Chuang, T. D., Chavez, J. A., et al. (2007). Rab10, a target of the AS160 Rab GAP, is required for insulin-stimulated translocation of GLUT4 to the adipocyte plasma membrane. *Cell Metabolism, 5*, 293–303.

Savage, D. A., Patterson, C. C., Deloukas, P., Whittaker, P., Morrison, J., Boulton, A. J., et al. (2008). Genetic association analyses of non-synonymous single nucleotide polymorphisms in diabetic nephropathy. *Diabetologia, 51*, 1998–2002.

Sewalem, A., Morrice, D. M., Law, A., Windsor, D., Haley, C. S., Ikeobi, C. O., et al. (2002). Mapping of quantitative trait loci for body weight at three, six, and nine weeks of age in a broiler layer cross. *Poultry Science*, *81*, 1775–1781.

Stenmark, H. (2009). Rab GTPases as coordinators of vesicle traffic. *Nature Reviews. Molecular Cell Biology*, *10*, 513–525.

Stone, S., Abkevich, V., Hunt, S. C., Gutin, A., Russell, D. L., Neff, C. D., et al. (2002). A major predisposition locus for severe obesity, at 4p15-p14. *The American Journal of Human Genetics*, *70*, 1459–1468.

Stone, S., Abkevich, V., Russell, D. L., Riley, R., Timms, K., Tran, T., et al. (2006). TBC1D1 is a candidate for a severe obesity gene and evidence for a gene/gene interaction in obesity predisposition. *Human Molecular Genetics*, *15*, 2709–2720.

Szekeres, F., Chadt, A., Tom, R. Z., Deshmukh, A. S., Chibalin, A. V., Björnholm, M., et al. (2012). The Rab-GTPase-activating protein TBC1D1 regulates skeletal muscle glucose metabolism. *American Journal of Physiology. Endocrinology and Metabolism*, *303*, E524–E533.

Taylor, E. B., An, D., Kramer, H. F., Yu, H., Fujii, N. L., Roeckl, K., et al. (2008). Discovery of TBC1D1 as an insulin-, AICAR-, and contraction-stimulated signaling nexus in mouse skeletal muscle. *The Journal of Biological Chemistry*, *283*, 9787–9796.

Veilleux, A., Houde, V. P., Bellmann, K., & Marette, A. (2010). Chronic inhibition of the mTORC1/S6K1 pathway increases insulin-induced PI3K activity but inhibits Akt2 and glucose transport stimulation in 3T3-L1 adipocytes. *Molecular Endocrinology*, *24*, 766–778.

Zhou, H., Deeb, N., Evock-Clover, C. M., Ashwell, C. M., & Lamont, S. J. (2006). Genome-wide linkage analysis to identify chromosomal regions affecting phenotypic traits in the chicken. I. Growth and average daily gain. *Poultry Science*, *85*, 1700–1711.

Zhou, Q. L., Jiang, Z. Y., Holik, J., Chawla, A., Hagan, G. N., Leszyk, J., et al. (2008). Akt substrate TBC1D1 regulates GLUT1 expression through the mTOR pathway in 3T3-L1 adipocytes. *The Biochemical Journal*, *411*, 647–655.

CHAPTER FIVE

Replication Initiator 1 in Adipose Tissue Function and Human Obesity

John T. Heiker[*,1], **Nora Klöting**[†]

[*]Department of Medicine, Endocrinology and Diabetes, Universität Leipzig, Leipzig, Germany
[†]IFB Adiposity Diseases, Junior Research Group 2 "Animal models of obesity" Universität Leipzig, Leipzig, Germany
[1]Corresponding author: e-mail address: john.heiker@medizin.uni-leipzig.de

Contents

Abstract

Replication initiator 1 (Repin1) was discovered as the replication initiation-region protein 60 kDa (RIP60) and characterized as a zinc finger protein involved in DNA binding and bending during initiation of DNA replication. As with many members of the zinc finger family, physiological functions and regulations of Repin1 as well as the underlying molecular mechanisms have remained elusive since its discovery in 1990. However, recent results have revealed interesting new insights into Repin1 function. Repin1 is highly expressed in liver and adipose tissue and it has been suggested as a candidate gene for obesity and related metabolic disorders in congenic and subcongenic rat strains. Repin1 seems to participate in the regulation of genes involved in adipogenesis, lipid droplet formation and fusion, as well as glucose and fatty acid transport in adipocytes. Given the observation that Repin1 expression in human adipose tissue significantly correlates with total body fat mass and adipocyte size, therapeutic strategies to reduce Repin1 action in adipose tissue may be a potential approach to tackle adipose tissue dysfunction in obesity.

Thus, human and *in vitro* data suggest a role of Repin1 in adipocyte function and could therefore emerge as an interesting therapeutic target in human obesity.

Vitamins and Hormones, Volume 91
ISSN 0083-6729
http://dx.doi.org/10.1016/B978-0-12-407766-9.00005-5

1. INTRODUCTION

Replication initiator 1 (Repin1) was initially discovered as replication initiation-region protein 60 kDa (RIP60) in a study investigating DNA-binding proteins to elucidate origin of replication activation of the Chinese hamster *dihydrofolate reductase* gene (*dhfr*) (Dailey, Caddle, Heintz, & Heintz, 1990). Repin1 binds to two ATT-rich sites in oriβ, a short region 3′ to the *dhfr* gene, acting as an enhancer of DNA bending during initiation of DNA synthesis (Caddle, Dailey, & Heintz, 1990; Schroll & Heintz, 2004), but plasmid replication assays have shown that it has only weak replication enhancing activity and thus may act as an accessory factor in origin recognition prior to the assembly of preinitiation complexes (Houchens et al., 2000). After it was first cloned in 2000, characterization of DNA-binding and bending properties revealed the first structural insight into Repin1/RIP60 function as a polydactyl zinc finger protein of the Cys_2-His_2 type (Houchens et al., 2000). However, as for most of polydactyl zinc finger proteins that have been sequenced, the cellular localization and function of Repin1 remain elusive.

Repin1 maps on human chromosome 7 within in a QTL region identified for altered total serum and lipoprotein lipid concentration as well as body weight regulation (Kovacs et al., 2001). Recently, it was demonstrated that there is an association of a SNP (449) in the coding region and a triplet repeat (TTT) in the 3′UTR-untranslated region of the *Repin1* gene with increased body weight, plasma glucose concentrations, fasting insulin, triglycerides, and cholesterol serum concentrations in congenic and subcongenic rat strains (Kloting, Wilke, & Kloting, 2007). The repeat size was different in several rat strains and increasing repeat size significantly reflected obesity and metabolic phenotype of the animals (Kloting et al., 2007). Despite its ubiquitous expression, Repin1 was found to be most abundantly expressed in the liver and adipose tissue (Bahr, Kloting, Kloting, & Follak, 2011; Kloting et al., 2007).

With the focus on these recent developments of Repin1 investigation, this chapter reviews the impact of *Repin1* on adipose tissue function *in vitro* and *in vivo*.

2. STRUCTURE OF REPIN1

Repin1 belongs to the most common Cys_2-His_2 (C2H2) type of zinc finger proteins containing 15 C2H2 zinc finger DNA motifs (ZF). These zinc finger domains self-fold into a $\beta\beta\alpha$ structure stabilized by the coordination of a zinc ion by the conserved cysteine and histidine residues that provide additional stability to the fold (Frankel, Berg, & Pabo, 1987; Lee, Gippert, Soman, Case, & Wright, 1989; Pavletich & Pabo, 1991). The zinc ion is chelated between two cysteine residues within the β-strands and two histidine residues in the α-helix (Lee et al., 1989). The ZFs within Repin1 are organized in three clusters, hands Z1 (ZF 1–5), Z2 (ZF 6–8), and Z3 (ZF 9–15). A proline-rich region between ZFs 8 and 9 separates Z2 and Z3. Strong specific DNA binding is mediated by Z1 and Z2, with Z2 contributing the highest binding strength (Houchens et al., 2000). Repin1 has been shown to bind DNA as a homodimer to two sets of sequences and subsequently inducing DNA bending and formation of the DNA loop (Large, Peroni, Letexier, Ray, & Beylot, 2004). Again, zinc finger cluster Z2 was found to be essential if not sufficient for Repin1 multimerization on DNA (Houchens et al., 2000). Investigating Repin1 activity, plasmid replication assays revealed only marginal replication enhancement and sequence analysis does not indicate structural transcription activator domains (Houchens et al., 2000). Thus, Repin1 seems not to be a functional transcription factor, but may act as an accessory factor in origin recognition prior to the assembly of preinitiation complexes (Houchens et al., 2000).

3. CELLULAR LOCALIZATION OF REPIN1

Investigation of Repin1 expression in 3T3-L1 cells revealed cytosolic localization (Fig. 5.1A) (Ruschke et al., 2010). Electron microscopy showed that Repin1 is also particularly located on ribosomes, the nucleus

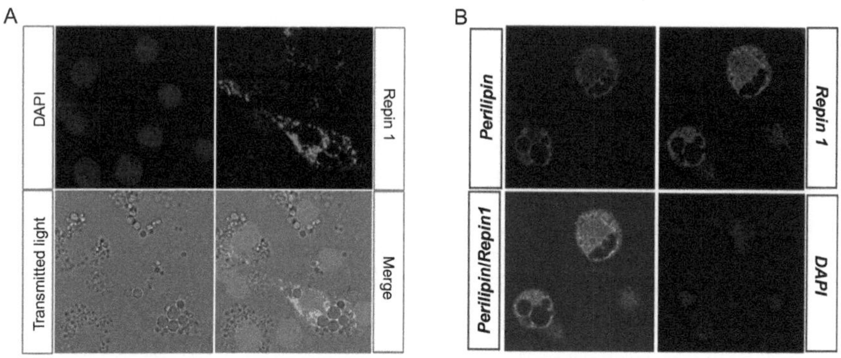

Figure 5.1 Cellular localization of Repin1. Antibody staining of mature 3T3-L1 cells against Repin1 (green). Nuclei were stained with DAPI (blue). (A) Repin1 is localized in the cytoplasm with contact to lipid droplets (see transmission picture). (B) Repin1 is in part colocalized with perilipin in 3T3-L1 cells. (See Color Insert.)

membrane, and the surface of intracellular lipid droplets of hepatocytes of C57BL/6NTac mice after a 12-week high-fat diet (Ruschke et al., 2010). Perilipin is one of the best characterized lipid droplet proteins and is localized surrounding neutral lipid storage droplets in adipocytes as a stabilizing structural protein (Ducharme & Bickel, 2008). This has led to the notion that it forms a hormonally regulated barrier between the cytosolic lipases and the neutral lipids within the droplet (Ducharme & Bickel, 2008). Confocal microscopy after immunofluorescence staining of 3T3-L1 adipocytes revealed extensive colocalization of Repin1 with endogenous perilipin at the surface of intracellular lipid droplets (Fig. 5.1B; Ruschke et al., 2010).

4. REPIN1 FUNCTION IN ADIPOSE TISSUE

4.1. Adipogenesis and lipid droplet size

Expression of most proteins required for adipocyte-specific functions is induced during the differentiation from preadipocytes into mature adipocytes (Ruschke et al., 2010). Within 3T3-L1 adipocytes, *Repin1* mRNA expression increases during adipogenesis from semiconfluent and confluent preadipocytes up to 9 days after induction of differentiation. *Repin1* mRNA expression is induced up to sevenfold during differentiation with the strongest increase beginning at day 4 after induction of adipogenesis (Fig. 5.2A). siRNA-mediated downregulation of *Repin1* mRNA expression resulted in considerably reduced lipid droplet size as reflected by a ~25% reduction of Oil-Red-O-staining intensity (Fig. 5.2B). The number of adipocytes containing lipid droplets with diameters >5 μm was decreased by about 85%. These data suggest a role of Repin1 in the development of adipose tissue with particular importance in the formation, growth, and/or fusion of intracellular lipid droplets (Ruschke et al., 2010).

Production of neutral lipids in the adipocytes and within the droplet is considered as one aspect of lipid droplet growth. The sources of triacylglycerol synthesis in adipocytes include fatty acid uptake from the plasma and *de novo* lipogenesis from nonlipids substrates, mainly carbohydrates (Large et al., 2004). *In vitro* fatty acid uptake experiments show that Repin1 is involved in the regulation of palmitate uptake. The reduced palmitate uptake upon downregulation of Repin1 in 3T3-L1 adipocytes was shown to be associated with reduced expression of fatty acid translocase *CD36*. However, reduced Repin1 expression in 3T3-L1 adipocytes did not cause alterations in mRNA expression of key genes of triglyceride

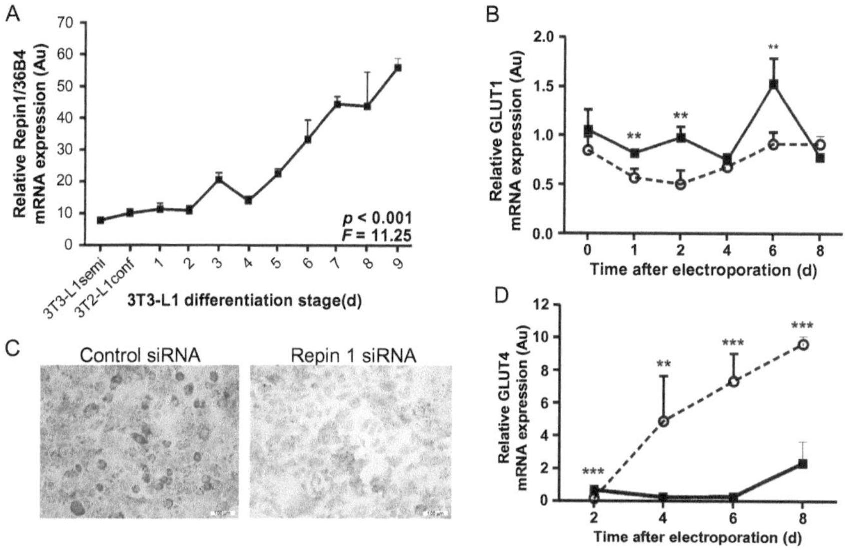

Figure 5.2 Repin1 during adipogenesis. (A) Expression of *Repin1* in 3T3-L1 adipocytes during differentiation. At the indicated time points (semiconfluent preadipocytes, confluent preadipocytes, day 1–9 after induction), total RNA was extracted and subjected to quantitative real-time PCR. *Repin1* is significantly upregulated during adipogenesis ($p < 0.001$) (magnification 630 ×). (B) Oil-Red-O staining of mature 3T3-L1 cells of controls and Repin1 siRNA-treated cells. (C) *GLUT1* and (D) *GLUT4* expression in 3T3-L1 cells during differentiation in controls and *Repin1* siRNA-treated cells. (See Color Insert.)

synthesis, indicating that Repin1 does not primarily affect triglyceride synthesis. Furthermore, gene expression of key enzymes of lipolysis was also unaffected by Repin1 knockdown *in vitro* (Ruschke et al., 2010).

Furthermore, fusion of smaller droplets to form larger droplets implicating SNARE proteins might contribute to droplet growth (Bostrom et al., 2007). In line with this, mRNA expression of important mediators of vesicle formation and fusion such as *vesicle-associated membrane protein 4* (*VAMP4*) and *synaptosomal-associated protein of 23 kDa* (*SNAP23*) was significantly reduced in *Repin1* siRNA knockdown cells (Ruschke et al., 2010). Taken together, *Repin1* mRNA expression increases during adipogenesis and siRNA-mediated knockdown results in reduced number and size of lipid droplets in differentiated adipocytes. These effects are most probably mediated by indirectly limiting lipid synthesis in the adipocytes or within droplets via decreased fatty acid uptake and inhibited droplet fusion preventing formation of larger droplets, but not by increased lipolysis.

4.2. Repin1 and adipocyte-specific glucose uptake

Repin1 has further been shown to affect adipocyte-specific glucose uptake. Knockdown of Repin1 in 3T3-L1 adipocytes resulted in reduced basal glucose uptake, while significantly increasing insulin-stimulated glucose uptake. These adversative effects on glucose uptake were associated with an altered protein and mRNA expression of glucose transporters *GLUT1* and *GLUT4* (Fig. 5.2C and D). Repin1 knockdown during adipogenesis reduced *GLUT1* mRNA expression and could thereby explain the lower basal glucose uptake (Fig. 5.2C). At the same time, loss of Repin1 considerably induced expression of *GLUT4* mRNA accounting for the improved insulin-stimulated glucose uptake in 3T3-L1 cells with reduced Repin1 expression (Ruschke et al., 2010). In congenic rat strains, *Repin1* was found as a candidate gene for deterioration of insulin sensitivity and lipid metabolism (Kloting et al., 2007). However, reduced Repin1 expression in 3T3-L1 adipocytes did not cause alterations in mRNA expression of key genes of the intracellular insulin signaling pathways (Ruschke et al., 2010).

The observation that Repin1 downregulation is associated with increased *GLUT4* expression is supported by a significant, BMI independent, negative correlation between *REPIN1* and *GLUT4* mRNA expression in human visceral ($r = -0.44$, $p < 0.001$) and subcutaneous ($r = -0.32$, $p < 0.01$) adipose tissue samples (Ruschke et al., 2010).

4.3. REPIN1 in human adipose tissue

REPIN1 expression is not gender and fat depot specific, as it has been shown in paired human adipose tissue samples [intraabdominal visceral (vis) and abdominal subcutaneous (sc)] (Ruschke et al., 2010). A correlation analysis revealed a significant correlation between visceral and subcutaneous *REPIN1* mRNA expression. Both, visceral and sc *REPIN1* mRNA expression, significantly correlate with total body fat mass (Fig. 5.3A) and maximal adipocyte size (Fig. 5.3B).

Noteworthy, multivariate linear regression analyses revealed that visceral *Repin1* mRNA expression is a significant predictor of visceral adipocyte size even after adjusting for age, gender, and % body fat. *REPIN1* expression is associated with determinants of human obesity (waist-to-hip ratio), fasting plasma insulin concentrations, 2-h OGTT glucose, glucose infusion rate during the steady state of an euglycemic–hyperinsulinemic clamp, serum LDL cholesterol, fasting serum triglycerides, circulating free fatty acid level, and hsCrP (Ruschke et al., 2010). However, these correlations did not remain significant after adjusting for age, gender, and BMI.

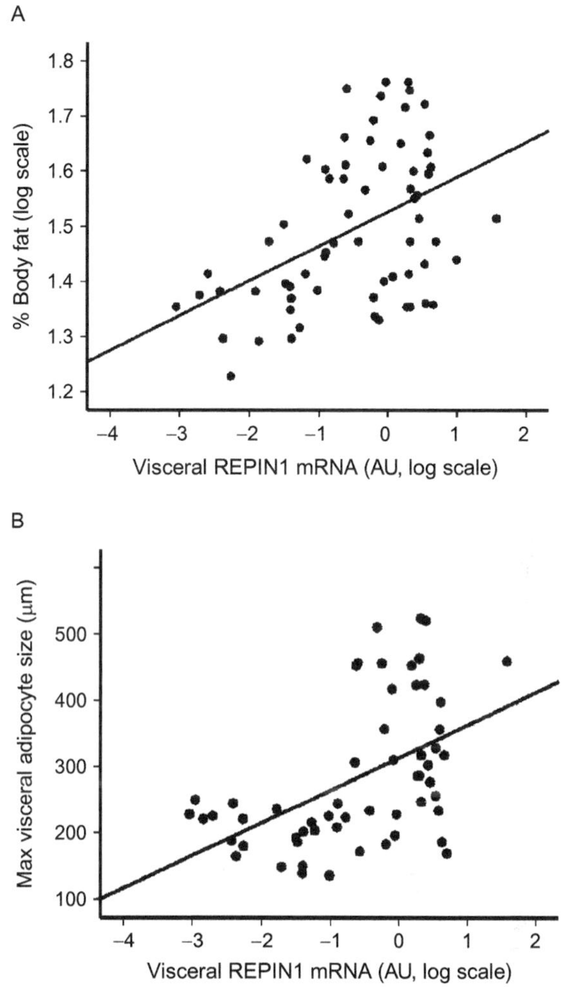

Figure 5.3 *Repin1* mRNA correlation with % body fat (A) and maximal adipocyte size (B) in human adipose tissue samples.

5. CONCLUDING REMARKS

In humans and rodents, Repin1 has been related to obesity and obesity-associated metabolic disorders. *In vitro* siRNA-mediated knock-down experiments as well as expression analysis in human adipose tissue samples indicate a potential role of Repin1 in adipose tissue development, fat mass, adipocyte size, and lipid storage (Fig. 5.4). Identification of target

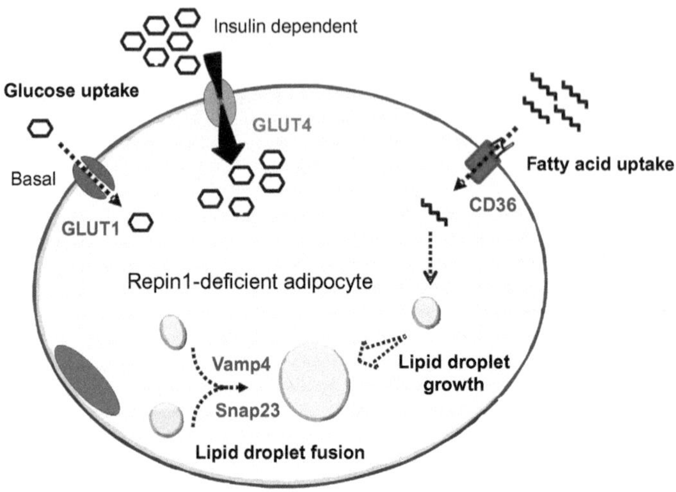

Figure 5.4 Schematic overview of Repin1 action in adipose tissue. Repin1 deficiency leads to downregulation (red) and upregulation (green) of proteins resulting in altered adipose tissue function. (See Color Insert.)

proteins which are regulated by Repin1 will lead to a better understanding of the underlying mechanisms of Repin1 action and enable therapeutic strategies to target Repin1 in obesity and adipose tissue dysfunction.

ACKNOWLEDGMENT

This work was supported by the Deutsche Forschungsgemeinschaft (DFG, KFO152, project KL-2346 to N. K.) and Federal Ministry of Education and Research (BMBF), Germany, FKZ: 01EO1001 (N. K.).

REFERENCES

Bahr, J., Kloting, N., Kloting, I., & Follak, N. (2011). Gene expression profiling supports the role of Repin1 in the pathophysiology of metabolic syndrome. *Endocrine*, *40*, 310–314.

Bostrom, P., Andersson, L., Rutberg, M., Perman, J., Lidberg, U., Johansson, B. R., et al. (2007). SNARE proteins mediate fusion between cytosolic lipid droplets and are implicated in insulin sensitivity. *Nature Cell Biology*, *9*, 1286–1293.

Caddle, M. S., Dailey, L., & Heintz, N. H. (1990). RIP60, a mammalian origin-binding protein, enhances DNA bending near the dihydrofolate reductase origin of replication. *Molecular and Cellular Biology*, *10*, 6236–6243.

Dailey, L., Caddle, M. S., Heintz, N., & Heintz, N. H. (1990). Purification of RIP60 and RIP100, mammalian proteins with origin-specific DNA-binding and ATP-dependent DNA helicase activities. *Molecular and Cellular Biology*, *10*, 6225–6235.

Ducharme, N. A., & Bickel, P. E. (2008). Lipid droplets in lipogenesis and lipolysis. *Endocrinology*, *149*, 942–949.

Frankel, A. D., Berg, J. M., & Pabo, C. O. (1987). Metal-dependent folding of a single zinc finger from transcription factor IIIA. *Proceedings of the National Academy of Sciences of the United States of America, 84*, 4841–4845.

Houchens, C. R., Montigny, W., Zeltser, L., Dailey, L., Gilbert, J. M., & Heintz, N. H. (2000). The dhfr oribeta-binding protein RIP60 contains 15 zinc fingers: DNA binding and looping by the central three fingers and an associated proline-rich region. *Nucleic Acids Research, 28*, 570–581.

Kloting, N., Wilke, B., & Kloting, I. (2007). Triplet repeat in the Repin1 3'-untranslated region on rat chromosome 4 correlates with facets of the metabolic syndrome. *Diabetes/Metabolism Research and Reviews, 23*, 406–410.

Kovacs, P., van den Brandt, J., Bonne, A. C., van Zutphen, L. F., van Lith, H. A., & Kloting, I. (2001). Congenic BB.SHR rat provides evidence for effects of a chromosome 4 segment (D4Mit6-Npy approximately 1 cm) on total serum and lipoprotein lipid concentration and composition after feeding a high-fat, high-cholesterol diet. *Metabolism, 50*, 458–462.

Large, V., Peroni, O., Letexier, D., Ray, H., & Beylot, M. (2004). Metabolism of lipids in human white adipocyte. *Diabetes & Metabolism, 30*, 294–309.

Lee, M. S., Gippert, G. P., Soman, K. V., Case, D. A., & Wright, P. E. (1989). Three-dimensional solution structure of a single zinc finger DNA-binding domain. *Science, 245*, 635–637.

Pavletich, N. P., & Pabo, C. O. (1991). Zinc finger-DNA recognition: Crystal structure of a Zif268-DNA complex at 2.1 A. *Science, 252*, 809–817.

Ruschke, K., Illes, M., Kern, M., Kloting, I., Fasshauer, M., Schon, M. R., et al. (2010). Repin1 maybe involved in the regulation of cell size and glucose transport in adipocytes. *Biochemical and Biophysical Research Communications, 400*, 246–251.

Schroll, A. L., & Heintz, N. H. (2004). Chemical footprinting of structural and functional elements of dhfr oribeta during the CHOC 400 cell cycle. *Gene, 332*, 139–147.

CHAPTER SIX

Adipokines in Childhood Obesity

Gabriel Ángel Martos-Moreno[*,†,‡], **Vicente Barrios**[*,‡], **Julie A. Chowen**[*,‡], **Jesús Argente**[*,†,‡,1]

[*]Department of Endocrinology, Hospital Infantil Universitario Niño Jesús, Instituto de Investigación La Princesa, Madrid, Spain
[†]Department of Pediatrics, Universidad Autónoma de Madrid, Madrid, Spain
[‡]CIBER Fisiopatología de la Obesidad y Nutrición (CIBERobn), Instituto Carlos III, Madrid, Spain
[1]Corresponding author: e-mail address: argentefen@terra.es

Contents

Abstract

The worldwide increase in the prevalence of obesity in children and adolescents during the past decades, in addition to mounting evidence indicating that obesity is associated with an increased incidence of comorbidities and the risk of premature death, resulting in a high economical impact, has stimulated obesity-focused research. These studies have highlightened the prominent endocrine activity of adipose tissue, which is exerted through the synthesis and secretion of a wide variety of peptides and cytokines, called adipokines.

In the present review, we have summarized the current knowledge and most relevant studies of adipokine dynamics and actions in children, focusing on the control of

Vitamins and Hormones, Volume 91
ISSN 0083-6729
http://dx.doi.org/10.1016/B978-0-12-407766-9.00006-7

energy homeostasis, metabolic regulation (particularly, carbohydrate metabolism), and inflammation. The particularities of adipose secretion and actions in healthy children, from birth to adolescence, and the modifications induced by early-onset obesity are highlighted.

1. INTRODUCTION

To date, there is no international agreement regarding the precise definition of obesity in children and adolescents, primarily due to the different population body mass index (BMI) references and cut-off thresholds suggested for the establishment of the pathological limit of excess white adipose tissue (WAT) accumulation (Martos-Moreno & Argente, 2011). In contrast, there is complete agreement in two epidemiological facts: (1) the accumulation of excess WAT in early ages increases the incidence of present and future comorbidities and the risk of premature death (Beilin & Huang, 2008; Camhi et al., 2010; Freedman et al., 2004); (2) The prevalence of obesity in children and adolescents has substantially increased worldwide during the past decades (Ogden, Carroll, Brian, & Flegal, 2012).

These facts, enhanced by their economical impact, have prompted an increase in the research focused on obesity and adipose tissue. One of the outcomes of this rise in obesity-focused investigation has been an increased understanding of the prominent endocrine activity of WAT, which is exerted through the synthesis and secretion of a wide variety of peptides and cytokines thus named adipokines (Ahima & Osei, 2008; Galic, Oakhill, & Steinberg, 2010; Kershaw & Flier, 2004). As for any pathophysiological condition, it must be taken into account that obesity during the earlier stages of life has some peculiarities that makes it different from the same disease in adults. These particularities can be extended to the production and actions of adipokines in fetal life, infancy, and adolescence, three periods of life characterized by intense growth, development, and a higher degree of structural and functional plasticity of tissues and organs, including WAT, compared to adulthood (Martos-Moreno & Argente, 2011).

The continuously growing list of adipokines extends its influence to almost every pathologic condition, making it an impossible mission to consider every peptide secreted by adipose tissue and all of their actions. Consequently, here we review what is currently known regarding the endocrine role of WAT focusing on the control of *energy homeostasis*, *metabolic regulation* (particularly, carbohydrate metabolism), and *inflammation*, highlighting the particularities in the secretion and actions of these adipokines in healthy children, from birth to adolescence, and the modifications associated with obesity onset and weight loss.

2. WHITE ADIPOSE TISSUE AS AN ENDOCRINE ORGAN: THE HYPOTHALAMIC–PITUITARY–ADIPOSE AXIS

Adipose tissue comprises two main components, adipocytes and the stromal-vascular fraction. Adipocytes are specific to adipose tissue and have the ability to store lipids. The stromal-vascular fraction contains a collagen matrix, nerves, blood and lymph vessels, and other cellular components such as preadipocytes, fibroblasts, monocytes, and macrophages. Adipocytes can already be identified in the sixth week of gestation in humans (Avram, Avram, & James, 2005; Symonds, Mostyn, Pearce, Budge, & Stephenson, 2003). At this time, the vast majority of fat is accumulated as brown adipose tissue (BAT) with its adipocytes designed for energy dissipation in contrast to adipocytes in WAT where energy is accumulated in the form of triglycerides (Avram et al., 2005). Throughout gestation, but mainly in the third trimester, both BAT and WAT accumulate in a maternal nutrition-dependent manner, with BAT playing an essential role in the adaptation of the newborn to the extrauterine environment at delivery (Symonds et al., 2003). After delivery, WAT adipocyte differentiation accelerates and BAT is almost completely substituted by WAT, with identified periods of rapid WAT accumulation from birth to 18 months of age, at mid-infancy, and in adolescence especially in women, but with WAT retaining the ability to recruit new adipocytes from preadipocytes throughout life (Prins & O'Rahilly, 1997).

The sequence of events leading to the development of differentiated adipose cells (adipocytes) from pluripotential undifferentiated precursors is called *adipogenesis* and differs between BAT and WAT. During WAT adipogenesis, embryonic pluripotential precursors elicit multipotential mesenchymal cells that after commitment to the adipogenic lineage results in the sequential formation of adipoblasts, type I preadipocytes and after clonal expansion type II preadipocytes. The growth arrest of type II preadipocytes is followed by the accumulation of lipid droplets and the formation of mature adipocytes. In this sequence of events, the ability to synthesize and secrete adipokines is almost exclusively restricted to mature adipocytes (Cristancho & Lazar, 2011; Feve, 2005; Majka, Barak, & Klemm, 2011) (Fig. 6.1).

As previously stated, the stromal matrix of adipose tissue also includes a number of cell types that are mainly derived from the

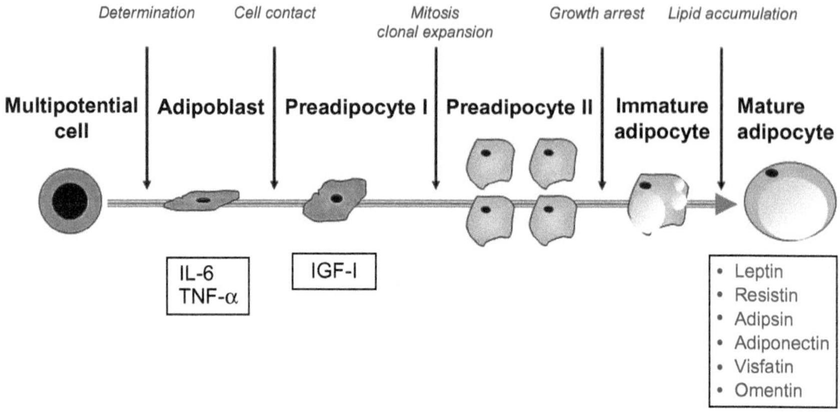

Figure 6.1 Adipogenic differentiation of multipotential mesenchymal cells and cell markers (in boxes). Of note, adipokine expression is restricted to the mature adipocyte, the end stage of this process. *Adapted from Feve (2005).* (See Color Insert.)

mononuclear–macrophage system (monocytes and macrophages). These cells possess the ability to produce and secrete cytokines (mainly proinflammatory) that in conjunction with the discovery of hormonal secretion from adipocytes resulted in WAT being considered as an endocrine organ (Kershaw & Flier, 2004). This concept was later supported by the demonstration that an effective bidirectional communication exists between adipocytes and the central nervous system (CNS), in particular, the hypothalamus, and the pituitary. Indeed, adipocytes express receptors for catecholamines and for most of the peptides and hormones produced in the hypothalamus and pituitary (Kershaw & Flier, 2004; Schäffler, Schölmerich, & Buechler, 2006) (Table 6.1).

Schäffler and coworkers proposed the term "*adipotropins*" to categorize those hormones and factors produced in the pituitary or hypothalamus with known effects on adipocytes (Schäffler et al., 2006). Among these, growth hormone (GH) plays a major role in adipose tissue accumulation and distribution, as well as in metabolic regulation by enhancing protein synthesis and lipolysis (Flint, Binart, Boumard, Kopchick, & Kelly, 2006) and directly antagonizing insulin's effects on adipocytes (del Rincon et al., 2007). Prolactin is also suggested to have a role in adipose tissue development, remodeling, and adipokine production (Flint et al., 2006). Functional thyrotropin (TSH) receptors are present in human preadipocytes and adipocytes (Sorisky, Bell, & Gagnon, 2000), with TSH apparently acting as a survival factor through reduction of apoptosis (Bell et al., 2002), in addition to enhancing leptin secretion (Menéndez et al., 2003) and IL-6 expression (Antunes, Gagnon,

Table 6.1 Neurotransmitter, hormone, and cytokine receptors expressed in the adipocyte (Kershaw & Flier, 2004)

Catecholamine receptors	α_1, α_2
	β_1, β_2, β_3
Hormone receptors	
Cell membrane	Insulin, glucagon
	GH, TSH, ACTH, CRH, prolactin
	Oxytocin
	Leptin, adiponectin
	Angiotensin II
Nucleus	Thyroid hormone
	Glucocorticoid
	Androgen, estrogen, progesterone
	Vitamin D
Cytokine receptors	IL-6
	TNF-α

ACTH, adrenocorticotropin; CRH, corticotropin releasing hormone; GH, growth hormone; IL-6, interleukin-6; TNF-α, tumoral necrosis factor alpha; TSH, thyrotropin.

Bell, & Sorisky, 2005) in human adipocytes. In contrast, there is limited information regarding the direct effect of adrenocorticotrophin (ACTH) on adipocytes, although the presence of ACTH receptors and the enhancement of proinflammatory adipokine production after ACTH stimulation have been demonstrated (Iwen et al., 2008).

3. CHANGES IN WHITE ADIPOSE TISSUE DUE TO OBESITY: IMPORTANCE OF THE AGE AT OBESITY ONSET

Obesity induces histological, metabolic, and endocrine changes in adipose tissue (Coppack, 2005; Fain, 2006; Hausman, DiGirolamo, Bartness, Hausman, & Martin, 2001; Skurk, Alberti-Huber, Herder, & Hausner, 2007; Wellen & Hotamisligil, 2003). These changes are determined by several factors: (1) the capacity of WAT to recruit new adipocytes from preadipocytes once the former have reached a critical size and after they modify their paracrine and endocrine secretion, with a prominent role of adipocyte hypertrophy, so-called the critical size hypothesis (Hausman et al., 2001); (2) the ability of adipose tissue to produce chemoattracting proteins (chemokines), which determines the increase in specific pro-inflammatory populations of monocytes and macrophages (Charo & Ransohoff, 2006; Lumeng, Deyoung, Bodzin, & Saltiel, 2007; Wellen & Hotamisligil, 2003), with this change in cellular profile substantially contributing to modifications in the adipokine secretion profile of WAT in obesity (Fain, 2006); and (3) the change in the pattern of adipokine secretion of hypertrophic adipocytes, compared to smaller ones (Skurk et al., 2007; Tsuchida, Yamauchi, & Kadowaki, 2005).

The impact of each of the above-mentioned factors on obesity changes during the different stages of human development. Both obese and lean adults maintain a fairly stable adipocyte population in their WAT due to proportional adipogenic and apoptotic rates. In contrast, children and adolescents progressively increase the number of adipocytes in their WAT, with a more rapid proliferation rate in obese compared to lean subjects (Spalding et al., 2008). This raises the concern that early-onset obesity induces a higher preadipocyte recruitment rate and an increase in the adipocyte population. This would allow for a lesser degree of adipocyte hypertrophy and abated impairment of the adipokine secretion profile during childhood, but with increased risk of severe obesity and comorbidity development in later stages of life (Freedman et al., 2004, Martos-Moreno, Barrios, Martínez, Hawkins, & Argente, 2010; Martos-Moreno, Chowen, & Argente, 2010 Spalding et al., 2008). This recalls the earlier concept of a "hyperplasic" model of obesity in children, with an increased number of normal size adipocytes versus a "hypertrophic" model with increased volume of preexisting adipocytes in adults, which would definitely exert differences in the dynamics of adipokine secretion throughout human life.

4. ENERGY HOMEOSTASIS CONTROL: LEPTIN AS A KEY PLAYER

The confirmation of the active involvement of adipose tissue in energy homeostasis control occurred when human *leptin* (Zhang et al., 1994) and its specific receptor genes (Tartaglia et al., 1995) were cloned and the leptin-hypothalamic signaling pathway was described as the main source of information to the CNS regarding energy storage in WAT in the form of triglycerides. Thus, a pathway was described by which WAT could influence growth, metabolism, and reproduction.

Leptin is a 16-kDa polypeptide adipokine mainly produced by differentiated adipocytes, although several other tissues also produce this peptide (Meier & Gressner, 2004). In humans (Gautron & Elmquist, 2011; Mantzoros et al., 2011) and other mammals (Denver, Bonett, & Boorse, 2011), leptin acts as an adiposity rather than a satiety signal, with its circulating levels directly correlated with body fat content and fluctuating in proportion to fat mass and triglyceride content, although with high interindividual variability for a given BMI (Gautron & Elmquist, 2011; Mantzoros et al., 2011). Leptin circulates in the bloodstream both free and bound to proteins, mainly to the soluble isoform of its specific receptor, with levels having a moderate circadian variation and being highest during the night (Meier & Gressner, 2004).

The leptin receptor (LEP-R) belongs to the superfamily of class I cytokine receptors and is thoroughly distributed throughout the body and allows for leptin transport across the blood–brain barrier (BBB). Its active (long) isoform is highly expressed in the hypothalamus. There is also a soluble isoform (sLEP-R) that binds leptin in an isomolecular manner, thus regulating its bioavailability and activities. Circulating levels of this isoform can be quantified and have been shown to change according to fat mass in the opposite direction to leptin (Korner, Kratzsch, & Kiess, 2005; Meier & Gressner, 2004).

Leptin modulates several neuronal populations in the CNS, especially in the hypothalamus and brainstem (Sone & Osamura, 2001). The hypothalamic nuclei most closely associated with food intake and body weight include the arcuate (ARC), paraventricular, and dorsomedial nuclei, as well as the ventromedial hypothalamus and lateral hypothalamus (Arch, 2005). The ARC contains two neuronal populations fundamental for metabolic control. One population stimulates food intake through the release of neuropeptide Y (NPY) and agouti-related peptide (AgRP) (Hahn, Breininger, Baskin, & Schwartz, 1998), whereas the other produces proopiomelanocortin (POMC) and cocaine- and amphetamine-regulated transcript (CART) and inhibits orexigenic stimuli (Ellacott & Cone, 2004). Both systems are direct targets for leptin, with this hormone-stimulating POMC and inhibiting NPY-producing neurons, thus exerting its main activity as a signal of energy sufficiency (Gautron & Elmquist, 2011; Mantzoros et al., 2011) (Fig. 6.2). However, the rise in free serum leptin as a result of the increase in leptin and decrease in sLEP-R observed in obesity is not reproduced in

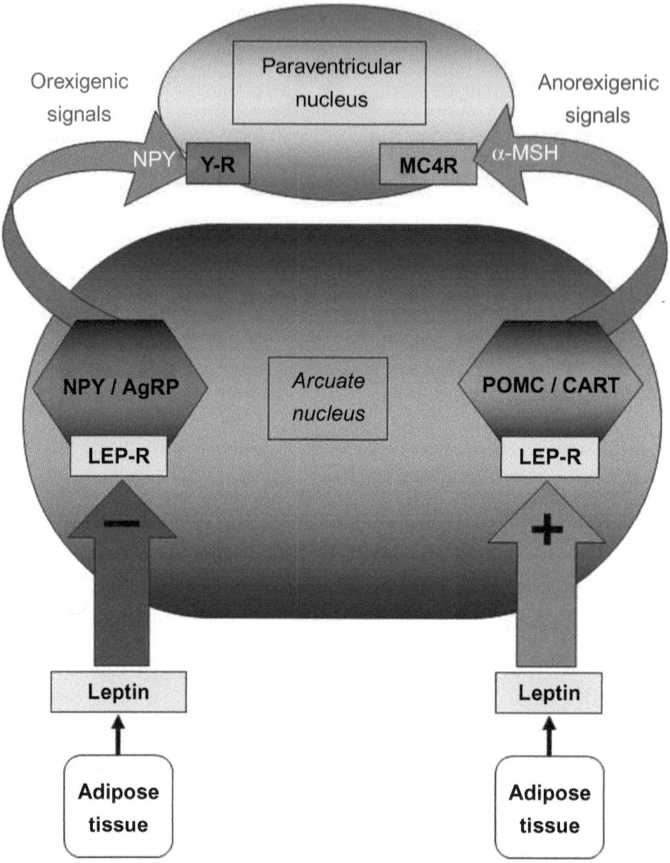

Figure 6.2 Schematic representation of the excitatory (+) and inhibitory (−) effects of leptin in NPY/AgRP and POMC/CART producing neurons in the hypothalamus. These neuronal groups reciprocally inhibit each other, through direct signaling, as well self-limiting their own activity to prevent overactivation. AgRP, agouti-related protein; CART, cocaine- and amphetamine-regulated transcript; LEP-R, leptin receptor; MC4R, melanocortin receptor number 4; NPY, neuropeptide Y; POMC, proopiomelanocortin; Y-R, Y receptor; α-MSH, alpha melanocyte-stimulating hormone. (See Color Insert.)

the spinal fluid. The saturable transport across the BBB, in addition to blunted leptin signaling in the hypothalamus have been described as a "leptin resistant state" that is reported to occur in most obese patients and be, at least partially, reversible after weight loss (Korner et al., 2005; Meier & Gressner, 2004).

Leptin also contributes to peripheral energy homeostasis control by stimulating AMP-dependent kinase in myocytes, thus inducing energy expenditure. In addition to its role in energy homeostasis, leptin is involved in growth promotion, bone metabolism, immune function, and cell mitosis and development in diverse organs, as well as in the control of almost every hypothalamic–pituitary axis (Denver et al., 2011; Gat-Yablonski & Phillip, 2008; Gautron & Elmquist, 2011; Mantzoros et al., 2011). A detailed description of the diverse actions of leptin is beyond this review and the reader is referred to the references cited above.

Several factors that change throughout human growth and development modulate leptin synthesis, including insulin, estrogens, and glucocorticoids, which stimulate leptin synthesis, and androgens, which inhibit leptin. This emphasizes the special interest that should be placed on the study of the dynamics and activities of leptin during childhood, and particularly, in childhood obesity.

Gestational age and birth weight are determinants for serum leptin levels and bioavailability in the newborn, with preterm and small for gestational age (SGA) infants showing lower leptin and higher sLEP-R levels in their cord blood than full-term and SGA newborns, respectively (Lo et al., 2002; Martos-Moreno et al., 2009). Conversely, macrosomic newborns have higher leptin levels (Shekhawat et al., 1998; Stoll-Becker et al., 2003). Birth size is suggested to be predictive of weight gain in infancy (Ong et al., 1999), with neonatal leptin levels being a better indicator of the amount of WAT in the newborn rather than of its level of maturation (Lepercq et al., 2001). Interestingly, a trend toward higher leptin levels and availability in full term, but not preterm female newborns compared to males has been reported (Martos-Moreno et al., 2009; Ong et al., 1999), suggesting that sexual dimorphism may exist from very early ages.

Leptin levels increase significantly throughout pubertal development in females, but decrease in the final stages of male puberty. In contrast, sLEP-R levels decrease in both sexes after pubertal onset, resulting in a puberty-related increase in free leptin that is most pronounced in adolescent females (Argente, Barrios, Chowen, Sinha, & Considine, 1997; Martos-Moreno, Barrios, & Argente, 2006). This has been interpreted as a signal to the CNS that metabolic conditions are adequate to undergo pubertal development (Landt, Parvin, & Wong, 2000; Martos-Moreno, Barrios, et al., 2010; Martos-Moreno, Chowen, et al., 2010). The leptin/receptor ratio is highest in adult females and is most likely the result of changes in fat mass and sex steroids, as BMI, percentage and distribution of body fat appear to be the most important predictors of free leptin levels (Misra et al., 2004; Wilasco et al., 2012). Additionally, sex steroids affect this ratio through regulating the amount and distribution of fat mass and directly modulating leptin transcriptional control, with testosterone inhibiting and estrogens stimulating leptin transcription and secretion (Wabitsch et al., 1997).

The impact of BMI and body fat mass on serum leptin levels and availability (Wilasco et al., 2012) is enhanced in the presence of malnutrition, with obese children showing increased circulating leptin levels and a prominent reduction in sLEP-R levels (Antunes, Santos, & Carvalho, 2008; Argente et al., 1997; Martos-Moreno, Barrios, et al., 2010; Martos-Moreno, Chowen, et al., 2010; Reinehr, Kratzsch, Kiess, & Andler, 2005). The inverse is observed in circumstances of undernutrition such as anorexia nervosa (Argente et al., 1997). These opposite changes in leptin and sLEP-R levels in childhood obesity impair leptin signaling as a consequence of LEP-R saturation by leptin (Landt, 2000). Additionally, a bidirectional influence between leptin and insulin exists, with hyperinsulinemia enhancing leptin production (Koutkia, Canavan, Johnson, DePaoli, & Grinspoon, 2003) and increased free leptin levels enhancing insulin resistance (IR) (Schwartz & Niswender, 2004).

The role of fat mass in determining serum leptin levels and availability is reinforced by the observation that leptin levels decrease and sLEP-R levels increase after weight reduction in children and adolescents (Argente et al., 1997; Martos-Moreno, Barrios, et al., 2010; Martos-Moreno, Chowen, et al., 2010; Reinehr et al., 2005) in a biphasic fashion (Holm et al., 2007). However, conversely to the "leptin resistance or insensitivity" status previously mentioned for adults, recent reports suggest that elevated baseline leptin levels in children can be followed by large leptin reductions after weight loss, that predicts short- and long-term loss of body fat, thus not necessarily reflecting the existence of leptin resistance at this age (Murer et al., 2011).

The definite confirmation of the importance of leptin in energy homeostasis control and pubertal development comes from the identification of human cases of congenital leptin (Montague et al., 1997) and LEP-R deficiency (Farooqi et al., 2007). These subjects present extremely severe, early-onset obesity and lack of pubertal development, which can be reversed by treatment with recombinant leptin (Farooqi et al., 2002).

The involvement of other adipose-secreted proteins, such as adiponectin, resistin, or interleukin-6 (IL-6), in energy homeostasis control is much less evident but cannot be ruled out. In this regard, adiponectin-specific receptors are widely distributed in the brain and adiponectin injection *icv* decreases body weight by stimulation of energy expenditure (Ahima, Qi, Singhal, Jackson, & Scherer, 2006). Similarly, inhibition of food intake after resistin administration *icv* has also been reported in mice (Ahima et al., 2006; Tovar et al., 2005), while *icv* infusion of IL-6 increases energy expenditure and decreases body fat in rats (Wallenius, Wallenius, Sunter, Dickson, & Jansson, 2002). However, the physiological role of these factors in energy homeostasis is far from being understood.

5. INSULIN SENSITIVITY: ROLE OF ADIPONECTIN, VISFATIN, AND VASPIN

Peripheral resistance to insulin action or "IR" was postulated by Gerald Reaven as the pathophysiological basis of the entire metabolic derangement observed in association with obesity and the so-called X or metabolic syndrome (Reaven, 1988). The role of the amount and, more importantly, the anatomical distribution of adipose tissue accumulated during the genesis of IR have been progressively elucidated, with visceral adipose tissue now known to be especially implicated in this process (Wajchenberg, 2000). One of the main determinants for the development of obesity-associated IR is the adipokine secretion profile of obese patients, especially of obese children. Among the adipose-derived peptides involved in producing IR are the proinflammatory cytokines including resistin and, particularly, adiponectin and the more recently characterized visfatin and vaspin.

5.1. Adiponectin

Adiponectin is a 30-kDa peptide with a collagen similar domain that allows for the formation of secondary and tertiary structures. It is produced exclusively in mature adipocytes, with higher synthesis in subcutaneous compared to visceral WAT. In the circulation, the monomers of adiponectin form homotrimers, which can sequentially build more complex structures such as hexamers of around 180 kDa (*low molecular weight*) and polymers (16–18 monomers) of around 400–600 kDa (HMW, *high molecular weight*). The HMW form is the most abundant in serum and is postulated to have differential activities, that is, insulin sensitivity correlates better with the ratio (S_A) of HMW–polymers to total (T) adiponectin levels than with T-adiponectin levels (Jeffery et al., 2008; Kadowaki & Yamauchi, 2005; Kershaw & Flier, 2004; Matsuzawa, 2010; Meier & Gressner, 2004; Pajvani et al., 2004).

This peptide has two specific receptors, adipoR1 and adipoR2, ubiquitously distributed in the human anatomy, but most highly expressed in muscle and liver, respectively. The stimulation of adipoR1 in muscle activates

AMP-dependent kinase (AMP-K) (as does leptin), which induces the expression of peroxisome proliferator activated receptor alpha (PPAR-α) and, as a consequence, the expression of enzymes involved in fatty acid catabolism and glucose uptake. Additionally, adiponectin enhances the expression and externalization of glucose transporter 4 (Glut-4) in myocytes and directly modulates insulin receptor activity. Activation of adipoR2 and liver AMP-K inhibits glyconeogenesis by modulating glucose-6-phosphatase and phosphoenolpyruvatecarboxykinase activity. Finally, WAT is also a target of adiponectin actions where it favors adipocyte differentiation and proliferation, as well as insulin-induced glucose uptake, fatty acid oxidation, and lipoprotein lipase activity (Chinetti-Gbaguidi, Fruchart, & Staels, 2005; Yu & Ginsberg, 2005). In summary, the effects of adiponectin on muscle, liver, and WAT result in an enhancement of insulin sensitivity and prevention of the development of type 2 diabetes and the promotion of fatty acid oxidation and a beneficial apolipoprotein profile (Kadowaki & Yamauchi, 2005; Kershaw & Flier, 2004; Matsuzawa, 2010; Meier & Gressner, 2004; Sowers, 2008; Yamauchi et al., 2003) (Fig. 6.3A and B).

Diverse evidence affirms the involvement of adiponectin in the modulation of glucose metabolism, including the decrease in adiponectin in type 2 diabetes, which is independent of adiposity. The genetic lack of adiponectin is associated with IR, and the experimental administration of adiponectin enhances insulin actions in animal models. Also, insulin appears to suppress adiponectin levels. However, although the influence of adiponectin on glucose metabolism appears evident, currently available data do not allow a clear relationship between adiponectin levels and insulin sensitivity to be established. Indeed, some authors suggest that adiponectin could merely be a marker, rather than an active player in the modulation of insulin sensitivity.

Conversely to leptin, serum adiponectin levels in adults are inversely related to the amount of WAT, with obese patients showing decreased circulating adiponectin levels (Arita et al., 1999). However, controversy exists regarding the influence of WAT distribution, that is, visceral versus subcutaneous depots, on circulating adiponectin levels (Jeffery et al., 2008). Interestingly, this inverse correlation between the amount of body fat and serum adiponectin levels does not exist at every stage of postnatal life and is also influenced by the conformational changes induced in adipose tissue by pharmacological treatments such as thiazolidinediones (Yu et al., 2002). In newborns, as for leptin, there is a positive correlation between adiponectin levels, gestational age, and birth weight (Kajantie, Hytinantti, Hovi, & Andersson, 2004; Martos-Moreno et al., 2009), which is opposite to the negative

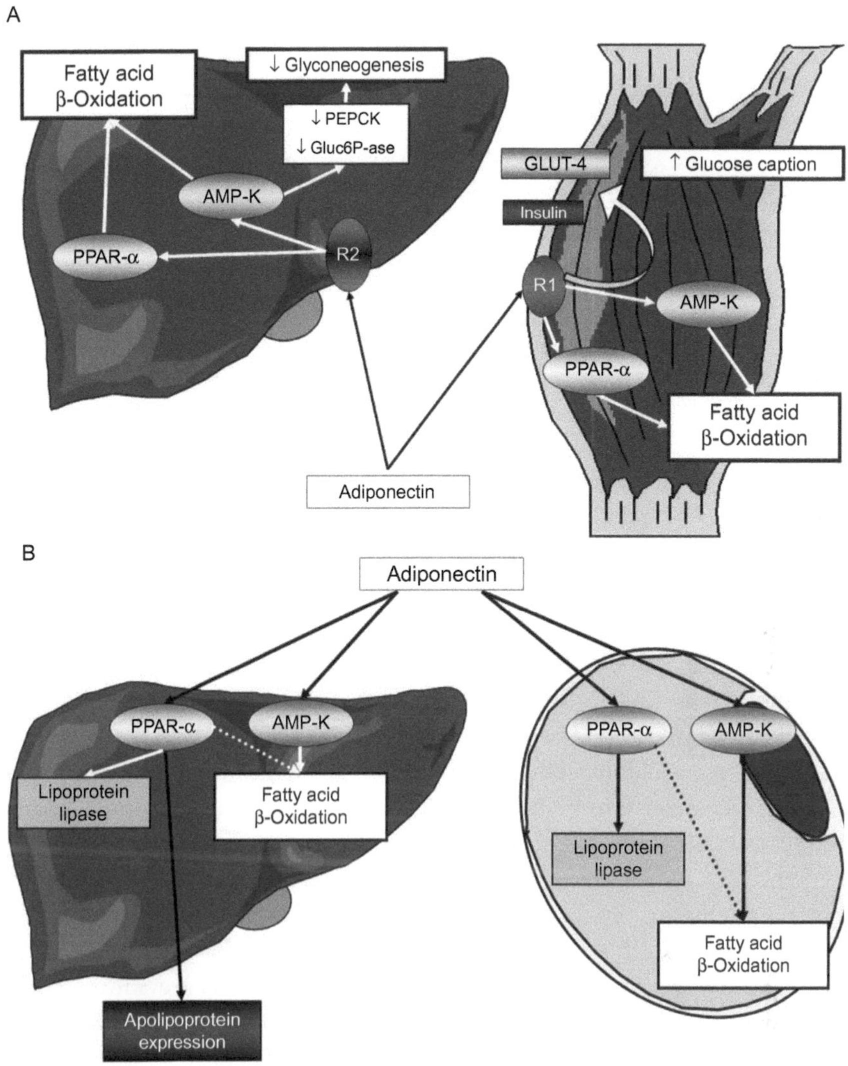

Figure 6.3 Schematic representation of the main metabolic effects of adiponectin on glucose metabolism in the liver and muscle (A) and on lipid metabolism in the liver and adipocytes (B). The stimulation of muscle adipoR1 determines the activation of AMP-dependent kinase (AMP-K) and induces the expression of the peroxisome proliferator activated receptor alpha (PPAR-α) and, as a consequence, the expression of enzymes involved in fatty acid catabolism and glucose uptake. Additionally, adiponectin enhances the expression and externalization of glucose transporter 4 (Glut-4) in myocytes and directly modulates insulin receptor activity. The activation of adipoR2 and liver AMP-K inhibits glyconeogenesis by modulating glucose-6-phosphatase (Gluc6P-ase) and phosphoenolpyruvatecarboxykinase (PEPCK) activity, enhancing lipoprotein lipase activity, and modulating apolipoprotein gene expression. Finally, adiponectin favors insulin-induced glucose uptake, fatty acid oxidation, and lipoprotein lipase activity in WAT adipocytes (Chinetti-Gbaguidi et al., 2005; Kadowaki & Yamauchi, 2005; Kershaw & Flier, 2004; Meier & Gressner, 2004; Yamauchi et al., 2003; Yu & Ginsberg, 2005). (See Color Insert.)

relationship between fat mass and adiponectin levels in adults. A sexual dimorphism in cord blood adiponectin levels, with higher concentrations in females, has also been reported (Li et al., 2009; Martos-Moreno et al., 2009; Lau et al., 2009). At birth, circulating adiponectin levels are two- to threefold those of adults, with an apparent decline and disappearance of the positive correlation between fat mass and adiponectin levels by 2 years of age, coincident with the increase in body fat (Jeffery et al., 2008).

There is also a sexual dimorphism in adiponectin levels during pubertal development. In prepubertal children, most studies report no differences between sexes (Jeffery et al., 2008; Martos-Moreno et al., 2006). However, boys show lower adiponectin levels from mid-puberty onwards, most probably due to the inhibitory effect of androgens and possibly of visceral fat accumulation on adiponectin secretion (Bottner et al., 2004; Jeffery et al., 2008; Martos-Moreno et al., 2006; Matsuzawa, 2010; Sowers, 2008).

In obese adolescents, there is an inverse correlation between adiponectin levels, body fat, and IR, similar to that reported for adults (Jeffery et al., 2008; Nishimura et al., 2007; Oh, Ciaraldi, & Henry, 2007). Conversely, whereas the negative association between adiponectin levels and IR markers remains in prepubertal children (Jeffery et al., 2008), there are contrasting results regarding the correlation between BMI and adiponectin levels. Some studies report low negative (Nishimura et al., 2007) or no correlation (Martos-Moreno, Barrios, et al., 2010) between BMI and adiponectin levels, with obese prepubertal children reported to have either reduced (Jeffery et al., 2008; Valle et al., 2005) or similar (Martos-Moreno, Barrios, et al., 2010) total adiponectin levels compared to lean counterparts. As previously stated, this could be due to the higher capacity of adipose tissue to recruit new adipocytes from preadipocytes at early ages, thus limiting their degree of hypertrophy and, consequently, reducing adiponectin secretion (Spalding et al., 2008; Tsuchida et al., 2005). However, studies of the HMW–adiponectin complexes have shown that, even at these early ages, obese children show impaired adiponectin synthesis, secretion, and post-translational processing that involves an increase in IR (Martos-Moreno, Barrios, et al., 2010). Interestingly, similar to what is observed in adults, weight loss increases adiponectin levels even at early ages (Jeffery et al., 2008; Martos-Moreno, Barrios, et al., 2010).

Adiponectin, which is produced by adipocytes in the perivascular adipose tissue, exerts its beneficial effects against atherosclerosis by several mechanisms that include modulation of endothelium-related vascular reactivity, inhibitory effects on expression of adhesion and inflammation molecules, and formation of lipid-rich foam cells (Kadowaki & Yamauchi, 2005).

Obesity-associated hypoadiponectinemia has also been suggested to be at least partially responsible for the chronic low-grade systemic inflammatory state. Adiponectin is postulated to play an anti-inflammatory role (Matsuzawa, 2010) and the hypoadiponectinemia, which generally courses with obesity, is associated with high C-reactive protein (CRP), IL-6, and tumoral necrosis factor alpha (TNF-α) levels (Sowers, 2008). In contrast, the evidence supporting an eventual protective role of adiponectin for the development of nonalcoholic fatty liver disease in children is inconclusive (Jeffery et al., 2008), which is in contrast to data in adults (Matsuzawa, 2010).

5.2. Visfatin/pre-B-cell colony enhancing factor/nicotinamide phosphoribosyltransferase

Nicotinamide phosphoribosyltransferase (NAMPT)/visfatin is a 52-kDa peptide that derives its name from the initial assumption that it was mainly produced by visceral adipose tissue. This protein is suggested to have a blood glucose lowering effect and *in vitro* studies have shown that it can induce the phosphorylation of the insulin receptor and its related substrates (IRS) 1 and 2 (Fukuhara et al., 2005). These characteristics predicted a pivotal role for visfatin in linking obesity (and specifically abdominal/visceral obesity) with carbohydrate metabolism impairment, which was supported by later reports demonstrating its role in the regulation of insulin secretion (Revollo et al., 2007). However, visfatin is found in low concentrations in the bloodstream and reports linking visfatin levels, type 2 diabetes and different states of IR are contradictory (Saddi-Rosa, Oliveira, Giuffrida, & Reis, 2010). Moreover, recent data have demonstrated that the main source of NAMPT/visfatin is not visceral WAT, but leukocytes, suggesting that the main role of this adipokine may involve the relationship between obesity and inflammation (Friebe et al., 2011). The multiple uncertainties existing regarding the pathophysiological significance of visfatin levels and its activities initially generated a certain degree of skepticism regarding its physiological significance (Rosen & Spiegelman, 2006), with some of these controversies remaining under discussion even to date.

A positive correlation between serum NAMPT/visfatin levels and body fat content and a decrease of its circulating levels after weight reduction have been described in adults (Haider, Holzer, et al., 2006). However, evidence indicating that its expression is higher in visceral than in subcutaneous adipose tissue is inconsistent (Berndt et al., 2005; Revollo et al., 2007), with some authors suggesting that this adipokine could merely be a marker of total WAT mass. Moreover, the effect of weight loss on serum visfatin levels in adults remains under discussion, as both a decrease (Manco et al., 2007) and an increase (Krzyzanowska, Mittermayer, Krugluger, Kopp, & Schernthaner, 2006) in circulating visfatin after extensive weight loss in obese adults following surgical procedures have been reported.

Several authors have studied NAMPT/visfatin levels during fetal life and in newborns, as well as their relationship with gestational age and birth weight. These reports have demonstrated high NAMPT/visfatin levels during fetal life, probably derived from placental expression and transfer of this adipokine to the fetus (Malamitsi-Puchner, Briana, et al., 2007), with a positive correlation between NAMPT/visfatin levels and birth weight, in both healthy preterm and full-term newborns (Malamitsi-Puchner, Briana, et al., 2007; Siahanidou, Margeli, Kappis, Papassotiriou, & Mandyla, 2011). This is in agreement with most (Cekmez et al., 2011; Meral et al., 2011), but not all (Evagelidou et al., 2010) studies showing higher NAMPT/visfatin levels in large for gestational age (LGA) newborns. The effect of intrauterine growth restriction on visfatin levels remains controversial, with SGA newborns reported to have similar (Mazaki-Tovi et al., 2010) or higher (Malamitsi-Puchner, Briana, et al., 2007) NAMPT/visfatin than those born with adequate anthropometry for their GA (AGA). One possible explanation for the observed differences is the influence of sex on these relationships as females are reported to have higher levels than males (Ibáñez et al., 2008).

Studies of NAMPT/visfatin levels in lean children are sparse (Dedoussis, Kapiri, Kalogeropoulos, et al., 2009; Friebe et al., 2011), with most data coming from lean subjects included as control groups to be compared with obese children (Evagelidou et al., 2010; Jin et al., 2008). These studies reveal a positive correlation between NAMPT/visfatin and BMI even in lean children (Dedoussis, Kapiri, Kalogeropoulos, et al., 2009), with no sexual dimorphism or changes during puberty in either sex (Friebe et al., 2011). In addition, no correlation of NAMPT/visfatin has been observed with height or surrogate markers of growth (IGF-I) or pubertal development (adrenal androgens, testosterone, and estradiol) (Dedoussis, Kapiri, Kalogeropoulos, et al., 2009; Friebe et al., 2011; Jin et al., 2008).

Higher circulating NAMPT/visfatin levels in obese children and their positive correlation with BMI and its surrogate markers (leptin and sLEP-R) have been reported in various series of children and adolescents (Araki et al., 2008; Friebe et al., 2011; Haider, Schindler, et al., 2006; Kolsgaard et al.,

2009), as well as in prepubertal children studied separately (Martos-Moreno, Kratzsch, et al., 2011). In this age range, the relationship between NAMPT/visfatin and visceral fat surrogates is uncertain, as its correlation with waist circumference disappears after controlling for the effect of BMI. This suggests that visfatin levels in obese children are influenced primarily by the total amount of body fat and not by its distribution (Haider, Schindler, et al., 2006; Martos-Moreno, Kratzsch, et al., 2011). Weight loss results in an initial decrease in NAMPT/visfatin levels in children and adolescents (Krzystek-Korpacka, Patryn, Bednarz-Misa, Hotowy, & Noczynska, 2011; Martos-Moreno, Kratzsch, et al., 2011), which remain stable after more extensive weight loss (Martos-Moreno, Kratzsch, et al., 2011).

Regarding the pathophysiological role of NAMPT/visfatin in childhood obesity, the relationship between visfatin levels and carbohydrate metabolism impairment is partially, but not completely, influenced by BMI. Thus, the decline in NAMPT/visfatin with weight loss, parallel to the decline in insulin and HOMA index of IR, is not exclusively driven by the decrease in body fat but also by the improvement in IR. Indeed, insulin is suggested to decrease visfatin levels (Bala et al., 2011), although most reports show that the correlation between visfatin and insulin and HOMA index in young obese patients disappears after controlling for the effect of BMI (Kolsgaard et al., 2009; Martos-Moreno, Kratzsch, et al., 2011). Supporting this, the decline in NAMPT/visfatin levels following OGTT seems to be mainly drive by insulin (Bala et al., 2011; Friebe et al., 2011), with this effect greatly enhanced by the increased insulin sensitization resulting from weight loss (Martos-Moreno, Kratzsch, et al., 2011). Taken together, these data suggest that, as young patients do not usually display any gross alteration in carbohydrate metabolism, the high basal visfatin levels could be a compensatory mechanism to maintain long-term glucose homeostasis, which would return to normality after weight loss. In contrast, rapid increases in glucose and insulin in the short term (OGTT) would decrease this proinsulin-sensitizing profile.

A relationship between NAMPT/visfatin levels and lipid metabolism has been suggested, with a direct correlation between circulating and tissue mRNA NAMPT/visfatin levels and cholesterol levels in lean and obese adults (Chang, Chang, Lee, & Chuang, 2010). Given the influence of body fat content on visfatin levels, it has been suggested that this could be an epiphenomenon driven by changes in BMI and/or carbohydrate metabolism. This is hypothesis is supported by the fact that in children, the observed correlations of NAMPT/visfatin with total and LDL cholesterol, but not with

HDL (Jin et al., 2008), lose their significance after controlling for both parameters (Martos-Moreno, Kratzsch, et al., 2011).

NAMPT/visfatin is predominantly produced by leucocytes both in the bloodstream (Friebe et al., 2011) and in the stromal-vascular fraction of the adipose tissue, especially by M1 subtype macrophages and other mononuclear cells infiltrating adipose tissue in obesity under the influence of inflammatory signals (Curat et al., 2006; Dedoussis, Kapiri, Samara, et al., 2009). This, in addition to the positive correlations between this adipokine and other proinflammatory factors such as IL-6 (Martos-Moreno, Kratzsch, et al., 2011; Romanowska & Lebensztejn, 2010), TNF-α (Dedoussis, Kapiri, Kalogeropoulos, et al., 2009), or resistin (Martos-Moreno, Kratzsch, et al., 2011), independently of BMI, emphasizes its contribution to the generation of the low-grade inflammation state associated with obesity.

5.3. Vaspin

Vaspin (visceral adipose tissue derived serine protease inhibitor) is a 415-amino acid peptide that belongs to the serpin (serine protease inhibitor) family, although its precise protease inhibitory activity is unknown (Li et al., 2008). In adult humans, both visceral and subcutaneous adipose tissue have been reported to express vaspin, with depot-specific regulation suggested to be controlled by either body fat content or insulin sensitivity (Blüher, 2012; Klöting et al., 2006). Vaspin administration reduces IR, thus improving glucose tolerance in obese mice (Hida et al., 2005), and reduces food intake. However, the precise mechanism by which vaspin influences glucose metabolism, its mode of action and molecular targets remain largely unknown (Blüher, 2012; Li et al., 2008). As a consequence, its relevance as an endogenous insulin sensitizer is under debate.

In human adults, vaspin levels are reported to be higher in females (Blüher, 2012; Youn et al., 2008) and to express a meal-related diurnal variation (Jeong et al., 2010). However, there are contrasting reports regarding the relationship between vaspin levels, BMI, and carbohydrate metabolism impairment in adults (von Loeffelholz et al., 2010; Ye, Hou, Pan, Lu, & Jia, 2009), as well as regarding the influence of physical training and weight reduction on its circulating levels (Chang, Lee, et al., 2010; Youn et al., 2008; Cho, 2010).

Information regarding vaspin in children is limited. The few available studies in newborns report that vaspin levels are detectable after birth and have no relationship with birth weight or insulin, with SGA, AGA, and LGA newborns showing similar serum vaspin concentrations (Briana et al., 2011; Cekmez et al., 2011). The sexual dimorphism in vaspin levels found in adults develops during puberty in lean healthy children, with levels remaining stable in boys and increasing progressively in girls and with no correlation between vaspin levels and BMI (Körner et al., 2011).

Two recent studies report increased vaspin levels in obese children and adolescents (Suleymanoglu et al., 2009) and a decrease after a short-term lifestyle modification program (Lee et al., 2010). However, results regarding the correlation of vaspin levels with insulin or HOMA index are conflicting, with both negative (Körner et al., 2011; Lee et al., 2010) and positive (Suleymanoglu et al., 2009) correlations being reported. In the prepubertal stage, vaspin levels are not sexual dimorphic, do not differ between controls and obese children, and are not modified after weight loss (Martos-Moreno, Kratzsch, et al., 2011). Furthermore, no significant correlations between vaspin, BMI, HOMA index, or insulin levels are reported in young children. These observations challenge, once again, the hypothetical relevance of vaspin in carbohydrate metabolism regulation. However, although these variables are reported to be unrelated to vaspin levels in nondiabetic humans (von Loeffelholz et al., 2010), changes are observed in patients with more severe impairment of carbohydrate metabolism (Chang, Lee, et al., 2010). This suggests that, as previously stated for visfatin, the lack of gross alterations in carbohydrate metabolism in young cohorts could be responsible for no correlation between vaspin and BMI or HOMA index being found.

Glucose ingestion decreases vaspin levels in hyperinsulinemic obese children and adolescents (Körner et al., 2011; Martos-Moreno, Kratzsch, et al., 2011), as well as in animal studies (González et al., 2009), and a postprandial decrease in serum vaspin levels occurs after every meal in healthy adults (Jeong et al., 2010). Interestingly, this glucose-induced drop in vaspin levels is absent in normoinsulinemic subjects (Körner et al., 2011), as well as in young obese patients after weight loss (Martos-Moreno, Kratzsch, et al., 2011), with the later showing a correlation between vaspin and insulin area under the curve in the OGTT performed after weight loss (Martos-Moreno, Kratzsch, et al., 2011). These observations support the hypothesis of a complex regulatory system involving the coordination between body fat content and insulin sensitivity status as the primary modulator of vaspin secretion, with any acute rise in glucose or insulin potentially exerting a suppressive effect on vaspin secretion.

5.4. Omentin

Omentin 1, also named intelectin (Schäffler et al., 2005), is a 313-amino acid adipokine preferentially produced and secreted by visceral adipose tissue, particularly by the stromal-vascular matrix cells (Yang et al., 2006). Its analog, omentin 2, shares 83% amino acid identity. Omentin 1 is proposed to exert three main functions: (1) enhancement of insulin-stimulated glucose uptake in adipocytes and of glucose metabolism via increasing Akt phosphorylation (Tan, Adya, & Randeva, 2010; Yang et al., 2006); (2) inhibition of CRP-induced angiogenesis and inflammation, potentially via the nuclear factor kappa beta signaling pathway (Tan et al., 2010); and (3) vasodilatation of blood vessels. In summary, this peptide is postulated to protect against metabolic syndrome.

In human adults, omentin 1 levels in plasma and gene expression are reported to be decreased in obesity (de Souza-Batista et al., 2007) and in situations of impaired glucose sensitivity (Pan, Guo, & Li, 2010) and to increase after weight loss (Moreno-Navarrete et al., 2010). In children and adolescents, the available information is minimal, but omentin 1 is present in the fetus and newborn, with no correlation with birth weight or insulin levels (Briana et al., 2011). The only study on omentin 1 levels in healthy prepubertal children ($n=161$) to date, surprisingly reports a positive association between omentin 1 levels and a poorer metabolic profile (higher HOMA, triglycerides, and blood pressure), independently of age, sex, or fat mass (Prats-Puig et al., 2011). This reinforces the particularity of this stage of life regarding the functional behavior of adipose tissue and adipokine secretions.

5.5. Other adipokines suggested to regulate glucose metabolism

Several other adipokines have been postulated to influence carbohydrate metabolism. Among them *apelin*, and its receptor the G-coupled receptor APJ ubiquitously distributed throughout the body, has suggested appetite and insulin sensitivity modulating activities (Castan-Laurell et al., 2011). However, results regarding its relationship with BMI and insulin sensitivity are conflicting (Castan-Laurell et al., 2011), with recent reports strongly challenging any pathophysiological relevance of this peptide in adults (Rittig et al., 2011) or in young children (Reinehr, Woelfle, & Roth, 2011). During the initial stages of life, apelin is already present in the fetus at higher levels than those in the maternal bloodstream with these levels increase even further after birth (Malamitsi-Puchner, Gourgiotis, et al., 2007). Reported studies are contradictory showing increased apelin levels in babies born LGA (Cekmez et al., 2011) or absolute lack of correlation between apelin and birth weight, sex, or insulin levels (Malamitsi-Puchner, Gourgiotis, et al., 2007). In young obese and lean children, evidence suggests a lack of any relationship between apelin levels and BMI or any metabolic parameter, with no effect of weight loss in obese children on serum apelin concentrations (Reinehr et al., 2011). The studies in obese adolescents are also conflicting and report higher, similar, and lower apelin levels in obese adolescents compared to controls (Castan-Laurell et al., 2011). In summary, the currently available data do not allow a clear pathophysiological effect of apelin related to obesity or IR in humans at any stage of life to be deduced, although the functional and metabolic effects of the activation of its receptor have been characterized (Castan-Laurell et al., 2011).

A potential role for *retinol binding protein 4* (RBP4) in glucose metabolism was suggested after its infusion was shown to induce IR and genetic knockout of RBP4 was shown to increase insulin sensitivity (Rabe, Lehrke, Parhofer, & Broedl, 2008; Rosen & Spiegelman, 2006). There are also numerous studies reporting both the existence and inexistence of an association between serum RBP4 levels and obesity or IR in adults as well as in children. Consequently, at present, these relationships and their clinical relevance remain uncertain (Rabe et al., 2008).

Chemerin is an essential adipokine for adipocyte differentiation that modulates the expression of several genes involved in glucose and lipid metabolism. It is reported to enhance glucose caption and IRS phosphorylation in adipocytes. However, the relationship between circulating chemerin levels, body fat content, and surrogate markers of IR yields controversial data in animal models and data in adult humans are minimal and absent in children (Rabe et al., 2008).

6. LOW-GRADE INFLAMMATORY ENVIRONMENT: PROINFLAMMATORY ADIPOKINES

In obesity, the excessive accumulation of WAT, the changes in WAT cell composition and morphology, and the impairment in adipokine secretion that modifies both their paracrine and endocrine actions all contribute to the generation of a proinflammatory environment that is ultimately involved in the associated metabolic derangement (Fain, 2006; Rabe et al., 2008; Tilg & Moschen, 2008). We have previously stated the pro- (NAMPT/visfatin) and anti-inflammatory (adiponectin) actions of the most clinically relevant adipokines. In addition, some of the proinflammatory cytokines produced by adipose tissue must also be highlighted. However, as these cytokines are also produced by other tissues, the diagnostic value of quantifying their circulating levels is limited.

6.1. Resistin

A few years ago, resistin would have undoubtedly been included in the previous subsection of this chapter devoted to insulin sensitivity. Its role in IR, as indicated by its name, in rodent models is indisputable, and this peptide was suggested to be the link between obesity and IR. Similar to that found for other adipokines, studies in humans regarding the relationship between resistin levels and adiposity and IR yielded discrepant results. This difference with murine models could result from the limited homology between rodent and human resistin molecules (around 60%) and, most importantly, from the differences in the main cellular source of resistin. Resistin is mainly produced by adipocytes in mice and circulating and stromal mononuclear cells in humans (Schwartz & Lazar, 2011). However, recent large case–control studies have shown an increased risk for the development of type 2 diabetes in subjects with elevated serum resistin levels, with the suggestion that at least part of the association between resistin levels, IR, and the risk of atherosclerosis and coronary disease could be mediated by other factors such as obesity and inflammation (Chen et al., 2009; Schwartz & Lazar, 2011).

In rodents, resistin is the adipocyte-specific member of the FIZZ (found in inflammatory zone) family and is also called FIZZ-3 or ADSF (adipocyte-specific secretory factor). Once secreted, resistin forms dimers followed by oligo- or polymerization processing, with this processing not only having functional implications but also interfering with its diagnostic utility (Gerber et al., 2005). No specific receptor has been identified for resistin, although it seems to share its signaling pathways with other proinflammatory molecules (Bokarewa, Nagaev, Dahlberg, Smith, & Tarkowski, 2005). Its parallel dynamics with other proinflammatory molecules in humans, such as IL-6, TNF-α, or CRP, have been widely reported (Korner et al., 2005; McTernan, Kusminski, & Kumar, 2006), as well as its decline after treatment with anti-TNF-α agents (infliximab) (Filková, Haluzík, Gay, & Senolt, 2009; Schwartz & Lazar, 2011).

A positive correlation between resistin and adiposity has been reported in obese adults, although not all authors are in agreement. Moreover, studies after intervention have reported decreases, increases, or no changes in resistin after weight loss (Schwartz & Lazar, 2011).

Resistin levels are higher in cord blood from preterm compared to full-term pregnancies (Martos-Moreno et al., 2009), which could be due to the higher prevalence of a proinflammatory environment in preterm pregnancies, as observed in cases with prelabor rupture of the placental membranes (Gursoy, Aliefendioglu, Caglayan, Aktas, & Ovali, 2011). This is supported by the positive correlation between resistin and proinflammatory cytokine levels in cord blood, as well as by the suppression of resistin observed in mothers receiving antenatal steroids (Gursoy et al., 2011; Martos-Moreno et al., 2009). Similar to adults, in newborns the relationship between resistin and inflammation is more robust than the correlation between resistin and fat mass, with studies reporting either positive (Gohlke et al., 2008), negative (Martos-Moreno et al., 2009; Wang et al., 2010), or no correlation between resistin levels and birth weight (Briana et al., 2008; Mamì et al., 2009). In newborns, no correlation between resistin and insulin levels has been reported (Briana et al., 2008; Mamì et al., 2009; Martos-Moreno et al., 2009).

In healthy children, serum resistin levels are higher in females than in males throughout pubertal development (Gerber et al., 2005; Martos-Moreno et al., 2006; Tadokoro et al., 2010; Yannakoulia et al., 2003), with resistin correlating with estradiol levels (Gerber et al., 2005) and the leptin/receptor ratio in females, but not in males. This suggests a possible link between resistin levels and female body fat content changes (Martos-Moreno et al., 2006). On the contrary, an independent correlation with waist circumference, not biased by BMI, in healthy children is lacking (Steene-Johannessen, Kolle, Reseland, Anderssen, & Andersen, 2010). In a large series of subjects pooling lean and obese children and adolescents, resistin was found to be a weak marker of obesity and inflammation and to have no relationship with IR indexes (Li et al., 2009).

Once more, when comparing resistin levels in obese children and controls the results are contradictory with similar (Gerber et al., 2005; Reinehr, Roth, Menke, & Andler, 2006) or higher resistin levels in obese children (Martos-Moreno, Barrios, et al., 2010; Rubin et al., 2008) being reported. Likewise, the changes in resistin levels in response to weight loss (Jones, Basilio, Brophy, McCammon, & Hickner, 2009; Reinehr et al., 2006) or exercise (Elloumi et al., 2009; Jones et al., 2009) are contradictory, with no correlation found between resistin and direct fat mass measurements or IR indexes in obese patients (Martos-Moreno, Barrios, et al., 2010).

6.2. IL-6 and TNF-α

IL-6 and TNF-α are the most studied proinflammatory adipokines. They are mainly secreted by monocytes and macrophages in the bloodstream and in the stromal-vascular matrix of WAT. Adipocytes not only produce both cytokines but also express their specific receptors, and IL-6 and TNF-α receptor expressions are enhanced in adipocytes in obesity. Additionally, IL-6 downregulates LEP-r expression in adipocytes and both cytokines regulate the early stages of adipogenesis (Feve, 2005; Hauner, 2004; Kershaw & Flier, 2004).

IL-6 and TNF-α production by adipocytes is also increased in human obesity (Fried, Bunkin, & Greenberg, 1998; Kershaw & Flier, 2004). TNF-α acts mainly in a paracrine fashion, circulates in low concentrations and is ubiquitously produced; thus, the question raises as to the influence of adipose TNF-α production on its circulating levels and the clinical utility in obesity of measuring its serum levels (Kershaw & Flier, 2004). In contrast, up to one-third of circulating IL-6 comes from WAT (mainly visceral), with reported correlations between IL-6 levels, BMI, and IR indexes (Kershaw & Flier, 2004; Stelzer et al., 2012). However, even in morbidly obese adults, IL-6 serum levels are close to the upper limit of normality, thus limiting its diagnostic utility (Kershaw & Flier, 2004). A predictive role of leptin levels on IL-6 levels has been recently reported in young, but not in older obese adults, thus suggesting a link between fat mass accumulation/leptin production and a rise in IL-6 in the initial phases of obesity-related inflammation (Stelzer et al., 2012).

As found with resistin, IL-6 levels are higher in preterm than in full-term newborns (Lo et al., 2002; Martos-Moreno et al., 2009), most probably due to the inflammatory environment in preterm deliveries as previously discussed. In contrast to resistin, the lower IL-6 levels observed in SGA compared to AGA infants might be related to their lower fat content (Martos-Moreno et al., 2009). Conversely, in the same study, TNF-α levels were not found to vary with gestational age or birth weight (Martos-Moreno et al., 2009). In addition to the preferentially paracrine secretion of TNF-α, the fact that the newborn immune system has an innately high IL-6 to TNF-α ratio, could explain the lack of correlation of circulating TNF-α levels with other factors (Angelone et al., 2006).

In healthy children, IL-6 levels are not different between the sexes and decrease in both boys and girls after pubertal onset, with IL-6 levels correlating negatively with sex steroids (testosterone in males and estradiol in females) and positively with sex hormone binding globulin levels. Indeed, both sex steroids inhibit IL-6 production (Bellido et al., 1995; Rachon, Mysliwska, Suchecka-Rachon, Wieckiewicz, & Mysliwski, 2002). Serum TNF-α levels show no differences between sexes or among pubertal stages (Martos-Moreno, Burgos-Ramos, et al., 2011).

Obese children have increased circulating IL-6 and TNF-α levels (Martos-Moreno, Barrios, et al., 2010), with IL-6 levels being increased in glucose intolerant obese subjects (Yeste et al., 2007). Both cytokines positively correlate with BMI with IL-6 levels decreasing after short-term BMI reduction and TNF-α exclusively after weight loss is sustained for at least 6 months. This observation indicates that long-term weight loss maintenance may be necessary in order to normalize circulating inflammatory markers in obese children. In contrast, no correlation was found between HOMA index and IL-6 or TNF-α levels (Martos-Moreno, Barrios, et al., 2010). In summary, as these inflammatory adipokines are involved in the innate immune response and their primary origin is mononuclear cells in humans, their usefulness as indicators of body fat content or carbohydrate metabolism impairment in obese children is limited.

Other proinflammatory adipokines have been studied in obesity, with reduced levels of IL-10 being reported in adults with obesity or type 2 diabetes and a negative correlation between this IL and BMI reported in children (Waters et al., 2007). This is consistent with its postulated insulin-sensitizing and anti-inflammatory roles through the antagonism of IL-6 and TNF-α activities (Gunnett, Heistad, & Faraci, 2002). Serum IL-1β and IL-8 levels are not sexually dimorphic in healthy children or adolescents, but IL-1β decreases and IL-8 increases at the termination of puberty (Martos-Moreno, Burgos-Ramos, et al., 2011), although the significance of these changes and their involvement in obesity are uncertain.

7. SUMMARY AND CONCLUSIONS

Here, we have summarized the current information regarding adipokine dynamics and activities in childhood, focusing on energy homeostasis balance, metabolic regulation, particularly carbohydrate metabolism, and inflammation. We have also discussed the particularities in the secretion and actions of these adipokines in healthy children, from birth to adolescence, in comparison to adults and the modifications induced by early-onset obesity.

We hope that after reading this review one comes to the obvious conclusion that for obesity, as for most physiological and pathological conditions, children are not just small size adults. In obesity, this difference is partially due to the extraordinary plasticity of their adipose tissue that confers upon them unique characteristics that evolve as longitudinal growth and puberty progress, with these changes often being influenced by their sex. Consequently, the age at obesity onset, as well as its severity, influences WAT structure and function, with adiponectin secretion and activity being an indisputable example.

Finally, documentation of the presence of adipokines during fetal life, their role in energy and metabolic homeostasis, and their permissive role for physiological processes, such as growth and puberty onset and development, has lead to the abolishment of the concept of adipose tissue as a passive "organ" devoted to energy storage and mechanical protection of more important anatomical structures. Taking into account the functional differences between the anatomically defined depots of WAT, it would be rather more appropriate to consider the existence of different "adipose organs" that communicate bidirectionally with the CNS, different hormonal axes, and with every organ in the body, and with an extraordinarily active role in the maintenance of whole body homeostasis.

ACKNOWLEDGMENTS

This work was supported by Fondo de Investigación Sanitaria (PI09/91060 and PI01/00747), CIBER de Fisiopatología de la Obesidad y Nutrición Instituto de Salud Carlos III, Fundación Mutua Madrileña (AP2561/2008), and Fundación Endocrinología y Nutrición. Gabriel Á. Martos-Moreno was a recipient of a fellowship (Rio Hortega) from the "Instituto de Salud Carlos III" (CM05/00100).

REFERENCES

Ahima, R. S., & Osei, S. Y. (2008). Adipokines in obesity. *Frontiers of Hormone Research*, *36*, 182–197.

Ahima, R. S., Qi, Y., Singhal, N. S., Jackson, M. B., & Scherer, P. E. (2006). Brain adipocytokine action and metabolic regulation. *Diabetes*, *55*, S145–S154.

Angelone, D. F., Wessels, M. R., Coughlin, M., Suter, E. E., Valentini, P., Kalish, L. A., et al. (2006). Innate immunity of the human newborn is polarized toward a high ratio of IL-6/TNF-alpha production in vitro and in vivo. *Pediatric Research*, *60*, 205–209.

Antunes, T. T., Gagnon, A., Bell, A., & Sorisky, A. (2005). Thyroid-stimulating hormone stimulates interleukin-6 release from 3T3-L1 adipocytes through a cAMP-protein kinase A pathway. *Obesity Research*, *13*, 2066–2071.

Antunes, H., Santos, C., & Carvalho, S. (2008). Serum leptin levels in overweight children and adolescents. *The British Journal of Nutrition*, *28*, 1–5.

Araki, S., Dobashi, K., Kubo, K., Kawagoe, R., Yamamoto, Y., Kawada, Y., et al. (2008). Plasma visfatin concentration as a surrogate marker for visceral fat accumulation in obese children. *Obesity (Silver Spring)*, *16*, 384–388.

Arch, J. R. (2005). Central regulation of energy balance: Inputs, outputs and leptin resistance. *The Proceedings of the Nutrition Society*, *64*, 39–46.

Argente, J., Barrios, V., Chowen, J. A., Sinha, M. K., & Considine, R. V. (1997). Leptin plasma levels in healthy Spanish children and adolescents, children with obesity, and adolescents with anorexia nervosa and bulimia nervosa. *The Journal of Pediatrics*, *131*, 833–838.

Arita, Y., Kihara, S., Ouchi, N., Takahashi, M., Maeda, K., Miyagawa, J., et al. (1999). Paradoxical decrease of an adipose-specific protein, adiponectin, in obesity. *Biochemical and Biophysical Research Communications*, *257*, 79–83.

Avram, A. S., Avram, M. M., & James, W. D. (2005). Subcutaneous fat in normal and diseased states: 2. Anatomy and physiology of white and brown adipose tissue. *Journal of the American Academy of Dermatology*, *53*, 671–683.

Bala, M., Martin, J., Kopp, A., Hanses, F., Buechler, C., & Schäffler, A. (2011). In vivo suppression of visfatin by oral glucose uptake: Evidence for a novel incretin-like effect by glucagon-like peptide-1 (GLP-1). *The Journal of Clinical Endocrinology and Metabolism*, *96*, 2493–2501.

Beilin, L., & Huang, R. C. (2008). Childhood obesity, hypertension, the metabolic syndrome and adult cardiovascular disease. *Clinical and Experimental Pharmacology and Physiology*, *35*, 409–411.

Bell, A., Gagnon, A., Dods, P., Papineau, D., Tiberi, M., & Sorisky, A. (2002). TSH signaling and cell survival in 3T3-L1 preadipocytes. *American Journal of Physiology. Cell Physiology*, *283*, C1056–C1064.

Bellido, T., Jilka, R. L., Boyce, B. F., Girasole, G., Broxmeyer, H., Dalrymple, S. A., et al. (1995). Regulation of interleukin-6, osteoclastogenesis, and bone mass by androgens. The role of the androgen receptor. *The Journal of Clinical Investigation*, *95*, 2886–2895.

Berndt, J., Klöting, N., Kralisch, S., Kovacs, P., Fasshauer, M., Schön, M. R., et al. (2005). Plasma visfatin concentrations and fat depot-specific mRNA expression in humans. *Diabetes*, *54*, 2911–2916.

Blüher, M. (2012). Vaspin in obesity and diabetes: Pathophysiological and clinical significance. *Endocrine*, *2012*(41), 176–182.

Bokarewa, M., Nagaev, I., Dahlberg, L., Smith, U., & Tarkowski, A. (2005). Resistin, an adipokine with potent proinflammatory properties. *The Journal of Immunology*, *174*, 5789–5795.

Bottner, A., Kratzsch, J., Muller, G., Kapellen, T. M., Bluher, S., Keller, E., et al. (2004). Gender differences of adiponectin levels develop during the progression of puberty and are related to serum androgen levels. *The Journal of Clinical Endocrinology and Metabolism*, *89*, 4053–4061.

Briana, D. D., Boutsikou, M., Baka, S., Gourgiotis, D., Marmarinos, A., Hassiakos, D., et al. (2008). Perinatal changes of plasma resistin concentrations in pregnancies with normal and restricted fetal growth. *Neonatology*, *93*, 153–157.

Briana, D. D., Boutsikou, M., Baka, S., Gourgiotis, D., Marmarinos, A., Liosi, S., et al. (2011). Omentin-1 and vaspin are present in the fetus and neonate, and perinatal concentrations are similar in normal and growth-restricted pregnancies. *Metabolism*, *60*, 486–490.

Camhi, S. M., Katzmarzyk, P. T., Broyles, S., Srinivasan, S. R., Chen, W., Bouchard, C., et al. (2010). Predicting adult body mass index-specific metabolic risk from childhood. *Metabolic Syndrome and Related Disorders*, *8*, 165–172.

Castan-Laurell, I., Dray, C., Attané, C., Duparc, T., Knauf, C., & Valet, P. (2011). Apelin, diabetes, and obesity. *Endocrine*, *40*, 1–9.

Cekmez, F., Canpolat, F. E., Pirgon, O., Çetinkaya, M., Aydinoz, S., Suleymanoglu, S., et al. (2011). Apelin, vaspin, visfatin and adiponectin in large for gestational age infants with insulin resistance. *Cytokine*, *56*, 387–391.

Chang, Y. C., Chang, T. J., Lee, W. J., & Chuang, L. M. (2010). The relationship of visfatin/pre-B-cell colony-enhancing factor/nicotinamide phosphoribosyltransferase in adipose tissue with inflammation, insulin resistance, and plasma lipids. *Metabolism*, *59*, 93–99.

Chang, H. M., Lee, H. J., Park, H. S., Isken, F., Spranger, J., Mai, K., et al. (2010). Effects of weight reduction on serum vaspin concentrations in obese subjects: Modification by insulin resistance. *Obesity*, *18*, 2105–2110.

Charo, I. F., & Ransohoff, R. M. (2006). The many roles of chemokines and chemokine receptors in inflammation. *The New England Journal of Medicine, 354*, 610–621.

Chen, B. H., Song, Y., Ding, E. L., Roberts, C. K., Manson, J. E., Rifai, N., et al. (2009). Circulating levels of resistin and risk of type 2 diabetes in men and women: Results from two prospective cohorts. *Diabetes Care, 32*, 329–334.

Chinetti-Gbaguidi, G., Fruchart, J. C., & Staels, B. (2005). Role of the PPAR family of nuclear receptors in the regulation of metabolic and cardiovascular homeostasis: New approaches to therapy. *Current Opinion in Pharmacology, 5*, 177–183.

Cho, J. K., Han, T. K., & Kang, H. S. (2010). Combined effects of body mass index and cardio/respiratory fitness on serum vaspin concentrations in Korean young men. *European Journal of Applied Physiology, 108*, 347–353.

Coppack, S. W. (2005). Adipose tissue changes in obesity. *Biochemical Society Transactions, 33*, 1049–1052.

Cristancho, A. G., & Lazar, M. A. (2011). Forming functional fat: A growing understanding of adipocyte differentiation. *Nature Reviews. Endocrinology, 12*, 722–734.

Curat, C. A., Wegner, V., Sengenès, C., Miranville, A., Tonus, C., Busse, R., et al. (2006). Macrophages in human visceral adipose tissue: Increased accumulation in obesity and a source of resistin and visfatin. *Diabetologia, 49*, 744–747.

Dedoussis, G. V., Kapiri, A., Kalogeropoulos, N., Samara, A., Dimitriadis, D., Lambert, D., et al. (2009). Adipokine expression in adipose tissue and in peripheral blood mononuclear cells in children. Correlation with BMI and fatty acid content. *Clinica Chimica Acta, 410*, 85–89.

Dedoussis, G. V., Kapiri, A., Samara, A., Dimitriadis, D., Lambert, D., Pfister, M., et al. (2009). Visfatin: The link between inflammation and childhood obesity. *Diabetes Care, 32*, e71.

del Rincon, J. P., Iida, K., Gaylinn, B. D., McCurdy, C. E., Leitner, J. W., Barbour, L. A., et al. (2007). Growth hormone regulation of p85alpha expression and phosphoinositide 3-kinase activity in adipose tissue: Mechanism for growth hormone-mediated insulin resistance. *Diabetes, 56*, 1638–1646.

Denver, R. J., Bonett, R. M., & Boorse, G. C. (2011). Evolution of leptin structure and function. *Neuroendocrinology, 94*, 21–38.

de Souza-Batista, C. M., Yang, R. Z., Lee, M. J., Glynn, N. M., Yu, D. Z., Pray, J., et al. (2007). Omentin plasma levels and gene expression are decreased in obesity. *Diabetes, 56*, 1655–1661.

Ellacott, K. L., & Cone, R. D. (2004). The central melanocortin system and the integration of short- and long-term regulators of energy homeostasis. *Recent Progress in Hormone Research, 59*, 395–408.

Elloumi, M., Ben-Ounis, O., Makni, E., Van Praagh, E., Tabka, Z., & Lac, G. (2009). Effect of individualized weight-loss programmes on adiponectin, leptin and resistin levels in obese adolescent boys. *Acta Paediatrica, 98*, 1487–1493.

Evagelidou, E. N., Giapros, V. I., Challa, A. S., Cholevas, V. K., Vartholomatos, G. A., Siomou, E. C., et al. (2010). Prothrombotic state, cardiovascular, and metabolic syndrome risk factors in prepubertal children born large for gestational age. *Diabetes Care, 33*, 2468–2470.

Fain, J. N. (2006). Release of interleukins and other inflammatory cytokines by human adipose tissue is enhanced in obesity and primarily due to the nonfat cells. *Vitamins and Hormones, 74*, 443–477.

Farooqi, I. S., Matarese, G., Lord, G. M., Keogh, J. M., Lawrence, E., Agwu, C., et al. (2002). Beneficial effects of leptin on obesity, T cell hyporesponsiveness, and neuroendocrine/metabolic dysfunction of human congenital leptin deficiency. *The Journal of Clinical Investigation, 110*, 1093–1103.

Farooqi, I. S., Wangensteen, T., Collins, S., Kimber, W., Matarese, G., Keogh, J. M., et al. (2007). Clinical and molecular genetic spectrum of congenital deficiency of the leptin receptor. *The New England Journal of Medicine, 356*, 237–247.

Feve, B. (2005). Adipogenesis: Cellular and molecular aspects. *Best Practice & Research. Clinical Endocrinology & Metabolism, 19*, 483–499.
Filková, M., Haluzík, M., Gay, S., & Senolt, L. (2009). The role of resistin as a regulator of inflammation: Implications for various human pathologies. *Clinical Immunology, 133*, 157–170.
Flint, D. J., Binart, N., Boumard, S., Kopchick, J. J., & Kelly, P. (2006). Developmental aspects of adipose tissue in GH receptor and prolactin receptor gene disrupted mice: Site-specific effects upon proliferation, differentiation and hormone sensitivity. *The Journal of Endocrinology, 191*, 101–111.
Freedman, D. S., Khan, L. K., Serdula, M. K., Dietz, W. H., Srinivasan, S. R., & Berenson, G. S. (2004). Inter-relationships among childhood BMI, childhood height, and adult obesity: The Bogalusa Heart Study. *International Journal of Obesity and Related Metabolic Disorders, 28*, 10–16.
Friebe, D., Neef, M., Kratzsch, J., Erbs, S., Dittrich, K., Garten, A., et al. (2011). Leucocytes are a major source of circulating nicotinamide phosphoribosyltransferase (NAMPT)/pre-B cell colony (PBEF)/visfatin linking obesity and inflammation in humans. *Diabetologia, 54*, 1200–1211.
Fried, S. K., Bunkin, D. A., & Greenberg, A. S. (1998). Omental and subcutaneous adipose tissues of obese subjects release interleukin-6: Depot difference and regulation by glucocorticoid. *The Journal of Clinical Endocrinology and Metabolism, 83*, 847–850.
Fukuhara, A., Matsuda, M., Nishizawa, M., Segawa, K., Tanaka, M., Kishimoto, K., et al. (2005). Visfatin: A protein secreted by visceral fat that mimics the effects of insulin. *Science, 307*, 426–430.
Galic, S., Oakhill, J. S., & Steinberg, G. R. (2010). Adipose tissue as an endocrine organ. *Molecular and Cellular Endocrinology, 316*, 129–139.
Gat-Yablonski, G., & Phillip, M. (2008). Leptin and regulation of linear growth. *Current Opinion in Clinical Nutrition and Metabolic Care, 11*, 303–308.
Gautron, L., & Elmquist, J. K. (2011). Sixteen years and counting: An update on leptin in energy balance. *The Journal of Clinical Investigation, 121*, 2087–2093.
Gerber, M., Boettner, A., Seidel, B., Lammert, A., Bär, J., Schuster, E., et al. (2005). Serum resistin levels of obese and lean children and adolescents: Biochemical analysis and clinical relevance. *The Journal of Clinical Endocrinology and Metabolism, 90*, 4503–4509.
Gohlke, B. C., Bartmann, P., Fimmers, R., Huber, A., Hecher, K., & Roth, C. L. (2008). Fetal adiponectin and resistin in correlation with birth weight difference in monozygotic twins with discordant growth. *Hormone Research, 69*, 37–44.
González, C. R., Caminos, J. E., Vázquez, M. J., Garcés, M. F., Cepeda, L. A., Angel, A., et al. (2009). Regulation of visceral adipose tissue-derived serine protease inhibitor by nutritional status, metformin, gender and pituitary factors in rat white adipose tissue. *The Journal of Physiology, 587*, 3741–3750.
Gunnett, C. A., Heistad, D. D., & Faraci, F. M. (2002). Interleukin-10 protects nitric oxide-dependent relaxation during diabetes: Role of superoxide. *Diabetes, 51*, 1931–1937.
Gursoy, T., Aliefendioglu, D., Caglayan, O., Aktas, A., & Ovali, F. (2011). Resistin levels in preterms: Are they influenced by fetal inflammatory course? *Journal of Perinatology, 31*, 171–175.
Hahn, T. M., Breininger, J. F., Baskin, D. G., & Schwartz, M. W. (1998). Coexpression of Agrp and NPY in fasting-activated hypothalamic neurons. *Nature Neuroscience, 1*, 271–272.
Haider, D. G., Holzer, G., Schaller, G., Weghuber, D., Widhalm, K., Wagner, O., et al. (2006). The adipokine visfatin is markedly elevated in obese children. *Journal of Pediatrics Gastroenterology and Nutrition, 43*, 548–549.
Haider, D. G., Schindler, K., Schaller, G., Prager, G., Wolzt, M., & Ludvik, B. (2006). Increased plasma visfatin concentrations in morbidly obese subjects are reduced after gastric banding. *The Journal of Clinical Endocrinology and Metabolism, 91*, 1578–1581.
Hauner, H. (2004). The new concept of adipose tissue function. *Physiology and Behavior, 83*, 653–658.

Hausman, D. B., DiGirolamo, M., Bartness, T. J., Hausman, G. J., & Martin, R. J. (2001). The biology of white adipocyte proliferation. *Obesity Reviews*, *2*, 239–254.

Hida, K., Wada, J., Eguchi, J., Zhang, H., Baba, M., Seida, A., et al. (2005). Visceral adipose tissue-derived serine protease inhibitor: A unique insulin-sensitizing adipocytokine in obesity. *Proceedings of the National Academy of Sciences of the United States of America*, *102*, 10610–10615.

Holm, J. C., Gamborg, M., Kaas-Ibsen, K., Gammeltoft, S., Ward, L., Heitmann, B. L., et al. (2007). Time course and determinants of leptin decline during weight loss in obese boys and girls. *Int J Pediatr Obes*, *2*, 2–10.

Ibáñez, L., Sebastiani, G., Lopez-Bermejo, A., Díaz, M., Gómez-Roig, M. D., & de Zegher, F. (2008). Gender specificity of body adiposity and circulating adiponectin, visfatin, insulin, and insulin growth factor-I at term birth: Relation to prenatal growth. *The Journal of Clinical Endocrinology and Metabolism*, *93*, 2774–2778.

Iwen, K. A., Senyaman, O., Schwartz, A., Drenckhan, M., Meier, B., Hadaschik, D., et al. (2008). Melanocortin crosstalk with adipose functions: ACTH directly induces insulin resistance, promotes a pro-inflammatory adipokine profile and stimulates UCP-1 in adipocytes. *The Journal of Endocrinology*, *196*, 465–472.

Jeffery, A. N., Murphy, M. J., Metcalf, B. S., Hosking, J., Voss, L. D., English, P., et al. (2008). Adiponectin in childhood. *International Journal of Pediatric Obesity*, *3*, 130–140.

Jeong, E., Youn, B. S., Kim, D. W., Kim, E. H., Park, J. W., Namkoong, C., et al. (2010). Circadian rhythm of serum vaspin in healthy male volunteers: Relation to meals. *The Journal of Clinical Endocrinology and Metabolism*, *95*, 1869–1875.

Jin, H., Jiang, B., Tang, J., Lu, W., Wang, W., Zhou, L., et al. (2008). Serum visfatin concentrations in obese adolescents and its correlation with age and high-density lipoprotein cholesterol. *Diabetes Research and Clinical Practice*, *79*, 412–418.

Jones, T. E., Basilio, J. L., Brophy, P. M., McCammon, M. R., & Hickner, R. C. (2009). Long-term exercise training in overweight adolescents improves plasma peptide YY and resistin. *Obesity (Silver Spring)*, *17*, 1189–1195.

Kadowaki, T., & Yamauchi, T. (2005). Adiponectin and adiponectin receptors. *Endocrine Reviews*, *26*, 439–451.

Kajantie, E., Hytinantti, T., Hovi, P., & Andersson, S. (2004). Cord plasma adiponectin: A 20-fold rise between 24 weeks gestation and term. *The Journal of Clinical Endocrinology and Metabolism*, *89*, 4031–4036.

Kershaw, E. E., & Flier, J. S. (2004). Adipose tissue as an endocrine organ. *The Journal of Clinical Endocrinology and Metabolism*, *89*, 2548–2556.

Klöting, N., Berndt, J., Kralisch, S., Kovacs, P., Fasshauer, M., Schön, M. R., et al. (2006). Vaspin gene expression in human adipose tissue: Association with obesity and type 2 diabetes. *Biochemical and Biophysical Research Communication*, *339*, 430–436.

Kolsgaard, M. L., Wangensteen, T., Brunborg, C., Joner, G., Holven, K. B., Halvorsen, B., et al. (2009). Elevated visfatin levels in overweight and obese children and adolescents with metabolic syndrome. *Scandinavian Journal of Clinical and Laboratory Investigation*, *69*, 858–864.

Korner, A., Kratzsch, J., & Kiess, W. (2005). Adipocytokines: Leptin—The classical, resistin—The controversial, adiponectin—The promising, and more to come. *Best Practice & Research. Clinical Endocrinology & Metabolism*, *19*, 525–546.

Körner, A., Neef, M., Friebe, D., Erbs, S., Kratzsch, J., Dittrich, K., et al. (2011). Vaspin is related to gender, puberty and deteriorating insulin sensitivity in children. *International Journal of Obesity*, *35*, 578–586.

Koutkia, P., Canavan, B., Johnson, M. L., DePaoli, A., & Grinspoon, S. (2003). Characterization of leptin pulse dynamics and relationship to fat mass, growth hormone, cortisol, and insulin. *American Journal of Physiology, Endocrinology and Metabolism*, *285*, E372–E379.

Krzystek-Korpacka, M., Patryn, E., Bednarz-Misa, I., Hotowy, K., & Noczynska, A. (2011). Visfatin in juvenile obesity—The effect of obesity intervention and sex. *European Journal of Clinical Investigation*, *41*, 1284–1291.

Krzyzanowska, K., Mittermayer, F., Krugluger, W., Kopp, H. P., & Schernthaner, G. (2006). Increase in visfatin after weight loss induced by gastroplastic surgery. *Obesity (Silver Spring)*, *14*, 1886–1889.

Landt, M. (2000). Leptin binding and binding capacity in serum. *Clinical Chemistry*, *46*, 379–384.

Landt, M., Parvin, C. A., & Wong, M. (2000). Leptin in cerebrospinal fluid from children: Correlation with plasma leptin, sexual dimorphism, and lack of protein binding. *Clinical Chemistry*, *46*, 854–858.

Lau, S. M., Hng, T. M., de Vries, B., McLean, M., & Cheung, N. W. (2009). Sexual dimorphism of high molecular weight adiponectin in cord blood. *Clinical Endocrinology*, *70*, 500–501.

Lee, M. K., Jekal, Y., Im, J. A., Kim, E., Lee, S. H., Park, J. H., et al. (2010). Reduced serum vaspin concentrations in obese children following short-term intensive lifestyle modification. *Clinica Chimica Acta*, *411*, 381–385.

Lepercq, J., Challier, J. C., Guerre-Millo, M., Cauzac, M., Vidal, H., & Hauguel-de Mouzon, S. (2001). Prenatal leptin production: Evidence that fetal adipose tissue produces leptin. *The Journal of Clinical Endocrinology and Metabolism*, *86*, 2409–2413.

Li, Q., Chen, R., Moriya, J., Yamakawa, J., Sumino, H., Kanda, T., et al. (2008). A novel adipocytokine, visceral adipose tissue-derived serine protease inhibitor (vaspin), and obesity. *The Journal of International Medical Research*, *36*, 625–629.

Li, M., Fisette, A., Zhao, X. Y., Deng, J. Y., Mi, J., & Cianflone, K. (2009). Serum resistin correlates with central obesity but weakly with insulin resistance in Chinese children and adolescents. *International Journal of Obesity*, *33*, 424–439.

Lo, H. C., Tsao, L. Y., Hsu, W. Y., Chen, H. N., Yu, W. K., & Chi, C. Y. (2002). Relation of cord serum levels of growth hormone, insulin-like growth factors, insulin-like growth factor binding proteins, leptin, and interleukin-6 with birth weight, birth length, and head circumference in term and preterm neonates. *Nutrition*, *18*, 604–608.

Lumeng, C. N., Deyoung, S. M., Bodzin, J. L., & Saltiel, A. R. (2007). Increased inflammatory properties of adipose tissue macrophages recruited during diet-induced obesity. *Diabetes*, *56*, 16–23.

Majka, S. M., Barak, Y., & Klemm, D. J. (2011). Concise review: Adipocyte origins: Weighing the possibilities. *Stem Cells*, *29*, 1034–1040.

Malamitsi-Puchner, A., Briana, D. D., Boutsikou, M., Kouskouni, E., Hassiakos, D., & Gourgiotis, D. (2007). Perinatal circulating visfatin levels in intrauterine growth restriction. *Pediatrics*, *119*, 1314–1318.

Malamitsi-Puchner, A., Gourgiotis, D., Boutsikou, M., Baka, S., Hassiakos, D., & Briana, D. D. (2007). Circulating apelin concentrations in mother/infant pairs at term. *Acta Paediatrica*, *96*, 1751–1754.

Mamì, C., Marseglia, L., Manganaro, R., Saitta, G., Martino, F., Gargano, R., et al. (2009). Serum levels of resistin and its correlation with adiponectin and insulin in healthy full term neonates. *Early Human Development*, *85*, 37–40.

Manco, M., Fernandez-Real, J. M., Equitani, F., Vendrell, J., Valera-Mora, M. E., Nanni, G., et al. (2007). Effect of massive weight loss on inflammatory adipocytokines and the innate immune system in morbidly obese women. *The Journal of Clinical Endocrinology and Metabolism*, *92*, 483–490.

Mantzoros, C. S., Magkos, F., Brinkoetter, M., Sienkiewicz, E., Dardeno, T. A., Kim, S. Y., et al. (2011). Leptin in human physiology and pathophysiology. *American Journal of Physiology, Endocrinology and Metabolism*, *301*, E567–E584.

Martos-Moreno, G. A., & Argente, J. (2011). Paediatric obesities: From childhood to adolescence. *Anales de Pediatría (Barcelona, Spain)*, *75*, e1–e23.

Martos-Moreno, G. A., Barrios, V., & Argente, J. (2006). Normative data for adiponectin, resistin, interleukin 6, and leptin/receptor ratio in a healthy Spanish pediatric population: Relationship with sex steroids. *European Journal of Endocrinology, 155*, 429–434.

Martos-Moreno, G. Á., Barrios, V., Martínez, G., Hawkins, F., & Argente, J. (2010). Effect of weight loss on high-molecular weight adiponectin in obese children. *Obesity (Silver Spring), 18*, 2288–2294.

Martos-Moreno, G. A., Barrios, V., Sáenz de Pipaón, M., Pozo, J., Dorronsoro, I., Martínez-Biarge, M., et al. (2009). Influence of prematurity and growth restriction on the adipokine profile, IGF1, and ghrelin levels in cord blood: Relationship with glucose metabolism. *European Journal of Endocrinology, 161*, 381–389.

Martos-Moreno, G. A., Burgos-Ramos, E., Canelles, S., Argente, J., & Barrios, V. (2011). Analysis of insulin and cytokines during development using a multiplexed immunoassay: Implications in pediatrics. *Anales de Pediatría (Barcelona, Spain), 74*, 356–362.

Martos-Moreno, G. A., Chowen, J. A., & Argente, J. (2010). Metabolic signals in human puberty: Effects of over and undernutrition. *Molecular and Cellular Endocrinology, 324*, 70–81.

Martos-Moreno, G. Á., Kratzsch, J., Körner, A., Barrios, V., Hawkins, F., Kiess, W., et al. (2011). Serum visfatin and vaspin levels in prepubertal children: Effect of obesity and weight loss after behavior modifications on their secretion and relationship with glucose metabolism. *International Journal of Obesity, 35*, 1355–1362.

Matsuzawa, Y. (2010). Adiponectin: A key player in obesity related disorders. *Current Pharmaceutical Design, 16*, 1896–1901.

Mazaki-Tovi, S., Vaisbuch, E., Romero, R., Kusanovic, J. P., Chaiworapongsa, T., Kim, S. K., et al. (2010). Maternal and neonatal circulating visfatin concentrations in patients with pre-eclampsia and a small-for-gestational age neonate. *The Journal of Maternal-Fetal & Neonatal Medicine, 23*, 1119–1128.

McTernan, P. G., Kusminski, C. M., & Kumar, S. (2006). Resistin. *Current Opinion in Lipidology, 17*, 170–175.

Meier, U., & Gressner, A. M. (2004). Endocrine regulation of energy metabolism: Review of pathobiochemical and clinical chemical aspects of leptin, ghrelin, adiponectin, and resistin. *Clinical Chemistry, 50*, 1511–1525.

Menéndez, C., Baldelli, R., Camiña, J. P., Escudero, B., Peino, R., Diéguez, C., et al. (2003). TSH stimulates leptin secretion by a direct effect on adipocytes. *The Journal of Endocrinology, 176*, 7–12.

Meral, C., Cekmez, F., Pirgon, O., Asya-Tanju, I., Metin-Ipcioglu, O., Karademir, F., et al. (2011). The relationship between serum visfatin, adiponectin, and insulin sensitivity markers in neonates after birth. *The Journal of Maternal-Fetal & Neonatal Medicine, 24*, 166–170.

Misra, M., Miller, K. K., Almazan, C., Ramaswamy, K., Aggarwal, A., Herzog, D. B., et al. (2004). Hormonal and body composition predictors of soluble leptin receptor, leptin, and free leptin index in adolescent girls with anorexia nervosa and controls and relation to insulin sensitivity. *The Journal of Clinical Endocrinology and Metabolism, 89*, 3486–3495.

Montague, C. T., Farooqi, I. S., Whitehead, J. P., Soos, M. A., Rau, H., Wareham, N. J., et al. (1997). Congenital leptin deficiency is associated with severe early-onset obesity in humans. *Nature, 387*, 903–908.

Moreno-Navarrete, J. M., Catalán, V., Ortega, F., Gómez-Ambrosi, J., Ricart, W., Frühbeck, G., et al. (2010). Circulating omentin concentration increases after weight loss. *Nutrition and Metabolism (London), 7*, 27.

Murer, S. B., Knöpfli, B. H., Aeberli, I., Jung, A., Wildhaber, J., Wildhaber-Brooks, J., et al. (2011). Baseline leptin and leptin reduction predict improvements in metabolic variables and long-term fat loss in obese children and adolescents: A prospective study of an inpatient weight-loss program. *The American Journal of Clinical Nutrition, 93*, 695–702.

Nishimura, R., Sano, H., Matsudaira, T., Miyashita, Y., Morimoto, A., Shirasawa, T., et al. (2007). Childhood obesity and its relation to serum adiponectin and leptin: A report from a population-based study. *Diabetes Research and Clinical Practice, 76*, 245–250.

Ogden, C. L., Carroll, M. D., Brian, K. K., & Flegal, K. M. (2012). Prevalence of obesity and trends in body mass index among US Children and Adolescents, 1999–2010. *Journal of the American Medical Association, 307*, 483–490.

Oh, D. K., Ciaraldi, T., & Henry, R. R. (2007). Adiponectin in health and disease. *Diabetes, Obesity & Metabolism, 9*, 282–289.

Ong, K. K., Ahmed, M. L., Sherriff, A., Woods, K. A., Watts, A., Golding, J., et al. (1999). Cord blood leptin is associated with size at birth and predicts infancy weight gain in humans. ALSPAC Study Team. Avon Longitudinal Study of Pregnancy and Childhood. *The Journal of Clinical Endocrinology and Metabolism, 84*, 1145–1148.

Pajvani, U. B., Hawkins, M., Combs, T. P., Rajala, M. W., Doebber, T., Berger, J. P., et al. (2004). Complex distribution, not absolute amount of adiponectin, correlates with thiazolidinedione-mediated improvement in insulin sensitivity. *The Journal of Biological Chemistry, 279*, 12152–12162.

Pan, H. Y., Guo, L., & Li, Q. (2010). Changes of serum omentin-1 levels in normal subjects and in patients with impaired glucose regulation and with newly diagnosed and untreated type 2 diabetes. *Diabetes Research and Clinical Practice, 88*, 29–33.

Prats-Puig, A., Bassols, J., Bargalló, E., Mas-Parareda, M., Ribot, R., Soriano-Rodríguez, P., et al. (2011). Toward an early marker of metabolic dysfunction: Omentin-1 in prepubertal children. *Obesity (Silver Spring), 19*, 1905–1907.

Prins, J. B., & O'Rahilly, S. (1997). Regulation of adipose cell number in man. *Clinical Science (London, England), 92*, 3–11.

Rabe, K., Lehrke, M., Parhofer, K. G., & Broedl, U. C. (2008). Adipokines and insulin resistance. *Molecular Medicine, 14*, 741–751.

Rachon, D., Mysliwska, J., Suchecka-Rachon, K., Wieckiewicz, J., & Mysliwski, A. (2002). Effects of oestrogen deprivation on interleukin-6 production by peripheral blood mononuclear cells of postmenopausal women. *The Journal of Endocrinology, 172*, 387–395.

Reaven, G. M. (1988). Role of insulin resistance in human disease. *Diabetes, 37*, 1595–1607 Banting lecture.

Reinehr, T., Kratzsch, J., Kiess, W., & Andler, W. (2005). Circulating soluble leptin receptor, leptin, and insulin resistance before and after weight loss in obese children. *International Journal of Obesity, 29*, 1230–1235.

Reinehr, T., Roth, C. L., Menke, T., & Andler, W. (2006). Resistin concentrations before and after weight loss in obese children. *International Journal of Obesity, 30*, 297–301.

Reinehr, T., Woelfle, J., & Roth, C. L. (2011). Lack of association between apelin, insulin resistance, cardiovascular risk factors, and obesity in children: A longitudinal analysis. *Metabolism, 60*, 1349–1354.

Revollo, J. R., Körner, A., Mills, K. F., Satoh, A., Wang, T., Garten, A., et al. (2007). Nampt/PBEF/Visfatin regulates insulin secretion in beta cells as a systemic NAD biosynthetic enzyme. *Cell Metabolism, 6*, 363–375.

Rittig, K., Hildebrandt, U., Thamer, C., Staiger, H., Peter, A., Stefan, N., et al. (2011). Apelin serum levels are not associated with early atherosclerosis or fat distribution in young subjects with increased risk for type 2 diabetes. *Experimental and Clinical Endocrinology & Diabetes, 119*, 358–361.

Romanowska, A., & Lebensztejn, D. M. (2010). Evaluation of serum visfatin concentrations in children with nonalcoholic fatty liver disease. *Polski Merkuriusz Lekarski, 28*, 459–461.

Rosen, E. D., & Spiegelman, B. M. (2006). Adipocytes as regulators of energy balance and glucose homeostasis. *Nature, 444*, 847–853.

Rubin, D. A., McMurray, R. G., Harrell, J. S., Hackney, A. C., Thorpe, D. E., & Haqq, A. M. (2008). The association between insulin resistance and cytokines in adolescents: The role of weight status and exercise. *Metabolism, 57*, 683–690.

Saddi-Rosa, P., Oliveira, C. S., Giuffrida, F. M., & Reis, A. F. (2010). Visfatin, glucose metabolism and vascular disease: A review of evidence. *Diabetology and Metabolic Syndrome, 2*, 21.

Schäffler, A., Neumeier, M., Herfarth, H., Fürst, A., Schölmerich, J., & Büchler, C. (2005). Genomic structure of human omentin, a new adipocytokine expressed in omental adipose tissue. *Biochimica et Biophysica Acta, 1732*, 96–102.

Schäffler, A., Schölmerich, J., & Buechler, C. (2006). The role of 'adipotropins' and the clinical importance of a potential hypothalamic-pituitary-adipose axis. *Nature Clinical Practice. Endocrinology & Metabolism, 2*, 374–383.

Schwartz, D. R., & Lazar, M. A. (2011). Human resistin: Found in translation from mouse to man. *Trends in Endocrinology and Metabolism, 22*, 259–265.

Schwartz, M. W., & Niswender, K. D. (2004). Adiposity signaling and biological defense against weight gain: Absence of protection or central hormone resistance? *The Journal of Clinical Endocrinology and Metabolism, 89*, 5889–5897.

Shekhawat, P. S., Garland, J. S., Shivpuri, C., Mick, G. J., Sasidharan, P., Pelz, C. J., et al. (1998). Neonatal cord blood leptin: Its relationship to birth weight, body mass index, maternal diabetes, and steroids. *Pediatric Research, 43*, 338–343.

Siahanidou, T., Margeli, A., Kappis, A., Papassotiriou, I., & Mandyla, H. (2011). Circulating visfatin levels in healthy preterm infants are independently associated with high-density lipoprotein cholesterol levels and dietary long-chain polyunsaturated fatty acids. *Metabolism, 60*, 389–393.

Skurk, T., Alberti-Huber, C., Herder, C., & Hausner, H. (2007). Relationship between adipocyte size and adipokine expression and secretion. *The Journal of Clinical Endocrinology and Metabolism, 92*, 1023–1033.

Sone, M., & Osamura, R. Y. (2001). Leptin and the pituitary. *Pituitary, 4*, 15–23.

Sorisky, A., Bell, A., & Gagnon, A. (2000). TSH receptor in adipose cells. *Hormone and Metabolic Research, 32*, 468–474.

Sowers, J. R. (2008). Endocrine functions of adipose tissue: Focus on adiponectin. *Clinical Cornerstone, 9*, 32–40.

Spalding, K. L., Arner, E., Westermark, P. O., Bernard, S., Buchholz, B. A., Bergmann, O., et al. (2008). Dynamics of fat cell turnover in humans. *Nature, 453*, 783–787.

Steene-Johannessen, J., Kolle, E., Reseland, J. E., Anderssen, S. A., & Andersen, L. B. (2010). Waist circumference is related to low-grade inflammation in youth. *International Journal of Pediatric Obesity, 5*, 313–319.

Stelzer, I., Zelzer, S., Raggam, R. B., Prüller, F., Truschnig-Wilders, M., Meinitzer, A., et al. (2012). Link between leptin and interleukin-6 levels in the initial phase of obesity related inflammation. *Translational Research, 159*, 118–124.

Stoll-Becker, S., Kreuder, J., Reiss, I., Etspüler, J., Blum, W. F., & Gortner, L. (2003). Influence of gestational age and intrauterine growth on leptin concentrations in venous cord blood of human newborns. *Klinische Pädiatrie, 215*, 3–8.

Suleymanoglu, S., Tascilar, E., Pirgon, O., Tapan, S., Meral, C., & Abaci, A. (2009). Vaspin and its correlation with insulin sensitivity indices in obese children. *Diabetes Research and Clinical Practice, 84*, 325–328.

Symonds, M. E., Mostyn, A., Pearce, S., Budge, H., & Stephenson, T. (2003). Endocrine and nutritional regulation of fetal adipose tissue development. *The Journal of Endocrinology, 179*, 293–299.

Tadokoro, N., Shinomiya, M., Yoshinaga, M., Takahashi, H., Matsuoka, K., Miyashita, Y., et al. (2010). Visceral fat accumulation in Japanese high school students and related atherosclerotic risk factors. *Journal of Atherosclerosis and Thrombosis, 17*, 546–557.

Tan, B. K., Adya, R., & Randeva, H. S. (2010). Omentin: A novel link between inflammation, diabesity, and cardiovascular disease. *Trends in Cardiovascular Medicine, 20*, 143–148.

Tartaglia, L. A., Dembski, M., Weng, X., Deng, N., Culpepper, J., Devos, R., et al. (1995). Identification and expression cloning of a leptin receptor, OB-R. *Cell, 83*, 1263–1271.

Tilg, H., & Moschen, A. R. (2008). Inflammatory mechanisms in the regulation of insulin resistance. *Molecular Medicine, 14*, 222–231.

Tovar, S., Nogueiras, R., Tung, L. Y., Castañeda, T. R., Vázquez, M. J., Morris, A., et al. (2005). Central administration of resistin promotes short-term satiety in rats. *European Journal of Endocrinology, 153*, R1–R5.

Tsuchida, A., Yamauchi, T., & Kadowaki, T. (2005). Nuclear receptors as targets for drug development: Molecular mechanisms for regulation of obesity and insulin resistance by peroxisome proliferator-activated receptor gamma, CREB-binding protein, and adiponectin. *Journal of Pharmacological Sciences, 97*, 164–170.

Valle, M., Martos, R., Gascón, F., Cañete, R., Zafra, M. A., & Morales, R. (2005). Low-grade systemic inflammation, hypoadiponectinemia and a high concentration of leptin are present in very young obese children, and correlate with metabolic syndrome. *Diabetes & Metabolism, 31*, 55–62.

von Loeffelholz, C., Möhlig, M., Arafat, A. M., Isken, F., Spranger, J., Mai, K., et al. (2010). Circulating vaspin is unrelated to insulin sensitivity in a cohort of nondiabetic humans. *European Journal of Endocrinology, 162*, 507–513.

Wabitsch, M., Blum, W. F., Muche, R., Braun, M., Hube, F., Rascher, W., et al. (1997). Contribution of androgens to the gender difference in leptin production in obese children and adolescents. *The Journal of Clinical Investigation, 100*, 808–813.

Wajchenberg, B. L. (2000). Subcutaneous and visceral adipose tissue: Their relation to the metabolic syndrome. *Endocrine Reviews, 21*, 697–738.

Wallenius, K., Wallenius, V., Sunter, D., Dickson, S. L., & Jansson, J. O. (2002). Intracerebroventricular interleukin-6 treatment decreases body fat in rats. *Biochemical and Biophysical Research Communication, 293*, 560–565.

Wang, J., Shang, L. X., Dong, X., Wang, X., Wu, N., Wang, S. H., et al. (2010). Relationship of adiponectin and resistin levels in umbilical serum, maternal serum and placenta with neonatal birth weight. *The Australian and New Zealand Journal of Obstetrics and Gynaecology, 50*, 432–438.

Waters, K. A., Mast, B. T., Vella, S., de la Eva, R., O'Brien, L. M., Bailey, S., et al. (2007). Structural equation modeling of sleep apnea, inflammation, and metabolic dysfunction in children. *Journal of Sleep Research, 16*, 388–395.

Wellen, K. E., & Hotamisligil, G. S. (2003). Obesity-induced inflammatory changes in adipose tissue. *The Journal of Clinical Investigation, 112*, 1785–1788.

Wilasco, M. I., Goldani, H. A., Dornelles, C. T., Maurer, R. L., Kieling, C. O., Porowski, M., et al. (2012). Ghrelin, leptin and insulin in healthy children: Relationship with anthropometry, gender, and age distribution. *Regulatory Peptides, 173*, 21–26.

Yamauchi, T., Kamon, J., Ito, Y., Tsuchida, A., Yokomizo, T., Kita, S., et al. (2003). Cloning of adiponectin receptors that mediate antidiabetic metabolic effects. *Nature, 423*, 762–769.

Yang, R. Z., Lee, M. J., Hu, H., Pray, J., Wu, H. B., Hansen, B. C., et al. (2006). Identification of omentin as a novel depot-specific adipokine in human adipose tissue: Possible role in modulating insulin action. *American Journal of Physiology, Endocrinology and Metabolism, 290*, E1253–E1261.

Yannakoulia, M., Yiannakouris, N., Bluher, S., Matalas, A. L., Klimis-Zacas, D., & Mantzoros, C. S. (2003). Body fat mass and macronutrient intake in relation to circulating soluble leptin receptor, free leptin index, adiponectin, and resistin concentrations in healthy humans. *The Journal of Clinical Endocrinology and Metabolism, 88*, 1730–1736.

Ye, Y., Hou, X. H., Pan, X. P., Lu, J. X., & Jia, W. P. (2009). Serum vaspin level in relation to postprandial plasma glucose concentration in subjects with diabetes. *Chinese Medical Journal*, *122*, 2530–2533.

Yeste, D., Vendrell, J., Tomasini, R., Broch, M., Gussinyé, M., Megia, A., et al. (2007). Interleukin-6 in obese children and adolescents with and without glucose intolerance. *Diabetes Care*, *30*, 1892–1894.

Youn, B. S., Klöting, N., Kratzsch, J., Lee, N., Park, J. W., Song, E. S., et al. (2008). Serum vaspin concentrations in human obesity and type 2 diabetes. *Diabetes*, *57*, 372–377.

Yu, Y. H., & Ginsberg, H. N. (2005). Adipocyte signaling and lipid homeostasis: Sequelae of insulin-resistant adipose tissue. *Circulation Research*, *96*, 1042–1052.

Yu, J. G., Javorschi, S., Hevener, A. L., Kruszynska, Y. T., Norman, R. A., Sinha, M., et al. (2002). The effect of thiazolidinediones on plasma adiponectin levels in normal, obese, and type 2 diabetic subjects. *Diabetes*, *51*, 2968–2974.

Zhang, Y., Proenca, R., Maffei, M., Barone, M., Leopold, L., & Friedman, J. M. (1994). Positional cloning of the mouse obese gene and its human homologue. *Nature*, *372*, 425–432.

CHAPTER SEVEN

Gut Hormones and Obesity: Physiology and Therapies

Rebecca Scott*,†,1, Tricia Tan*, Stephen Bloom*

*Division of Diabetes, Endocrinology, Metabolism, Hammersmith Hospital, Imperial College London, London, United Kingdom

†Queen Elizabeth II Hospital, Howlands, Welwyn Garden City, Hertfordshire, United Kingdom

[1]Corresponding author: e-mail address: rebeccavscott@gmail.com

Contents

Vitamins and Hormones, Volume 91
ISSN 0083-6729
http://dx.doi.org/10.1016/B978-0-12-407766-9.00007-9

Abstract

Over the past 30 years, it has been established that hormones produced by the gut, pancreas, and adipose tissue are key players in the control of body weight. These hormones act through a complex neuroendocrine system, including the hypothalamus, to regulate metabolism and energy homeostasis. In obesity, this homeostatic balance is disrupted, either through alterations in the levels of these hormones or through resistance to their actions. Alterations in gut hormone secretion following gastric bypass surgery are likely to underlie the dramatic and persistent loss of weight following this procedure, as well as the observed amelioration in type 2 diabetes mellitus. Medications based on the gut hormone GLP-1 are currently in clinical use to treat type 2 diabetes mellitus and have been shown to produce weight loss. Further therapies for obesity based on other gut hormones are currently in development.

1. INTRODUCTION

It is well established that obesity is a multinational, and growing, problem. Worldwide, the number of people suffering from obesity has doubled since 2008; at least 2.8 million adults die each year as a result of being overweight or obese (WHO, 2012). In order to devise safe, effective, and rational therapies to treat this pandemic, an understanding of the physiological mechanisms of energy balance and specifically the complex regulatory interactions between the brain, the gut, the liver, and adipose tissue is required. As the route for nutrient absorption, the gut is a natural point for regulation of energy balance and is indeed considered the largest endocrine organ in the body (Ahlman & Nilsson, 2001). The range of gut hormones that are thought to be involved in regulation of energy balance is continually increasing, as is knowledge about their actions and interactions that enable them to regulate energy homeostasis (Field et al., 2010; Harrold et al., 2012; Ladenheim, 2012; Woods et al., 2006; Zac-Varghese et al., 2010). There is also greater knowledge about how the normal functions of these hormones are subverted to cause and perpetuate obesity, and therefore how they may also be used in the treatment of obesity (Suzuki et al., 2010). This chapter starts by looking at the physiology of the hormones produced by the gut that regulate body weight and energy intake, followed by how these are affected by obesity. We will conclude by looking at current therapies targeting gut hormones to treat obesity.

2. CENTRAL STRUCTURES INVOLVED IN THE CONTROL OF FOOD INTAKE

Regulation of energy balance involves a number of different processes, including homeostatic control of energy intake; processing of gustatory, visual, and olfactory sensation; determining the rewarding/emotional aspects of food; higher cortical functions involved in motivation and impulse control; and control of energy expenditure through metabolic rate and activity. These different processes need to be coordinated, and this occurs in the hypothalamus and the brainstem.

The role of the hypothalamus in controlling feeding was first established in the 1950s (Stellar, 1954). It is well situated for this as it possesses reciprocal links with the brainstem and higher cortical areas, as well as neuroendocrine projections that can manipulate a number of hormonal axes – notably the hypothalamic-pituitary-thyroid (HPT) axis (Berthoud, 2002). The blood–brain barrier is incomplete at the median eminence of the hypothalamus, enabling the hypothalamus to directly sense circulating substances such as gut hormones (Gross, 1992).

The arcuate nucleus of the hypothalamus is particularly important in the control of feeding. It contains two distinct populations of neurons with opposing effects on food intake. The first coexpresses neuropeptide Y (NPY) and Agouti-related peptide (AgRP; Aponte, Atasoy, & Sternson, 2011; Gropp et al., 2005) and stimulates feeding: that is, they are orexigenic. The other population containing proopiomelanocortin (POMC) and cocaine- and amphetamine-related transcript (CART), has anorectic effects (Aponte et al., 2011).

NPY is a member of the pancreatic polypeptide (PP)-fold family and is related to gut hormones PP and peptide YY (PYY). Central administration of NPY produces hyperphagia in rodents (Raposinho et al., 2001; Sainsbury, Rohner-Jeanrenaud, Grouzmann, & Jeanrenaud, 1996), and mice that overexpress NPY are obese (Kaga et al., 2001). NPY acts through a number of members of the Y receptor superfamily: it activates Y1 and Y5 receptors to produce hyperphagia, while activation of the Y2 inhibitory autoreceptors reduces feeding (Henry, Ghibaudi, Gao, & Hwa, 2005). The POMC/CART neurons, which oppose the NPY/AgRP neurones, produce the neuropeptide α-melanocortin-stimulating hormone (αMSH), which is derived from POMC by posttranslational cleavage. αMSH acts as a ligand for the melanocortin receptors Mc3r and Mc4r (Chen et al., 2000; Huszar et al., 1997; Marsh et al., 1999). Blockage of either receptor increases food intake (Chen et al., 2000; Giraudo, Billington, & Levine, 1998; Huszar et al., 1997). AgRP is an antagonist of αMSH at Mc3r and Mc4r (Pritchard & White, 2005), and rats that overexpress AgRP in the hypothalamus eat more and have reduced energy expenditure, causing weight gain (De Backer et al., 2011).

These two populations of neurons within the arcuate synapse on each other (Broberger, 2005; Sainsbury & Zhang, 2010) and also project to other hypothalamic nuclei (Bai et al., 1985). Activation of the NPY/AgRP neurons counteracts the effects of the POMC/CART neurons via AgRP's antagonistic effects at the Mc3r and Mc4r (Pritchard & White, 2005) and also through direct inhibition of the POMC/CART neurons via GABA release (Tong, Ye, Jones, Elmquist, & Lowell, 2008; Roseberry et al., 2004). Activation of NPY neurons is caused by signals of energy deficiency, such as high levels of ghrelin, stress, and glucocorticoids (Guan, Li, & Jiang, 2010; Pralong, Corder, & Gaillard, 1993), while signals of energy repletion, such as leptin and insulin, suppress activity in the NPY neurons (Ahima et al., 1996; Schwartz et al., 1992; Zhang, Bijker, & Herzog, 2011).

Other hypothalamic nuclei are also involved in controlling food intake. The lateral hypothalamic area was initially thought of as a "hunger center" as lesions here produced hypophagic animals with reduced body fat (Bernardis, Awad, Fink, & Bellinger, 1992), while the ventromedial hypothalamic nucleus (VMN) was thought of as a "satiety center," with stimulation of the area reducing food intake in animals (Wyrwicka & Dobrzecka, 1960). The dorsomedial nucleus is also responsible for maintaining food intake (Bellinger & Bernardis, 2002), and lesions of the paraventricular nucleus cause increased food intake and obesity (Leibowitz, Hammer, & Chang, 1981).

The brainstem is also involved in control of energy homeostasis. The areas within the brainstem that regulate energy balance include the area postrema, the nucleus tractus solitarius (NTS), and the dorsal vagal complex (DVC). The brainstem acts as an interface between the vagus nerve, and the hypothalamus and higher cortical areas. As the area postrema lies outside the blood–brain barrier, the brainstem can also directly sense circulating hormones and substances within the plasma.

3. GUT HORMONES

Because the gut is directly exposed to nutrients, it is the obvious place to signal the current state of energy balance within the body. These signals are relayed centrally, and one way this happens is through hormones produced by the gut and pancreas. The variety of hormones produced, and their varying actions, can provide a detailed picture of current nutrient status. The signals involved are discussed below.

3.1. Peptide YY

This 36-amino acid hormone is a member of the PP-fold family, sharing sequence homology and tertiary structure with PP and the neurotransmitter NPY (Tatemoto, Carlquist, & Mutt, 1982; Tatemoto, 1982). PYY

(1–36) is produced by the L cells of the intestine but is converted by dipeptidyl peptidase-4 (DPP-4) to PYY(3–36), the dominant form in the circulation (Mentlein, Dahms, Grandt, & Krüger, 1993). The PP-fold family act through the Y receptor superfamily. There are five Y receptors: Y1, Y2, Y4, Y5, and Y6, which are G-protein-coupled receptors (Blomqvist & Herzog, 1997). Receptor homology is low and their function varied, and they may be found both pre- and postsynaptically (Zhang et al., 2011). Affinity of the differing PP-fold peptides varies for the different receptors (reviewed by Blomqvist & Herzog, 1997). With regards to PYY, the cleavage of PYY(1–36) to PYY(3–36) changes the peptide affinity from binding to Y1, Y2, and Y5 receptors to being a more specific Y2 agonist (Keire, Bowers, Solomon, & Reeve, 2002; Mentlein et al., 1993). It is through this receptor PYY exerts its satiating effects.

PYY(3–36) secretion is stimulated by food intake, peaking 1–2 h after a meal and persisting for several hours postprandially (Adrian et al., 1985), suggesting it has a role as a satiety signal. Commensurate with this, PYY-null mice are hyperphagic and obese (Batterham et al., 2006). Peripheral injection of PYY(3–36) in rodents reduces food intake; similarly, infusion of PYY(3–36) to mimic postprandial levels reduces caloric intake in both lean and obese humans (Batterham et al., 2002; Batterham, Cohen, et al., 2003; Chelikani et al., 2006).

PYY secretion varies with nutrient ingestion: release is greater following a high-protein meal than a high-fat or high-carbohydrate meal, which may explain the finding that volunteers report greatest satiety after a high-protein meal (Batterham et al., 2006). Prolonged exposure to a high-protein diet also increases basal and postprandial PYY levels compared to normal protein diets (Batterham et al., 2006).

PYY(3–36) produces its anorectic effect through the arcuate nucleus of the hypothalamus: peripherally administered PYY(3–36) increases immunohistochemical staining for c-*fos* (a marker of neuronal activation) within the POMC neurons, while NPY mRNA levels were reduced (Abbott et al., 2005; Batterham et al., 2002; Koda et al., 2005). Direct injection of PYY(3–36) into the arcuate reduces feeding, while injection into the PVN does not, confirming the anatomical specificity of PYY (3–36) action (Batterham et al., 2002). PYY(3–36) activates the Y2 receptor, and the anorectic effects of PYY(3–36) are abolished in Y2-null mice (Batterham et al., 2002). The anorectic effects are also abolished by vagotomy, suggesting that the vagus nerve conveys the effect of PYY(3–36) from the periphery (Abbott et al., 2005; Koda et al., 2005).

3.2. Pancreatic polypeptide

Another member of the PP-fold peptide family, this 36-amino acid polypeptide is secreted by the PP cells of the endocrine pancreas with small amounts from the distal gut (Adrian et al., 1976). PP binds to and activates the Y4 receptor (Gehlert et al., 1996). These receptors are widely distributed through the brain, notably, the hypothalamus, forebrain, and brainstem (Whitcomb, Puccio, Vigna, Taylor, & Hoffman, 1997), although the most significant binding of PP appears to be in the area postrema (Dumont, Moyse, Fournier, & Quirion, 2007).

PP is released in response to oral but not intravenous nutrients (Adrian et al., 1977). PP slows gastric emptying (Asakawa et al., 2003; Schmidt et al., 2005). It also acts as a satiety signal, and peripheral administration of PP reduces appetite and produces a sustained reduction in food intake in both humans and rodents (Asakawa et al., 1999, 2003; Batterham, Le Roux, et al., 2003). The anorectic effect has been confirmed in mice that overexpress PP in the pancreatic islets: they eat less and gain less weight than controls, to the extent that there is a lethal lack of milk intake in half of the mouse pups (Ueno et al., 1999).

The route of administration of PP determines its effects on food intake: peripheral administration reduces intake, but direct intracerebroventricular injections stimulate feeding (Asakawa et al., 1999). As *in vivo* PP appears to bind mostly to the area postrema (Dumont et al., 2007), it is unlikely that the activity produced by intracerebroventricular injection is of physiological significance (Field, Chaudhri, & Bloom, 2009; Whitcomb, Taylor, & Vigna, 1990).

Peripheral PP, binding to the area postrema, reduces the expression of the orexigenic neuropeptides NPY and orexin in the hypothalamus, but only in fasted and not fed mice (Asakawa et al., 2003). However, vagotomy prevents the effect of peripherally administered PP, implying that the effects of PP are relayed to the central nervous system via the vagus nerve as well (Asakawa et al., 2003; McTigue & Rogers, 1995). There is a reciprocal relationship with the vagus nerve, as vagal stimulation of the pancreas causes release of PP (Schwartz, 1983; Schwartz et al., 1978).

3.3. Insulin

Insulin is produced by the beta cells in the pancreas. The main role of insulin is the regulation of blood glucose; it causes efficient disposal of plasma glucose by stimulating GLUT4 expression and translocation to the plasma membrane, allowing uptake of glucose into skeletal muscle and adipose tissue. Insulin also causes lipogenesis and stimulates fatty acid uptake into adipose tissue. It can therefore act as a signal of food intake, and in a homeostatic model, it should prompt a reduction in food intake. Administration of insulin in rats has been shown to reduce meal size (Oetting & Vanderweele, 1985; VanderWeele, 1994), while treatment with diazoxide, which suppresses insulin levels, results in increased food intake (VanderWeele, 1994). However, these findings have not been replicated consistently, with some studies showing insulin causes an initial transient increase in food intake (Oetting & Vanderweele, 1985), and yet others showing a persistent increase in food intake and body weight in both rats and humans after insulin administration (Rodin, Wack, Ferrannini, & DeFronzo, 1985; Willing, Walls, & Koopmans, 1994). The cause of these differing results cannot adequately be explained.

It was initially assumed that insulin would only work in the periphery, as it is too large to cross the blood–brain barrier. Subsequent studies have, however, shown an insulin receptor-facilitated, saturable transport mechanism that enables insulin to cross the blood–brain barrier (Baura et al., 1993). Central administration of insulin reduces food intake in baboons, rats, and humans (Hallschmid et al., 2004; McGowan, Andrews, & Grossman, 1992; Van Dijk et al., 1997; Woods, Lotter, McKay, & Porte, 1979), while blockage of central insulin receptors causes hyperphagia and increased body fat in rats (McGowan et al., 1992; Obici, Feng, Karkanias, Baskin, & Rossetti, 2002). The site of insulin action in the central nervous system is the arcuate nucleus of the hypothalamus (Marks, Porte, Stahl, & Baskin, 1990; McGowan et al., 1992). Transport of insulin across the blood–brain barrier only occurs in energy replete states, meaning that, in energy-deficient states, insulin has less of a satiating effect (Strubbe, Porte, & Woods, 1988).

Central insulin administration also produces effects outside the central nervous system. Rats given intracerebroventricular insulin required a higher rate of peripheral glucose infusion to maintain euglycemia, compared to control animals not given insulin centrally, though peripheral plasma insulin levels were the same (Obici, Zhang, Karkanias, & Rossetti, 2002). In an insulin-clamp experiment, hyperinsulinemia reduced hepatic glucose production to a lesser extent after central insulin receptors were blocked than in control mice (Obici, Feng, et al., 2002). These data imply that central insulin administration reduces peripheral insulin resistance and that hypothalamic insulin resistance could contribute to hyperglycemia seen in type 2 diabetes (Obici, Feng, et al., 2002; Obici, Zhang, et al., 2002). Some of the peripheral action of insulin is mediated through the vagus nerve, as the effects of centrally administered insulin on hepatic glucose production are also abolished by vagotomy (Pocai et al., 2005).

3.4. Glucagon

Shortly after the discovery of insulin, an antagonistic hyperglycaemic factor was postulated and named glucagon (Kimball & Murlin, 1923), though it was not purified until 1955 (Staub, Sinn, & Behrens, 1955). This 29-amino acid polypeptide is produced from the cleavage of proglucagon in the alpha cells of the islets of Langerhans. Glucagon acts on a class 2 G-protein-coupled receptor which is structurally related but distinct to those stimulated by the other products of proglucagon cleavage, glucagon-like peptide-1 (GLP-1) and glucagon-like peptide-2 (GLP-2) (Mayo et al., 2003).

Glucagon opposes the effects of insulin by increasing hepatic glucose output via hepatic glycogenolysis or gluconeogenesis. Glucagon release is stimulated by hypoglycemia (Havel & Valverde, 1996), and also by amino acids (Rocha, Faloona, & Unger, 1972), though it is suppressed by fatty acids (Gerich et al., 1976).

Peripheral administration of glucagon reduces food intake in humans and rodents (Geary, Kissileff, Pi-Sunyer, & Hinton, 1992; Le Sauter & Geary, 1991; Martin & Novin, 1977; Schulman et al., 1957), while immunoneutralization of glucagon increases meal size in rats (Geary, Le Sauter, & Noh, 1993; Le Sauter, Noh, & Geary, 1991). Exogenously administered glucagon also increases energy expenditure (Tan et al., 2012). Glucagon-receptor knockout mice have reduced adiposity, though the effect on food intake and energy expenditure in such animals is variable (Conarello et al., 2007; Gelling et al., 2003). The anorectic effect of glucagon may appear counterintuitive for a hormone that normally acts to increase blood glucose, but it is compatible with glucagon's role as a stress hormone, where an anorectic effect might be advantageous as part of a "fight or flight" response (Jones, Tan, & Bloom, 2012).

The satiating effect of glucagon occurs both centrally and peripherally. Central administration of glucagon suppresses feeding to a greater degree than peripheral administration in rats, chicks, and sheep (Honda et al., 2007; Inokuchi, Oomura, & Nichimura, 1984; Kurose et al., 2009). The central site of glucagon action is the hypothalamus, where glucagon inhibits neurons within multiple hypothalamic nuclei (Inokuchi et al., 1986). In the periphery, glucagon acts on the liver: infusion of glucagon into the hepatic portal vein suppresses feeding at levels that fail to do so when infused into the inferior vena cava (Geary et al., 1993). The effect of peripheral glucagon is also conveyed centrally: vagotomy prevents the suppression of feeding caused by peripherally administered glucagon (Geary et al., 1993; Le Sauter & Geary, 1990; Martin, Novin, & Vanderweele, 1978), as do lesions to the area postrema and the NTS (Weatherford & Ritter, 1988). The vagus nerve also stimulates glucagon release through its innervation of the pancreas (Ahrén & Taborsky, 1986); the cephalic phase of glucagon release is lost from transplanted pancreata, due to loss of these innervations (Secchi et al., 1995).

3.5. Glucagon-like peptide-1

GLP-1 is a post-translational product of the proglucagon precursor in L cells of the intestine. GLP-1 exists in two forms in the plasma: GLP(1–37) and GLP(7–37). Only the latter has biological activity (Mosjov, Weir, & Habener, 1987; Komatsu et al., 1989). Degradation of GLP-1 occurs very rapidly *in vivo* by DPP-4 (Kieffer, McIntosh, & Pederson, 1995).

Release of GLP-1 is partly activated by direct contact of nutrients with the L cells in the intestine through taste transduction systems (Jang et al., 2007) and partly through a complex neuroendocrine circuit involving the vagus nerve and another gut hormone, glucose-dependent insulinotropic peptide (GIP) (Roberge & Brubaker, 1993; Rocca & Brubaker, 1999). Leptin also stimulates GLP-1 release (Anini & Brubaker, 2003).

Oral glucose induces a much larger rise in insulin (between two- and three-fold) than with the same amount of glucose given intravenously: this is termed the incretin effect (Perley & Kipnis, 1967; Vilsbøll & Holst, 2004). GLP-1, together with GIP (see below), causes the incretin effect. In addition to directly stimulating insulin release from the pancreas (Mosjov et al., 1987; Van Dijk et al., 1996), GLP-1 alters insulin gene expression, enhances beta cell proliferation, and reduces cell apoptosis (Drucker, Philippe, Moisoy, Chick, & Habener, 1987; Farilla et al., 2002). It also suppresses glucagon release, but only in euglycemia (Nauck et al., 2002; Van Dijk et al., 1996).

GLP-1 also regulates energy intake and satiety. An infusion of GLP-1 into normal-weight humans causes an increase in the subjective sensation of fullness after a set meal and reduces food intake (Flint, Raben, Astrup, & Holst, 1998); it also reduces gastric emptying (Willms et al., 1996). The mechanism by which GLP-1 causes satiety and reduces food intake is uncertain, but postulated mechanisms include through activation of vagal afferents, through direct activation of central nervous system targets, and even through induction of nausea (Chaudhri et al., 2006; Hayes et al., 2011; Kanoski, Rupprecht, Fortin, De Jonghe, & Hayes, 2012; Parkinson et al., 2009; Rocca & Brubaker, 1999; Kanoski et al., 2011).

3.6. Glucose-dependent insulinotropic polypeptide

Initially called gastric inhibitory polypeptide due to its inhibitory effects on gastric acid secretion in dogs (Brown, Dryburgh, Ross, & Dupré, 1975), this 42-amino acid polypeptide has been renamed glucose-dependent insulinotropic polypeptide to reflect its role in glucose metabolism. GIP is released from K cells, found mainly in the jejunum (Buchan, Polak, Capella, Solcia, & Pearse, 1978; Mortensen, Christensen, Holst, & Orskov, 2003), and acts upon G-protein-coupled receptors found on the beta cells of the pancreas and in small numbers on many other tissues including bone, brain, stomach, thyroid, and adipose tissue (Baggio & Drucker, 2007; Girard, 2008). As with GLP-1, GIP is inactivated very rapidly *in vivo* by DPP-4.

Release of GIP is triggered by K cell sensing of nutrients in the intestinal lumen (Cataland, Crockett, Brown, & Mazzaferri, 1974). The nutrients which trigger GIP release are species specific: in humans, it is predominantly in response to carbohydrates and lipids (Elliott et al., 1993; Rijkelijkhuizen et al., 2010). The vagus nerve does not appear to affect GIP release in animals (Berthoud, Trimble, & Moody, 1982; Ohneda et al., 1985); in humans, there is some evidence that vagal stimulation may cause GIP release, but this has not been consistently demonstrated (Lauritsen, Moesgaard, Nielsen, & Jensen, 1982; Taylor & Feldman, 1982).

As an incretin, GIP potentiates glucose-stimulated insulin secretion; it also promotes insulin biosynthesis, pancreatic beta cell proliferation, and survival (Drucker, 2006; Ehses et al., 2003). GIP also affects lipid metabolism by stimulating lipoprotein lipase to release fatty acids from chylomicron particles carrying ingested fats, which can then be incorporated into adipose tissue (Eckel, Fujimoto, & Brunzell, 1979; Girard, 2008). Consistent with this, GIP-receptor knockout mice are resistant to diet-induced obesity (Miyawaki et al., 2002). Through these effects on lipid metabolism, GIP indirectly influences energy balance and body weight; however, direct infusion of GIP into healthy male subjects does not affect appetite, energy intake, or energy expenditure (Asmar et al., 2010).

3.7. Oxyntomodulin

Oxyntomodulin is a 37-amino acid polypeptide, consisting of the glucagon molecule plus a C-terminal extension (Bataille et al., 1981). It is produced by ileal L cells after cleavage of proglucagon by prohormone convertase 1 (Dhanvantari, Seidah, & Brubaker, 1996); mice lacking prohormone convertase 1 have a reduction in oxyntomodulin, GLP-1, GLP-2, and glicentin levels (Ugleholdt et al., 2004). As with GLP-1, oxyntomodulin is inactivated by DPP-4 (Zhu et al., 2003).

Endogenous oxyntomodulin appears to work as a satiety signal, rising after a meal (Ghatei, Uttenthal, Christofides, Bryant, & Bloom, 1983; Le Quellec, Kervran, Blache, Ciurana, & Bataille, 1992). Exogenous oxyntomodulin transiently reduces food intake in both rodents and humans (Baggio, Huang, Brown, & Drucker, 2004; Cohen et al., 2003; Dakin et al., 2002; Parkinson et al., 2009). The central mediation of oxyntomodulin's effect is through the arcuate and paraventricular nuclei of the hypothalamus (Baggio et al., 2004; Dakin et al., 2004). Oxyntomodulin also reduces circulating ghrelin levels; it may be this interaction that is behind the postprandial fall in ghrelin levels (Cohen et al., 2003; Dakin et al., 2002; Patterson et al., 2009).

Oxyntomodulin works via both the GLP-1 and glucagon receptors (Baggio et al., 2004), though with much lower potency than either cognate ligand (Dakin et al., 2001). This dual-agonism may explain the range of actions of oxyntomodulin. Oxyntomodulin is insulinotropic (Schjoldager, Baldissera, Mortensen, Holst, & Christiansen, 1988), and it blunts the rise in plasma glucose levels that occurs in response to either an oral or an intravenous glucose challenge (Maida, Lovshin, Baggio, & Drucker, 2008; Parlevliet et al., 2008). This effect is lost in GLP-1-receptor knockout mice, suggesting it is through the GLP-1 receptor that oxyntomodulin exerts this effect (Maida et al., 2008). However, when the structure of oxyntomodulin is altered to lose its affinity for the glucagon receptor, the weight loss in rodents is reduced, though food intake remains the same (Kosinksi, Hubert, Carrington, et al., 2012). This suggests that the glucagon receptor is involved in weight loss effects of oxyntomodulin, probably through increasing energy expenditure. Peripheral administration of oxyntomodulin and GLP-1 also produces distinct patterns of brain activity on manganese-enhanced MRI (MEMRI), suggesting the activation of different pathways (Parkinson et al., 2009).

3.8. Amylin

Also known as islet amyloid polypeptide, this 37-amino acid peptide is cosecreted from the beta cells with insulin, though in much smaller amounts. It is derived from cleavage of an 89-residue prohormone (Sanke, Bell, Sample, Rubenstein, & Steiner, 1988) by prohormone convertases PC2 and PC3 (Higham et al., 2000).

Unsurprisingly for a hormone cosecreted with insulin, amylin has glucoregulatory effects, stimulating glycogenolysis and inhibiting glucose uptake, and reducing glucagon secretion (Silvestre et al., 2001; Young, 2005; Young,

Deems, Deacon, McIntosh, & Foley, 1990; Young et al., 1993). Amylin additionally acts as a satiety signal: infusion into both lean and obese rodents causes a reduction in food intake and a loss of body weight through reduction in adipose tissue (Roth, Hughes, Kendall, Baron, & Anderson, 2006).

Amylin acts on the brain stem, activating the NTS and the area postrema (Mack et al., 2010; Riediger, Zuend, Becskei, & Lutz, 2004); this activity is then transmitted to the arcuate nucleus, where anorexigenic POMC/CART neurons are stimulated (Roth et al., 2006).

3.9. Cholecystokinin

Secreted by the I cells of the proximal intestine, cholecystokinin (CCK) is actually a family of peptides produced from preprocholecystokinin, the product of the *CCK* gene. Each member is denoted by the number of amino acids it contains, and CCK8, CCK22, CCK33, and CCK 58 are the predominant forms found in the plasma (Zwirska-Korczala et al., 2007). CCK has two seven-transmembrane domain G-protein-coupled receptors, denoted in various ways, but most simply, CCK1R and CCK2R (Dufresne, Seva, & Fourmy, 2006).

CCK affects multiple physiological processes, including stimulating pancreatic secretions, gallbladder contraction, gastric emptying, and colonic motility (Berna & Jensen, 2007). CCK is secreted in response to lipid ingestion (Pilichiewicz et al., 2006) and has a role in lipid processing: CCK-knockout mice not only are resistant to diet-induced obesity but also demonstrate defects in fat absorption (Lo et al., 2010). CCK also acts as a satiety signal. Exogenous administration suppresses feeding in rats (Gibbs, Young, & Smith, 1973; Hewson, Leighton, Hill, & Hughes, 1988), while antibodies and antagonists to CCK lead to increased food intake and body weight in rodent models (Hewson et al., 1988; McLaughlin, Baile, & Buonomo, 1985; Meereis-Schwanke, Klonoswski-Stumpe, Herberg, & Niederau, 1998).

CCK acts both centrally and peripherally. CCK activates the vagus nerve (Schwartz & Moran, 1994; Zwirska-Korczala et al., 2007), and leptin works through the same fibers to enhance the satiating effect of CCK (Peters, Simasko, & Ritter, 2006). Direct injections of CCK8 into the hypothalamus and into the hindbrain also suppress food intake (Blevins, Stanley, & Reidleberger, 2000), while in vagotomized rats, a CCK antagonist will stimulate feeding but only if the antagonist can cross the blood–brain barrier (Reidleberger, Hernandez, Fritzsch, & Hulce, 2004). The anorectic effects of CCK are mediated via the CCK-1 receptor (Reidleberger et al., 2004).

3.10. Leptin

Leptin, the product of the *obese (ob)* gene, is produced predominantly by adipose tissue and is more accurately described as an adipokine than a gut hormone. Serum levels of leptin are higher in obese compared to lean subjects, and leptin levels fall with weight loss (Considine et al., 1996; Hamilton, Paglia, Kwan, & Deitel, 1995; Maffei et al., 1995). Leptin has an important role in obesity, and its interactions with gut hormones ensure it merits mention in this discussion of obesity.

Leptin was initially thought to act as a satiety signal after parabiotic experiments (Coleman, 1973), and administration of exogenous leptin does reduce feeding in normal-weight subjects (Flynn, Scott, Pritchard, & Plata-Salamán, 1998). Mutations in the *ob* gene cause a small number of genetically determined cases of obesity associated with hyperphagia (Farooqi, Wangensteen, et al., 2007; Montague et al., 1997; Zhang et al., 1994). Leptin also affects food preference, reducing carbohydrate and protein intake, and increasing fat intake (Licinio et al., 2007; Wetzler, Dumaz, Goubern, Tomé, & Larue-Achagiotis, 2004). However, leptin levels do not change postprandially (Drewes et al., 1997), arguing against leptin being a short-term satiety signal. Rather, it is a marker of adiposity and informs the body of long-term energy supply. It has a particular role in mediating the physiological response to starvation, during which there is downregulation of the adrenal, thyroid, and gonadal axes to conserve energy. As starvation resolves, leptin increases and switches back on these downregulated processes. Experimentally, administering leptin to starving mice prevents these normal adaptive responses, implying it acts as a signal of adequate energy stores (Ahima et al., 1996). A relative deficiency in leptin may be responsible for the hypothalamic hypogonadism seen in patients with anorexia nervosa, or who have low body weights due to exercise (Blüher & Mantzoros, 2007; Scheid & De Souza, 2010).

Leptin acts through a single membrane-spanning receptor similar to the class 1 cytokine receptor family (Tartaglia et al., 1995), encoded by the *diabetes (db)* gene. Leptin receptors are found in multiple brain areas, including the choroid plexus, hindbrain, cortex, hypothalamus, and leptomeninges (Mercer et al., 1996; Scott, Williams, Rossi, Lee, & Elmquist, 2011; Tartaglia et al., 1995). Within the arcuate nucleus, leptin stimulates anorexigenic POMC/CART neurons and reduces activity in orexigenic NPY/AgRP neurones (Cowley et al., 2001; Wang et al., 2008). Leptin receptors are also found on the vagus nerve (Buyse et al., 2001), and leptin activates vagal afferents (Peters et al., 2006), while the reduction in food intake caused by peripheral leptin administration is stopped by vagotomy in rats (Peters et al., 2006).

3.11. Ghrelin

Only one peripheral hormone is known to stimulate food intake: ghrelin, a 28-amino acid peptide that is *n*-octanoylated on its third residue (Kojima et al., 1999). It is produced by X/A-like cells of the stomach (Date et al., 2000), the octanoylation being catalyzed by the enzyme ghrelin-O-acyltransferase (GOAT; Yang, Brown, Liang, Grishin, & Goldstein, 2008). Ghrelin was originally characterized on the basis of its growth hormone secretagogue activity and acts as the endogenous ligand to the GHS receptor, a G-protein-coupled receptor, found throughout the body, including the pancreas, intestine, reproductive system, pituitary, and spinal cord (Ueberberg, Unger, Saeger, Mann, & Petersenn, 2009). Ghrelin circulates in two forms, the *n*-octanoylated form (ghrelin) and a des-*n*-octanyl form (des-acyl ghrelin); it is only ghrelin with its *n*-octanoylated residue that activates the GHS receptor (Hosoda, Kojima, Matsuo, & Kangawa, 2000).

Endogenous ghrelin acts as a trigger to start eating, rising in fasting to peak just before meals, and falling postprandially (Cummings, Frayo, Marmonier, Aubert, & Chapelot, 2004; Cummings et al., 2001; Tschöp, Wawarta, et al., 2001). Exogenous administration also induces feeding and increases appetite, which leads to increased body weight, in both rodents and humans (Wren, Seal, et al., 2001; Wren, Small, et al., 2001; Wren et al., 2000; Nakazato et al., 2001; Neary et al., 2004).

As with leptin, ghrelin is essential for survival in starvation. The rise in ghrelin when fasting triggers growth hormone release with a subsequent protection of plasma glucose levels. Without the rise in ghrelin, blood glucose falls to dangerously low levels in starvation (Goldstein et al., 2011; Zhao et al., 2010). Ghrelin further protects against an energy-deficient environment by reducing spontaneous physical activity, and therefore energy output (Pfluger et al., 2011).

When given centrally, ghrelin acts predominantly through the arcuate nucleus of the hypothalamus (Wren, Small, et al., 2001); however, vagotomy prevents the increase in food intake seen after peripheral ghrelin administration, suggesting that the vagus nerve transmits ghrelin's signals from the periphery (Le Roux et al., 2005).

4. INTEGRATION OF GUT HORMONE SIGNALS

There are therefore a multitude of gut hormones that inform the body of its current nutritional status. These signals need to be integrated to provide a coordinated drive to seek and consume food when in an energy-deficient

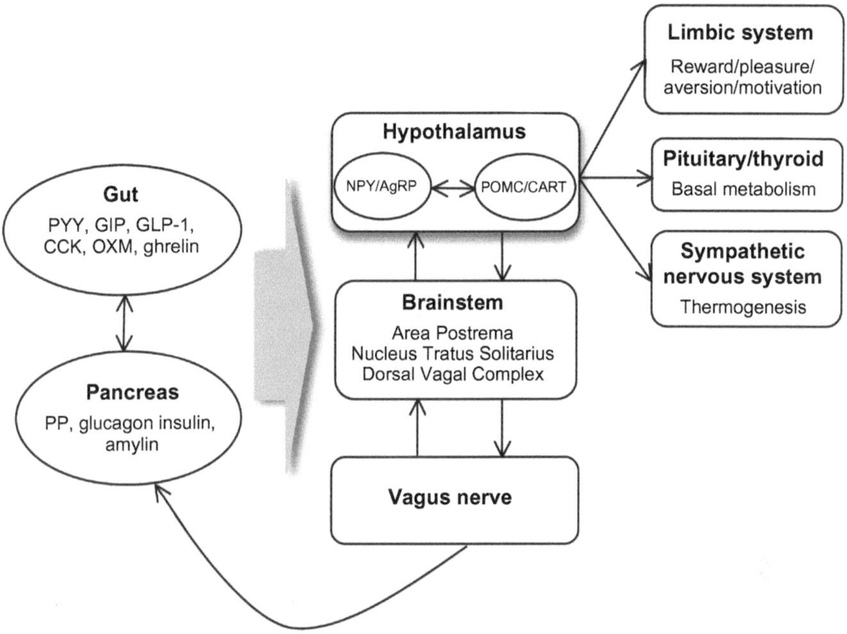

Figure 7.1 How gut hormones can influence energy balance. Gut hormones are produced by the intestine, stomach, and pancreas: PP, amylin, insulin, oxyntomodulin (OXM), GLP-1, GIP, PYY, and CCK are released in response to nutrient ingestion; ghrelin and glucagon in response to fasting. They alter food intake through actions on the vagus nerve, the brainstem, and the hypothalamus. They also influence food preference and the rewarding properties of food. Response to these signals is not only through direct alteration in food intake but also through affects on metabolic rate and through vagal stimulation of different intraabdominal organs.

state or to stop eating when the body is energy replete. This integration occurs in different levels: direct interactions between the gut hormones; at the level of the vagus nerve; and in the CNS, in particular, the hypothalamus and brainstem (Fig. 7.1).

4.1. The effect of gut hormones on the production and release of each other

Gut hormones can affect the production and release of other hormones. This can reinforce or suppress hormonal actions on energy balance and does not always take the form of a reciprocal relationship.

The best characterized network of interactions concerns insulin. Insulin reduces blood glucose levels, an effect counteracted by glucagon. Insulin acts on the alpha cells of the pancreas to suppress glucagon synthesis and release

(Maruyama, Hisatomi, Orci, Grodsky, & Unger, 1984; Philippe, 1989); in contrast, glucagon stimulates insulin production within the beta cells (Moens et al., 1996), an adaptive response in fasting to ensure efficient disposal of any glucose that becomes available.

The incretins, and to an extent, amylin, also stimulate insulin production and release in response to food ingestion and the presence of nutrients within the gut lumen (Drucker et al., 1987; Farilla et al., 2002; Fehmann et al., 1991; Mosjov et al., 1987; Van der Burg, Guicherit, Frölich, & Gooszen, 1995). This allows efficient disposal of intraintestinal nutrients, and the satiety effect of the incretins augments that of insulin. GLP-1 also increases insulin gene expression and enhances pancreatic cell proliferation and survival (Drucker et al., 1987; Farilla et al., 2002). CCK also stimulates insulin release from beta cells (Fehmann et al., 1991; Verspohl, Ammon, Williams, & Goldfine, 1986), while CCK-knockout mice show a reduced insulin response to glucose (Lo et al., 2011). Insulin sensitizes the body to the satiety effects of CCK, causing a greater suppression of food intake in rats when insulin and CCK8 are coadministered, than following CCK8 alone (Riedy, Chavez, Figlewicz, & Woods, 1995).

In contrast to these signals of energy sufficiency, ghrelin, a signal of energy deficiency, reduces insulin release (Date et al., 2002; Dezaki et al., 2004, 2006), while ghrelin receptor antagonists increase insulin release in response to glucose (Dezaki et al., 2004, 2006). As ghrelin's effects on insulin secretion are dose dependent, it reduces the satiating effects of insulin only in an environment of energy deficiency when endogenous ghrelin levels are high (Salehi, Dornonville de la Cour, Håkanson, & Lundquist, 2004).

4.2. Integration of information by the vagus nerve

Peripheral information about energy status needs to be integrated and conveyed to the brain to alter behavior. This information includes not only gut hormone levels but also the degree of intestinal distension, the extent of liver stores, and gustatory inputs. The vagus nerve innervates the intra-abdominal viscera as well as the oropharyngeal cavity, allowing all this peripheral information to be relayed to the brain. Efferent branches of the vagus can also produce a peripheral response to central brain signals. It is therefore the ideal structure to integrate gut hormone activity.

Many of the gut hormones modulate vagal nerve activity. This includes CCK (Dockray, 2012; Peters et al., 2006), leptin (Peters et al., 2006), and PP (Asakawa et al., 2003). Many of the physiological effects of gut hormones are abolished by vagotomy, including the satiating effects of CCK8 in rats (Smith, Jerome, Cushin, Eterno, & Simansky, 1981); the reduction in food intake and activity in the brain after injection of PYY(3–36) or GLP-1 (Abbott et al., 2005; Koda et al., 2005); the anorectic effects of PP (Asakawa et al., 2003); and the suppression of food intake in rats given insulin (VanderWeele, 1994) and glucagon (Geary et al., 1993; Le Sauter & Geary, 1990; Martin et al., 1978).

There is also a direct interaction between gut hormones at the level of the vagus. Leptin potentiates the ability of CCK to stimulate the transcription factor early-growth response factor-1 (EGFR1) within the vagus, which subsequently increases transcription of the anorectic neuropeptide CART (De Lartigue et al., 2010); in contrast, ghrelin inhibits the increase in EGFR1 in response to CCK (De Lartigue et al., 2010).

The vagus is also an effector of central control of energy homeostasis. The vagus nerve stimulates release of satiety signals in response to a meal, including PYY(3–36), PP, and GLP-1 (McTigue & Rogers, 1995; Roberge & Brubaker, 1993; Rocca & Brubaker, 1999; Schwartz, 1983). It is also responsible for the cephalic phase of glucagon secretion, a phenomenon lost in pancreas transplant patients due to denervation of the pancreas (Secchi et al., 1995). The vagus is also involved in complex neuroendocrine loops involving gut hormones, for example, mediating the GIP-induced stimulation of GLP-1 release (Rocca & Brubaker, 1999). Furthermore, it mediates the effect of centrally administered insulin on hepatic glucose production (Pocai et al., 2005).

Obesity affects the function of the vagus nerve: the vagus is less responsive to the effects of CCK, leptin, and intestinal distension in obese subjects (Daly, Park, Valinsky, & Beyak, 2011; De Lartigue, Barbier de la Serre, Espero, Lee, & Raybould, 2011). This alteration in responsiveness may underlie the development of relative hyperphagia in obesity (Dockray, 2012).

4.3. Integration of gut hormone signals within central structures

As mentioned above, the hypothalamus, in particular the arcuate nucleus, is an important structure in controlling energy balance. How the hypothalamus is affected by gut hormones has been widely investigated, through functional imaging as well as immunohistochemical and behavioral studies, which shall be discussed in the following examples.

Central injection of ghrelin increases the expression of NPY and AgRP mRNA within the arcuate nucleus (Kamegai et al., 2001; Nakazato et al., 2001). Intraperitoneal injections of ghrelin similarly induce expression of c-*fos* in the arcuate, PVN, and DMH (Kobelt et al., 2008), and functional imaging with MEMRI showed *in vivo* activation of the arcuate, PVN, and VMN by ghrelin (Kuo et al., 2007). Ghrelin therefore increases the activity of the orexigenic NPY/AgRP neurones relative to the anorexigenic POMC/CART neurones within the arcuate nucleus.

Studies with MEMRI have shown that the anorexigenic hormones PP and PYY(3–36) reduce activity in the arcuate nucleus (Hankir et al., 2011; Kuo et al., 2007), with additional reduction in signal intensity in the PVN and VMH after PP administration (Hankir et al., 2011). Peripheral administration of PYY(3–36) increases c-*fos* within the arcuate but reduces NPY mRNA, and direct injection of PYY(3–36) into the arcuate inhibits food intake (Batterham et al., 2002). PYY(3–36) therefore appears to swing the balance of activity in favor of the anorexigenic POMC/CART neurones as opposed to the NPY/AgRP neurones. Interestingly, injecting PYY (3–36) into the paraventricular nucleus stimulates feeding in a similar way to NPY administration (Stanley, Daniel, Chin, & Leibowitz, 1985)—it may be that, in this situation, PYY(3–36) is acting on Y1 rather than Y2 receptors and therefore stimulating feeding.

Intracerebroventricular administration of GLP-1 inhibits food intake by attenuating the normal activation of orexigenic NPY/AgRP and inhibition of anorexigenic POMC/CART neurons during fasting (Seo, Ju, Chung, Lee, & Park, 2008). Functional imaging using PET scanning shows that in humans fed to satiety, the peak in endogenous GLP-1 levels correlates with increased blood flow in the hypothalamus and the dorsolateral prefrontal cortex (Pannacciulli et al., 2007). In rodent MEMRI studies, GLP-1 increases activity within the VMN but reduces signal intensities within the PVN (Chaudhri et al., 2006; Parkinson et al., 2009). Oxyntomodulin produces similar expression of c-*fos* in the PVN as GLP-1, consistent with its activity at the GLP-1 receptor (Baggio et al., 2004), but in functional imaging, oxyntomodulin administered to fasted mice showed reduction in signal intensity in the arcuate and supraoptic nuclei of the hypothalamus, rather than the PVN or the VMN (Parkinson et al., 2009); thus, even though these structurally related hormones have similar behavioral effects, they work through distinct pathways within the hypothalamus.

The brainstem is also involved in the integration of gut hormone signals. Injection of CCK8 directly into the brainstem reduces feeding (Blevins et al., 2000), while lesions of the NTS abolish the reduction in food intake caused by amylin in rats (Lutz, Mollet, Rushing, Riediger, & Scharrer, 2001). Higher doses of exogenously administered amylin do still have a satiating effect, suggesting that amylin works through multiple pathways (Lutz et al., 2001). Exogenous administration of the satiety signals GLP-1 and oxyntomodulin both increase activity within the area postrema on MEMRI (Parkinson et al., 2009), while transection of the brainstem prevents the reduction in feeding normally seen after administration of GLP-1 and PYY(3–36) (Abbott et al., 2005). Interaction occurs between gut hormones in the brainstem as well, with leptin receptors being found on GLP-1-expressing neurons within the NTS; deletion of these receptors results in a hyperphagic, obese mouse (Scott et al., 2011). Ghrelin receptors are also found throughout the brainstem (Zigman, Jones, Lee, Saper, & Elmquist, 2006), and indeed, ghrelin is produced within the cells of the DVC, expression of which is increased during fasting (Huang et al., 2011).

5. INFLUENCE OF GUT HORMONE SIGNALS ON OTHER ASPECTS OF ENERGY HOMEOSTASIS

5.1. Modification of nutrient preferences by gut hormones

Obese patients appear to have a preference for high-fat foods, a phenomenon confirmed in monozygotic twin studies (Rissanen et al., 2002). After gastric bypass surgery, patients typically show an aversion for high-calorie, high-fat, and sweet foods (Halmi, Mason, Falk, & Stunkard, 1981; Kenler, Brolin, & Cody, 1990). Gut hormones may underlie both these phenomena, as there is increasing evidence that gut hormones can modulate taste preference.

As may be expected for a hormone, the role of which is to drive feeding, infusion of ghrelin into rats increases the consumption of sweet substances, even liquid sucrose over solid chow, a preference abolished in animals lacking ghrelin receptors or who have received a GHSR antagonist (Disse et al., 2010; Egecioglu et al., 2010). How ghrelin modulates food preferences is, however, not understood.

GLP-1 and glucagon are both found within the taste cells of mice, rats, and monkeys, and GLP-1 receptors are found on afferent nerve fibers close to the taste cells (Shin et al., 2008). GLP-1-knockout mice show hypersensitivity to aversive stimuli but reduced sensitivity to two different sweeteners, suggesting GLP-1 also modulates preference for sweet food (Shin et al., 2008; Martin et al., 2009). Taste cells are also found lower down in the gut where they are responsible for nutrient-dependent hormone release: the taste protein gustducin, found in intestinal L cells, is responsible for glucose-related GLP-1 and PYY release (Jang et al., 2007; Steinert et al., 2011). CCK and NPY are also found within oropharyngeal taste cells in rats; within these, NPY enhances an inward-rectifying potassium current, an action that is antagonized by CCK (Herness, Zhao, Lu, Kaya, & Shen, 2002; Shen et al., 2005; Zhao et al., 2005). Thus, the different gut hormones can stimulate different taste experiences and modulate nutrient preferences.

5.2. Effects on energy expenditure

Energy balance is that which is struck between calories consumed and energy expended. Energy expenditure consists of three main processes: basal metabolic rate, diet-induced thermogenesis, and physical activity (Westerterp, 2004). Each of these processes is affected by gut hormones.

Diet-induced thermogenesis is the phenomenon of heat production within brown adipose tissue, in response to food intake, and is produced by sympathetic activation of the brown adipose tissue (Landsberg, Saville, & Young, 1984). This has been demonstrated in rats and humans (Rothwell & Stock, 1979; Schutz, Bessard, & Jéquier, 1984). Diet-induced thermogenesis is blunted in obesity (Bessard, Schutz, & Jéquier, 1983; Schutz et al., 1984). Whether reduced thermogenesis is a cause or consequence of obesity has not yet been established, but as studies have shown in some women that postprandial thermogenesis does not improve with weight loss (Bessard et al., 1983; Jéquier & Schutz, 1985), it may be that abnormalities in diet-induced thermogenesis predispose people to obesity.

Satiety signals increase diet-induced thermogenesis, and thus energy expenditure: transgenic mice that overexpress PYY have significantly higher body temperatures than wild-type mice, which cause a reduction in body fat despite no change in food intake (Boey et al., 2008), and administration of PP and oxyntomodulin produced thermogenesis (Asakawa et al., 2003; Dakin et al., 2002). Leptin also increases activation within the sympathetic innervation of brown adipose tissue (Haynes, Morgan, Djalali, Sivitz, & Mark, 1999; Haynes, Morgan, Walsh, Mark, & Sivitz, 1997). Glucagon also increases thermogenesis, an effect seen in both rodent and man (Cockburn, Hull, & Walton, 1967; Joel, 1966; Nair,1987; Tan et al., 2012). This effect on energy expenditure is blunted by concurrent infusion of insulin (Calles-Escandón, 1994).

While satiety signals increase thermogenesis, ghrelin reduces thermogenesis to preserve energy levels. This has been confirmed *in vivo*, as higher endogenous ghrelin levels in healthy women are inversely correlated with the magnitude of diet-induced thermogenesis (St-Pierre et al., 2004), while mice lacking both ghrelin and the GHS receptor have an increased core temperature (Pfluger et al., 2008).

Energy expenditure through physical activity is also influenced by gut hormones. Central administration of both PYY(3–36) and NPY transiently increases spontaneous physical activity in rats (Pfluger et al., 2011); in humans, oxyntomodulin increases activity-related energy expenditure, which contributes to its weight loss effects (Wynne et al., 2006). In contrast, administration of ghrelin reduces spontaneous physical activity both acutely and chronically, commensurate with its role in preserving the body's energy reserves (Pfluger et al., 2011; Tang-Christensen et al., 2004).

The HPT axis also regulates energy expenditure and basal metabolic rate. Thyroid hormones increase the metabolic rate of most tissues, increasing energy expenditure and thermogenesis (Kim, 2008). In physiological states, the HPT axis also responds to the energy status of the body: high-fat diets increase activity of the HPT axis to increase energy expenditure (Boey et al., 2008), and there is downregulation of the HPT axis in starvation to reduce metabolic rate (Ahima et al., 1996). The arcuate nucleus is implicated in modulation of the HPT axis, as lesions of the arcuate prevent the normal fasting-related changes in the HPT axis (Legradi et al., 1998).

The HPT axis is altered by gut hormones. In rats, ghrelin suppresses the thyroid axis and thus reduces energy expenditure (Sosić-Jurjević, Stevanović, Milosević, Sekulić, & Starcević, 2009). In humans, ghrelin produces an acute increase in thyroxine and reduction in TSH levels (Kluge et al., 2010). This suggests that, at least initially, ghrelin activates the HPT axis; however, due to the short length of the study, the investigators could not exclude that in the long term, the reduction in TSH following ghrelin would suppress the HPT axis (Kluge et al., 2010). NPY, through Y1 and Y5 receptors, also suppresses the HPT axis and reduces peripheral thyroxine levels, fitting with its role as a signal of energy deficit (Costa-e-Sousa et al., 2011; Fekete et al., 2001, 2002). Conversely, signals of energy excess activate the thyroid axis. PYY(3–36) increases TSH release when administered to rats (Oliveira et al., 2006), and leptin also activates the HPT axis (Ahima et al., 1996; Fekete et al., 2006).

5.3. The role of gut hormones in the hedonistic drive to obesity

The hypothalamus mediates the homeostatic control of energy balance. However, this model does not account for the desirability of food. There is increasing understanding that the hedonic qualities of eating can drive obesity (Stice et al., 2009). Implicated in this are the dopaminergic reward pathways involving the ventral tegmental area (VTA), the nucleus accumbens, the amygdala, and the orbitofrontal cortex (Fulton, 2010). The importance of the reward pathways in obesity is illustrated by the following: dopamine, a sign of reward, increases in the brains of normal-weight rats in response to both standard chow and fatty foods, whereas obese rats only show dopamine release in response to the high-fat diet (Geiger et al., 2009). Furthermore rats fed a high-fat diet show reduced learning of cocaine self-administration, a marker of reward-pathway activity, compared to those fed standard chow (Wellman, Nation, & Davis, 2007). In humans, the relationship between obesity and sensitivity to the rewarding aspects of food is not straightforward: people who are overweight have an increased sensitivity to food reward compared to lean subjects, but those with moderate or morbid obesity have reduced sensitivity to food reward (Davis & Fox, 2008; Davis et al., 2004).

There is now evidence that gut hormones modulate some of these hedonic properties. In healthy volunteers shown pictures of food, administration of ghrelin increases activity on functional MRI in areas of the brain associated with determining the incentive values of food, such as the orbitofrontal cortex and amygdala (Malik, McGlone, Bedrossian, & Dagher, 2008). Direct injection of ghrelin into the VTA in rodents stimulates eating of rewarding food (Egecioglu et al., 2010; Naleid, Grace, Cummings, & Levine, 2005), and ghrelin antagonists reduce the intake of rewarding food in a free choice (standard chow/rewarding food) experimental paradigm (Egecioglu et al., 2010). Infusion of ghrelin also increases the amount that rats are willing to work for a sugar reward, while in rats with high endogenous ghrelin levels after a fast, ghrelin antagonists reduce the work they will do for a sucrose reward (Skibicka, Hansson, Egecioglu, & Dickson, 2012).

Satiety signals also affect the reward pathways of the brain (Figlewicz and Benoit, 2009). GLP-1, when injected into the nucleus accumbens in rats, suppresses feeding (Dossat, Lilly, Kay, & Williams, 2011). Insulin receptors are found within the reward pathway (Figlewicz, Evans, Murphy, Hoen, & Baskin, 2003), and insulin administered into the VTA reduces feeding (Figlewicz, Bennett, Aliakbari, Zavosh, & Sipols, 2008) and elevates brain reward thresholds (Bruijnzeel, Corrie, Rogers, & Yamada, 2011). A similar effect on brain reward thresholds has been found with leptin administration (Bruijnzeel et al., 2011), and leptin receptors are also found on neurons in areas associated with reward processing within the brain (Farooqi et al., 2007; Hommel et al., 2006). Leptin-deficient *ob/ob* mice have reduced dopamine expression and function within the nucleus accumbens and VTA (Fulton et al., 2006), though humans with leptin deficiency have a greater "liking" for all food stuffs, which is reduced when they are given leptin replacement (Farooqi, Bullmore, et al., 2007). This suggests that their ability to differentiate the rewarding properties of different food stuffs is reduced in leptin deficiency and leads to hyperphagia. It may be that leptin resistance, and even insulin resistance, reduces the response to the hedonic properties of food in normal obese subjects, who eat more in order to gain the same level of reward (Figlewicz, Bennett, Naleid, Davis, & Grimm, 2006).

PYY(3–36) also affects the rewarding aspects of feeding. Compared to a saline infusion, infusion of PYY(3–36) leads to reduced activity in a number of brain regions, including the amygdala, nucleus accumbens, and orbitofrontal cortex, when fasted subjects are shown food images (De Silva et al., 2011). A similar pattern of activation is seen in fed subjects when shown the same images, as they have high endogenous PYY(3–36) levels. Using blood oxygen level-dependent functional MRI, Batterham et al. (2007) showed that in the fasted state, it is activity in the hypothalamus which correlates with how much the subject will subsequently eat. After an injection of PYY(3–36), activity in the orbitofrontal cortex determines calorie intake. Therefore, the homeostatic feeding center—the hypothalamus—influences feeding in the fasted state; but in the presence of a satiety signal, the areas involved in reward and motivation determine feeding. Interestingly, higher PYY(3–36) levels are associated with features of disordered eating independent of BMI (Lawson et al., 2011).

Further studies into the method by which the reward system causes and maintains obesity are required; they may provide important targets for the treatment of obesity.

6. ALTERATIONS IN GUT HORMONES AND THEIR FUNCTION IN OBESITY

How gut hormones normally affect energy balance has been discussed. However, there is increasing evidence that the function and efficacy of gut hormones change in the obese patient, including after weight loss; this might perpetuate the state of obesity and make weight loss difficult.

Fasting CCK levels initially increase in obesity, but in morbid obesity, they actually fall, and with it the satiating effect (Li, Ma, & Wang, 2011; Zwirska-Korczala et al., 2007). The normal postprandial rise in CCK is also lower in morbidly obese subjects compared with either obese or lean controls (Zwirska-Korczala et al., 2007), and the pattern of CCK release is altered, with loss of the normal biphasic CCK release pattern (Zwirska-Korczala et al., 2007). In a mouse model, this is mediated through a reduction in vagal afferent stimulation in obesity (Daly et al., 2011; Nefti, Chaumontet, Fromentin, Tomé, & Darcel, 2009). The satiating effects of exogenously administered CCK are also reduced in obese subjects (McLaughlin & Baile, 1979; Nefti et al., 2009). These observations suggest that morbidly obese subjects have disordered satiety feedback signals from CCK. This disorder can appear very early on, as overfed weaning rats have reduced expression of CCK receptors in the hypothalamus compared to normally fed controls (Plagemann et al., 1998), implying that infant obesity may affect life-long body habitus.

Fasting levels of PYY(3–36) are reduced in obese humans to a degree inversely proportional to body mass, although there is no difference in fasting PYY(3–36) levels between obese and lean mice (Gueugnon et al., 2012; Roth et al., 2005; Xu, McNearney, & Chen, 2011). In both species, the postprandial rise in PYY(3–36) is blunted in obesity, along with the satiating effects (Batterham et al., 2006 Le Roux et al., 2006; Xu, McNearney, et al., 2011). Indeed, obese humans need to consume twice as many calories to achieve the same plasma PYY(3–36) levels as lean controls (Le Roux et al., 2006). Therefore, PYY(3–36) deficiency may maintain obesity. Endogenous PYY levels may also predict long-term changes in body weight, as they increase with weight loss (and decline in weight gain), and higher baseline PYY(3–36) levels correlate with greater weight loss in obese subjects who undergo a weight-reducing lifestyle modification program (Gueugnon et al., 2012; Roth et al., 2005). Although endogenous levels of PYY(3–36) are reduced in obesity, obese humans and rodents remain sensitive to the effects of exogenous PYY(3–36) (Batterham, Cohen, et al., 2003; Le Roux et al., 2006; Pittner et al., 2004), which makes PYY (3–36) an attractive target for antiobesity therapies.

Similar to the blunted release of PYY(3–36) seen in obesity, in humans with Prader-Willi syndrome, a genetic form of obesity, the stimulated release of PP in response to meals is reduced compared to lean controls. However, this blunting has not been consistently replicated in other forms of human obesity (Glaser, Zoghlin, Pienta, & Vinik, 1988; Zipf, O'Dorisio, Cataland, & Sotos, 1981). As with PYY(3–36), sensitivity to PP's anorectic effects is not affected by obesity (Asakawa et al., 2003), which again makes it attractive as a target for antiobesity therapies.

The interaction of obesity and insulin is complex. Insulin causes adiposity by stimulating fatty acid production and uptake into adipose tissue. Obesity, in return, is associated with insulin resistance and hyperinsulinemia

(Baranowska, Wolińska-Witort, Martyńska, Chmielowska, & Baranowska-Bik, 2005; Björntorp, 1996; Peiris et al., 1989; Polonsky, Given, & Van Cauter, 1988; Frayn, 2001; Santiago et al., 1977). Insulin resistance occurs both at a peripheral and central level, with skeletal muscles and glucose-responsive neurons in the hypothalamus being less responsive to insulin in obese versus lean subjects (Rabinowitz, 1968; Spanswick, Smith, Mirshamsi, Routh, & Ashford, 2000). The pathophysiological link between adiposity and insulin resistance is complex but involves the effects of excess free fatty acids, enlarged adipose cells, adipokine release, and the production of a proinflammatory state (Benito, 2011; Boden, 2011; Gregor & Hotamisligil, 2011; Kim & Reaven, 2010; Boden, 1997; Golay et al., 1986). It is this combination of hyperinsulinemia and resistance to the actions of insulin that can lead to type 2 diabetes in susceptible individuals.

Resistance to the action of leptin is also seen in obesity. Leptin replacement therapy can dramatically improve body weight in patients with obesity due to abnormalities in the leptin gene (Farooqi et al., 1999), and recombinant leptin therapy can cause weight loss in lean subjects. However, it is less effective in obese subjects (Heymsfield et al., 1999; Seth, Knight, & Overton, 2011). The pathophysiological basis of leptin resistance is twofold: first, impaired leptin transport across the blood–brain barrier and second, impaired leptin receptor function and signaling (Oswal & Yeo, 2010). Ultimately, once leptin resistance is established, the anorectic effects of the hormone are lost, which exacerbates diet-induced obesity and further perpetuates leptin resistance (Scarpace & Zhang, 2009).

In contrast, obesity increases sensitivity to ghrelin. While basal ghrelin levels are lower in obese compared with lean individuals (Tschöp, Weyer, et al., 2001), food intake is significantly higher in obese subjects compared with normal-weight humans when low doses of ghrelin are given (Druce et al., 2005). Endogenous ghrelin is also affected, and the normal postprandial reduction in ghrelin is reduced in both human and rodent models of obesity (Xu, McNearney, et al., 2011; Zwirska-Korczala et al., 2007).

Incretin levels change in obesity. Fasting GLP-1 levels are reduced, and the response to an oral glucose load is blunted (Anini & Brubaker, 2003; Ranganath et al., 1996; Knop et al., 2011). However, exogenous GLP-1 still reduces food intake and hunger ratings in obese subjects (Näslund et al., 1999). Interestingly, in obese subjects fed to satiety, activity in the dorsolateral prefrontal cortex is much less than in lean subjects (Le et al., 2009); activity in this area correlates with GLP-1 levels (Pannacciulli et al., 2007). Why the GLP-1 response is reduced is unknown but may be due to the lack of leptin-induced GLP-1 release in obese subjects with leptin resistance (Anini & Brubaker, 2003). The reduction in GLP-1 may contribute to the impaired glucose tolerance that occurs in obesity. Changes in GIP levels in obesity have not been consistently observed: different studies have shown that levels are either unchanged or increased, both after fasting and in response to food ingestion; this makes the exact role of GIP in obesity unclear (Ranganath et al., 1996; Verdich, Toubro, et al., 2001; Vilsbøll et al., 2003). However, any increase in GIP may, in theory, propagate obesity by increasing fatty acid uptake into adipose tissue.

Alterations in gut hormones not only occur in obesity but also following diet and lifestyle induced weight loss. The levels of many satiating hormones decline with weight loss, including amylin and PP (Sumithran et al., 2011), while orexigenic NPY and ghrelin, plus its activating enzyme GOAT, increase after weight loss (Hansen et al., 2002; Reimer, Maurer, Lau, & Auer, 2010; Sumithran et al., 2011; Yu, Deng, & Huang, 2009). Thus, after weight loss, subjects are likely to feel hungrier, and find it harder to become sated: these counter-regulatory hormonal changes contribute to the difficulty in maintaining weight loss produced by diet and lifestyle changes.

In summary, while the effects of some gut hormones persist in obesity, the overall change in hormonal environment is to one where satiety signals are less efficacious and are produced at lower levels while orexigenic stimuli are more efficacious. Furthermore, once weight is lost, counter-regulatory alterations in gut hormone secretion tend to return the body to its previous weight, making sustained weight loss very difficult.

7. TARGETING OF GUT HORMONES IN OBESITY

After the withdrawal of two drug from the market (Rimonbant and Sibutramine), only two drugs have been licensed for the specific treatment of obesity. These are orlistat (*Xenical*©) and Qsymia© (at present this is only licensed for use in the USA but not Europe). Orlistat is a tetrahydrolipstatin which prevents absorption of fat from the intestine, whilst Qsymia© is a combination of phentermine, a psychostimulant, and the anti-epilepsy drug topiramate. The weight loss of these is limited to usually less than 10% (Padwal, Li, & Lau, 2004; Allison et al., 2012; Garvey et al., 2012), and in the case of orlistat, there are some unpleasant side effects such as fecal leakage (Padwal, Li, & Lau, 2004).

Current evidence suggests that bariatric surgery is the most effective method to produce significant and sustained weight loss (Colquitt, Picot, Loveman, & Clegg, 2009; Picot et al., 2009; Simpson, Shaw, &

McNamara, 2011), and it causes undoubted long-term improvements in mortality, principally through reductions in cardiovascular events (Sjöström et al., 2012). The maximum weight loss achieved in this way is 20–30% of total weight or 40–70% of excess weight, depending on the bariatric procedure used (Buchwald et al., 2004; Sjöström et al., 2007). In the UK, bariatric surgery is currently first line treatment for obesity in those with a BMI of greater than 50 kg/m^2 and should be considered in those with a BMI over 40 or between 35 and 40 with medical comorbidities that may be improved with weight loss NICE (2006). However, perioperative mortality, while small, remains significant at approximately 1.1% (Buchwald et al., 2004), and the requirement for specialist facilities and surgeons means that bariatric surgery is available only to the few.

Bariatric surgical procedures can be divided into those which are purely restrictive, for example, gastric banding, sleeve gastrectomy, and those with an element of bypass, for example, Roux-en-Y gastric bypass (RYGB), biliopancreatic diversion, and duodenal switch. Bypass procedures are more effective in causing weight loss (Colquitt et al., 2009; Padwal et al., 2011; Picot et al., 2009; Simpson et al., 2011), and alterations in gut hormone secretion after gastric bypass are thought to be responsible for this (Le Roux et al., 2011). Bypass surgery also substantially improves glucose tolerance and, in many cases, causes diabetes remission (Buchwald et al., 2004; Inabnet, Winegar, Sherif, & Sarr, 2012). This occurs within days of the procedure, before weight loss can be the cause (Pories, 2004); rather, alteration in the incretin profile seems the likely underlying mechanism. Subjects also have altered food preference after surgery, reporting a decreased desire to eat sweet and high-fat foods (Le Roux et al., 2011; Shin, Zheng, Townsend, Sigalet, & Berthoud, 2010; Shin and Berthoud, 2011); as seen previously, gut hormones affect food preference, and alterations in gut hormone levels (e.g., elevations in GLP-1) may be responsible for this postsurgical change.

Gut hormone changes following bypass produce a more anorexigenic environment, with an increase in satiety signals. Fasting PYY(3–36) levels increase (Shin et al., 2010), and following a meal or glucose tolerance test, the PYY(3–36) and amylin responses are much greater after bypass than in sham operated, obese, or lean controls (Bose, Machineni, et al., 2010; Oliván et al., 2009; Shin et al., 2010). These changes are not seen in weight loss following diet, or following the purely restrictive procedure of gastric banding (Bose, Machineni, et al., 2010; Oliván et al., 2009).

Incretin responses also change after gastric bypass. Fasting GIP levels are the same as in controls, but the peak following an OGTT dramatically increases. The GLP-1 rise in response to oral glucose also increases post-RYGB (Bose, Machineni, et al., 2010; Shin et al., 2010). These changes occur rapidly after surgery, and persist for at least 1 year, and may be responsible for the rapid remission of diabetes seen after bypass surgery, as well as the prolonged weight loss.

Orexigenic stimuli are reduced following bypass surgery. Fasting ghrelin levels do not change (Bose, Machineni, et al., 2010; Cummings et al., 2002; Oliván et al., 2009), but the normal preprandial rise in ghrelin is lost (Cummings et al., 2002). Also, the postprandial fall in ghrelin is greater post-bypass than after gastric banding in humans (Bose, Machineni, et al., 2010; Shin et al., 2010).

The clear success of bariatric surgery in causing weight loss has understandably led to substantial interest in trying to mimic the changes in gut hormone profile seen, but without surgery. This is most simply accomplished by administration of gut hormones such as GLP-1, PYY, and oxyntomodulin. Infusion of GLP-1 reduces food intake and body weight (Tang-Christensen et al., 1996; Verdich, Flint, et al., 2001). However, use of native GLP-1 as therapy is limited as it is rapidly degraded within the plasma by DPP-4. To turn GLP-1 into a practical proposition as a treatment for diabetes, DPP-4-resistant analogues have been developed. The first to be marketed, exenatide (*Byetta*®, a.k.a. Exendin-4) is a DPP-4-resistant peptide found in the saliva of the lizard *Heloderma suspectum* (the Gila monster) which is a potent agonist at the GLP-1 receptor (Göke et al., 1993). Another long-acting GLP-1 analogue, liraglutide (*Victoza*®), is derived from human GLP-1 but is coupled to a fatty acid chain binding to serum albumin to extend half-life. Both exenatide and liraglutide cause significant improvements in long-term glycemic control and significant weight loss (Vilsbøll, Christensen, Junker, Knop, & Gluud, 2012). However, all GLP-1 analogues are currently only licensed for the treatment of type 2 diabetes, rather than as stand-alone antiobesity therapy. Liraglutide has been studied in obese nondiabetic patients and has been shown to cause significant weight loss of up to 7 kg, but at the expense of considerable nausea and vomiting (Astrup et al., 2009) limiting its utility in this context. It is also known that GLP-1 has a tendency to reduce energy expenditure, reducing its ability to produce weight loss (Flint, Raben, Rehfeld, Holst, & Astrup, 2000).

A more attractive target for antiobesity therapy is oxyntomodulin. Obese subjects remain sensitive to oxyntomodulin: after 28 days of oxyntomodulin injections, subjects lost 2.4% body weight and had reduced food intake during a test meal (Wynne et al., 2005, 2006).

It also appears to have few side effects (Wynne et al., 2005, 2006). Oxyntomodulin is of particular interest because of its dual agonist action at both glucagon and GLP-1 receptors. Although glucagon has a hyperglycaemic effect, the simultaneous activation of GLP-1 receptors can counterbalance this due to their insulinotropic action. In addition, as previously mentioned, GLP-1 and glucagon receptor agonism may be synergistic in terms of suppression of food intake. Lastly, glucagon receptor activation increases energy expenditure which should accelerate weight loss. Indeed, in animal studies, dual agonists at the GLP-1 and glucagon receptors have been shown to be much more efficacious at weight loss than simple GLP-1 receptor agonists (Day et al., 2009; Pocai et al., 2009). In these animal models of obesity, glucose tolerance is improved, though how these dual agonists will affect subjects with known impaired glucose tolerance, or with diabetes, has not been established. Long-lasting analogues of oxyntomodulin such as TKS-1225 are currently being developed as potential therapies for obesity (Druce et al., 2009).

PYY(3–36) is another attractive gut hormone target, especially as obese subjects remain sensitive to its effects (Batterham, Cohen, et al., 2003; Batterham et al., 2007; Reidleberger, Haver, Chelikani, & Buescher, 2008). Early attempts to deliver PYY(3–36) intranasally led to problematic nausea, consequent on rapid absorption and development of supraphysiological levels of PYY (Gantz et al., 2007; le Roux et al., 2008). These can be overcome by modification of the pharmacokinetics of PYY, and degradation-resistant analogues of PYY are in current development as antiobesity therapies. PYY's cousin, PP, is also able to suppress appetite (Batterham, Le Roux, et al., 2003; Jesudason et al., 2007), and these satiating effects are still present in obese volunteers (Tan, unpublished data), confirming the results found in mice (Asakawa et al., 2003). A degradation-resistant analogue of PP intended for obesity treatment, PP 1420, has recently concluded Phase I studies and has been found to be safe and well tolerated (Tan et al., 2012).

Another treatment that has been developed for type 2 diabetes, with the added benefit of promoting and maintaining weight loss, is pramlintide (*Symlin*®), a nonamyloidogenic analogue of amylin. It has been licensed in the United States for the treatment of both type 1 and type 2 diabetes in patients using insulin. In addition to beneficial effects on overall glycaemic control, pramlintide has been shown to achieve and maintain weight loss (Smith et al., 2008); however, as with the GLP-1 analogues, pramlintide is not licensed for therapy of obesity without diabetes.

Although the GLP-1 analogues and synthetic amylin show that gut hormone-based therapy can have weight loss as a useful side effect, trials of other gut hormones have shown the pitfalls of developing gut hormones as weight loss therapies.

In the few cases of obesity determined by abnormalities in the leptin gene, therapy is relatively simple: obesity responds dramatically to leptin replacement therapy (Farooqi et al., 1999). However, its efficacy is reduced in the majority of obese subjects because of leptin resistance (Heymsfield et al., 1999; Seth et al., 2011). Studies have looked at reducing leptin resistance, as a means to restoring the anorectic effects of leptin, but unfortunately, the signaling pathways involved affect a number of other physiological processes, and attempts to target these pathways have so far been either ineffective at reducing leptin resistance or have been fatal in rodent models (Oswal & Yeo, 2010).

Another difficulty in the use of gut hormones as antiobesity therapy is that some of their effects are subject to tachyphylaxis. Exogenous CCK is effective at reducing meal length and overall food intake, and increasing satiety, even in obese individuals (Lieverse, Jansen, Masclee, & Lamers, 1995; Pi-Sunyer, Kissileff, Thornton, & Smith, 1982). However, it has a very short half-life, and with repeated injections, or continuous CCK infusions, the satiating effect is rapidly lost and body weight is regained (Covasa, Marcuson, & Ritter, 2001; Crawley & Beinfeld, 1983; West, Fey, & Woods, 1984). Attempts have been made, nevertheless, to devise agonists that may be suitable therapeutically; however, a lack of efficacy has stymied their development (Elliott et al., 2010).

8. CONCLUSION

Gut hormones clearly play a fundamental role in the cause of obesity, both from a homeostatic drive to eat: "eat to live," and through the hedonistic value of food: "live to eat." With the increasing financial and social burden worldwide due to obesity, new strategies must be used to tackle this epidemic. Therapies involving gut hormones have already been shown to have beneficial effects on type 2 diabetes, with welcome loss of weight. Current research efforts are in progress to develop gut hormone analogues for the specific treatment of obesity.

REFERENCES

Abbott, C. R., Monteiro, M., Small, C. J., Sajedi, A., Smith, K. L., Parkinson, J. R., et al. (2005). The inhibitory effects of peripheral administration of peptide YY(3–36) and glucagon-like peptide 1 on food intake are attenuated by ablation of the vagal-brainstem-hypothalamic pathway. *Brain Research*, *1044*, 127–131.

Adrian, T. E., Besterman, H. S., Cooke, T. J., Bloom, S. R., Barnes, A. J., & Russell, R. C. (1977). Mechanism of pancreatic polypeptide release in man. *Lancet*, *1*, 161–163.

Adrian, T. E., Bloom, S. R., Bryant, M. G., Polak, J. M., Heitz, P. H., & Barnes, A. J. (1976). Distribution and release of human pancreatic polypeptide. *Gut*, *17*, 940–944.

Adrian, T. E., Savage, A. P., Sagor, G. R., Allen, J. M., Bacarese-Hamilton, A. J., Tatemoto, K., et al. (1985). Effect of peptide YY on gastric, pancreatic, and biliary function in humans. *Gastroenterology*, *89*, 494–499.

Ahima, R. S., Prabakaran, D., Mantzoros, C., Qu, D., Lowell, B., Maratos-Flier, E., et al. (1996). Role of leptin in the neuroendocrine response to fasting. *Nature*, *382*, 250–252.

Ahlman, H., & Nilsson, O. (2001). The gut as the largest endocrine organ in the body. *Annals of Oncology*, *12*(Suppl. 2), 63–68.

Ahrén, B., & Taborsky, G. J., Jr. (1986). The mechanism of vagal nerve stimulation of glucagon and insulin secretion in the dog. *Endocrinology*, *118*, 1551–1557.

Allison, D. B., Gadde, K. M., Garvey, W. T., Peterson, C. A., Schwiers, M. L., & Najarian, T. (2012). Controlled-release phentermine/topiramate in severely obese adults: a randomized controlled trial (EQUIP). *Obesity*, *20*(2), 330–342.

Anini, Y., & Brubaker, P. L. (2003). Role of leptin in the regulation of glucagon-like peptide-1 secretion. *Diabetes*, *52*, 252–259.

Aponte, Y., Atasoy, D., & Sternson, S. M. (2011). AGRP neurons are sufficient to orchestrate feeding behaviour rapidly and without training. *Nature Neuroscience*, *14*, 351–355.

Asakawa, A., Inui, A., Ueno, N., Fukimiya, M., Fujino, M. A., & Kasuga, M. (1999). Mouse pancreatic polypeptide modulates food intake, while not influencing anxiety in mice. *Peptides*, *20*, 1445–1448.

Asakawa, A., Inui, A., Yuzuriha, H., Ueno, N., Katsuura, G., Fujimiya, M., et al. (2003). Characterization of the effects of pancreatic polypeptide in the regulation of energy balance. *Gastroenterology*, *124*, 1325–1336.

Asmar, M., Tangaa, W., Madsbad, S., Hare, K., Astrup, A., Flint, A., et al. (2010). On the role of glucose-dependent insulinotropic polypeptide in postprandial metabolism in humans. *American Journal of Physiology. Endocrinology and Metabolism*, *298*, 614–621.

Astrup, A., Rössner, S., Van Gaal, L., Rissanen, A., Niskanen, L., Al Hakim, M., et al. (2009). Effects of liraglutide in the treatment of obesity: A randomised, double-blind, placebo-controlled study. *Lancet*, *374*, 1606–1616.

Baggio, L. L., & Drucker, D. J. (2007). Biology of incretins: GLP-1 and GIP. *Gastroenterology*, *132*, 2131–2157.

Baggio, L. L., Huang, Q., Brown, T. J., & Drucker, D. J. (2004). Oxyntomodulin and glucagon-like peptide-1 differentially regulate murine food intake and energy expenditure. *Gastroenterology*, *127*, 546–558.

Bai, F. L., Yamano, M., Shiotani, Y., Emson, P. C., Smith, A. D., Powell, J. F., et al. (1985). An arcuato-paraventricular and -dorsomedial hypothalamic neuropeptide Y-containing system which lacks noradrenaline in the rat. *Brain Research*, *331*, 172–175.

Baranowska, B., Wolińska-Witort, E., Martyńska, L., Chmielowska, M., & Baranowska-Bik, A. (2005). Plasma orexin A, orexin B, leptin, neuropeptide Y (NPY) and insulin in obese women. *Neuroendocrinology Letters*, *26*, 293–296.

Bataille, D., Gespach, C., Tatemoto, K., Marie, J. C., Coudray, A. M., Rosselin, G., et al. (1981). Bioactive enteroglucagon (oxyntomodulin): Present knowledge on its chemical structure and its biological activities. *Peptides*, *2*(Suppl. 2), 41–44.

Batterham, R. L., Cohen, M. A., Ellis, S. M., LeRoux, C. W., Withers, D. J., Frost, G. S., et al. (2003). Inhibition of food intake in obese subjects by peptide YY3-36. *The New England Journal of Medicine*, *349*, 941–948.

Batterham, R. L., Cowley, M. A., Small, C. J., Herzog, H., Cohen, M. A., Dakin, C. L., et al. (2002). Gut hormone PYY(3–36) physiologically inhibits food intake. *Nature*, *418*, 650–654.

Batterham, R. L., Ffytche, D. H., Rosenthal, J. M., Zelaya, F. O., Barker, G. J., Withers, D. J., et al. (2007). PYY modulation of cortical and hypothalamic brain areas predicts feeding behaviour in humans. *Nature, 450*, 106–109.

Batterham, R. L., Heffron, H., Kapoor, S., Chivers, J. E., Chandarana, K., Herzog, H., et al. (2006). Critical role for peptide YY in protein-mediated satiation and body weight regulation. *Cell Metabolism, 4*, 223–333.

Baura, G. D., Foster, D. M., Porte, D., Jr., Kahn, S. E., Bergman, R. N., Cobelli, C., et al. (1993). Saturable transport of insulin from plasma in to the central nervous system of dogs in vivo. A mechanism for regulated insulin delivery to the brain. *The Journal of Clinical Investigation, 92*, 1824–1830.

Bellinger, L. L., & Bernardis, L. L. (2002). The dorsomedial hypothalamic nucleus and its role in ingestive behaviour and body weight regulation: Lessons learned from lesioning studies. *Physiology & Behavior, 76*, 431–442.

Benito, M. (2011). Tissue specificity on insulin action and resistance: Past to recent mechanisms. *Acta Physiologica, 201*, 297–312.

Berna, M. J., & Jensen, R. T. (2007). Role of CCK/gastrin receptors in gastrointestinal/metabolic diseases and results of human studies using gastrin/CCK receptor agonists/antagonists in these diseases. *Current Topics in Medicinal Chemistry*, 7, 1211–1231.

Bernardis, L. L., Awad, A., Fink, C., & Bellinger, L. L. (1992). Metabolic and neuroendocrine indices one month after lateral hypothalamic area lesions. *Physiology & Behavior, 52*, 133–139.

Berthoud, H. R. (2002). Multiple neural systems controlling food intake and body weight. *Neuroscience and Biobehavioral Reviews, 26*, 393–428.

Berthoud, H. R., Trimble, E. R., & Moody, A. J. (1982). Lack of gastric inhibitory polypeptide (GIP) response to vagal stimulation in the rat. *Peptides, 3*, 907–912.

Bessard, T., Schutz, Y., & Jéquier, E. (1983). Energy expenditure and postprandial thermogenesis in obese women before and after weight loss. *The American Journal of Clinical Nutrition, 38*, 680–693.

Björntorp, P. (1996). The regulation of adipose tissue distribution in humans. *International Journal of Obesity and Related Metabolic Disorders: Journal of the International Association for the Study of Obesity, 20*, 291–302.

Blevins, J. E., Stanley, B. G., & Reidleberger, R. D. (2000). Brain regions where cholecystokinin suppresses feeding in rats. *Brain Research, 860*, 1–10.

Blomqvist, A. G., & Herzog, H. (1997). Y-receptor subtypes—How many more? *Trends in Neurosciences, 20*, 294–298.

Blüher, S., & Mantzoros, C. S. (2007). Leptin in reproduction. *Current Opinion in Endocrinology, Diabetes, and Obesity, 14*, 458–464.

Boden, G. (1997). Role of fatty acids in the pathogenesis of insulin resistance and NIDDM. *Diabetes, 46*, 3–10.

Boden, G. (2011). Obesity, insulin resistance and free fatty acids. *Current Opinion in Endocrinology, Diabetes, and Obesity, 18*, 139–143.

Boey, D., Lin, S., Enriquez, R. F., Lee, N. J., Slack, K., Couzens, M., et al. (2008). PYY transgenic mice are protected against diet-induced and genetic obesity. *Neuropeptides, 42*, 19–30.

Bose, M., Machineni, S., Oliván, B., Teixeira, J., McGinty, J. J., Bawa, B., et al. (2010). Superior appetite hormonal profile after equivalent weight loss by gastric bypass compared to gastric banding. *Obesity (Silver Spring), 18*, 1085–1091.

Broberger, C. (2005). Brain regulation of food intake and appetite: Molecules and networks. *Journal of Internal Medicine, 258*, 301–327.

Brown, J. C., Dryburgh, J. R., Ross, S. A., & Dupré, J. (1975). Identification and actions of gastric inhibitory polypeptide. *Recent Progress in Hormone Research, 31*, 487–532.

Bruijnzeel, A. W., Corrie, L. W., Rogers, J. A., & Yamada, H. (2011). Effects of insulin and leptin in the ventral tegmental area and arcuate hypothalamic nucleus on food intake and brain reward function in female rats. *Behavioural Brain Research*, *219*, 254–264.

Buchan, A. M., Polak, J. M., Capella, C., Solcia, E., & Pearse, A. G. (1978). Electronimmunocytochemical evidence of the K cell localisation of gastric inhibitory polypeptide (GIP) in man. *Histochemistry*, *56*, 37–44.

Buchwald, H., Avidor, Y., Braunwald, E., Jensen, M. D., Pories, W., Fahrbach, K., et al. (2004). Bariatric surgery: A systematic review and meta-analysis. *Journal of the American Medical Association*, *292*, 1724–1737.

Buyse, M., Ovesjö, M. L., Goïot, H., Guilmeau, S., Péranzi, G., Moizo, L., et al. (2001). Expression and regulation of leptin receptor proteins in afferent and efferent neurons of the vagus nerve. *The European Journal of Neuroscience*, *14*, 64–72.

Calles-Escandón, J. (1994). Insulin dissociates hepatic glucose cycling and glucagon-induced thermogenesis in man. *Metabolism*, *43*, 1000–1005.

Cataland, S., Crockett, S. E., Brown, J. C., & Mazzaferri, E. L. (1974). Gastric inhibitory polypeptide (GIP) stimulation by oral glucose in man. *The Journal of Clinical Endocrinology and Metabolism*, *39*, 223–228.

Chaudhri, O. B., Parkinson, J. R., Kuo, Y. T., Druce, M. R., Herlihy, A. H., Bell, J. D., et al. (2006). Differential hypothalamic neuronal activation following peripheral injection of GLP-1 and oxyntomodulin in mice detected by manganese-enhanced magnetic resonance imaging. *Biochemical and Biophysical Research Communications*, *350*, 298–306.

Chelikani, P. K., Haver, A. C., Reeve, J. R., Jr., Keire, D. A., & Reidelberger, R. D. (2006). Daily, intermittent intravenous infusion of peptide YY(3–36) reduces daily food intake and adiposity in rats. *American Journal of Physiology. Regulatory, Integrative and Comparative Physiology*, *290*, 298–305.

Chen, A. S., Marsh, D. J., Trumbauer, M. E., Frazier, E. G., Guan, X. M., Yu, H., et al. (2000). Inactivation of the mouse melanocortin-3 receptor results in increased fat mass and reduced lean body mass. *Nature Genetics*, *26*, 97–102.

Cockburn, F., Hull, D., & Walton, I. (1967). The effect of lipolytic hormones and theophylline on heart production in brown adipose tissue in vivo. *British Journal of Pharmacology and Chemotherapy*, *31*, 568–577.

Cohen, M. A., Ellis, S. M., Le Roux, C. W., Batterham, R. L., Park, A., Patterson, M., et al. (2003). Oxyntomodulin suppresses appetite and reduces food intake in humans. *The Journal of Clinical Endocrinology and Metabolism*, *88*, 4696–4701.

Coleman, D. L. (1973). Effects of parabiosis of obese with diabetes and normal mice. *Diabetologia*, *9*, 294–298.

Colquitt, J. L., Picot, J., Loveman, E., & Clegg, A. J. (2009). Surgery for obesity. *Cochrane Database of Systematic Reviews (Online)*, (2), CD003641.

Conarello, S. L., Jiang, G., Mu, J., Li, Z., Woods, J., Zycband, E., et al. (2007). Glucagon receptor knockout mice are resistance to diet-induced obesity and streptozocin-mediated beta cell loss and hyperglycaemia. *Diabetologia*, *50*, 142–150.

Considine, R. V., Sinha, M. K., Heiman, M. L., Kriauciunas, A., Stephens, T. W., Nyce, M. R., et al. (1996). Serum immunoreactive-leptin concentrations in normal-weight and obese humans. *The New England Journal of Medicine*, *334*, 292–295.

Costa-e-Sousa, R. H., Souza, L. L., Calviño, C., Cabanelas, A., Almeida, N. A., Oliveira, K. J., et al. (2011). Central NPY-Y5 receptors activation plays a major role in fasting-induced pituitary-thyroid axis suppression in adult rat. *Regulatory Peptides*, *171*, 43–47.

Covasa, M., Marcuson, J. K., & Ritter, R. C. (2001). Diminished satiation in rats exposed to elevated levels of endogenous or exogenous cholecystokinin. *American Journal of Physiology. Regulatory, Integrative and Comparative Physiology*, *208*, 331–337.

Cowley, M. A., Smart, J. L., Rubinstein, M., Cerdán, M. G., Diano, S., Horvath, T. L., et al. (2001). Leptin activates anorexigenic POMC neurons through a neural network in the arcuate nucleus. *Nature*, *411*, 480–484.
Crawley, J. N., & Beinfeld, M. C. (1983). Rapid development of tolerance to the behavioural actions of cholecystokinin. *Nature*, *302*, 703–706.
Cummings, D. E., Frayo, R. S., Marmonier, C., Aubert, R., & Chapelot, D. (2004). Plasma ghrelin levels and hunger scores in humans initiating meals voluntarily without time- and food-related cues. *American Journal of Physiology. Endocrinology and Metabolism*, *287*, 297–304.
Cummings, D. E., Purnell, J. Q., Frayo, R. S., Schmidova, K., Wisse, B. E., & Weigle, D. S. (2001). A pre-prandial rise in ghrelin levels suggests a role in meal initiation in humans. *Diabetes*, *50*, 1714–1719.
Cummings, D. E., Weigle, D. S., Frayo, R. S., Breen, P. A., Ma, M. K., Dellinger, E. P., et al. (2002). Plasma ghrelin levels after diet-induced weight loss or gastric bypass surgery. *The New England Journal of Medicine*, *23*, 1623–1630.
Dakin, C. L., Gunn, I., Small, C. J., Edwards, C. M., Hay, D. L., Smith, D. M., et al. (2001). Oxyntomodulin inhibits food intake in the rat. *Endocrinology*, *142*, 4244–4250.
Dakin, C. L., Small, C. J., Batterham, R. L., Neary, N. M., Cohen, M. A., Patterson, M., et al. (2004). Peripheral oxyntomodulin reduces food intake and body weight gain in rats. *Endocrinology*, *145*, 2687–2695.
Dakin, C. L., Small, C. J., Park, A. J., Seth, A., Ghatei, M. A., & Bloom, S. R. (2002). Repeated ICV administration of oxyntomodulin causes a greater reduction in body weight gain than in pair-fed rats. *American Journal of Physiology. Endocrinology and Metabolism*, *283*, 1173–1177.
Daly, D. M., Park, S. J., Valinsky, W. C., & Beyak, M. J. (2011). Impaired intestinal afferent nerve satiety signalling and vagal afferent excitability in diet induced obesity in the mouse. *The Journal of Physiology*, *589*, 2857–2870.
Date, Y., Kojima, M., Hosoda, H., Sawaguchi, A., Mondal, M. S., Suganuma, T., et al. (2000). Ghrelin, a novel growth hormone-releasing acylated peptide, is synthesized in a distinct endocrine cell type in the gastrointestinal tracts of rats and humans. *Endocrinology*, *141*, 4255–4261.
Date, Y., Nakazato, M., Hashiguchi, S., Dezaki, K., Mondal, M. S., Hosoda, H., et al. (2002). Ghrelin is present in pancreatic alpha-cells of humans and rats and stimulates insulin secretion. *Diabetes*, *51*, 124–129.
Davis, C., & Fox, J. (2008). Sensitivity to reward and body mass index (BMI): Evidence for a non-linear relationship. *Appetite*, *50*, 43–49.
Davis, C., Strachan, S., & Berkson, M. (2004). Sensitivity to reward: Implications for overeating and overweight. *Appetite*, *42*, 131–138.
Day, J. W., Ottaway, N., Patterson, J. T., Gelfanov, V., Smiley, D., Gidda, J., et al. (2009). A new glucagon and GLP-1 co-agonist eliminates obesity in rodents. *Nature Chemical Biology*, *5*, 749–757.
De Backer, M. W., la Fleur, S. E., Brans, M. A., van Rozen, A. J., Luijendijk, M. C., Merkestein, M., et al. (2011). Melanocortin receptor-mediated effects on obesity are distributed over specific hypothalamic regions. *International Journal of Obesity*, *35*, 629–641.
De Lartigue, G., Barbier de la Serre, C., Espero, E., Lee, J., & Raybould, H. E. (2011). Diet-induced obesity leads to the development of leptin resistance in vagal afferent neurons. *American Journal of Physiology. Endocrinology and Metabolism*, *301*, 187–195.
De Lartigue, G., Lur, G., Dimaline, R., Varro, A., Raybould, H., & Dockray, G. J. (2010). EGR1 is a target for cooperative interactions between cholecystokinin and leptin, and inhibition by ghrelin, in vagal afferent neurons. *Endocrinology*, *151*, 3589–3599.
De Silva, A., Salem, V., Long, C. J., Makwana, A., Newbould, R. D., Rabiner, E. A., et al. (2011). The gut hormones PYY3-36 and GLP-1 7–36 amide reduce food

intake and modulate brain activity in appetite centres in humans. *Cell Metabolism, 14*, 700–706.

Dezaki, K., Hosoda, H., Kakei, M., Hashiguchi, S., Watanabe, M., Kangawa, K., et al. (2004). Endogenous ghrelin in pancreatic islets restricts insulin release by attenuating Ca2+ signalling in beta-cells: Implication in the glycaemic control in rodents. *Diabetes, 53*, 3142–3151.

Dezaki, K., Sone, H., Koizumi, M., Nakata, M., Kakei, M., Nagai, H., et al. (2006). Blockade of pancreatic islet-derived ghrelin enhances insulin secretion to prevent high-fat diet-induced glucose intolerance. *Diabetes, 55*, 3486–3493.

Dhanvantari, S., Seidah, N. G., & Brubaker, P. L. (1996). Role of prohormone convertases in the tissue-specific processing of proglucagon. *Molecular Endocrinology, 10*, 342–355.

Disse, E., Bussier, A. L., Veyrat-Durebex, C., Deblon, N., Pfluger, P. T., Tschöp, M. H., et al. (2010). Peripheral ghrelin enhances sweet taste food consumption and preference, regardless of its caloric content. *Physiology and Behavior, 101*, 277–281.

Dockray, G. J. (2012). Cholecystokinin. *Current Opinion in Endocrinology, Diabetes, and Obesity, 19*, 8–12.

Dossat, A. M., Lilly, N., Kay, K., & Williams, D. L. (2011). Glucagon-like peptide 1 receptors in nucleus accumbens affect food intake. *The Journal of Neuroscience, 31*, 14453–14457.

Drewes, C., Nauck, M. A., Horn, R., Holst, J., Schmiegel, W., & Brabant, G. (1997). A liquid mixed meal or exogenous glucagon-like peptide 1 (GLP-1) do not alter plasma leptin concentrations in healthy volunteers. *Acta Diabetologica, 34*, 230–234.

Druce, M. R., Minnion, J. S., Field, B. C., Patel, S. R., Shillito, J. C., Tilby, M., et al. (2009). Investigation of structure-activity relationships of Oxyntomodulin (Oxm) using Oxm analogs. *Endocrinology, 150*, 1712–1722.

Druce, M. R., Wren, M. A., Park, A. J., Milton, J. E., Patterson, M., Frost, G., et al. (2005). Ghrelin increases food intake in obese as well as lean subjects. *International Journal of Obesity, 29*, 1130–1136.

Drucker, D. J. (2006). The biology of incretin hormones. *Cell Metabolism, 3*, 153–165.

Drucker, D. J., Philippe, J., Moisoy, S., Chick, W. L., & Habener, J. F. (1987). Glucagon-like peptide-1 stimulates insulin gene expression and increases cyclic AMP levels in a rat islet cell line. *Proceedings of the National Academy of Sciences of the United States of America, 84*, 3434–3438.

Dufresne, M., Seva, C., & Fourmy, D. (2006). Cholecystokinin and gastrin receptors. *Physiological Reviews, 86*, 805–847.

Dumont, Y., Moyse, E., Fournier, A., & Quirion, R. (2007). Distribution of peripherally injected peptide YY ([125I] PYY (3–36)) and pancreatic polypeptide ([125I] hPP) in the CNS: Enrichment in the area postrema. *Journal of Molecular Neuroscience, 33*, 294–304.

Eckel, R. H., Fujimoto, W. Y., & Brunzell, J. D. (1979). Gastric inhibitory polypeptide enhanced lipoprotein lipase activity in cultured preadipocytes. *Diabetes, 28*, 1411–1412.

Egecioglu, E., Jerlhag, E., Salomé, N., Skibicka, K. P., Haage, D., Bohlooly-Y, M., et al. (2010). Ghrelin increases intake of rewarding food in rodents. *Addiction Biology, 15*, 304–311.

Ehses, J. A., Casilla, V. R., Doty, T., Pospisilik, J. A., Winter, K. D., Demuth, H. U., et al. (2003). Glucose-dependent insulinotropic polypeptide promotes beta-(INS-1) cell survival via cyclic adenosine monophosphate-mediated caspase-3 inhibition and regulation of p38 mitogen-activated protein kinase. *Endocrinology, 144*, 4433–4445.

Elliott, R. L., Cameron, K. O., Chin, J. E., Bartlett, J. A., Beretta, E. E., Chen, Y., et al. (2010). Discovery of N-benzyl-2-[(4S)-4-(1H-indol-3-ylmethyl)-5-oxo-1-phenyl-4,5-dihydro-6H-[1,2,4]triazolol[4,3-a][1,5]benzodiazepine-6-y]-N-isopropylacetamide, an orally active, gut selective CCK1 receptor agonist for the potential treatment of obesity. *Bioorganic & Medicinal Chemistry Letters, 20*, 6797–6801.

Elliott, R. M., Morgan, L. M., Tredger, J. A., Deacon, S., Wright, J., & Marks, V. (1993). Glucagon-like peptide-1 (7–36)amide and glucose-dependent insulinotropic polypeptide secretion in response to nutrient ingestion in man: Acute post-prandial and 24-h secretion patterns. *The Journal of Endocrinology, 138*, 159–166.

Farilla, L., Hui, H., Bertolotto, C., Kang, E., Bulotta, A., Di Mario, U., et al. (2002). Glucagon-like peptide-1 promotes islet cell growth and inhibits apoptosis in Zucker diabetic rats. *Endocrinology, 143*, 4397–4408.

Farooqi, I. S., Bullmore, E., Keogh, J., Gillard, J., O'Rahilly, S., & Fletcher, P. C. (2007). Leptin regulates striatal regions and human eating behaviour. *Science, 317*, 1355.

Farooqi, I. S., Jebb, S. A., Langmack, G., Lawrence, E., Cheetham, C. H., Prentice, A. M., et al. (1999). Effects of recombinant leptin therapy in a child with congenital leptin deficiency. *The New England Journal of Medicine, 341*, 879–884.

Farooqi, I. S., Wangensteen, T., Collins, S., Kimber, W., Matarese, G., Keogh, J. M., et al. (2007). Clinical and molecular genetic spectrum of congenital deficiency of the leptin receptor. *The New England Journal of Medicine, 356*, 237–247.

Fehmann, H. C., Göke, R., Göke, B., Bächle, R., Wagner, B., & Arnold, R. (1991). Priming effect of glucagon-like peptide-1 (7–36) amide, glucose-dependent insulinotropic polypeptide and cholecystokinin-8 at the isolated perfused rat pancreas. *Biochimica et Biophysica Acta, 1091*, 356–363.

Fekete, C., Kelly, J., Mihály, E., Sarkar, S., Rand, W. M., Légrádi, G., et al. (2001). Neuropeptide Y has a central inhibitory action on the hypothalamic-pituitary-thyroid axis. *Endocrinology, 142*, 2606–2613.

Fekete, C., Sarkar, S., Rand, W. M., Harney, J. W., Emerson, C. H., Bianco, A. C., et al. (2002). Neuropeptide Y1 and Y5 receptors mediate the effects of neuropeptide Y on the hypothalamic-pituitary-thyroid axis. *Endocrinology, 143*, 4513–4519.

Fekete, C., Singru, P. S., Sanchez, E., Sarkar, S., Christoffolete, M. A., Riberio, R. S., et al. (2006). Differential effects of central leptin, insulin or glucose administration during fasting on the hypothalamic-pituitary-thyroid axis and feeding-related neurons in the arcuate nucleus. *Endocrinology, 147*, 520–529.

Field, B. C., Chaudhri, O. B., & Bloom, S. R. (2009). Obesity treatment: Novel peripheral targets. *British Journal of Clinical Pharmacology, 68*, 830–843.

Field, B. C., Chaudhri, O. B., & Bloom, S. R. (2010). Bowels control brain: Gut hormones and obesity. *Nature Reviews. Endocrinology, 6*, 444–453.

Figlewicz, D. P., Bennett, J. L., Aliakbari, S., Zavosh, A., & Sipols, A. J. (2008). Insulin acts at different CNS sites to decrease acute sucrose intake and sucrose self-administration in rats. *American Journal of Physiology. Regulatory, Integrative and Comparative Physiology, 295*, 388–394.

Figlewicz, D. P., Bennett, J. L., Naleid, A. M., Davis, C., & Grimm, J. W. (2006). Intraventricular insulin and leptin decreases sucrose self-administration in rats. *Physiology and Behavior, 89*, 611–616.

Figlewicz, D. P., & Benoit, S. C. (2009). Insulin, leptin and food reward: Update 2008. *American Journal of Physiology. Regulatory, Integrative and Comparative Physiology, 296*, 9–19.

Figlewicz, D. P., Evans, S. B., Murphy, J., Hoen, M., & Baskin, D. G. (2003). Expression of receptors for insulin and leptin in the ventral tegmental area/substantia nigra (VTA/SN) of the rats. *Brain Research, 964*, 107–115.

Flint, A., Raben, A., Astrup, A., & Holst, J. J. (1998). Glucagon-like peptide 1 promotes satiety and suppresses energy intake in humans. *The Journal of Clinical Investigation, 101*, 515–520.

Flint, A., Raben, A., Rehfeld, J. F., Holst, J. J., & Astrup, A. (2000). The effect of glucagon-like peptide-1 on energy expenditure and substrate metabolism in humans. *International Journal of Obesity and Related Metabolic Disorders: Journal of the International Association for the Study of Obesity, 24*, 288–290.

Flynn, M. C., Scott, T. R., Pritchard, T. C., & Plata-Salamán, C. R. (1998). Mode of action of OB protein (leptin) on feeding. *American Journal of Physiology. Regulatory, Integrative and Comparative Physiology*, *275*, 174–179.

Frayn, K. N. (2001). Adipose tissue and the insulin resistance syndrome. *The Proceedings of the Nutrition Society*, *60*, 375–380.

Fulton, S. (2010). Appetite and reward. *Frontiers in Neuroendocrinology*, *31*, 85–103.

Fulton, S., Pissios, P., Manchon, R. P., Stiles, L., Frank, L., Pothos, E. N., et al. (2006). Leptin regulation of the mesoaccumbens dopamine pathway. *Neuron*, *51*, 811–822.

Gantz, I., Erondu, N., Mallick, M., Musser, B., Krishna, R., Tanaka, W. K., et al. (2007). Efficacy and safety of intranasal peptide YY3-36 for weight reduction in obese adults. *The Journal of Clinical Endocrinology and Metabolism*, *92*, 1754–1757.

Garvey, W. T., Ryan, D. H., Look, M., Gadde, K. M., Allison, D. B., & Peterson, C. A. (2012). Two-year sustained weight loss and metabolic benefits with controlled-release phentermine/topiramate in obese and overweight adults (SEQUEL): a randomized, placebo-controlled, phase 3 extension study. *American Journal of Nutrition*, *95*(2), 297–308.

Geary, N., Kissileff, H. R., Pi-Sunyer, F. X., & Hinton, V. (1992). Individual, but not simultaneous, glucagon and cholecystokinin infusions inhibit feeding in men. *American Journal of Physiology. Regulatory, Integrative and Comparative Physiology*, *262*, 975–980.

Geary, N., Le Sauter, J., & Noh, U. (1993). Glucagon acts in the liver to control spontaneous meal size in rats. *American Journal of Physiology. Regulatory, Integrative and Comparative Physiology*, *264*, 116–122.

Gehlert, D. R., Schober, D. A., Beavers, L., Gadski, R., Hoffman, J. A., Smiley, D. L., et al. (1996). Characterization of the peptide binding requirements for the cloned human pancreatic polypeptide-preferring receptor. *Molecular Pharmacology*, *50*, 112–118.

Geiger, B. M., Haburcak, M., Avena, N. M., Moyer, M. C., Hoebel, B. G., & Pothos, E. N. (2009). Deficits of mesolimbic dopamine neurotransmission in rat dietary obesity. *Neuroscience*, *159*, 1193–1199.

Gelling, R. W., Du, X. Q., Dichmann, D. S., Romer, J., Huang, H., Cui, L., et al. (2003). Lower blood glucose, hyperglucagonemia, and pancreatic alpha cell hyperplasia in glucagon receptor knockout mice. *Proceedings of the National Academy of Sciences of the United States of America*, *100*, 1438–1443.

Gerich, J. E., Langlois, M., Noacco, C., Lorenzi, M., Karam, J. H., & Korsham, P. H. (1976). Comparison of the suppressive effects of elevated plasma glucose and free fatty acid levels on glucagon secretion in normal and insulin-dependent diabetic. Evidence for selective alpha-cell insensitivity to glucose in diabetes mellitus subjects. *The Journal of Clinical Investigation*, *58*, 320–325.

Ghatei, M. A., Uttenthal, L. O., Christofides, N. D., Bryant, M. G., & Bloom, S. R. (1983). Molecular forms of human enteroglucagon in tissue and plasma: Plasma responses to nutrient stimuli in health and in disorders of the upper gastrointestinal tract. *The Journal of Clinical Endocrinology and Metabolism*, *57*, 488–495.

Gibbs, J., Young, R. C., & Smith, G. P. (1973). Cholecystokinin elicits satiety in rats with open gastric fistulas. *Nature*, *245*, 323–325.

Girard, J. (2008). The incretins: From the concept to their use in the treatment of type 2 diabetes. Part A: Incretins: Concept and physiological functions. *Diabetes & Metabolism*, *34*, 550–559.

Giraudo, S. Q., Billington, C. J., & Levine, A. S. (1998). Feeding effects of hypothalamic injection of melanocortin 4 receptor ligands. *Brain Research*, *809*, 302–306.

Glaser, B., Zoghlin, G., Pienta, K., & Vinik, A. I. (1988). Pancreatic polypeptide response to secretin in obesity: Effects of glucose intolerance. *Hormone and Metabolic Research*, *20*, 288–292.

Göke, R., Fehmann, H. C., Linn, T., Schmidt, H., Krause, M., Eng, J., et al. (1993). Exendin-4 is a high potent agonist and truncated exendin-(9–39)-amide an antagonist at the glucagon-like peptide 1-(7–36)-amide receptor of insulin-secreting beta-cells. *The Journal of Biological Chemistry*, *268*, 19650–19655.

Golay, A., Swislocki, A. L., Chen, Y. D., Jaspan, J. B., & Reaven, G. M. (1986). Effect of obesity on ambient plasma glucose, free fatty acid, insulin, growth hormone, and glucagon concentrations. *The Journal of Clinical Endocrinology and Metabolism*, *63*, 481–484.

Goldstein, J. L., Zhao, T. J., Li, R. L., Sherbet, D. P., Liang, G., & Brown, M. S. (2011). Surviving starvation: Essential role of the ghrelin-growth hormone axis. *Cold Spring Harbor Symposia on Quantitative Biology*, *76*, 121–127.

Gregor, M. F., & Hotamisligil, G. S. (2011). Inflammatory mechanisms in obesity. *Annual Review of Immunology*, *29*, 415–445.

Gropp, E., Shanabrough, M., Borok, E., Xu, A. W., Janoschek, R., Buch, T., et al. (2005). Agouti-related peptide-expressing neurons are mandatory for feeding. *Nature Neuroscience*, *8*, 1289–1291.

Gross, P. M. (1992). Circumventricular organ capillaries. *Progress in Brain Research*, *91*, 219–233.

Guan, H. Z., Li, Q. C., & Jiang, Z. Y. (2010). Ghrelin acts on rat dorsal vagal complex to stimulate feeding via arcuate neuropeptide Y/agouti-related peptide neurons activation. *Sheng Li Xue Bao*, *62*, 357–364.

Gueugnon, C., Mougin, F., Nguyen, N. U., Bouhaddi, M., Nicolet-Guénet, M., & Dumoulin, G. (2012). Ghrelin and PYY levels in adolescents with severe obesity: Effects of weight loss induced by long-term exercise training and modified food habits. *European Journal of Applied Physiology*, *112*, 1797–1805.

Hallschmid, M., Benedict, C., Schultes, B., Fehm, H. L., Born, J., & Kern, W. (2004). Intranasal insulin reduces body fat in men but not in women. *Diabetes*, *53*, 3024–3029.

Halmi, K. A., Mason, E., Falk, J. R., & Stunkard, A. (1981). Appetitive behaviour after gastric bypass for obesity. *International Journal of Obesity*, *5*, 457–464.

Hamilton, B. S., Paglia, D., Kwan, A. Y., & Deitel, M. (1995). Increased obese mRNA expression in mental fat cells from massively obese humans. *Nature Medicine*, *1*, 953–956.

Hankir, M. K., Parkinson, J. R., Minnion, J. S., Addison, M. L., Bloom, S. R., & Bell, J. D. (2011). Peptide YY3-36 and pancreatic polypeptide differentially regulate hypothalamic neuronal activity in mice in vivo as measured by manganese-enhanced magnetic resonance imaging. *Journal of Neuroendocrinology*, *23*, 371–380.

Hansen, T. K., Dall, R., Hosoda, H., Kojima, M., Kangawa, K., Christiansen, J. S., et al. (2002). Weight loss increases circulating levels of ghrelin in human obesity. *Clinical Endocrinology (Oxford, UK)*, *56*, 203–206.

Harrold, J. A., Dovey, T. M., Blundell, J. E., & Halford, J. C. (2012). CNS regulation of appetite. *Neuropharmacology*, *63*, 3–17.

Havel, P. J., & Valverde, C. (1996). Autonomic mediation of glucagon secretion during insulin-induced hypoglycaemia in rhesus monkeys. *Diabetes*, *45*, 960–966.

Hayes, M. R., Kanoski, S. E., De Jonghe, B. C., Leichner, T. M., Alhadeff, A. L., Fortin, S. M., et al. (2011). The common hepatic branch of the vagus is not required to mediate the glycaemic and food intake suppressive effects of glucagon-like-peptide-1. *American Journal of Physiology. Regulatory, Integrative and Comparative Physiology*, *301*, 1479–1485.

Haynes, W. G., Morgan, D. A., Djalali, A., Sivitz, W. I., & Mark, A. L. (1999). Interactions between the melanocortin system and leptin in control of sympathetic nerve traffic. *Hypertension*, *33*, 542–547.

Haynes, W. G., Morgan, D. A., Walsh, S. A., Mark, A. L., & Sivitz, W. I. (1997). Receptor-mediated regional sympathetic nerve activation by leptin. *The Journal of Clinical Investigation*, *100*, 270–278.

Henry, M., Ghibaudi, L., Gao, J., & Hwa, J. J. (2005). Energy metabolic profile of mice after chronic activation of central NPY Y1, Y2 or Y5 receptors. *Obesity Research*, *13*, 36–47.

Herness, S., Zhao, F. L., Lu, S. G., Kaya, N., & Shen, T. (2002). Expression and physiological actions of cholecystokinin in rat taste receptor cells. *The Journal of Neuroscience*, *22*, 10018–10029.

Hewson, G., Leighton, G. E., Hill, R. G., & Hughes, J. (1988). The cholecystokinin receptor antagonist L364,718 increases food intake in the rat by attenuation of the action of endogenous cholecystokinin. *British Journal of Pharmacology*, *93*, 79–84.

Heymsfield, S. B., Greenberg, A. S., Fujioka, K., Dixon, R. M., Kushner, R., Hunt, T., et al. (1999). Recombinant leptin for weight loss in obese and lean adults. *JAMA*, *282*, 1568–1575.

Higham, C. E., Hull, R. L., Lawrie, L., Shennan, K. I., Morris, J. F., Birch, N. P., et al. (2000). Processing of synthetic pro-islet amyloid polypeptide (proIAPP) 'amylin' by recombinant prohormone convertase enzymes, PC2 and PC3, in vitro. *European Journal of Biochemistry*, *267*, 4998–5004.

Hommel, J. D., Trinko, R., Sears, R. M., Georgescu, D., Liu, Z. W., Gao, X. B., et al. (2006). Leptin receptor signalling in midbrain dopamine neurons regulates feeding. *Neuron*, *51*, 801–810.

Honda, K., Kamisoyama, H., Saito, N., Kurose, Y., Sugahara, K., & Hasegawa, S. (2007). Central administration of glucagon suppresses food intake in chicks. *Neuroscience Letters*, *416*, 198–201.

Hosoda, H., Kojima, M., Matsuo, H., & Kangawa, K. (2000). Ghrelin and des-acyl ghrelin: Two major forms of rat ghrelin peptide in gastrointestinal tissue. *Biochemical and Biophysical Research Communications*, *279*, 909–913.

Huang, L., Qiu, B., Yuan, L., Zheng, L., Li, Q., & Zhu, S. (2011). Influence of fasting, insulin and glucose on ghrelin in the dorsal vagal complex in rats. *The Journal of Endocrinology*, *211*, 257–262.

Huszar, D., Lynch, C. A., Fairchild-Huntress, V., Dunmore, J. H., Fang, Q., Berkemeier, L. R., et al. (1997). Targeted disruption of the melanocortin-4 receptor results in obesity in mice. *Cell*, *88*, 131–141.

Inabnet, W. B., 3rd, Winegar, D. A., Sherif, B., & Sarr, M. G. (2012). Early outcomes of bariatric surgery in patients with metabolic syndrome: An analysis of the bariatric outcomes longitudinal database. *Journal of the American College of Surgeons*, *214*, 550–556.

Inokuchi, A., Oomura, Y., & Nichimura, H. (1984). Effect of intracerebroventricularly infused glucagon on feeding behaviour. *Physiology and Behavior*, *33*, 397–400.

Inokuchi, A., Oomura, Y., Shimizu, N., & Yamamoto, T. (1986). Central action of glucagon in rat hypothalamus. *American Journal of Physiology. Regulatory, Integrative and Comparative Physiology*, *250*, 120–126.

Jang, H. J., Kokrashvili, Z., Theodorakis, M. J., Carlson, O. D., Kim, B. J., Zhou, J., et al. (2007). Gut-expressed gustducin and taste receptors regulate secretion of glucagon-like peptide-1. *Proceedings of the National Academy of Sciences of the United States of America*, *104*, 15069–15074.

Jéquier, E., & Schutz, Y. (1985). New evidence for a thermogenic defect in human obesity. *International Journal of Obesity*, *9*(Suppl. 2), 1–7.

Jesudason, D. R., Monteiro, M. P., McGowan, B. M., Neary, N. M., Park, A. J., Philippou, E., et al. (2007). Low-dose pancreatic polypeptide inhibits food intake in man. *The British Journal of Nutrition*, *97*, 426–429.

Joel, C. D. (1966). Stimulation of metabolism of rat brown adipose tissue by addition of lipolytic hormones in vitro. *The Journal of Biological Chemistry*, *241*, 814–821.

Jones, B. J., Tan, T., & Bloom, S. R. (2012). Minireview: Glucagon in stress and energy homeostasis. *Endocrinology*, *153*, 1049–1054.

Kaga, T., Inui, A., Okita, M., Asakawa, A., Ueno, N., Kasuga, M., et al. (2001). Modest overexpression of neuropeptide Y in the brain leads to obesity after high-sucrose feeding. *Diabetes, 50*, 1206–1210.

Kamegai, J., Tamura, H., Shimizu, T., Ishii, S., Sugihara, H., & Wakabayashi, I. (2001). Chronic central infusion of ghrelin increases hypothalamic neuropeptide Y and agouti-related protein mRNA levels and body weight in rats. *Diabetes, 50*, 2438–2443.

Kanoski, S. E., Fortin, S. M., Arnold, M., Grill, H. J., & Hayes, M. R. (2011). Peripheral and central GLP-1 receptor populations mediate the anorectic effects of peripherally administered GLP-1 agonists, liraglutide and exendin-4. *Endocrinology, 152*, 3103–3112.

Kanoski, S. E., Rupprecht, L. E., Fortin, S. M., De Jonghe, B. C., & Hayes, M. R. (2012). The role of nausea in food intake and body weight suppression by peripheral GLP-1 receptor agonists, exendin-4 and liraglutide. *Neuropharmacology, 62*, 1916–1927.

Keire, D. A., Bowers, C. W., Solomon, T. W., & Reeve, J. R., Jr. (2002). Structure and receptor binding of PYY analogues. *Peptides, 23*, 305–321.

Kenler, H. A., Brolin, R. E., & Cody, R. P. (1990). Changes in eating behaviour after horizontal gastroplasty and Roux-en-Y gastric bypass. *The American Journal of Clinical Nutrition, 52*, 87–92.

Kieffer, T. J., McIntosh, C. H., & Pederson, R. A. (1995). Degradation of glucose-dependent insulinotropic polypeptide and truncated glucagon-like peptide 1 in vitro and vivo by dipeptidyl peptidase IV. *Endocrinology, 136*, 3585–3596.

Kim, B. (2008). Thyroid hormone as a determinant of energy expenditure and the basal metabolic rate. *Thyroid, 18*, 141–144.

Kim, S. H., & Reaven, G. (2010). Obesity and insulin resistance: An ongoing saga. *Diabetes, 59*, 2105–2106.

Kimball, C. P., & Murlin, J. R. (1923). Aqueous extracts of pancreas III. Some precipitation reactions of insulin. *The Journal of Biological Chemistry, 58*, 337–348.

Kluge, M., Rield, S., Uhr, M., Schmidt, D., Zhang, X., Yassouridis, A., et al. (2010). Ghrelin affects the hypothalamus-pituitary-thyroid axis in humans by increasing free thyroxin and decreasing TSH in plasma. *European Journal of Endocrinology, 162*, 1059–1065.

Knop, F. K., Aaboe, K., Vilsbøll, T., Vølund, A., Holst, J. H., & Krarup, T. (2012). Impaired incretin effect and fasting hyperglucagonaemia characterizing type 2 diabetic subjects are early sings of dysmetabolism in obesity. *Diabetes, Obesity & Metabolism, 14*, 500–510.

Kobelt, P., Wisser, A. S., Stengel, A., Goebel, M., Inhoff, T., Noetzel, S., et al. (2008). Peripheral injection of ghrelin induces *fos* expression in the dorsomedial hypothalamic nucleus in rats. *Brain Research, 1204*, 77–86.

Koda, S., Date, Y., Murakami, N., Shimbara, T., Hanada, T., Toshinai, K., et al. (2005). The role of the vagal nerve in peripheral PYY3-36-induced feeding reduction in rats. *Endocrinology, 146*, 2369–2375.

Kojima, M., Hosoda, H., Date, Y., Nakazato, M., Matsuo, H., & Kangawa, K. (1999). Ghrelin is a growth-hormone-releasing acylated peptide from the stomach. *Nature, 402*, 656–660.

Komatsu, R., Matsuyama, T., Namba, M., Watanabe, N., Itoh, H., Kono, N., et al. (1989). Glucagonostatic and insulinotropic action of glucagon-like peptide 1-(7–36)-amide. *Diabetes, 38*, 902–905.

Kuo, Y. T., Parkinson, J. R., Chaudhri, O. B., Herlihy, A. H., So, P. W., Dhillo, W. S., et al. (2007). The temporal sequence of gut peptide CNS interactions tracked in vivo by magnetic resonance imaging. *The Journal of Neuroscience, 27*, 12341–12348.

Kurose, Y., Kamisoyama, H., Honda, K., Azuma, Y., Sugahara, K., Hasegawa, S., et al. (2009). Effects of central administration of glucagon on feed intake and endocrine responses in sheep. *Animal Science Journal, 80*, 686–690.

Ladenheim, E. E. (2012). Gastrointestinal regulatory peptides and central nervous system mechanisms of weight control. *Current Opinion in Endocrinology, Diabetes, and Obesity*, *19*, 13–18.
Landsberg, L., Saville, M. E., & Young, J. B. (1984). Sympathoadrenal system and regulation of thermogenesis. *American Journal of Physiology. Endocrinology and Metabolism*, *247*, 181–189.
Lauritsen, K. B., Moesgaard, F., Nielsen, M. L., & Jensen, S. L. (1982). Release of gastric inhibitory polypeptide and insulin in response to intrajejunal glucose in duodenal ulcer patients before and after truncal vagotomy. *Scandinavian Journal of Gastroenterology*, *17*, 507–511.
Lawson, E. A., Eddy, K. T., Donoho, D., Misra, M., Miller, K. K., Meenaghan, E., et al. (2011). Appetite-regulating hormones cortisol and peptide YY are associated with disordered eating psychopathology, independent of body mass index. *European Journal of Endocrinology*, *164*, 253–261.
Le Quellec, A., Kervran, A., Blache, P., Ciurana, A. J., & Bataille, D. (1992). Oxyntomodulin-like immunoreactivity: Diurnal profile of a new potential enterogastrone. *The Journal of Clinical Endocrinology and Metabolism*, *74*, 1405–1409.
Le Roux, C. W., Batterham, R. L., Aylwin, S. J., Patterson, M., Borg, C. M., Wynne, K. J., et al. (2006). Attenuated peptide YY release in obese subjects is associated with reduced satiety. *Endocrinology*, *147*, 3–8.
le Roux, C. W., Borg, C. M., Murphy, K. G., Vincent, R. P., Ghatei, M. A., & Bloom, S. R. (2008). Supraphysiological doses of intravenous PYY3-36 cause nausea, but no additional reduction in food intake. *Annals of Clinical Biochemistry*, *45*, 93–95.
Le Roux, C. W., Bueter, M., Theis, N., Werling, M., Ashrafian, H., Löwenstein, C., et al. (2011). Gastric bypass reduces fat intake and preference. *American Journal of Physiology. Regulatory, Integrative and Comparative Physiology*, *301*, 1057–1066.
Le Roux, C. W., Neary, N. M., Halsey, T. J., Small, C. J., Martinez-Isla, A. M., Ghatei, M. A., et al. (2005). Ghrelin does not stimulate food intake in patients with surgical procedures involving vagotomy. *The Journal of Clinical Endocrinology and Metabolism*, *90*, 4521–4524.
Le Sauter, J., & Geary, N. (1990). Redundant vagal mediation of the synergist satiety effect of pancreatic glucagon and cholecystokinin in sham feeding rats. *Journal of the Autonomic Nervous System*, *30*, 13–22.
Le Sauter, J., & Geary, N. (1991). Hepatic portal glucagon infusion decreases spontaneous meal size in rats. *American Journal of Physiology. Regulatory, Integrative and Comparative Physiology*, *261*, 154–161.
Le Sauter, J., Noh, U., & Geary, N. (1991). Hepatic portal infusion of glucagon antibodies increases spontaneous meal size in rats. *American Journal of Physiology. Regulatory, Integrative and Comparative Physiology*, *261*, 162–165.
Le, D. S., Chen, K., Pannacciulli, N., Gluck, M., Reiman, E. M., & Krakoff, J. (2009). Reanalysis of the obesity-related attenuation in the left dorsolateral prefrontal cortex response to a satiating meal using gyral regions-of-interest. *Journal of the American College of Nutrition*, *28*, 667–673.
Legradi, G., Emerson, C. H., Ahima, R. S., Rand, W. M., Flier, J. S., & Lechan, R. M. (1998). Arcuate nucleus ablation prevents fasting-induced suppression of ProTRH mRNA in the hypothalamic paraventricular nucleus. *Neuroendocrinology*, *68*, 89–97.
Leibowitz, S. F., Hammer, N. J., & Chang, K. (1981). Hypothalamic paraventricular nucleus lesions produce overeating and obesity in the rat. *Physiology and Behavior*, *27*, 1031–1040.
Li, J., Ma, W., & Wang, S. (2011). Slower gastric emptying in high-fat diet induced obese rats is associated with attenuated plasma ghrelin and elevated plasma leptin and cholecystokinin concentrations. *Regulatory Peptides*, *171*, 53–57.
Licinio, J., Ribeiro, L., Busnello, J. V., Delibasi, T., Thakur, S., Elashoff, R. M., et al. (2007). Effects of leptin replacement on macro- and micronutrient preferences. *International Journal of Obesity*, *31*, 1859–1863.

Lieverse, R. J., Jansen, J. B., Masclee, A. A., & Lamers, C. B. (1995). Satiety effects of a physiological dose of cholecystokinin in humans. *Gut, 36,* 176–179.
Lo, C. M., King, A., Samuelson, L. C., Kindel, T. L., Rider, T., Jandacek, R. J., et al. (2010). Cholecystokinin knockout mice are resistant to high-fat diet-induced obesity. *Gastroenterology, 138,* 1997–2005.
Lo, C. M., Obici, S., Dong, H. H., Haas, M., Lou, D., Kim, D. H., et al. (2011). Impaired insulin secretion and enhanced insulin sensitivity in cholecystokinin-deficient mice. *Diabetes, 60,* 2000–2007.
Lutz, T. A., Mollet, A., Rushing, P. A., Riediger, T., & Scharrer, E. (2001). The anorectic effect of a chronic peripheral infusion of amylin is abolished in area postrema/nucleus of the solitary tract (AP/NTS) lesioned rats. *International Journal of Obesity and Related Metabolic Disorders: Journal of the International Association for the Study of Obesity, 25,* 1005–1011.
Mack, C. M., Soares, C. J., Wilson, J. K., Athanacio, J. R., Turek, V. F., Trevaskis, J. L., et al. (2010). Davalintide (AC2307), a novel amylin-mimetic peptide: Enhanced pharmacological properties over native amylin to reduce food intake and body weight. *International Journal of Obesity, 34,* 385–395.
Maffei, M., Halaas, J., Ravussin, E., Pratley, R. E., Lee, G. H., Zhang, Y., et al. (1995). Leptin levels in human and rodent: Measurement of plasma leptin and ob RNA in obese and weight-reduced subjects. *Nature Medicine, 1,* 1155–1161.
Maida, A., Lovshin, J. A., Baggio, L. L., & Drucker, D. J. (2008). The glucagon-like peptide-1 receptor agonist oxyntomodulin enhances beta-cell function but does not inhibit gastric emptying in mice. *Endocrinology, 149,* 5670–5678.
Malik, S., McGlone, F., Bedrossian, D., & Dagher, A. (2008). Ghrelin modulates brain activity in areas that control appetitive behaviour. *Cell Metabolism, 7,* 400–409.
Marks, J. L., Porte, D., Jr., Stahl, W. L., & Baskin, D. G. (1990). Localization of insulin receptor mRNA in rat brain by in situ hybridization. *Endocrinology, 127,* 3234–3236.
Marsh, D. J., Hollopeter, G., Huszar, D., Laufer, R., Yagaloff, K. A., Fisher, S. L., et al. (1999). Response of melanocortin-4 receptor-deficient mice to anorectic and orexigenic peptides. *Nature Genetics, 21,* 119–122.
Martin, B., Dotson, C. D., Shin, Y. K., Ji, S., Drucker, D. J., Maudsley, S., et al. (2009). Modulation of taste sensitivity by GLP-1 signalling in taste buds. *Annals of the New York Academy of Sciences, 1170,* 98–101.
Martin, J. R., & Novin, D. (1977). Decreased feeding in rats following hepatic-portal infusion of glucagon. *Physiology and Behavior, 19,* 461–466.
Martin, J. R., Novin, D., & Vanderweele, D. A. (1978). Loss of glucagon suppression of feeding after vagotomy in rats. *American Journal of Physiology. Endocrinology and Metabolism, 234,* 314–318.
Maruyama, H., Hisatomi, A., Orci, L., Grodsky, G. M., & Unger, R. H. (1984). Insulin within islets is a physiologic glucagon release inhibitor. *The Journal of Clinical Investigation, 74,* 2296–2299.
Mayo, K. E., Miller, L. J., Bataille, D., Dalle, S., Göke, B., Thorens, B., et al. (2003). International Union of Pharmacology XXV. The glucagon receptor family. *Pharmacological Reviews, 55,* 167–194.
McGowan, M. K., Andrews, K. M., & Grossman, S. P. (1992). Chronic intrahypothalamic infusions of insulin or insulin antibodies alter body weight and food intake in the rat. *Physiology and Behavior, 51,* 753–766.
McLaughlin, C. L., & Baile, C. A. (1979). Cholecystokinin, amphetamine and diazepam and feeding in lean and obese Zucker rats. *Pharmacology, Biochemistry, and Behavior, 10,* 87–93.
McLaughlin, C. L., Baile, C. A., & Buonomo, F. C. (1985). Effect of CCK antibodies on food intake and weight gain in Zucker rats. *Physiology and Behavior, 34,* 277–282.

McTigue, D. M., & Rogers, R. C. (1995). Pancreatic polypeptide stimulates gastric acid secretion through a vagal mechanism in rats. *American Journal of Physiology. Regulatory, Integrative and Comparative Physiology, 269*, 983–987.
Meereis-Schwanke, K., Klonoswski-Stumpe, H., Herberg, L., & Niederau, C. (1998). Long term effects of CCK-agonist and -antagonist on food intake and body weight in Zucker lean and obese rats. *Peptides, 19*, 291–299.
Mentlein, R., Dahms, P., Grandt, D., & Krüger, R. (1993). Proteolytic processing of neuropeptide Y and peptide YY by dipeptidyl peptidase IV. *Regulatory Peptides, 49*, 133–144.
Mercer, J. G., Hoggard, N., Williams, L. M., Lawrence, C. B., Hannah, L. T., & Trayhurn, P. (1996). Localization of leptin receptor mRNA and the long form splice variant (Ob-Rb) in mouse hypothalamus and adjacent brain regions by in situ hybridization. *FEBS Letters, 387*, 113–116.
Miyawaki, K., Yamada, Y., Ban, N., Ihara, Y., Tsukiyama, K., Zhou, H., et al. (2002). Inhibition of gastric inhibitory polypeptide signalling prevents obesity. *Nature Medicine, 8*, 738–742.
Moens, K., Heimberg, H., Flamez, D., Huypens, P., Quartier, E., Ling, Z., et al. (1996). Expression and functional activity of glucagon, glucagon-like peptide 1, and glucose-dependent insulinotropic peptide receptors in rat pancreatic islet cells. *Diabetes, 45*, 257–261.
Montague, C. T., Farooqi, I. S., Whitehead, J. P., Soos, M. A., Rau, H., Wareham, N. J., et al. (1997). Congenital leptin deficiency is associated with severe early-onset obesity in humans. *Nature, 387*, 903–908.
Mortensen, K., Christensen, L. L., Holst, J. J., & Orskov, C. (2003). GLP-1 and GIP are colocalised to a subset of endocrine cells in the small intestine. *Regulatory Peptides, 114*, 189–196.
Mosjov, S., Weir, G. C., & Habener, J. F. (1987). Insulinotropin: Glucagon-like peptide I (7–37) co-encoded in the glucagon gene is a potent stimulator of insulin release in the perfused rat pancreas. *The Journal of Clinical Investigation, 79*, 616–619.
Nair, K. S. (1987). Hyperglucagonemia increases resting metabolic rate in man during insulin deficiency. *The Journal of Clinical Endocrinology and Metabolism, 64*, 896–901.
Nakazato, M., Murakami, N., Date, Y., Kojima, M., Matsuo, H., Kangawa, K., et al. (2001). A role for ghrelin in the central regulation of feeding. *Nature, 409*, 194–198.
Naleid, A. M., Grace, M. K., Cummings, D. E., & Levine, A. S. (2005). Ghrelin induces feeding in the mesolimbic reward pathway between the ventral tegmental area and the nucleus accumbens. *Peptides, 26*, 2274–2279.
Näslund, E., Barkeling, B., King, N., Gutniak, M., Blundell, J. E., Holst, J. J., et al. (1999). Energy intake and appetite are suppressed by glucagon-like peptide-1 (GLP-1) in obese men. *International Journal of Obesity and Related Metabolic Disorders: Journal of the International Association for the Study of Obesity, 23*, 304–311.
Nauck, M. A., Heimesaat, M. M., Behle, K., Holst, J. J., Nauck, M. S., Ritzel, R., et al. (2002). Effects of glucagon-like peptide 1 on counterregulatory hormone responses, cognitive functions, and insulin secretion during hyperinsulinemic, stepped hypoglycemic clamp experiments in healthy volunteers. *The Journal of Clinical Endocrinology and Metabolism, 87*, 1239–1246.
Neary, N. M., Small, C. J., Wren, A. M., Lee, J. L., Druce, M. R., Palmieri, C., et al. (2004). Ghrelin increases energy intake in cancer patients with impaired appetite: Acute, randomised, placebo-controlled trial. *The Journal of Clinical Endocrinology and Metabolism, 89*, 2832–2836.
Nefti, W., Chaumontet, C., Fromentin, G., Tomé, D., & Darcel, N. (2009). A high-fat diet attenuates the central response to within-meal satiation signals and modifies the receptor

expression of vagal afferents in mice. *American Journal of Physiology. Regulatory, Integrative and Comparative Physiology*, *296*, 1681–1686.

NICE, (2006). Obesity: the prevention, identification, assessment and management of overweight and obesity in adults and children. *Clinical Guidance*, *43*, http://www.nice.org.uk/cg043.

Obici, S., Feng, Z., Karkanias, G., Baskin, D. G., & Rossetti, L. (2002). Decreasing hypothalamic insulin receptors causes hyperphagia and insulin resistance in rats. *Nature Neuroscience*, *5*, 566–572.

Obici, S., Zhang, B. B., Karkanias, G., & Rossetti, L. (2002). Hypothalamic insulin signalling is required for inhibition of glucose production. *Nature Medicine*, *8*, 1376–1382.

Oetting, R. L., & Vanderweele, D. A. (1985). Insulin suppresses intake without inducing illness in sham feeding rats. *Physiology and Behavior*, *34*, 557–562.

Ohneda, A., Kobayashi, T., Nihei, J., Imamura, M., Naito, H., & Tsuchiya, T. (1985). Role of vagus nerve in secretion of gastric inhibitory polypeptide in dogs. *The Tohoku Journal of Experimental Medicine*, *147*, 183–190.

Oliván, B., Teixeira, J., Bose, M., Bawa, B., Chang, T., Summe, H., et al. (2009). Effect of weight loss by diet or gastric bypass surgery on peptide YY3-36 levels. *Annals of Surgery*, *249*, 948–953.

Oliveira, K. J., Paula, G. S., Costa-e-Sousa, R. H., Souza, L. L., Moraes, D. C., Curty, F. H., et al. (2006). Peptide YY (PYY)3-36 modulates thyrotropin secretion in rats. *The Journal of Endocrinology*, *191*, 459–463.

Oswal, A., & Yeo, G. (2010). Leptin and the control of body weight: A review of its diverse central targets, signalling mechanisms, and role in the pathogenesis of obesity. *Obesity*, *18*, 221–229.

Padwal, R., Klarenbach, S., Wiebe, N., Birch, D., Karmali, D., Manns, B., et al. (2011). Bariatric surgery: A systematic review and network meta-analysis of randomized trials. *Obesity Reviews*, *12*, 602–612.

Padwal, R., Li, S. K., & Lau, D. C. (2004). Long-term pharmacotherapy for obesity and overweight. *Cochrane Database of Systematic Reviews (Online)*, (3), CD004094.

Pannacciulli, N., Le, D. S., Salbe, A. D., Chen, K., Reiman, E. M., Tataranni, P. A., & Krakoff, J. (2007). Postprandial glucagon-like peptide-1 (GLP-1) response is positively associated with changes in neuronal activity of brain areas implicated in satiety and food intake regulation in humans. *NeuroImage*, *35*, 511–517.

Parkinson, J. R., Chaudhri, O. B., Kuo, Y. T., Field, B. C., Herlihy, A. H., Dhillo, W. S., et al. (2009). Differential patterns of neuronal activation in the brainstem and hypothalamus following peripheral injection of GLP-1, oxyntomodulin and lithium chloride in mice detected by manganese-enhanced magnetic resonance imaging (MEMRI). *NeuroImage*, *44*, 1022–1031.

Parlevliet, E. T., Heijober, A. C., Schröder-van der Elst, J. P., Haveskes, L. M., Romijn, J. A., Pijl, H., et al. (2008). Oxyntomodulin ameliorates glucose intolerance in mice fed a high-fat diet. *American Journal of Physiology. Endocrinology and Metabolism*, *294*, 142–147.

Patterson, M., Murphy, K. G., Patel, S. R., Patel, N. A., Greenwood, H. C., Cooke, J. H., et al. (2009). Hypothalamic injection of oxyntomodulin suppresses circulating ghrelin-like immunoreactivity. *Endocrinology*, *150*, 3513–3520.

Peiris, A. N., Sothmann, M. S., Hennes, M. I., Lee, M. B., Wilson, C. R., Gustafson, A. B., et al. (1989). Relative contribution of obesity and body fat distribution to alterations in glucose insulin homeostasis: Predictive values of selected indices in premenopausal women. *The American Journal of Clinical Nutrition*, *49*, 758–764.

Perley, M. J., & Kipnis, D. M. (1967). Plasma insulin responses to oral and intravenous glucose: Studies in normal and obese subjects. *The Journal of Clinical Investigation*, *46*, 1954–1962.

Peters, J. H., Simasko, S. M., & Ritter, R. C. (2006). Modulation of vagal afferent excitation and reduction of food intake by leptin and cholecystokinin. *Physiology and Behavior, 89*, 477–485.

Pfluger, P. T., Castañeda, T. R., Heppner, K. M., Strassburg, S., Kruthaupt, T., Chaudhary, N., et al. (2011). Ghrelin, peptide YY and their hypothalamic targets differentially regulate spontaneous physical activity. *Physiology and Behavior, 105*, 52–61.

Pfluger, P. T., Kirchner, H., Günnel, S., Schrott, B., Perez-Tilve, D., Fu, S., et al. (2008). Simultaneous deletion of ghrelin and its receptor increases motor activity and energy expenditure. *American Journal of Physiology. Gastrointestinal and Liver Physiology, 294*, 610–618.

Philippe, J. (1989). Glucagon gene transcription is negatively regulated by insulin in a hamster islet cell line. *The Journal of Clinical Investigation, 84*, 672–677.

Picot, J., Jones, J., Colquitt, J. L., Gospodarevskava, E., Loveman, E., Baxter, L., et al. (2009). The clinical effectiveness and cost-effectiveness of bariatric (weight loss) surgery for obesity: A systematic review and economic evaluation. *Health Technology Assessment (Winchester, England), 13*, 1–190 215–357.

Pilichiewicz, A. N., Little, T. J., Brennan, I. M., Meyer, J. H., Wishart, J. M., Otto, B., et al. (2006). Effects of load, and duration, of duodenal lipid on antropyloroduodenal motility, plasma CCK and PYY, and energy intake in healthy men. *American Journal of Physiology. Regulatory, Integrative and Comparative Physiology, 290*, 668–677.

Pi-Sunyer, X., Kissileff, H. R., Thornton, J., & Smith, G. P. (1982). C-terminal octapeptide of cholecystokinin decreases food intake in obese man. *Physiology and Behavior, 29*, 627–630.

Pittner, R. A., Moore, C. X., Bhavsar, S. P., Gedulin, B. R., Smith, P. A., Jodka, C. M., et al. (2004). Effects of PYY[3–36] in rodent models of diabetes and obesity. *International Journal of Obesity and Related Metabolic Disorders: Journal for the International Association for the Study of Obesity, 28*, 963–971.

Plagemann, A., Rake, A., Harder, T., Melchior, K., Rohde, W., & Dörner, G. (1998). Reduction of cholecystokinin-8S-neurons in the paraventricular hypothalamic nucleus of neonatally overfed weanling rats. *Neuroscience Letters, 258*, 13–16.

Pocai, A., Carrington, P. E., Adams, J. R., Wright, M., Eiermann, G., Zhu, L., et al. (2009). Glucagon-like peptide 1/glucagon receptor dual agonism reverses obesity in mice. *Diabetes, 58*, 2258–2266.

Pocai, A., Lam, T. K., Gutierrez-Juarez, R., Obici, S., Schwartz, G. J., Bryan, J., et al. (2005). Hypothalamic K(ATP) channels control hepatic glucose production. *Nature, 434*, 1026–1031.

Polonsky, K. S., Given, B. D., & Van Cauter, E. (1988). Twenty-four-hour profiles and pulsatile patterns of insulin secretion in normal and obese subjects. *The Journal of Clinical Investigation, 81*, 442–448.

Pories, W. J. (2004). Diabetes: The evolution of a new paradigm. *Annals of Surgery, 239*, 12–13.

Pralong, F. P., Corder, R., & Gaillard, R. C. (1993). The effects of chronic glucocorticoid excess, adrenalectomy and stress on neuropeptide Y in individual rat hypothalamic nuclei. *Neuropeptides, 25*, 223–231.

Pritchard, L. E., & White, A. (2005). Agouti-related protein: More than a melanocortin-4 receptor antagonist? *Peptides, 26*, 1759–1770.

Rabinowitz, D. (1968). Hormonal profile and forearm metabolism in human obesity. *The American Journal of Clinical Nutrition, 21*, 1438–1444.

Ranganath, L. R., Beety, J. M., Morgan, L. M., Wright, J. W., Howland, R., & Marks, V. (1996). Attenuated GLP-1 secretion in obesity: Cause or consequence? *Gut, 38*, 916–919.

Raposinho, P. D., Pierroz, D. D., Broqua, P., Whie, R. B., Pedrazzini, T., & Aubert, M. L. (2001). Chronic administration of neuropeptide Y into the lateral ventricle of C57BL/6J male mice produces an obesity syndrome including hyperphagia, hyperleptinaemia, insulin resistance, and hypogonadism. *Molecular and Cellular Endocrinology*, *185*, 195–204.

Reidleberger, R. D., Haver, A. C., Chelikani, P. K., & Buescher, J. L. (2008). Effects of different intermittent peptide YY(3–36) dosing strategies on food intake, body weight, and adiposity in diet-induced obese rats. *American Journal of Physiology. Regulatory, Integrative and Comparative Physiology*, *295*, 449–458.

Reidleberger, R. D., Hernandez, J., Fritzsch, B., & Hulce, M. (2004). Abdominal vagal mediation of the satiety effects of CCK in rats. *American Journal of Physiology. Regulatory, Integrative and Comparative Physiology*, *286*, 1005–1012.

Reimer, R. A., Maurer, A. D., Lau, D. C., & Auer, R. N. (2010). Long-term dietary restriction influences plasma ghrelin and GOAT mRNA level in rats. *Physiology and Behavior*, *99*, 605–610.

Riediger, T., Zuend, D., Becskei, C., & Lutz, T. A. (2004). The anorectic hormone amylin contributes to feeding-related changes of neuronal activity in key structures in the gut-brain axis. *American Journal of Physiology. Regulatory, Integrative and Comparative Physiology*, *286*, 114–122.

Riedy, C. A., Chavez, M., Figlewicz, D. P., & Woods, S. C. (1995). Central insulin enhances sensitivity to cholecystokinin. *Physiology and Behavior*, *58*, 755–760.

Rijkelijkhuizen, J. M., McQuarrie, K., Girman, C. J., Stein, P. P., Mari, A., Holst, J. J., et al. (2010). Effects of meal size and composition on incretin, alpha-cell, and beta-cell responses. *Metabolism*, *59*, 502–511.

Rissanen, A., Hakala, P., Lissner, L., Mattlar, C. E., Koskenvuo, M., & Rönnemaa, T. (2002). Acquired preference especially for dietary fat and obesity: A study of weight-discordant monozygotic twin pairs. *International Journal of Obesity and Related Metabolic Disorders: Journal of the International Association for the Study of Obesity*, *26*, 973–977.

Roberge, J. N., & Brubaker, P. L. (1993). Regulation of intestinal proglucagon-derived peptide secretion by glucose-dependent insulinotropic peptide in a novel enteroendocrine loop. *Endocrinology*, *133*, 233–240.

Rocca, A. S., & Brubaker, P. L. (1999). Role of the vagus nerve in mediating proximal nutrient-induced glucagon-like peptide-1 secretion. *Endocrinology*, *140*, 1687–1694.

Rocha, D. M., Faloona, G. R., & Unger, R. H. (1972). Glucagon-stimulating activity of 20 amino acids in dogs. *The Journal of Clinical Investigation*, *51*, 2346–2351.

Rodin, J., Wack, J., Ferrannini, E., & DeFronzo, R. A. (1985). Effect of insulin and glucose on feeding behaviour. *Metabolism*, *34*, 826–831.

Roseberry, A. G., Liu, H., Jackson, A. C., Cai, X., & Friedman, J. M. (2004). Neuropeptide Y-mediated inhibition of proopiomelanocortin neurons in the arcuate nucleus shows enhanced desensitization in ob/ob mice. *Neuron*, *41*, 711–722.

Roth, C. L., Enriori, P. J., Harz, K., Woelfle, J., Cowley, M. A., & Reinehr, T. (2005). Peptide YY is a regulator of energy homeostasis in obese children before and after weight loss. *The Journal of Clinical Endocrinology and Metabolism*, *90*, 6386–6391.

Roth, J. D., Hughes, H., Kendall, E., Baron, A. D., & Anderson, C. M. (2006). Anti-obesity effects of the beta-cell hormone amylin in diet-induced obese rats: Effects on food intake, body weight, composition, energy expenditure, and gene expression. *Endocrinology*, *147*, 5855–5864.

Rothwell, N. J., & Stock, M. J. (1979). A role for brown adipose tissue in diet-induced thermogenesis. *Nature*, *281*, 31–35.

Sainsbury, A., Rohner-Jeanrenaud, F., Grouzmann, E., & Jeanrenaud, B. (1996). Acute intracerebroventricular administration of neuropeptide Y stimulates corticosterone output and feeding but not insulin output in normal rats. *Neuroendocrinology*, *63*, 318–326.

Sainsbury, A., & Zhang, L. (2010). Role of the arcuate nucleus of the hypothalamus in regulation of body weight during energy deficit. *Molecular and Cellular Endocrinology, 316*, 109–119.
Salehi, A., Dornonville de la Cour, C., Håkanson, R., & Lundquist, I. (2004). Effects of ghrelin on insulin and glucagon secretion: A study of isolated pancreatic islets and intact mice. *Regulatory Peptides, 118*, 143–150.
Sanke, T., Bell, G. I., Sample, C., Rubenstein, A. H., & Steiner, D. F. (1988). An islet amyloid peptide is derived from an 89-amino acid precursor by proteolytic processing. *The Journal of Biological Chemistry, 263*, 17243–17246.
Santiago, J. V., Haymond, M. W., Clarke, W. L., & Pagliara, A. S. (1977). Glucagon, insulin and glucose response to physiologic testing in normal and massively obese adults. *Metabolism, 26*, 1115–1122.
Scarpace, P. J., & Zhang, Y. (2009). Leptin resistance: A predisposing factor for diet-induced obesity. *American Journal of Physiology. Regulatory, Integrative and Comparative Physiology, 296*, 493–500.
Scheid, J. L., & De Souza, M. J. (2010). Menstrual irregularities and energy deficiency in physically active women: The role of ghrelin PYY and adipocytokines. *Medicine and Sport Science, 55*, 82–102.
Schjoldager, B. T., Baldissera, F. G., Mortensen, P. E., Holst, J. J., & Christiansen, J. (1988). Oxyntomodulin: A potential hormone from the distal gut. Pharmacokinetics and effects on gastric acid and insulin secretion in man. *European Journal of Clinical Investigation, 18*, 499–503.
Schmidt, P. T., Näslund, E., Grybäck, P., Jacobsson, H., Holst, J. J., Hilsted, L., et al. (2005). A role for pancreatic polypeptide in the regulation of gastric emptying and short-term metabolic control. *The Journal of Clinical Endocrinology and Metabolism, 90*, 5241–5246.
Schulman, J. L., Carleton, J. L., Whitney, G., & Whitehorn, J. C. (1957). Effect of glucagon on food intake and body weight in man. *Journal of Applied Physiology, 11*, 419–421.
Schutz, Y., Bessard, T., & Jéquier, E. (1984). Diet-induced thermogenesis measured over a whole day in obese and non-obese women. *The American Journal of Clinical Nutrition, 40*, 542–552.
Schwartz, T. W. (1983). Pancreatic polypeptide: A unique model for vagal control of endocrine systems. *Journal of the Autonomic Nervous System, 9*, 99–111.
Schwartz, T. W., Holst, J. J., Fahrenkrug, J., Jensen, S. L., Nielsen, O. V., Rehfeld, J. F., et al. (1978). Vagal, cholinergic regulation of pancreatic polypeptide secretion. *The Journal of Clinical Investigation, 61*, 781–789.
Schwartz, G. J., & Moran, T. H. (1994). CCK elicits and modulates vagal afferent activity arising from gastric and duodenal sites. *Annals of the New York Academy of Sciences, 23*, 121–128.
Schwartz, M. W., Sipols, A. J., Marks, J. L., Sanacora, G., White, J. D., Scheurink, A., et al. (1992). Inhibition of hypothalamic neuropeptide Y gene expression by insulin. *Endocrinology, 130*, 3608–3616.
Scott, M. M., Williams, K. W., Rossi, J., Lee, C. E., & Elmquist, J. K. (2011). Leptin receptor expression in hindbrain GLP-1 neurons regulates food intake and energy balance in mice. *The Journal of Clinical Investigation, 121*, 2413–2421.
Secchi, A., Caldara, R., Caumo, A., Monti, L. D., Bonfatti, D., Di Carlo, V., et al. (1995). Cephalic-phase insulin and glucagon release in normal subjects and in patients receiving pancreas transplantation. *Metabolism, 44*, 1153–1158.
Seo, S., Ju, S., Chung, H., Lee, D., & Park, S. (2008). Acute effects of glucagon-like peptide-1 on hypothalamic neuropeptide and AMP activated kinase expression in fasted rats. *Endocrine Journal, 55*, 867–874.
Seth, R., Knight, W. D., & Overton, J. M. (2011). Combined amylin-leptin treatment lowers blood pressure and adiposity in lean and obese rats. *International Journal of Obesity, 35*, 1183–1192.

Shen, T., Kaya, N., Zhao, F. L., Lu, S. G., Cao, Y., & Herness, S. (2005). Co-expression patterns of the neuropeptides vasoactive intestinal peptide and cholecystokinin with the transduction molecules alpha-gustducin and T1R2 in rat taste receptor cells. *Neuroscience, 130*, 229–238.

Shin, A. C., & Berthoud, H. R. (2011). Food reward functions as affected by obesity and bariatric surgery. *International Journal of Obesity, 35*(Suppl. 3), 40–44.

Shin, Y. K., Martin, B., Golden, E., Dotson, C. D., Maudsley, S., Kim, W., et al. (2008). Modulation of taste sensitivity by GLP-1 signalling. *Journal of Neurochemistry, 106*, 455–463.

Shin, A. C., Zheng, H., Townsend, R. L., Sigalet, D. L., & Berthoud, H. R. (2010). Meal-induced hormone responses in a rat model of Roux-en-Y gastric bypass surgery. *Endocrinology, 151*, 1588–1597.

Silvestre, R. A., Rodríguez-Gallardo, J., Jodka, C., Parkes, D. G., Pittner, R. A., Young, A. A., et al. (2001). Selective amylin inhibition of the glucagon response to arginine is extrinsic to the pancreas. *American Journal of Physiology. Endocrinology and Metabolism, 280*, 443–449.

Simpson, S. A., Shaw, C., & McNamara, R. (2011). What is the most effective way to maintain weight loss in adults? *BMJ, 343*, 8042.

Sjöström, L., Narbro, K., Sjöström, C. D., Karason, K., Larsson, B., Wedel, H., et al. (2007). Effects of bariatric surgery on mortality in Swedish obese subjects. *The New England Journal of Medicine, 357*, 741–752.

Sjöström, L., Peltonen, M., Jacobson, P., Sjöström, C. D., Karason, K., Wedel, H., et al. (2012). Bariatric surgery and long-term cardiovascular events. *Journal of the American Medical Association, 307*, 56–65.

Skibicka, K. P., Hansson, C., Egecioglu, E., & Dickson, S. L. (2012). Role of ghrelin in food reward: Impact of ghrelin on sucrose self-administration and mesolimbic dopamine and acetylcholine receptor gene expression. *Addiction Biology, 17*, 95–107.

Smith, S. R., Aronne, L. J., Burns, C. M., Kesty, N. C., Halseth, A. E., & Weyer, C. (2008). Sustained weight loss following 12-month pramlintide treatment as an adjunct to lifestyle intervention in obesity. *Diabetes Care, 31*, 1816–1823.

Smith, G. P., Jerome, C., Cushin, B. J., Eterno, R., & Simansky, K. J. (1981). Abdominal vagotomy blocks the satiety effect of cholecystokinin in the rat. *Science, 213*, 1036–1037.

Sosić-Jurjević, B., Stevanović, D., Milosević, V., Sekulić, M., & Starcević, V. (2009). Central ghrelin affects pituitary-thyroid axis: Histomorphological and hormonal study in rats. *Neuroendocrinology, 89*, 327–336.

Spanswick, D., Smith, M. A., Mirshamsi, S., Routh, V. H., & Ashford, M. L. (2000). Insulin activates ATP-sensitive K+ channels in hypothalamic neurons of lean, but not obese, rats. *Nature Neuroscience, 3*, 757–758.

Stanley, B. G., Daniel, D. R., Chin, A. S., & Leibowitz, S. F. (1985). Paraventricular nucleus injection of peptide YY and neuropeptide Y preferentially enhance carbohydrate ingestion. *Peptides, 6*, 1205–1211.

Staub, A., Sinn, L., & Behrens, O. K. (1955). Purification and crystallization of glucagon. *The Journal of Biological Chemistry, 214*, 619–632.

Steinert, R. E., Gerspach, A. C., Gutmann, H., Asarian, L., Drewe, J., & Beglinger, C. (2011). The functional involvement of gut-expressed sweet taste receptors in glucose-stimulated secretion of glucagon-like peptide-1 (GLP-1) and peptide YY (PYY). *Clinical Nutrition, 30*, 524–532.

Stellar, E. (1954). The physiology of motivation. *Psychology Review, 61*, 5–22.

Stice, E., Spoor, S., Ng, J., & Zald, D. H. (2009). Relation of obesity to consummatory and anticipatory food reward. *Physiology and Behavior, 97*, 551–560.

St-Pierre, D. H., Karelis, A. D., Cianflone, K., Conus, F., Mignault, D., Rabasa-Lhoret, R., et al. (2004). Relationship between ghrelin and energy expenditure in healthy young women. *The Journal of Clinical Endocrinology and Metabolism, 89*, 5993–5997.

Strubbe, J. H., Porte, D., Jr., & Woods, S. C. (1988). Insulin responses and glucose levels in plasma and cerebrospinal fluid during fasting and refeeding in the rat. *Physiology and Behavior, 44*, 205–208.

Sumithran, P., Prendergast, L. A., Delbridge, E., Purcell, K., Shulkes, A., Kriketos, A., et al. (2011). Long-term persistence of hormonal adaptations to weight loss. *The New England Journal of Medicine, 365*, 1597–1604.

Suzuki, K., Simpson, K. A., Minnion, J. S., Shillito, J. C., & Bloom, S. R. (2010). The role of gut hormones and the hypothalamus in appetite regulation. *Endocrine Journal, 57*, 359–372.

Tan, T. M., Field, B. C., McCullough, K. A., Troke, R. C., Chambers, E. S., & Salem, V. (2012). Coadministration of Glucagon-Like Peptide-1 During Glucagon Infusion in Man Results in Increased Energy Expenditure and Amelioration of Hyperglycemia. *Diabetes*, Epub ahead of print.

Tan, T. M., Field, B. C., Minnion, J. S., Cuenco-Shillito, J., Chambers, E. S., Zac-Varghese, S., et al. (2012). Pharmacokinetics, adverse effects and tolerability of a novel analogue of human pancreatic polypeptide, PP 1420. *British Journal of Clinical Pharmacology, 73*, 232–239.

Tang-Christensen, M., Larsen, P. J., Göke, R., Fink-Jensen, A., Jessop, D. S., Møller, M., et al. (1996). Central administration of GLP-1-(7–36) amide inhibits food and water intake in rats. *American Journal of Physiology. Regulatory, Integrative and Comparative Physiology, 271*, 848–856.

Tang-Christensen, M., Vrang, N., Ortmann, S., Bidlingmaier, M., Horvath, T. L., & Tschöp, M. (2004). Central administration of ghrelin and agouti-related protein (83–132) increases food intake and decreases spontaneous locomotor activity in rats. *Endocrinology, 145*, 4645–4652.

Tartaglia, L. A., Dembski, M., Weng, X., Deng, N., Culpepper, J., Devos, R., et al. (1995). Identification and expression cloning of a leptin receptor, OB-R. *Cell, 83*, 1263–1271.

Tatemoto, K. (1982). Isolation and characterization of peptide YY (PYY), a candidate gut hormone that inhibits pancreatic exocrine secretion. *Proceedings of the National Academy of Sciences of the United States of America, 79*, 2514–2518.

Tatemoto, K., Carlquist, M., & Mutt, V. (1982). Neuropeptide Y—A novel brain peptide with structure similarities to peptide YY and pancreatic polypeptide. *Nature, 296*, 659–660.

Taylor, I. L., & Feldman, M. (1982). Effect of cephalic-vagal stimulation on insulin, gastric inhibitory polypeptide, and pancreatic polypeptide release in humans. *The Journal of Clinical Endocrinology and Metabolism, 55*, 1114–1117.

Tong, Q., Ye, C. P., Jones, J. E., Elmquist, J. K., & Lowell, B. B. (2008). Synaptic release of GABA by AgRP neurons is required for normal regulation of energy balance. *Nature Neuroscience, 11*, 998–1000.

Tschöp, M., Wawarta, R., Riepl, R. L., Friedrich, S., Bidlingmaier, M., Landgraf, R., et al. (2001). Post-prandial decrease of circulating human ghrelin levels. *Journal of Endocrinological Investigation, 24*, 19–21.

Tschöp, M., Weyer, C., Tataranni, P. A., Devanarayan, V., Ravussin, E., & Heiman, M. L. (2001). Circulating ghrelin levels are decreased in human obesity. *Diabetes, 50*, 707–709.

Ueberberg, B., Unger, N., Saeger, W., Mann, K., & Petersenn, S. (2009). Expression of ghrelin and its receptor in human tissues. *Hormone and Metabolic Research, 41*, 814–821.

Ueno, N., Inui, A., Iwamoto, M., Kaga, T., Asakawa, A., Okita, M., et al. (1999). Decreased food intake and body weight in pancreatic polypeptide-overexpressing mice. *Gastroenterology, 117*, 1427–1432.

Ugleholdt, R., Zhu, X., Deacon, C. F., Ørskov, C., Steiner, D. F., & Holst, J. J. (2004). Impaired intestinal proglucagon processing in mice lacking prohormone convertase 1. *Endocrinology, 145*, 1349–1355.

Van der Burg, M. P., Guicherit, O. R., Frölich, M., & Gooszen, H. G. (1995). Insulinotropic effects of cholecystokinin, gastric inhibitory polypeptide and glucagon-like peptide-1 during perfusion of short-term cultured canine isolated islets. *Regulatory Peptides*, *60*, 61–67.

Van Dijk, G., de Groote, C., Chavez, M., van der Werf, Y., Steffens, A. B., & Strubbe, J. H. (1997). Insulin in the arcuate nucleus of the hypothalamus reduces fat consumption in rats. *Brain Research*, 777, 147–152.

Van Dijk, G., Lindskog, S., Holst, J. J., Steffens, A. B., & Ahrén, B. (1996). Effects of glucagon-like peptide-1 on glucose turnover in rats. *American Journal of Physiology. Endocrinology and Metabolism*, *270*, 1015–1021.

VanderWeele, D. A. (1994). Insulin is a prandial satiety hormone. *Physiology and Behavior*, *56*, 619–622.

Verdich, C., Flint, A., Gutzwiller, J. P., Näslund, E., Beglinger, C., Hellström, P. M., et al. (2001). A meta-analysis of the effect of glucagon-like peptide-1(7–36) amide on ad libitum energy intake in humans. *The Journal of Clinical Endocrinology and Metabolism*, *86*, 4382–4389.

Verdich, C., Toubro, S., Buemann, B., Lysgård Madsen, J., Juul Holst, J., & Astrup, A. (2001). The role of postprandial releases of insulin and incretin hormones in meal-induced satiety—Effect of obesity and weight reduction. *International Journal of Obesity and Related Metabolic Disorders: Journal of the International Association for the Study of Obesity*, *25*, 1206–1214.

Verspohl, E. J., Ammon, H. P., Williams, J. A., & Goldfine, I. D. (1986). Evidence that cholecystokinin interacts with specific receptors and regulates insulin release in isolate rat islets of Langerhans. *Diabetes*, *35*, 38–43.

Vilsbøll, T., Christensen, M., Junker, A. E., Knop, F. K., & Gluud, L. L. (2012). Effects of glucagon-like peptide-1 receptor agonists on weight loss: Systematic review and meta-analyses of randomised controlled trials. *BMJ*, *344*, 7771.

Vilsbøll, T., & Holst, J. J. (2004). Incretins, insulin secretion and Type 2 diabetes mellitus. *Diabetologia*, *47*, 357–366.

Vilsbøll, T., Krarup, T., Sonne, J., Madsbad, S., Vølund, A., Juul, A. G., et al. (2003). Incretin secretion in relation to meal size and body weight in healthy subjects and people with type 1 and type 2 diabetes mellitus. *The Journal of Clinical Endocrinology and Metabolism*, *88*, 2706–2713.

Wang, J. H., Wang, F., Yang, M. J., Yu, D. F., Wu, W. N., Liu, J., et al. (2008). Leptin regulated calcium channels of neuropeptide Y and proopiomelanocortin neurons by activation of different signal pathways. *Neuroscience*, *156*, 89–98.

Weatherford, S. C., & Ritter, S. (1988). Lesion of vagal afferent terminals impairs glucagon-induced suppression of food intake. *Physiology and Behavior*, *43*, 645–650.

Wellman, P. J., Nation, J. R., & Davis, K. W. (2007). Impairment of acquisition of cocaine self-administration in rats maintained on a high-fat diet. *Pharmacology, Biochemistry, and Behavior*, *88*, 89–93.

West, D. B., Fey, D., & Woods, S. C. (1984). Cholecystokinin persistently suppresses meal size but not food intake in free-feeding rats. *American Journal of Physiology. Regulatory, Integrative and Comparative Physiology*, *246*, 776–787.

Westerterp, K. R. (2004). Diet induced thermogenesis. *Nutrition and Metabolism (London)*, *18*, 5.

Wetzler, S., Dumaz, V., Goubern, M., Tomé, D., & Larue-Achagiotis, C. (2004). Intraperitoneal leptin modifies macronutrient choice in self-selecting rats. *Physiology and Behavior*, *83*, 65–72.

Whitcomb, D. C., Puccio, A. M., Vigna, S. R., Taylor, I. L., & Hoffman, G. E. (1997). Distribution of pancreatic polypeptide receptors in the rat brain. *Brain Research*, *760*, 137–149.

Whitcomb, D. C., Taylor, I. L., & Vigna, S. R. (1990). Characterization of saturable binding sites for circulating pancreatic polypeptide in rat brain. *American Journal of Physiology. Gastrointestinal and Liver Physiology, 259*, 687–691.

Willing, A. E., Walls, E. K., & Koopmans, H. S. (1994). Insulin increases the daily food intake of diabetic rats on high and low fat diets. *Physiology and Behavior, 56*, 983–991.

Willms, B., Werner, J., Holst, J. J., Orskov, C., Creutzfeldt, W., & Nauck, M. A. (1996). Gastric emptying, glucose responses, and insulin secretion after a liquid test meal: Effects of exogenous glucagon-like peptide-1 (GLP-1)-(7–36)amide in type 2 (noninsulin-dependent) diabetic patients. *The Journal of Clinical Endocrinology and Metabolism, 81*, 327–332.

Woods, S. C., Lotter, E. C., McKay, L. D., & Porte, D., Jr. (1979). Chronic intracerebroventricular infusion of insulin reduces food intake and body weight of baboons. *Nature, 282*, 503–505.

Woods, S. C., Lutz, T. A., Geary, N., & Langhans, W. (2006). Pancreatic signals controlling food intake; insulin, glucagon and amylin. *Philosophical Transactions of the Royal Society B: Biological Sciences, 361*, 1219–1235.

World Health Organisation (2012) Obesity and Overweight (2012) *Factsheet 311* http://www.who.int/mediacentre/factsheets/fs311/en/

Wren, A. M., Seal, L. J., Cohen, M. A., Brynes, A. E., Frost, G. S., Murphy, K. G., et al. (2001). Ghrelin enhances appetite and increases food intake in humans. *The Journal of Clinical Endocrinology and Metabolism, 86*, 5992.

Wren, A. M., Small, C. J., Abbott, C. R., Dhillo, W. S., Seal, L. J., Cohen, M. A., et al. (2001). Ghrelin causes hyperphagia and obesity in rats. *Diabetes, 50*, 2540–2547.

Wren, A. M., Small, C. J., Ward, H. L., Murphy, K. G., Dakin, C. L., Taheri, S., et al. (2000). The novel hypothalamic peptide ghrelin stimulates food intake and growth hormone secretion. *Endocrinology, 141*, 4325–4328.

Wynne, K., Park, A. J., Small, C. J., Meeran, K., Ghatei, M. R., Frost, G. S., et al. (2006). Oxyntomodulin increase energy expenditure in addition to decreasing energy intake in overweight and obese humans: A randomised controlled trial. *International Journal of Obesity, 30*, 1729–1736.

Wynne, K., Park, A. J., Small, C. J., Patterson, M., Ellis, S. M., Murphy, K. G., et al. (2005). Subcutaneous oxyntomodulin reduces body weight in overweight and obese subjects: A double-blind, randomized, controlled trial. *Diabetes, 54*, 2390–2395.

Wyrwicka, W., & Dobrzecka, C. (1960). Relationship between feeding and satiation centres of the hypothalamus. *Science, 132*, 805–806.

Xu, J., McNearney, T. A., & Chen, J. D. (2011). Impaired postprandial release/syntheses of ghrelin and PYY(3–36) and blunted responses to exogenous ghrelin and PYY(3–36) in a rodent model of diet-induced obesity. *Journal of Gastroenterology and Hepatology, 26*, 700–705.

Yang, J., Brown, M. S., Liang, G., Grishin, N. V., & Goldstein, J. L. (2008). Identification of the acyltransferase that octanoylates ghrelin, an appetite-stimulating peptide hormone. *Cell, 132*, 387–396.

Young, A. (2005). Effects on plasma glucose and lactate. *Advances in Pharmacology (San Diego, CA, US), 52*, 193–208.

Young, A. A., Carlo, P., Smith, P., Wolfe-Lopez, D., Pittner, R., Wang, M. W., et al. (1993). Evidence for release of free glucose from muscle during amylin-induced glycogenolysis in rats. *FEBS Letters, 334*, 317–321.

Young, D. A., Deems, R. O., Deacon, R. W., McIntosh, R. H., & Foley, J. E. (1990). Effects of amylin on glucose metabolism and glycogenolysis in vivo and in vitro. *American Journal of Physiology. Endocrinology and Metabolism, 259*, 457–461.

Yu, Y., Deng, C., & Huang, X. F. (2009). Obese reversal by a chronic energy restricted diet leaves an increased Arc NPY/AgRP but no alteration in POMC/CART, mRNA expression in diet-induced obese mice. *Behavioural Brain Research, 205*, 50–56.

Zac-Varghese, S., Tan, T., & Bloom, S. (2010). Hormonal interactions between gut and brain. *Discovery Medicine, 10*, 543–552.

Zhang, L., Bijker, M. S., & Herzog, H. (2011). The neuropeptide Y system: Pathophysiological and therapeutic implications in obesity and cancer. *Pharmacology & Therapeutics, 131*, 91–113.

Zhang, Y., Proenca, R., Maffei, M., Barone, M., Leopold, L., & Friedman, J. M. (1994). Positional cloning of the mouse obese gene and its human homologue. *Nature, 372*, 425–432.

Zhao, T. J., Liang, G., Li, R. L., Xie, X., Sleeman, M. W., Murphy, A. J., et al. (2010). Ghrelin O-acyltransferase (GOAT) is essential for growth-hormone mediated survival of calorie-restricted mice. *Proceedings of the National Academy of Sciences of the United States of America, 107*, 7467–7472.

Zhao, F. L., Shen, T., Kaya, N., Lu, S. G., Cao, Y., & Herness, S. (2005). Expression, physiological action, and co-expression patterns of neuropeptide Y in rat taste-bud cells. *Proceedings of the National Academy of Sciences of the United States of America, 102*, 11100–11105.

Zhu, L., Tamyakopoulos, C., Xie, D., Dragovic, J., Shen, X., Fenyk-Melody, J. E., et al. (2003). The role of dipeptidyl peptidase IV in the cleavage of glucagon family peptides: in vivo metabolism of pituitary adenylate cyclase activating polypeptide-(1–38). *The Journal of Biological Chemistry, 278*, 22418–22423.

Zigman, J. M., Jones, J. E., Lee, C. E., Saper, C. B., & Elmquist, J. K. (2006). Expression of ghrelin receptor mRNA in the rat and mouse brain. *The Journal of Comparative Neurology, 494*, 528–548.

Zipf, W. B., O'Dorisio, T. M., Cataland, S., & Sotos, J. (1981). Blunted pancreatic polypeptide response in children with obesity of Prader-Willi syndrome. *The Journal of Clinical Endocrinology and Metabolism, 52*, 1264–1266.

Zwirska-Korczala, K., Konturek, S. J., Sodowski, M., Wylezol, M., Kuka, D., Sowa, P., et al. (2007). Basal and post-prandial levels of PYY, ghrelin, cholecystokinin, gastrin, and insulin in women with moderate and morbid obesity and metabolic syndrome. *Journal of Physiology and Pharmacology, 58*(Suppl. 1), 13–35.

CHAPTER EIGHT

Neuroinflammation in Overnutrition-Induced Diseases

Dongsheng Cai[1]

Department of Molecular Pharmacology, Institute of Aging, Diabetes Research Center, Albert Einstein College of Medicine, Bronx, New York, USA

[1]Corresponding author: e-mail address: dongsheng.cai@einstein.yu.edu

Contents

Abstract

Inflammation is a biological response mounted by the immune system against dangerous assaults that threaten the integrity and normal physiology of an organism. During the past decades, cross-disciplinary research from immunology and endocrinology has much broadened this knowledge by demonstrating that chronic conditions of nutritional excess constitute an independent category of inflammatory activators, and the resulting chronic and low-grade inflammation is an important characteristic of overnutrition-induced diseases. A large body of research has demonstrated that these diseases are pathogenically associated with the local, negative actions of inflammation in peripheral tissues predominantly including the liver, muscle, and fat. In this research background, more recent research has advanced to a new level, with the important discoveries showing that overnutrition-induced inflammation occurs in the brain and thus plays a broad and leadership role in overnutrition-induced diseases. While much more research establishments are expected in this emerging and quickly expanding research, the appreciated understandings have been mainly based on proinflammatory IKKβ/NF-κB pathway and related molecules in the hypothalamus. In this chapter, the author focuses on describing IKKβ/NF-κB-induced neural inflammation in the context of overnutrition-induced metabolic inflammation and especially the central roles of this neural inflammation in the development of a spectrum of overnutrition-related diseases.

Vitamins and Hormones, Volume 91
ISSN 0083-6729
http://dx.doi.org/10.1016/B978-0-12-407766-9.00008-0

ABBREVIATIONS

AGRP agouti-related peptide
CNS central nervous system
CVD cardiovascular diseases
ER endoplasmic reticulum
HFD high-fat diet
IKKβ IκB kinase beta
JNK c-Jun N-terminal kinase
MBH mediobasal hypothalamus
NF-κB nuclear factor kappa B
POMC proopiomelanocortin
ROS reactive oxygen species
SOCS3 suppressor of cytokine signaling-3
T2D type-2 diabetes
TLR toll-like receptor
TNF-α tumor necrosis factor alpha

1. INTRODUCTION

Inflammation is a biological response mounted by the immune system against dangerous assaults that threaten the integrity and normal physiology of an organism. In response to pathogens or injuries, pattern recognition receptors of the innate immune system are activated, which, in turn, initiates proinflammatory signaling cascades to produce cytokines, chemokines, and other inflammatory products in order to eradicate inflammatory insults and perform tissue repair. IκB kinase beta (IKKβ) and its downstream effector nuclear factor kappa B (NF-κB) form a key proinflammatory axis to pivotally mediate a broad network of inflammatory reactions (Baeuerle & Baltimore, 1996; Hayden & Ghosh, 2008; Karin & Lin, 2002). Activation of NF-κB leads to the expression of a large battery of inflammatory genes that promote inflammation and also direct proliferation of involved immune cells. Many other inflammatory signaling pathways, such as c-Jun N-terminal kinases (JNKs) and downstream transcription factor activator protein 1, can interact with IKKβ/NF-κB pathway to mediate inflammation (Zhang & Kaufman, 2008). In the past decades, cross-disciplinary studies at the interface of immunology and endocrinology have revealed negative relationships between immunity and metabolic regulation. It has been known that overnutrition-induced metabolic diseases and, in particular, obesity and type 2 diabetes (T2D) are pathogenically associated with immune changes in the circulation as well as metabolic tissues including liver, muscle, and fat (Cai, 2009; Cai & Liu, 2011, 2012; Gregor & Hotamisligil, 2011; Lehrke & Lazar, 2004; Lumeng & Saltiel, 2011; Schenk, Saberi, & Olefsky, 2008; Shoelson & Goldfine, 2009). In contrast to classical inflammation seen in diseases, such as infections, trauma, cancers, and autoimmune diseases, the type of inflammation in obesity and T2D is often devoid of pathogen invasion or physical injuries but instead primarily results from various metabolic abnormalities. The onset of systemic and tissue inflammation in obesity and T2D is chronically low grade and atypical in many aspects, and "metabolic inflammation" (Cai, 2009; Cai & Liu, 2011, 2012) or "metaflammation" (Gregor & Hotamisligil, 2011) is specifically used to reflect its biological characteristics. In-depth research has revealed that under proobesity/T2D conditions IKKβ/NF-κB pathway can induce metabolic inflammation to mediate insulin resistance and glucose/lipid disorders and thus mechanistically contribute to the development and progression of T2D and relevant metabolic diseases (Cai, 2009; Cai & Liu, 2011, 2012; Shoelson & Goldfine, 2009).

More recently, metabolic inflammation was discovered also to significantly exist in the brain and was further identified as a root and baseline cause for many overnutrition-induced diseases (Cai, 2009; Cai & Liu, 2011, 2012; Thaler & Schwartz, 2010). In these studies, it was shown that excessive nutrients in the form of glucose and fatty acids can directly activate the innate immune system of the central nervous system (CNS), and the resulting central metabolic inflammation disrupts the central neuroendocrine and neural regulations of metabolic-related physiology. Recently, nonneuronal glial cells including astrocytes and microglia have been recognized to be as an integral part of central inflammation (Horvath et al., 2010; Thaler et al., 2012). Molecular research had led to the elucidation that IKKβ/NF-κB in the hypothalamus can be activated by excessive nutrients via receptor-independent intracellular organelle stresses as well as via receptor-dependent pathways through such as toll-like receptor (TLR) and cytokine receptors. To date, IKKβ/NF-κB-directed metabolic inflammation in the CNS has been studied mostly in the mediobasal hypothalamus (MBH)—the hypothalamic region containing first-order metabolic sensing neurons which are known to primarily regulate feeding, energy balance, and other metabolic activities. Metabolic inflammation in these neurons has been linked to their neuronal dysfunctions, leading to energy imbalance and other pre-T2D changes including insulin resistance and glucose intolerance. Furthermore, metabolic inflammation in the MBH was found to interfere with the central control of cardiovascular homeostasis to mediate the development of obesity-related hypertension (Purkayastha, Zhang, & Cai, 2011; Purkayastha, Zhang, Zhang, et al., 2011). In addition, although still limitedly studied, central metabolic inflammation is relevant to neurodegeneration which might be a basis for the epidemic correlation between obesity/T2D and neurodegenerative diseases including Alzheimer's disease, Parkinson's disease, and Huntington's disease. Very recently, obesity-related neural inflammation was discovered to impair neural stem cells (NSC) to mediate neurodegenerative mechanism of obesity and pre-T2D (Li, Tang, & Cai, 2012). Altogether, stemming from obesity/T2D-associated peripheral inflammation, neural inflammation has now become an emerging topic and an expanding research area in studying various overnutrition-induced diseases and disorders. Herein, in the context of overnutrition-associated inflammation, this chapter will focus on describing the involvement of IKKβ/NF-κB-mediated neural metabolic inflammation in a spectrum of overnutrition-related diseases.

2. IKKβ/NF-κB AND METABOLIC INFLAMMATION

The NF-κB family transcription factors play central roles in immunity and inflammation (Baeuerle & Baltimore, 1996; Hayden & Ghosh, 2008; Karin & Lin, 2002). Mammalian NF-κB family proteins have five members, and they have Rel-homology domain for protein dimerization and DNA binding. Three members including p65/RelA, RelB, and cRel have C-terminal transcription activation domain which is required for exerting transcriptional regulation. Two other members, p105 and p100, are inert precursor proteins with C-terminal ankyrin repeats which inhibit DNA binding, and they need proteolytic processing to produce functional proteins p50 and p52. Distinct members of NF-κB proteins can form homo- or heterodimeric complexes and bind to κB site of target genes for transcriptional activation. In quiescent state, canonical NF-κB protein activity is inhibited through being maintained in the cytoplasm due to the binding of IκB proteins. Canonical activation of NF-κB pathway is induced through proinflammatory receptors such as TNF receptor members, cytokine receptors, and TLRs, and this process is crucially mediated by IKK complex which consists of three subunits, the catalytic IKKα and IKKβ subunits and the regulatory IKKγ subunit. IKKβ is known to be mainly accountable for inflammatory response, while IKKα can be involved in alternative functions of NF-κB such as tissue development. Upon activation of receptors, IKK complex is activated to induce phosphorylation and then ubiquitin-dependent degradation of IκB proteins, leading to NF-κB nuclear translocation for its transcriptional activities. Activated

IKKβ/NF-κB induces the expression of a large number of genes, and many of them are involved in inducing and regulating immune reaction, inflammation, and related molecular and cellular events.

The IKKβ/NF-κB pathway plays a central role in metabolic inflammation. It has been long appreciated that in chronic obesity and obesity-related diseases, several metabolic tissues including fat, liver, and muscle undergo inflammatory changes. In response to metabolic overload, immune cells such as macrophages invade metabolic tissues, and for example, in the case of adipose tissue, infiltration of M1 population of macrophages is proinflammatory (Weisberg et al., 2003; Xu et al., 2003). Within adipose tissue, resident T cells were found to have compositional changes (Feuerer et al., 2009; Nishimura et al., 2009; Winer et al., 2009), where the number of immunosuppressive T regulatory cells decreases and the ratio of $CD8^+$ to $CD4^+$ T cells increases. All these factors coherently produce a proinflammatory milieu, which also divergently occurs in other types of metabolic tissues. At molecular level, proinflammatory cytokines such as tumor necrosis factor alpha (TNF-α) and interleukins due to IKKβ/NF-κB activity can upactivate IKKβ/NF-κB signaling via TNF-α and interleukin receptors, which leads to the sustained production of proinflammatory cytokines, thus formulating a feed-forward inflammatory loop to maintain metabolic inflammation in metabolic tissues. Chronically elevated inflammatory molecules in metabolic tissues can substantially impair their physiological functions to promote the development of obesity and T2D. The recent discovery of metabolic inflammation in the hypothalamus is leading metabolic research to a new area. Among these new levels of research, IKKβ/NF-κB pathway has been identified as a pivotal mediator for neural metabolic inflammation in overnutrition-induced diseases (Kleinridders et al., 2009; Milanski et al., 2009; Posey et al., 2009; Zhang et al., 2008). In summary, IKKβ/NF-κB is a pivotal mediator of metabolic inflammation in not only peripheral tissues but also the CNS, and because of the brain's leadership position in controlling the whole body physiology, neural metabolic inflammation has been conceived to play broad and fundamental roles in the development of various overnutrition-induced diseases.

3. MEDIATORS OF BRAIN METABOLIC INFLAMMATION

Obesity and T2D-associated inflammation in peripheral tissues has predominantly been studied in terms of cytokine receptor-mediated NF-κB activation. In contrast, nutrient overload in the brain was reported to directly activate IKKβ/NF-κB and induce neural inflammation via receptor-independent intracellular stresses, putting forth a concept that excessive nutrients under proobesity/T2D conditions *per se* represent a class of independent stimuli for brain IKKβ/NF-κB pathway (Meng & Cai, 2011; Purkayastha, Zhang, & Cai, 2011; Purkayastha, Zhang, Zhang, et al., 2011; Zhang et al., 2008). Along this line, mounting evidences have demonstrated that excessive nutrients in the brain present a type of metabolic stresses to neurons in the hypothalamus or the brain. The intracellular consequences can be multiple, and so far, researchers have paid attention to a few organelles including mitochondria, endoplasmic reticulum (ER), and autophagosomes. As revealed, neurons seem to be very sensitive to increased oxidative activities induced by increased amount of nutrient influx, which puts great pressure on intracellular oxidative machinery, leading to mitochondrial dysfunction and oxidative stress. At the same time, increased cellular oxidative workload drives cellular protein synthesis to higher levels and thus promotes unfolded protein response of the ER which can result in ER stress response. While overnutrition-induced cellular oxidative stress and ER stress persist, accumulation of defective macromolecules (e.g., misfolded proteins and dysfunctional mitochondria and ER) can negatively affect macroautophagic protein degradation machineries and leads to development of overnutrition-induced autophagic defect.

From the molecular signaling perspective, cellular oxidative stress can activate NF-κB via reactive oxygen species (ROS)-induced alternative phosphorylation of IκBα to abolish its inhibitory effect on NF-κB (Schieven, Kirihara, Myers, Ledbetter, & Uckun, 1993; Schoonbroodt et al., 2000; Tanaka et al., 2000) or through oxidative inactivation of IKKβ/NF-κB pathway inhibitors (Herscovitch et al., 2008; Kamata, Manabe, Oka, Kamata, & Hirata, 2002; Lee et al., 2002). Reversely, activation of NF-κB has reinforcing effects on cellular oxidative stress by producing cytotoxic products (Pahl, 1999). In addition, as established in recent research (Deng et al., 2004; Hu, Han, Couvillon, Kaufman, & Exton, 2006; Yamazaki et al., 2009), ER stress can activate NF-κB via signaling cross talk between IKKβ/NF-κB pathway and three UPR pathways which are mediated by PKR-like ER kinase, inositol requiring enyzyme-1, and activating transcription factor-6. Under overnutrition, activation of IKKβ/NF-κB and induction of neuronal ER stress indeed promote each other (Purkayastha, Zhang, Zhang, et al., 2011; Zhang et al., 2008). Furthermore, prolonged overnutrition can lead to brain autophagy defect which further promotes IKKβ/NF-κB-dependent inflammation in the hypothalamus (Meng

& Cai, 2011). The underlying mechanism that links neural autophagic defect to hypothalamic inflammation remains unexplored, but theoretically, it may work at least through NLRP3 inflammasomes (Zhou, Yazdi, Menu, & Tschopp, 2011), as NLRP3 inflammasomes can activate IKKβ/NF-κB pathway through the comprised inflammatory IL-1β and IL-18 release (Strowig, Henao-Mejia, Elinav, & Flavell, 2012).

In addition to intracellular organelle stresses, receptor-dependent signaling pathways can be parallely or secondarily involved in the formation of neural inflammation under chronic conditions of overnutrition. An important candidate is TLR pathway, which was initially found to be involved in obesity/T2D-associated peripheral inflammation. Though classically known as mediators of innate immune defense by recognizing pathogens (Beutler, 2004; Takeuchi & Akira, 2010), researchers found that in models of dietary obesity, high circulating levels of fatty acids can activate TLR2 and TLR4 to induce inflammation in muscle and adipose tissues as well as macrophages (Caricilli et al., 2008; Shi et al., 2006). More recently, researchers discovered that high-fat diet (HFD) feeding or central administration of fatty acids can activate brain TLR4 to induce neural inflammation (Milanski et al., 2009). Consistently, brain-specific inhibition of TLR4 signaling significantly attenuates NF-κB-mediated central metabolic inflammation in mice under chronic HFD feeding (Kleinridders et al., 2009). These findings indicated that TLR pathway is linked to NF-κB activation in the context of metabolic inflammation.

A number of relevant signaling pathways have been significantly related to metabolic inflammation in the brain. JNK pathway is known to be closely interacted with IKKβ/NF-κB, and in terms of metabolic inflammation, JNK pathway can respond to lipid overexposure, ER stress, and cellular oxidative stress to mediate systemic inflammation which contributes to the development of insulin resistance and dyslipidemia (Hirosumi et al., 2002; Sabio et al., 2009, 2008, 2010; Solinas et al., 2007; Vallerie, Furuhashi, Fucho, & Hotamisligil, 2008). Recent studies demonstrated that JNK-mediated inflammation in the brain can contribute to the development of obesity and insulin resistance under dietary overnutrition (Belgardt et al., 2010; Sabio et al., 2010; Unger, Piper, Olofsson, & Xu, 2010). MyD88, an adaptor protein in IKKβ/NF-κB signaling, has also been investigated in the brain, and it was found that mice with brain-specific MyD88 knockout were protected from developing HFD-induced obesity (Kleinridders et al., 2009). Regarding the downstream molecules of IKKβ/NF-κB, suppressor of cytokine signaling-3 (SOCS3) has been examined, and brain-specific SOCS3 knockout was demonstrated to normalize central regulation despite overnutrition and thus prevent against the induction of obesity by overnutrition (Mori et al., 2004). Moreover, SOCS3 overexpression in the hypothalamus reduces the antiobesity effect of centrally inhibiting IKKβ/NF-κB, indicating that SOCS3 is a molecular mediator for the effect of IKKβ/NF-κB-induced central inflammation in obesity and T2D (Zhang et al., 2008).

Nonneuronal glia cells including microglia and astrocytes are also significantly involved in overnutrition-induced central inflammation. Newborn rats that were exposed to HFD feeding at 60-day age showed prominent CNS microglial activation and increased local production of IL-6, followed by hyperleptinemia and weight gain (Tapia-Gonzalez et al., 2011), demonstrating that nutrient overload can lead to central metabolic deregulation via microglial inflammation. In another study, researchers reported that central administration of IL-4 to induce microglial activation resulted in elevated hypothalamic inflammation and weight gain in HFD-fed rats (Oh et al., 2010). In addition to microglia, astrocytes can mediate central metabolic inflammation, which agrees with the observation that astrocytes are susceptible to the influences of various proinflammatory cytokines (Belanger, Allaman, & Magistretti, 2011). A recent study found that saturated fatty acids activate inflammatory signaling in astrocytes and trigger astrocyte release of proinflammatory cytokines TNF-α and IL-6 in rats (Gupta, Knight, Gupta, Keller, & Bruce-Keller, 2012). Thus, the cross talk between neurons and nonneuronal cells is another integral part of central metabolic inflammation.

4. METABOLIC INFLAMMATION AND NEURAL DYSREGULATION

The CNS's metabolic regulatory functions largely reside in the hypothalamus, a midbrain structure that consists of functionally distinct nuclei and forms extensive neural connections with limbic forebrain and hindbrain structures. The hypothalamus is the "headquarters" for controlling energy and metabolic homeostasis through sensing metabolic signals and accordingly regulating appetite, energy expenditure, and nutrient metabolism (Coll, Farooqi, & O'Rahilly, 2007; Cone, 2005; Flier, 2006; Lam, Schwartz, & Rossetti, 2005; Morton, Cummings, Baskin, Barsh, & Schwartz, 2006; Sandoval, Cota, & Seeley, 2008). The hypothalamus exerts its metabolic regulatory functions based on a variety of metabolic signals sent to the brain, which include circulating hormones leptin, insulin, gut hormones as well as nutrients (Belgardt & Bruning, 2010; Kahn, Alquier, Carling, & Hardie,

2005; Murphy & Bloom, 2006; Myers, Cowley, & Munzberg, 2008; Zhang, Zhang, Gonzalez, Park, & Cai, 2011). The arcuate nucleus in the MBH is the first-order metabolic regulatory center. Agouti-related peptide (AGRP) and proopiomelanocortin (POMC) neurons in the arcuate nucleus receive metabolic signals such as circulating leptin, insulin, and nutrient moieties to interpret the body's energy statuses. When energy is in positive balance, increased circulating levels of insulin, leptin, and other metabolic clues can activate POMC neurons by promoting gene expression and secretion of two anorexigenic neuropeptides including α-melanocyte-stimulating hormone and cocaine and amphetamine-regulated transcript, and on the other hand, these metabolic signals inhibit AGRP neurons through reducing gene expression and release of two orexigenic neuropeptides including AGRP and neuropeptide Y (Belgardt & Bruning, 2010; Morton et al., 2006; Myers et al., 2008). The synergistic effects of these neuroendocrine regulations ensure balanced energy intake and expenditure and therefore body weight balance. In parallel, hypothalamic regions such as arcuate nucleus and paraventricular nucleus can use neural pathways to regulate various metabolic physiological activities.

At the pathophysiological level, the neuroendocrine and neural pathways can be negatively impacted by overnutrition-induced central metabolic inflammation. It has been shown that metabolic inflammation in first-order metabolic sensing neurons in the MBH interferes with the comprised neuronal leptin and insulin signaling, resulting in central leptin and insulin resistance to cause feeding and energy imbalance (Zhang et al., 2008). In parallel, as these neurons also exert the functions of regulating systemic glucose and blood pressure homeostasis via central–peripheral neuroendocrine axes and neural circuitries, metabolic inflammation can negatively impact these neuroendocrine and neural networks to potentiate the development of diabetes and related complications. For example, monosodium L-glutamate-induced neural inflammation was linked to the suppression of hypothalamic–pituitary–thyroid axis and upregulation of hypothalamic–pituitary–adrenal axis to cause metabolic problems (Castrogiovanni, Gaillard, Giovambattista, & Spinedi, 2008). In addition to the hypothalamus-regulated neuroendocrine pathways, the sympathetic nervous system modulated by hypothalamic neurons represents another mechanistic avenue in the central inflammatory mechanism of overnutrition-induced diseases. Recent literature has demonstrated that overnutrition-induced hypothalamic metabolic inflammation can increase the sympathetic outflow into peripheral organs (Purkayastha, Zhang, & Cai, 2011; Purkayastha, Zhang, Zhang, et al., 2011) and cause obesity-related hypertension and glucose disorders (Arruda et al., 2011; Muse, Lam, Scherer, & Rossetti, 2007; Purkayastha, Zhang, Zhang, et al., 2011; Romanatto et al., 2007; Singhal, Lazar, & Ahima, 2007). Altogether, overnutrition-induced neural inflammation can affect multiple neuroendocrine and neural regulatory functions to comprehensively mediate the development of overnutrition-induced diseases.

5. BRAIN METABOLIC INFLAMMATION AND OBESITY

Obesity is a pathological state of chronic positive energy balance cumulatively resulting from excessive energy intake and/or decreased energy expenditure. According to World Health Organization guideline, the body mass index greater than or equal to 25 is defined as overweight, and if it is greater than or equal to 30, obesity is then diagnosed. Based on this standard, the obesity incidence has reached an alarming rate in today's obesogenic environment, with 300 million women and 200 million men affected worldwide (World Health Organization, 2011). Obesity is a core and baseline component of metabolic syndrome and is a major risk factor for many diseases such as T2D, hypertension, stroke, and neurodegenerative diseases. The hypothalamus is the master regulator of whole body energy balance, and metabolic sensing center in the hypothalamus critically governs food intake and energy expenditure via combined neuroendocrine and neural mechanisms. Research to study the neural inflammatory mechanism of obesity has largely focused on identifying inflammatory molecules. Using HFD feeding or intracerebroventricular lipid infusion in rodents to model overnutrition, researchers found that lipid excess can induce metabolic inflammation in the hypothalamus via IKKβ/NF-κB activation (Kleinridders et al., 2009; Milanski et al., 2009; Posey et al., 2009; Zhang et al., 2008), which mechanistically impairs hypothalamic leptin and insulin signaling (Zhang et al., 2008). Brain-specific ablation of IKKβ (Zhang et al., 2008) or MyD88 (Kleinridders et al., 2009), MBH-specific inhibition of autophagy defect (Meng & Cai, 2011), and whole body deficiency of NF-κB subunit p50 (Gao et al., 2009) all have protective effects against dietary obesity. The role of hypothalamic IKKβ/NF-κB in obesity development is significantly related to ER stress (Zhang et al., 2008), and indeed, central induction of ER stress via genetic inactivation of X-box binding protein-1 can render mice highly susceptible to central leptin resistance and dietary obesity (Ozcan et al., 2009). Researchers also studied IKKβ/NF-κB-related signaling pathways such as JNK1. It was shown that

JNK1-mediated central metabolic inflammation plays a role in the pathogenesis of obesity. Through analyzing different tissue-specific deletion of JNK1, only brain-specific JNK1 deletion exhibited antiobesity effect (Belgardt et al., 2010; Sabio, Cavanagh-Kyros, et al., 2010) which was similarly seen in the whole body JNK1 knockout mice (Hirosumi et al., 2002), indicating that JNK1-mediated neural inflammation causes energy imbalance to induce obesity. These mice with brain-specific JNK1 deletion manifested decreased food intake, increased energy expenditure, and increased physical activity (Belgardt et al., 2010), all of which are indicative of brain regulatory improvements. SOCS3 (Howard & Flier, 2006; Myers et al., 2008) is a well-established negative regulator of leptin and insulin signaling, and it was found that SOCS3 haploinsufficiency (Howard et al., 2004) or brain-specific SOCS3 knockout (Mori et al., 2004) can enhance central leptin sensitivity and counteract HFD-induced obesity. Research proved that SOCS3 is a key effector downstream of IKKβ/NF-κB in the neural inflammatory mechanism of dietary obesity, as obesity-associated hypothalamic SOCS3 upregulation requires IKKβ/NF-κB signaling, and transgenic overexpression of SOCS3 in the MBH abolishes the antiobesity effect of central IKKβ inactivation (Zhang et al., 2008). The receptor signaling upstream of IKKβ/NF-κB has also been investigated, and in this regard, mice deficient of TLR2 (Caricilli et al., 2008; Himes & Smith, 2010) or TLR4 (Milanski et al., 2009; Tsukumo et al., 2007) have reduced HFD-induced inflammation and are strongly protected from HFD-induced obesity.

To summarize, metabolic inflammation is a prominent feature of obesity, and central metabolic inflammation forms the pathogenic root of overnutrition-induced obesity. This central inflammatory mechanism of obesity is critically mediated by IKKβ/NF-κB, its upstream and downstream components, and related signaling pathways. From the therapeutic perspective, these molecular mediators in the CNS can be potential therapeutic targets for combating obesity.

6. BRAIN METABOLIC INFLAMMATION AND DIABETES

Diabetes is a medical condition arising from the body's inability to lower blood sugar to the normal range, and chronic hyperglycemia and related pathophysiological changes can lead to serious, life-threatening complications. Insulin is a hormone produced by pancreatic β cells to be responsible for lowering blood glucose through insulin's actions in metabolic tissues including liver, muscle, and adipocytes. Peripheral insulin resistance in these metabolic tissues is a major etiological basis for the development of T2D. Interestingly, recent research advances have established the role of the brain and especially the hypothalamus in regulation and dysregulation of glucose homeostasis. As elucidated in the literature, the hypothalamus can employ multiple neuroendocrine and neural mechanisms to exert the central control over glucose homeostasis (Elmquist, Coppari, Balthasar, Ichinose, & Lowell, 2005; Herman & Kahn, 2006; Morton, 2007; Plum, Belgardt, & Bruning, 2006). Metabolic signals such as circulating insulin, leptin, gut hormones, and nutrients are converged in certain types of hypothalamic neurons, and through secreting neuropeptides and/or neurotransmitters, these neurons send metabolic orders to downstream neuroendocrine networks and hindbrain autonomic sites (e.g., dorsal motor nucleus of the vagus nerve and nucleus tractus solitarius in the brain stem) to regulate peripheral glucose metabolism. Under physiological conditions, central insulin (Obici, Zhang, Karkanias, & Rossetti, 2002) and leptin (Asilmaz et al., 2004) signaling have hypoglycemic effects via suppressing hepatic glucose production and promoting peripheral glucose uptake.

It has been long known that obesity-induced inflammation in the peripheral metabolic sites contributes to insulin signaling defect and the development of insulin resistance (Cai, 2009; Hotamisligil, 2010; Lumeng & Saltiel, 2011). Recently, research has begun to demonstrate that obesity-related metabolic inflammation can impact the CNS, leading to the central dysregulation of glucose homeostasis. While the molecular mechanism of this central dysregulation still largely remains to be investigated, it includes inflammation-induced central insulin and leptin resistance. The relationship between T2D and obesity-related inflammation, particularly in the CNS, has been considered as a basis for developing novel anti-T2D strategies. Given that ER stress is a critical inducer of obesity-related central metabolic inflammation, ER stress inhibitors can be a class of optional therapeutic targets which can inhibit inflammation in not only peripheral tissues but also the CNS to treat diabetes. Indeed, brain-specific inhibition of hypothalamic ER stress can improve overnutrition-induced systemic insulin resistance and glucose abnormality without involving body weight changes (Purkayastha, Zhang, Zhang, et al., 2011). The finding suggested that the antidiabetic effect of systemically delivered ER stress inhibitors shown in the literature (Ozcan et al., 2006) might involve anti-ER stress/inflammation actions in the CNS. Thus, while anti-ER stress therapeutic strategy is driving new drug discovery efforts to treat T2D (Kim, Xu, & Reed, 2008), it is an important research area to investigate the roles of such designs in manipulating neural ER stress and inflammation to treat systemic glucose disorder and insulin resistance in diabetes.

7. BRAIN METABOLIC INFLAMMATION IN HYPERTENSION AND STROKE

Obesity-related hypertension is a frequent and serious cardiovascular complication of obesity. Due to its prominent role in promoting deleterious CVDs such as atherosclerosis and stroke, obesity-induced hypertension is an important epidemic factor for CVDs. Blood pressure is a physiological output of multiple factors such as cardiac output, body fluid and salt balance, and vascular compliance, and all of these functions are under the neural control of the CNS. Earlier studies have provided evidences showing that obesity-related inflammation in endothelial cells, smooth muscles, and vasculature-resident macrophages contribute to the development of obesity-related hypertension (Duan, Usher, & Mortensen, 2009; Harrison, Guzik, Goronzy, & Weyand, 2008; Savoia & Schiffrin, 2006). More recently, obesity-induced central neuroendocrine and neural changes, such as overactivation of the melanocortin system (Rahmouni, Correia, Haynes, & Mark, 2005; Rahmouni, Haynes, & Mark, 2002) and sympathetic nervous system (Purkayastha, Zhang, & Cai, 2011; Purkayastha, Zhang, Zhang, et al., 2011), have been identified to underlie the pathogenesis of obesity-related hypertension. In this context, it was discovered that central metabolic inflammation is a critical pathogenic link between obesity and sympathetic upregulation of blood pressure (Purkayastha, Zhang, & Cai, 2011; Purkayastha, Zhang, Zhang, et al., 2011). In these studies, researchers found that activation of IKKβ/NF-κB pathway by overnutrition in the MBH or, more specifically, the POMC neurons can upregulate sympathetic nervous activity to cause hypertension. On the other hand, systemic or CNS-targeted anti-inflammatory interventions have demonstrated antihypertensive therapeutic effects. For example, in experimental rodent models, genetic inactivation of IKKβ/NF-κB pathway in POMC neurons (Purkayastha, Zhang, & Cai, 2011) or pharmacological inhibition of brain ER stress upstream of NF-κB activation (Purkayastha, Zhang, Zhang, et al., 2011) both significantly intercept the development of obesity-related hypertension. Hence, suppression of central metabolic inflammation holds two converging benefits in combating not only obesity but also obesity-related hypertension in an obesity-independent manner. It is also worth mentioning that due to CNS's important role in blood pressure control,

NF-κB-induced central inflammation has also been causally involved in many other forms of hypertension such as essential hypertension, spontaneous hypertension, and angiotensin II-induced hypertension (Kang et al., 2009; Qadri, Hauser, Johren, & Dominiak, 2002; Wu & Schmid-Schonbein, 2011), and it should be an interesting topic to explore if similar central inflammatory mechanism is widely shared between these hypertension and obesity-related hypertension.

Stroke is a devastating cerebrovascular disease featuring rapid loss of brain function(s) due to interruption of cerebral blood supply. Worldwidely, stroke is second only to cardiac ischemia in causing nonaccidental death. Based on the cause of disturbance of brain blood supply, there is ischemic stroke in which brain blood vessels are blocked, such as thrombosis, and hemorrhagic stroke in which brain arteries rupture due to pathological conditions such as hypertension, atheroma, and brain aneurysm. Thus, obesity-related disorders such as hypertension, diabetes, dyslipidemia, and atherosclerosis all become risk factors of stroke (Donnan, Fisher, Macleod, & Davis, 2008). It has been clear that brain inflammation plays critical roles in stroke pathogenesis. First, brain inflammation can generally mediate stroke development through promoting the above-mentioned obesity-related metabolic disorders. Second, brain inflammation can directly induce ischemic brain damage (Iadecola & Anrather, 2011). In an ischemic event, brain neurons that are at the center of ischemic region quickly die from energy failure-induced degeneration and necrosis. Neurons in the ischemic penumbra, though remain viable, are under severe stress and vulnerable to adverse events such as postischemic inflammatory insults. Meanwhile, hypoxia-induced cell death or injury at early postischemic stage leads to release of a host of proinflammatory factors to trigger postischemic inflammation. This inflammation leads to microvascular occlusions and exacerbated ischemia at the intravascular site and recruits perivascular macrophages and mast cells to reinforce postischemic inflammation and tissue damage at the perivascular site. The signaling mediators of inflammatory activation in stroke can include many molecules, such as TLRs which respond to dead or damaged cells, ROS that result from hypoxia, and cytokines and chemokines produced by primary inflammatory response. Brain IKKβ is found to be activated in mouse stroke model, while inhibition of brain IKKβ/NF-κB can protect against stroke (Cheng et al., 2010; Herrmann et al., 2005; Jiang et al., 2012; Liu, Zhang, Yang, & Fan, 2009). In addition to directly inhibiting IKKβ/NF-κB, other relevant anti-inflammation methods have been explored to treat ischemic brain damage and stroke (Wang, 2005). For example, pharmacological anti-inflammatory agents such as granulocyte colony-stimulating factor, tetramethylpyrazine, and picroside 2 (Guo, Xu, Li, Li, & Du, 2010; Liao et al., 2004; Lu & Xiao, 2007) or that promote expression of anti-inflammatory cytokines TGFβ and IL-10 (Becker et al., 1997; Cheyuo et al., 2012) all exhibited neuroprotective effects in animal stroke models. Clinically, IL-1 receptor blockade treatment has proven to markedly decrease neurological impairments in stroke patients (Emsley et al., 2005). On the other hand, while anti-inflammation can effectively reduce ischemic brain lesions, its application in stroke should bear cautions, because stroke is known to be associated with postischemia immune suppression of lymphoid organs and increased risk for opportunistic infections (Meisel, Schwab, Prass, Meisel, & Dirnagl, 2005). Therefore, selective antineuroinflammation without compromising the function of systemic immune system may represent an optimal strategy for treating stroke.

8. METABOLIC INFLAMMATION AND NEURODEGENERATION

Neurodegenerative diseases are a group of severe health problems characterized by loss of neurons and neural cells, for example, Parkinson's disease, Alzheimer's disease, and Huntington's disease. Though their symptoms differ from each other, these diseases have a common basis in pathology, that is, excessive intracellular accumulation of misfolded proteins such as α-synuclein aggregates in Parkinson's disease, mutant Huntingtin protein aggregates in Huntington's disease, and tau protein aggregates in addition to abnormal formation of extracellular β-amyloid plaques in Alzheimer's disease (Taylor, Hardy, & Fischbeck, 2002). Accumulation of these toxic proteins is attributed to intracellular changes including organelle membrane damage, oxidative stress, mitochondrial dysfunctions, and defective protein degradation machinery, and such cellular proteinopathy can lead to neural cell apoptosis and thus neurodegenerative diseases. In addition to the appreciated genetic etiology, overnutrition-related environmental factors are important for these neurodegenerative diseases. For example, obesity and insulin resistance are known to be two major risk factors of developing neurodegeneration. Along this line, these neurodegenerative diseases have been causally related to neural oxidative stress and mitochondrial dysfunction (Cui et al., 2006; Dawson & Dawson, 2003; DiMauro & Schon, 2008; Lin & Beal, 2006), neural ER stress (Tabas & Ron, 2011; Uehara et al.,

2006; Xu, Bailly-Maitre, & Reed, 2005), and neural autophagy defect (Komatsu et al., 2006; Lee & Gao, 2009; Martinez-Vicente & Cuervo, 2007), and it should be noted that all these neural stresses can be induced by overnutrition. Also importantly, the interactive relationships between intracellular stresses and neural IKKβ/NF-κB activation have been observed in neurodegenerative diseases, for example, oxidative stress (Bazan, Palacios-Pelaez, & Lukiw, 2002; Lezoualc'h & Behl, 1998; Yamashima, 2012), ER stress (Braun & Zischka, 2008; Giraud, Caron, Pham-Dinh, Kitabgi, & Nicot, 2010) and autophagy defect (Lim & Lim, 2011) have all been related to neural NF-κB activation in these diseases. Therapeutically, anti-inflammatory agents were shown to effectively treat animals with neurodegenerative diseases (Hu et al., 2010; Kawamata & Shimohama, 2011; Zhu & Qian, 2006), and moreover, clinical studies have shown promises in testing the potential use of antineuroinflammation in treating these diseases in humans (Choi, Aid, & Bosetti, 2009; Dinarello, 2010; Glass, Saijo, Winner, Marchetto, & Gage, 2010). Altogether, based on evidences in the literature, metabolic neuroinflammation can be significantly accountable for the susceptibility of developing neurodegenerative diseases under conditions of obesity and T2D.

Very recently, research has begun to appreciate that obesity and T2D also suffer from the problem of neurodegeneration. Most recent publications have interestingly demonstrated that chronic HFD feeding can lead to a reduction of certain hypothalamic neurons (Li, Tang, & Cai, 2012; McNay, Briancon, Kokoeva, Maratos-Flier, & Flier, 2012; Thaler et al., 2012). Based on more in-depth research findings, obesity-related metabolic inflammation is sufficient to disrupt NSC in the brain and especially in the hypothalamus to mediate a neurodegenerative mechanism of obesity-related diseases (Li, Tang, & Cai, 2012). In this research, it was shown that chronic HFD feeding activates IKKβ/NF-κB pathway in adult NSC, leading to not only dramatic apoptotic depletion of NSCs but also their impaired neuronal differentiation, and over a long period, animals under chronic overnutrition had poor neural regeneration which results in reduced numbers of functional neurons (Li, Tang, & Cai, 2012). Though the focus of this research was on hypothalamic NSC, it is likely that overnutrition and obesity employ the inflammatory mechanism to induce neurodegeneration in other brain regions. Thus, to study the involvement of inflammation in neural regeneration and degeneration will be much required, which may represent an important future research area in order to understand the neural inflammation in overnutrition-induced diseases.

9. CONCLUSIONS

Research over the past decades has significantly unveiled the pathogenic mechanisms of overnutrition-induced metabolic diseases including obesity, insulin resistance, T2D, CVD, and neurodegenerative diseases. A milestone understanding is that chronic overnutrition causes an atypical form of inflammation, namely, metabolic inflammation to mediate the development and progression of these diseases. While this type of inflammation occurs in the peripheral sites to locally affect individual metabolic functions, it is significantly induced in the CNS and the hypothalamus, leading to the central dysregulation of metabolic physiology. Because the CNS is in the leadership position in regulating physiology including energy and nutrient metabolism, metabolic inflammation in the brain has widespread involvement in various aspects of metabolic diseases ranging from feeding disorder, energy imbalance, obesity, glucose intolerance, blood pressure increase to neurodegeneration. IKKβ/NF-κB pathway is a major molecular mediator for neural metabolic inflammation and represents an anti-inflammatory target in order to treat various overnutrition-induced diseases. Despite these understandings, the knowledge on neural inflammation in metabolic diseases is still limited in general, and future research is much needed to delineate the involved molecular and cellular network and complex actions, which will help developing more specific and effective interventional strategies to conquer related diseases.

REFERENCES

Arruda, A. P., Milanski, M., Coope, A., Torsoni, A. S., Ropelle, E., Carvalho, D. P., et al. (2011). Low-grade hypothalamic inflammation leads to defective thermogenesis, insulin resistance, and impaired insulin secretion. *Endocrinology, 152*, 1314–1326.

Asilmaz, E., Cohen, P., Miyazaki, M., Dobrzyn, P., Ueki, K., Fayzikhodjaeva, G., et al. (2004). Site and mechanism of leptin action in a rodent form of congenital lipodystrophy. *The Journal of Clinical Investigation, 113*, 414–424.

Baeuerle, P. A., & Baltimore, D. (1996). NF-kappa B: Ten years after. *Cell, 87*, 13–20.

Bazan, N. G., Palacios-Pelaez, R., & Lukiw, W. J. (2002). Hypoxia signaling to genes: Significance in Alzheimer's disease. *Molecular Neurobiology, 26*, 283–298.

Becker, K. J., McCarron, R. M., Ruetzler, C., Laban, O., Sternberg, E., Flanders, K. C., et al. (1997). Immunologic tolerance to myelin basic protein decreases stroke size after transient focal cerebral ischemia. *Proceedings of the National Academy of Sciences of the United States of America, 94*, 10873–10878.

Belanger, M., Allaman, I., & Magistretti, P. J. (2011). Differential effects of pro- and anti-inflammatory cytokines alone or in combinations on the metabolic profile of astrocytes. *Journal of Neurochemistry, 116*, 564–576.

Belgardt, B. F., & Bruning, J. C. (2010). CNS leptin and insulin action in the control of energy homeostasis. *Annals of the New York Academy of Sciences, 1212*, 97–113.

Belgardt, B. F., Mauer, J., Wunderlich, F. T., Ernst, M. B., Pal, M., Spohn, G., et al. (2010). Hypothalamic and pituitary c-Jun N-terminal kinase 1 signaling coordinately regulates glucose metabolism. *Proceedings of the National Academy of Sciences of the United States of America*, *107*, 6028–6033.

Beutler, B. (2004). Inferences, questions and possibilities in Toll-like receptor signalling. *Nature*, *430*, 257–263.

Braun, R. J., & Zischka, H. (2008). Mechanisms of Cdc48/VCP-mediated cell death: From yeast apoptosis to human disease. *Biochimica et Biophysica Acta*, *1783*, 1418–1435.

Cai, D. (2009). NFkappaB-mediated metabolic inflammation in peripheral tissues versus central nervous system. *Cell Cycle*, *8*, 2542–2548.

Cai, D., & Liu, T. (2011). Hypothalamic inflammation: A double-edged sword to nutritional diseases. *Annals of the New York Academy of Sciences*, *1243*, E1–E39.

Cai, D., & Liu, T. (2012). Inflammatory cause of metabolic syndrome via brain stress and NF-kappaB. *Aging (Albana, NY)*, *4*, 98–115.

Caricilli, A. M., Nascimento, P. H., Pauli, J. R., Tsukumo, D. M., Velloso, L. A., Carvalheira, J. B., et al. (2008). Inhibition of toll-like receptor 2 expression improves insulin sensitivity and signaling in muscle and white adipose tissue of mice fed a high-fat diet. *The Journal of Endocrinology*, *199*, 399–406.

Castrogiovanni, D., Gaillard, R. C., Giovambattista, A., & Spinedi, E. (2008). Neuroendocrine, metabolic, and immune functions during the acute phase response of inflammatory stress in monosodium L-glutamate-damaged, hyperadipose male rat. *Neuroendocrinology*, *88*, 227–234.

Cheng, C. C., Chen, Y. H., Chang, W. L., Yang, S. P., Chang, D. M., Lai, J. H., et al. (2010). Phytoestrogen bavachin mediates anti-inflammation targeting Ikappa B kinase-I kappaB alpha-NF-kappaB signaling pathway in chondrocytes in vitro. *European Journal of Pharmacology*, *636*, 181–188.

Cheyuo, C., Jacob, A., Wu, R., Zhou, M., Qi, L., Dong, W., et al. (2012). Recombinant human MFG-E8 attenuates cerebral ischemic injury: Its role in anti-inflammation and anti-apoptosis. *Neuropharmacology*, *62*, 890–900.

Choi, S. H., Aid, S., & Bosetti, F. (2009). The distinct roles of cyclooxygenase-1 and -2 in neuroinflammation: Implications for translational research. *Trends in Pharmacological Sciences*, *30*, 174–181.

Coll, A. P., Farooqi, I. S., & O'Rahilly, S. (2007). The hormonal control of food intake. *Cell*, *129*, 251–262.

Cone, R. D. (2005). Anatomy and regulation of the central melanocortin system. *Nature Neuroscience*, *8*, 571–578.

Cui, L., Jeong, H., Borovecki, F., Parkhurst, C. N., Tanese, N., & Krainc, D. (2006). Transcriptional repression of PGC-1alpha by mutant huntingtin leads to mitochondrial dysfunction and neurodegeneration. *Cell*, *127*, 59–69.

Dawson, T. M., & Dawson, V. L. (2003). Molecular pathways of neurodegeneration in Parkinson's disease. *Science*, *302*, 819–822.

Deng, J., Lu, P. D., Zhang, Y., Scheuner, D., Kaufman, R. J., Sonenberg, N., et al. (2004). Translational repression mediates activation of nuclear factor kappa B by phosphorylated translation initiation factor 2. *Molecular and Cellular Biology*, *24*, 10161–10168.

DiMauro, S., & Schon, E. A. (2008). Mitochondrial disorders in the nervous system. *Annual Review of Neuroscience*, *31*, 91–123.

Dinarello, C. A. (2010). Anti-inflammatory agents: Present and future. *Cell*, *140*, 935–950.

Donnan, G. A., Fisher, M., Macleod, M., & Davis, S. M. (2008). Stroke. *The Lancet*, *371*, 1612–1623.

Duan, S. Z., Usher, M. G., & Mortensen, R. M. (2009). PPARs: The vasculature, inflammation and hypertension. *Current Opinion in Nephrology and Hypertension*, *18*, 128–133.

Elmquist, J. K., Coppari, R., Balthasar, N., Ichinose, M., & Lowell, B. B. (2005). Identifying hypothalamic pathways controlling food intake, body weight, and glucose homeostasis. *The Journal of Comparative Neurology*, *493*, 63–71.

Emsley, H. C., Smith, C. J., Georgiou, R. F., Vail, A., Hopkins, S. J., Rothwell, N. J., et al. (2005). A randomised phase II study of interleukin-1 receptor antagonist in acute stroke patients. *Journal of Neurology, Neurosurgery & Psychiatry*, *76*, 1366–1372.

Feuerer, M., Herrero, L., Cipolletta, D., Naaz, A., Wong, J., Nayer, A., et al. (2009). Lean, but not obese, fat is enriched for a unique population of regulatory T cells that affect metabolic parameters. *Nature Medicine*, *15*, 930–939.

Flier, J. S. (2006). Neuroscience. Regulating energy balance: The substrate strikes back. *Science*, *312*, 861–864.

Gao, Z., Yin, J., Zhang, J., He, Q., McGuinness, O. P., & Ye, J. (2009). Inactivation of NF-kappaB p50 leads to insulin sensitization in liver through post-translational inhibition of p70S6K. *The Journal of Biological Chemistry*, *284*, 18368–18376.

Giraud, S. N., Caron, C. M., Pham-Dinh, D., Kitabgi, P., & Nicot, A. B. (2010). Estradiol inhibits ongoing autoimmune neuroinflammation and NFkappaB-dependent CCL2 expression in reactive astrocytes. *Proceedings of the National Academy of Sciences of the United States of America*, *107*, 8416–8421.

Glass, C. K., Saijo, K., Winner, B., Marchetto, M. C., & Gage, F. H. (2010). Mechanisms underlying inflammation in neurodegeneration. *Cell*, *140*, 918–934.

Gregor, M. F., & Hotamisligil, G. S. (2011). Inflammatory mechanisms in obesity. *Annual Review of Immunology*, *29*, 415–445.

Guo, Y., Xu, X., Li, Q., Li, Z., & Du, F. (2010). Anti-inflammation effects of picroside 2 in cerebral ischemic injury rats. *Behavioral and Brain Functions*, *6*, 43.

Gupta, S., Knight, A. G., Gupta, S., Keller, J. N., & Bruce-Keller, A. J. (2012). Saturated long-chain fatty acids activate inflammatory signaling in astrocytes. *Journal of Neurochemistry*, *120*, 1060–1071.

Harrison, D. G., Guzik, T. J., Goronzy, J., & Weyand, C. (2008). Is hypertension an immunologic disease? *Current Cardiology Reports*, *10*, 464–469.

Hayden, M. S., & Ghosh, S. (2008). Shared principles in NF-kappaB signaling. *Cell*, *132*, 344–362.

Herman, M. A., & Kahn, B. B. (2006). Glucose transport and sensing in the maintenance of glucose homeostasis and metabolic harmony. *The Journal of Clinical Investigation*, *116*, 1767–1775.

Herrmann, O., Baumann, B., de Lorenzi, R., Muhammad, S., Zhang, W., Kleesiek, J., et al. (2005). IKK mediates ischemia-induced neuronal death. *Nature Medicine*, *11*, 1322–1329.

Herscovitch, M., Comb, W., Ennis, T., Coleman, K., Yong, S., Armstead, B., et al. (2008). Intermolecular disulfide bond formation in the NEMO dimer requires Cys54 and Cys347. *Biochemical and Biophysical Research Communications*, *367*, 103–108.

Himes, R. W., & Smith, C. W. (2010). Tlr2 is critical for diet-induced metabolic syndrome in a murine model. *The FASEB Journal*, *24*, 731–739.

Hirosumi, J., Tuncman, G., Chang, L., Gorgun, C. Z., Uysal, K. T., Maeda, K., et al. (2002). A central role for JNK in obesity and insulin resistance. *Nature*, *420*, 333–336.

Horvath, T. L., Sarman, B., Garcia-Caceres, C., Enriori, P. J., Sotonyi, P., Shanabrough, M., et al. (2010). Synaptic input organization of the melanocortin system predicts diet-induced hypothalamic reactive gliosis and obesity. *Proceedings of the National Academy of Sciences of the United States of America*, *107*, 14875–14880.

Hotamisligil, G. S. (2010). Endoplasmic reticulum stress and the inflammatory basis of metabolic disease. *Cell*, *140*, 900–917.

Howard, J. K., Cave, B. J., Oksanen, L. J., Tzameli, I., Bjorbaek, C., & Flier, J. S. (2004). Enhanced leptin sensitivity and attenuation of diet-induced obesity in mice with haploinsufficiency of Socs3. *Nature Medicine*, *10*, 734–738.

Howard, J. K., & Flier, J. S. (2006). Attenuation of leptin and insulin signaling by SOCS proteins. *Trends in Endocrinology and Metabolism, 17*, 365–371.
Hu, P., Han, Z., Couvillon, A. D., Kaufman, R. J., & Exton, J. H. (2006). Autocrine tumor necrosis factor alpha links endoplasmic reticulum stress to the membrane death receptor pathway through IRE1alpha-mediated NF-kappaB activation and down-regulation of TRAF2 expression. *Molecular and Cellular Biology, 26*, 3071–3084.
Hu, L. F., Lu, M., Tiong, C. X., Dawe, G. S., Hu, G., & Bian, J. S. (2010). Neuroprotective effects of hydrogen sulfide on Parkinson's disease rat models. *Aging Cell, 9*, 135–146.
Iadecola, C., & Anrather, J. (2011). The immunology of stroke: From mechanisms to translation. *Nature Medicine, 17*, 796–808.
Jiang, T., Gao, L., Guo, J., Lu, J., Wang, Y., & Zhang, Y. (2012). Suppressing inflammation by inhibiting NF-kappaB pathway contributes to the neuroprotection ofik Angiotensin-(1-7) in rats with permanent cerebral ischemia. *British Journal of Pharmacology, 167*, 1520–1532.
Kahn, B. B., Alquier, T., Carling, D., & Hardie, D. G. (2005). AMP-activated protein kinase: Ancient energy gauge provides clues to modern understanding of metabolism. *Cell Metabolism, 1*, 15–25.
Kamata, H., Manabe, T., Oka, S., Kamata, K., & Hirata, H. (2002). Hydrogen peroxide activates IkappaB kinases through phosphorylation of serine residues in the activation loops. *FEBS Letters, 519*, 231–237.
Kang, Y. M., Ma, Y., Zheng, J. P., Elks, C., Sriramula, S., Yang, Z. M., et al. (2009). Brain nuclear factor-kappa B activation contributes to neurohumoral excitation in angiotensin II-induced hypertension. *Cardiovascular Research, 82*, 503–512.
Karin, M., & Lin, A. (2002). NF-kappaB at the crossroads of life and death. *Nature Immunology, 3*, 221–227.
Kawamata, J., & Shimohama, S. (2011). Stimulating nicotinic receptors trigger multiple pathways attenuating cytotoxicity in models of Alzheimer's and Parkinson's diseases. *Journal of Alzheimer's Disease, 24*(Suppl. 2), 95–109.
Kim, I., Xu, W., & Reed, J. C. (2008). Cell death and endoplasmic reticulum stress: Disease relevance and therapeutic opportunities. *Nature Reviews. Drug Discovery*, 7, 1013–1030.
Kleinridders, A., Schenten, D., Konner, A. C., Belgardt, B. F., Mauer, J., Okamura, T., et al. (2009). MyD88 signaling in the CNS is required for development of fatty acid-induced leptin resistance and diet-induced obesity. *Cell Metabolism, 10*, 249–259.
Komatsu, M., Waguri, S., Chiba, T., Murata, S., Iwata, J., Tanida, I., et al. (2006). Loss of autophagy in the central nervous system causes neurodegeneration in mice. *Nature, 441*, 880–884.
Lam, T. K., Schwartz, G. J., & Rossetti, L. (2005). Hypothalamic sensing of fatty acids. *Nature Neuroscience, 8*, 579–584.
Lee, J. A., & Gao, F. B. (2009). Inhibition of autophagy induction delays neuronal cell loss caused by dysfunctional ESCRT-III in frontotemporal dementia. *The Journal of Neuroscience, 29*, 8506–8511.
Lee, S. R., Yang, K. S., Kwon, J., Lee, C., Jeong, W., & Rhee, S. G. (2002). Reversible inactivation of the tumor suppressor PTEN by H_2O_2. *The Journal of Biological Chemistry, 277*, 20336–20342.
Lehrke, M., & Lazar, M. A. (2004). Inflamed about obesity. *Nature Medicine, 10*, 126–127.
Lezoualc'h, F., & Behl, C. (1998). Transcription factor NF-kappaB: Friend or foe of neurons? *Molecular Psychiatry, 3*, 15–20.
Li, J., Tang, Y., & Cai, D. (2012). IKKbeta/NF-kappaB disrupts adult hypothalamic neural stem cells to mediate a neurodegenerative mechanism of dietary obesity and pre-diabetes. *Nature Cell Biology, 14*, 999–1012.
Liao, S. L., Kao, T. K., Chen, W. Y., Lin, Y. S., Chen, S. Y., Raung, S. L., et al. (2004). Tetramethylpyrazine reduces ischemic brain injury in rats. *Neuroscience Letters, 372*, 40–45.

Lim, K. L., & Lim, G. G. (2011). K63-linked ubiquitination and neurodegeneration. *Neurobiology of Disease, 43*, 9–16.

Lin, M. T., & Beal, M. F. (2006). Mitochondrial dysfunction and oxidative stress in neurodegenerative diseases. *Nature, 443*, 787–795.

Liu, Y., Zhang, X. J., Yang, C. H., & Fan, H. G. (2009). Oxymatrine protects rat brains against permanent focal ischemia and downregulates NF-kappaB expression. *Brain Research, 1268*, 174–180.

Lu, C. Z., & Xiao, B. G. (2007). Neuroprotection of G-CSF in cerebral ischemia. *Frontiers in Bioscience, 12*, 2869–2875.

Lumeng, C. N., & Saltiel, A. R. (2011). Inflammatory links between obesity and metabolic disease. *The Journal of Clinical Investigation, 121*, 2111–2117.

Martinez-Vicente, M., & Cuervo, A. M. (2007). Autophagy and neurodegeneration: When the cleaning crew goes on strike. *Lancet Neurology, 6*, 352–361.

McNay, D. E., Briancon, N., Kokoeva, M. V., Maratos-Flier, E., & Flier, J. S. (2012). Remodeling of the arcuate nucleus energy-balance circuit is inhibited in obese mice. *The Journal of Clinical Investigation, 122*, 142–152.

Meisel, C., Schwab, J. M., Prass, K., Meisel, A., & Dirnagl, U. (2005). Central nervous system injury-induced immune deficiency syndrome. *Nature Reviews. Neuroscience, 6*, 775–786.

Meng, Q., & Cai, D. (2011). Defective hypothalamic autophagy directs the central pathogenesis of obesity via IKK-beta/NF-kappaB pathway. *The Journal of Biological Chemistry, 286*, 32324–32332.

Milanski, M., Degasperi, G., Coope, A., Morari, J., Denis, R., Cintra, D. E., et al. (2009). Saturated fatty acids produce an inflammatory response predominantly through the activation of TLR4 signaling in hypothalamus: Implications for the pathogenesis of obesity. *The Journal of Neuroscience, 29*, 359–370.

Mori, H., Hanada, R., Hanada, T., Aki, D., Mashima, R., Nishinakamura, H., et al. (2004). Socs3 deficiency in the brain elevates leptin sensitivity and confers resistance to diet-induced obesity. *Nature Medicine, 10*, 739–743.

Morton, G. J. (2007). Hypothalamic leptin regulation of energy homeostasis and glucose metabolism. *The Journal of Physiology, 583*, 437–443.

Morton, G. J., Cummings, D. E., Baskin, D. G., Barsh, G. S., & Schwartz, M. W. (2006). Central nervous system control of food intake and body weight. *Nature, 443*, 289–295.

Murphy, K. G., & Bloom, S. R. (2006). Gut hormones and the regulation of energy homeostasis. *Nature, 444*, 854–859.

Muse, E. D., Lam, T. K., Scherer, P. E., & Rossetti, L. (2007). Hypothalamic resistin induces hepatic insulin resistance. *The Journal of Clinical Investigation, 117*, 1670–1678.

Myers, M. G., Cowley, M. A., & Munzberg, H. (2008). Mechanisms of leptin action and leptin resistance. *Annual Review of Physiology, 70*, 537–556.

Nishimura, S., Manabe, I., Nagasaki, M., Eto, K., Yamashita, H., Ohsugi, M., et al. (2009). CD8+ effector T cells contribute to macrophage recruitment and adipose tissue inflammation in obesity. *Nature Medicine, 15*, 914–920.

Obici, S., Zhang, B. B., Karkanias, G., & Rossetti, L. (2002). Hypothalamic insulin signaling is required for inhibition of glucose production. *Nature Medicine, 8*, 1376–1382.

Oh, I., Thaler, J. P., Ogimoto, K., Wisse, B. E., Morton, G. J., & Schwartz, M. W. (2010). Central administration of interleukin-4 exacerbates hypothalamic inflammation and weight gain during high-fat feeding. *American Journal of Physiology, Endocrinology and Metabolism, 299*, E47–E53.

Ozcan, L., Ergin, A. S., Lu, A., Chung, J., Sarkar, S., Nie, D., et al. (2009). Endoplasmic reticulum stress plays a central role in development of leptin resistance. *Cell Metabolism, 9*, 35–51.

Ozcan, U., Yilmaz, E., Ozcan, L., Furuhashi, M., Vaillancourt, E., Smith, R. O., et al. (2006). Chemical chaperones reduce ER stress and restore glucose homeostasis in a mouse model of type 2 diabetes. *Science, 313*, 1137–1140.

Pahl, H. L. (1999). Activators and target genes of Rel/NF-kappaB transcription factors. *Oncogene, 18*, 6853–6866.

Plum, L., Belgardt, B. F., & Bruning, J. C. (2006). Central insulin action in energy and glucose homeostasis. *The Journal of Clinical Investigation, 116*, 1761–1766.

Posey, K. A., Clegg, D. J., Printz, R. L., Byun, J., Morton, G. J., Vivekanandan-Giri, A., et al. (2009). Hypothalamic proinflammatory lipid accumulation, inflammation, and insulin resistance in rats fed a high-fat diet. *American Journal of Physiology, Endocrinology and Metabolism, 296*, E1003–E1012.

Purkayastha, S., Zhang, G., & Cai, D. (2011). Uncoupling the mechanisms of obesity and hypertension by targeting hypothalamic IKK-beta and NF-kappaB. *Nature Medicine, 17*, 883–887.

Purkayastha, S., Zhang, H., Zhang, G., Ahmed, Z., Wang, Y., & Cai, D. (2011). Neural dysregulation of peripheral insulin action and blood pressure by brain endoplasmic reticulum stress. *Proceedings of the National Academy of Sciences of the United States of America, 108*, 2939–2944.

Qadri, F., Hauser, W., Johren, O., & Dominiak, P. (2002). Kinin B1 and B2 receptor mRNA expression in the hypothalamus of spontaneously hypertensive rats. *Canadian Journal of Physiology and Pharmacology, 80*, 258–263.

Rahmouni, K., Correia, M. L., Haynes, W. G., & Mark, A. L. (2005). Obesity-associated hypertension: New insights into mechanisms. *Hypertension, 45*, 9–14.

Rahmouni, K., Haynes, W. G., & Mark, A. L. (2002). Cardiovascular and sympathetic effects of leptin. *Current Hypertension Reports, 4*, 119–125.

Romanatto, T., Cesquini, M., Amaral, M. E., Roman, E. A., Moraes, J. C., Torsoni, M. A., et al. (2007). TNF-alpha acts in the hypothalamus inhibiting food intake and increasing the respiratory quotient—Effects on leptin and insulin signaling pathways. *Peptides, 28*, 1050–1058.

Sabio, G., Cavanagh-Kyros, J., Barrett, T., Jung, D. Y., Ko, H. J., Ong, H., et al. (2010). Role of the hypothalamic-pituitary-thyroid axis in metabolic regulation by JNK1. *Genes & Development, 24*, 256–264.

Sabio, G., Cavanagh-Kyros, J., Ko, H. J., Jung, D. Y., Gray, S., Jun, J. Y., et al. (2009). Prevention of steatosis by hepatic JNK1. *Cell Metabolism, 10*, 491–498.

Sabio, G., Das, M., Mora, A., Zhang, Z., Jun, J. Y., Ko, H. J., et al. (2008). A stress signaling pathway in adipose tissue regulates hepatic insulin resistance. *Science, 322*, 1539–1543.

Sabio, G., Kennedy, N. J., Cavanagh-Kyros, J., Jung, D. Y., Ko, H. J., Ong, H., et al. (2010). Role of muscle c-Jun NH2-terminal kinase 1 in obesity-induced insulin resistance. *Molecular and Cellular Biology, 30*, 106–115.

Sandoval, D., Cota, D., & Seeley, R. J. (2008). The integrative role of CNS fuel-sensing mechanisms in energy balance and glucose regulation. *Annual Review of Physiology, 70*, 513–535.

Savoia, C., & Schiffrin, E. L. (2006). Inflammation in hypertension. *Current Opinion in Nephrology and Hypertension, 15*, 152–158.

Schenk, S., Saberi, M., & Olefsky, J. M. (2008). Insulin sensitivity: Modulation by nutrients and inflammation. *The Journal of Clinical Investigation, 118*, 2992–3002.

Schieven, G. L., Kirihara, J. M., Myers, D. E., Ledbetter, J. A., & Uckun, F. M. (1993). Reactive oxygen intermediates activate NF-kappa B in a tyrosine kinase-dependent mechanism and in combination with vanadate activate the p56lck and p59fyn tyrosine kinases in human lymphocytes. *Blood, 82*, 1212–1220.

Schoonbroodt, S., Ferreira, V., Best-Belpomme, M., Boelaert, J. R., Legrand-Poels, S., Korner, M., et al. (2000). Crucial role of the amino-terminal tyrosine residue 42 and

the carboxyl-terminal PEST domain of I kappa B alpha in NF-kappa B activation by an oxidative stress. *The Journal of Immunology*, *164*, 4292–4300.
Shi, H., Kokoeva, M. V., Inouye, K., Tzameli, I., Yin, H., & Flier, J. S. (2006). TLR4 links innate immunity and fatty acid-induced insulin resistance. *The Journal of Clinical Investigation*, *116*, 3015–3025.
Shoelson, S. E., & Goldfine, A. B. (2009). Getting away from glucose: Fanning the flames of obesity-induced inflammation. *Nature Medicine*, *15*, 373–374.
Singhal, N. S., Lazar, M. A., & Ahima, R. S. (2007). Central resistin induces hepatic insulin resistance via neuropeptide Y. *The Journal of Neuroscience*, *27*, 12924–12932.
Solinas, G., Vilcu, C., Neels, J. G., Bandyopadhyay, G. K., Luo, J. L., Naugler, W., et al. (2007). JNK1 in hematopoietically derived cells contributes to diet-induced inflammation and insulin resistance without affecting obesity. *Cell Metabolism*, *6*, 386–397.
Strowig, T., Henao-Mejia, J., Elinav, E., & Flavell, R. (2012). Inflammasomes in health and disease. *Nature*, *481*, 278–286.
Tabas, I., & Ron, D. (2011). Integrating the mechanisms of apoptosis induced by endoplasmic reticulum stress. *Nature Cell Biology*, *13*, 184–190.
Takeuchi, O., & Akira, S. (2010). Pattern recognition receptors and inflammation. *Cell*, *140*, 805–820.
Tanaka, Y., Guhde, G., Suter, A., Eskelinen, E. L., Hartmann, D., Lullmann-Rauch, R., et al. (2000). Accumulation of autophagic vacuoles and cardiomyopathy in LAMP-2-deficient mice. *Nature*, *406*, 902–906.
Tapia-Gonzalez, S., Garcia-Segura, L. M., Tena-Sempere, M., Frago, L. M., Castellano, J. M., Fuente-Martin, E., et al. (2011). Activation of microglia in specific hypothalamic nuclei and the cerebellum of adult rats exposed to neonatal overnutrition. *Journal of Neuroendocrinology*, *23*, 365–370.
Taylor, J. P., Hardy, J., & Fischbeck, K. H. (2002). Toxic proteins in neurodegenerative disease. *Science*, *296*, 1991–1995.
Thaler, J. P., & Schwartz, M. W. (2010). Minireview: Inflammation and obesity pathogenesis: The hypothalamus heats up. *Endocrinology*, *151*, 4109–4115.
Thaler, J. P., Yi, C. X., Schur, E. A., Guyenet, S. J., Hwang, B. H., Dietrich, M. O., et al. (2012). Obesity is associated with hypothalamic injury in rodents and humans. *The Journal of Clinical Investigation*, *122*, 153–162.
Tsukumo, D. M., Carvalho-Filho, M. A., Carvalheira, J. B., Prada, P. O., Hirabara, S. M., Schenka, A. A., et al. (2007). Loss-of-function mutation in Toll-like receptor 4 prevents diet-induced obesity and insulin resistance. *Diabetes*, *56*, 1986–1998.
Uehara, T., Nakamura, T., Yao, D., Shi, Z. Q., Gu, Z., Ma, Y., et al. (2006). S-nitrosylated protein-disulphide isomerase links protein misfolding to neurodegeneration. *Nature*, *441*, 513–517.
Unger, E. K., Piper, M. L., Olofsson, L. E., & Xu, A. W. (2010). Functional role of c-Jun-N-terminal kinase in feeding regulation. *Endocrinology*, *151*, 671–682.
Vallerie, S. N., Furuhashi, M., Fucho, R., & Hotamisligil, G. S. (2008). A predominant role for parenchymal c-Jun amino terminal kinase (JNK) in the regulation of systemic insulin sensitivity. *PLoS One*, *3*, e3151.
Wang, X. (2005). Investigational anti-inflammatory agents for the treatment of ischaemic brain injury. *Expert Opinion on Investigational Drugs*, *14*, 393–409.
Weisberg, S. P., McCann, D., Desai, M., Rosenbaum, M., Leibel, R. L., & Ferrante, A. W., Jr. (2003). Obesity is associated with macrophage accumulation in adipose tissue. *The Journal of Clinical Investigation*, *112*, 1796–1808.
Winer, S., Chan, Y., Paltser, G., Truong, D., Tsui, H., Bahrami, J., et al. (2009). Normalization of obesity-associated insulin resistance through immunotherapy. *Nature Medicine*, *15*, 921–929.
World Health Organization. (2011). Media centre: Obesity and overweight.

Wu, K. I., & Schmid-Schonbein, G. W. (2011). Nuclear factor kappa B and matrix metalloproteinase induced receptor cleavage in the spontaneously hypertensive rat. *Hypertension, 57*, 261–268.

Xu, C., Bailly-Maitre, B., & Reed, J. C. (2005). Endoplasmic reticulum stress: Cell life and death decisions. *The Journal of Clinical Investigation, 115*, 2656–2664.

Xu, H., Barnes, G. T., Yang, Q., Tan, G., Yang, D., Chou, C. J., et al. (2003). Chronic inflammation in fat plays a crucial role in the development of obesity-related insulin resistance. *The Journal of Clinical Investigation, 112*, 1821–1830.

Yamashima, T. (2012). Hsp70.1 and related lysosomal factors for necrotic neuronal death. *Journal of Neurochemistry, 120*, 477–494.

Yamazaki, H., Hiramatsu, N., Hayakawa, K., Tagawa, Y., Okamura, M., Ogata, R., et al. (2009). Activation of the Akt-NF-kappaB pathway by subtilase cytotoxin through the ATF6 branch of the unfolded protein response. *The Journal of Immunology, 183*, 1480–1487.

Zhang, K., & Kaufman, R. J. (2008). From endoplasmic-reticulum stress to the inflammatory response. *Nature, 454*, 455–462.

Zhang, H., Zhang, G., Gonzalez, F. J., Park, S. M., & Cai, D. (2011). Hypoxia-inducible factor directs POMC gene to mediate hypothalamic glucose sensing and energy balance regulation. *PLoS Biology, 9*, e1001112.

Zhang, X., Zhang, G., Zhang, H., Karin, M., Bai, H., & Cai, D. (2008). Hypothalamic IKKbeta/NF-kappaB and ER stress link overnutrition to energy imbalance and obesity. *Cell, 135*, 61–73.

Zhou, R., Yazdi, A. S., Menu, P., & Tschopp, J. (2011). A role for mitochondria in NLRP3 inflammasome activation. *Nature, 469*, 221–225.

Zhu, F., & Qian, C. (2006). Berberine chloride can ameliorate the spatial memory impairment and increase the expression of interleukin-1beta and inducible nitric oxide synthase in the rat model of Alzheimer's disease. *BMC Neuroscience, 7*, 78.

CHAPTER NINE

p66^{ShcA}: Linking Mammalian Longevity with Obesity-Induced Insulin Resistance

Sofia Chiatamone Ranieri*, Salvatore Fusco†, Giovambattista Pani†,1

*Clinical Chemistry, Department of Laboratory Medicine, Laboratory and Endocrinology Unit, Azienda Ospedaliera ASMN, Istituto di Ricovero e Cura a Carattere Scientifico, Reggio Emilia, Italy

†Laboratory of Cell Signaling, Institute of General Pathology, Catholic University Medical School, Rome, Italy

[1]Corresponding author: e-mail address: gpani@rm.unicatt.it

Contents

Abstract

Traditionally linked to an accumulation of cell and tissue oxidative damage, the aging process is currently viewed as the result of a chronic metabolic imbalance and deranged body response to nutrients. This is in keeping with the pivotal role of the insulin/IGF pathway in longevity determination from model organisms to primates, with the association between obesity and age-related disorders, and with the dramatic acceleration of senescence by diabetes. Importantly, metabolic and oxidative damages concomitantly occur in aging as consequences of tissue hyperstimulation by insulin.

With its double identity of generator of mitochondrial oxidant species and of signaling adaptor in the insulin receptor cascade, the 66Kd Shc protein (p66Shc) has drawn major attention as a negative determinant of life span and healthy longevity in mammals. We have demonstrated that these effects are related, at least in part, to p66Shc effect on the mTOR/S6K cascade that promotes obesity and insulin resistance and shortens life span.

Vitamins and Hormones, Volume 91
ISSN 0083-6729
http://dx.doi.org/10.1016/B978-0-12-407766-9.00009-2

1. AGING: OXIDATIVE STRESS VERSUS METABOLIC UNBALANCE

In the twenty-first century, human aging and age-associated diseases will become one of the biggest challenges faced by both rich and emerging countries.

Aging is a complex biological process characterized by progressive functional and structural deterioration of multiple organ systems (primary aging) but also a result of external influences, including diseases (secondary aging): in the past decades, life expectancy has significantly increased, leading to a prepotent rise of age-associated metabolic disorders (including obesity, insulin resistance, type 2 diabetes, and atherosclerosis) and cancer. Successful interventions aimed at delaying the development and progression of age-related diseases are expected to further increase average life span.

Calorie restriction is the only known intervention that can slow primary aging and protect against secondary aging. The most important calorie restriction-induced antiaging mechanisms are thought to be neuroendocrine adaptations and protection against oxidative stress damage (Fontana, 2009). There is a close relationship between IGF-1/insulin signaling activity, stress resistance, and longevity: metabolic effects of attenuated insulin signaling are usually accompanied by changes in mitochondrial activity and a decrease in ROS (reactive oxygen species) production that, coupled with enhanced antioxidant defense capacity, leads to lower levels of oxidative damage. These changes are often concurrently reported in calorie-restricted and mutant long-living organisms.

The initial idea of a link between oxidative stress resistance and longevity was theorized by Harman (1956) in his free radical theory on aging. Free radicals produced as by-products of normal metabolism can damage cells and induce the physiological decline of the organism. These molecules can adversely affect DNA, proteins, and lipids, and this accumulation of damage leads, in a rather consolidated view, to aging (Brown-Borg, 2006). As most of the endogenous cellular ROS are generated in mitochondria as a consequence of oxidative metabolism, the antiaging action of calorie restriction could be easily explained in the framework of Harman's theory. In recent years, however, reduced activity of insulin/insulin-like growth factor signaling (IIS) has emerged as another finding consistently associated with extended longevity in organisms ranging from yeast to mammals, and potentially related to nutrient intake and energy metabolism. These observations, now largely confirmed in transgenic murine systems, have initially come from genetic studies conducted on model invertebrate organisms that, in reason of their short life spans, represent powerful systems to investigate molecular pathways that may influence longevity. Aging studies on genetic mutants in both *Caenorhabditis elegans* and *Drosophila melanogaster* have demonstrated that a reduction in intracellular signaling pathways homologous or similar to that triggered by insulin or IGF-1 signaling (IIS) in mammals increases longevity (Clancy et al., 2001; Kenyon, 2001; Kenyon, Chang, Gensch, Rudner, & Tabtiang, 1993; Tatar et al., 2001). In *C. elegans*, a set of genes involved in IIS cascades is able to influence longevity: it includes daf-2 (insulin/IGF-1-like receptor), pdk-1, akt-1, akt-2, sgk-1 (serine–threonine kinases PDK-1, SGK-1, Akt-1, and Akt-2), age-1 (catalytic subunit of phosphatidylinositol-3 kinase (PI3K)), as well as daf-16 (forkhead transcription factor and the major target of insulin-like signaling in this organism), and daf-18 (PTEN, a phosphatase that dephosphorylates the PI3K lipid products PtdIns(3,4)P_2 and PtdIns(3,4,5)P_3 and inhibits the AKT signaling pathway) (Berryman, Christiansen, Johannsson, Thorner, & Kopchick, 2008). Likewise, in Drosophila mutations of genes encoding insulin-like receptor, insulin receptor substrate (Chico), PI3K, PTEN, Akt/PKB, and FOXO modify life span.

All of these genes and the encoded proteins and functions are phylogenetically preserved also in mammals. Accordingly, common phenotypes among long-lived mice, that likely contribute to increased health and longevity, include, besides improved stress resistance, and protection from oxidative damage and carcinogenesis, also inhibition of IGF-1 signaling and enhanced insulin sensitivity. In fact, among the several long-lived mouse mutants that have been discovered or created and that exhibit extended life span are (a) mice with defects in anterior pituitary function (Snell and Ames dwarf mice), (b) mice with selective defects of growth hormone (GH) function (GH receptor knockout), and (c) mice with defects in IGF-1 or IGF-1 receptor (IGF-1 receptor heterozygote) or in downstream IGF-1 signaling molecules (p66Shc knockout and Klotho transgenic mouse).

Ames and Snell dwarf mice live 42–68% longer than wild type (Brown-Borg, Borg, Meliska, & Bartke, 1996; Flurkey, Papaconstantinou, & Harrison, 2002): they have a point mutation in a gene, respectively, Prop-1 and Pit1, that both determine a primary pituitary deficiency (resulting in the absence of GH, prolactin, and thyrotropin).

Homozygous deletion of GH receptors or GH brings to reduction of circulating IGF-1 and insulin with significant delay of aging and extension of life (Coschigano, Clemmons, Bellush, & Kopchick, 2000; Zhou et al., 1997). Although IGF-1 receptor knockout mice are not viable, a partial reduction of IGF-1 activity (IGF-1-R heterozygous) increases life span and resistance to oxidative stress (assayed by paraquat toxicity), although only in females (Holzenberger et al., 2003).

In keeping with Harman's theory, resistance to oxidative stress is currently considered one of the most important mechanisms by which these mutations delay aging in all organisms, from yeast to mammals. Fibroblasts derived from long-lived Snell dwarf, Ames dwarf, and GHRKO mice are more resistant to a variety of oxidative, cytotoxic, and metabolic stresses (Harper et al., 2006). Moreover, the reduction of insulin levels and the enhancement of insulin sensitivity are among the most significant endocrine features shared by Ames dwarf, Snell dwarf, and GHRKO mice (Bartke, 2006).

A general explanation for the relationship between reduced growth factors (GFs)/nutrient signaling and oxidative stress resistance could be that a lower demand for protein synthesis causes a shift in resources from growth and proliferation toward cytoprotection. Genetic models of delayed aging and extended life span have recently provided a molecular framework for the cross talk between the insulin/IGF-1 and ROS signaling pathways. Some key regulators of these biological connections are, in fact, been identified in the FOXO transcription factors: FOXO transcriptional activity is inhibited by IGF-1/insulin signaling as part of its mitogenic and anabolic effects in the context of nutrient replenishment, whereas nutrient or GFs deprivation or oxidative challenge promotes the nuclear translocation of the factor, where it transactivates a number of antioxidant genes (Catalase, SOD2, GPX) thus increasing cell resistance to oxidative stress.

Thus, FOXO may be the factor that integrates nutritional and stress signals to regulate growth, cellular proliferation, and stress resistance (Accili & Arden, 2004).

Two other long-lived mouse mutants, Klotho and p66Shc KO (the latter discussed in detail below), link the role of oxidative stress ROS signaling to aging and longevity.

The Klotho gene encodes a transmembrane protein with an extracellular soluble peptide detected in blood and cerebrospinal fluid, implying it may also act as a hormone (Imura et al., 2004).

Overexpression of Klotho in mice extends life span and reduces oxidative stress. Interestingly, Klotho also decreases insulin signaling by suppressing tyrosine phosphorylation of the insulin and IGF-1 receptors (Kurosu et al., 2005). This in turn reduces IRS protein association with

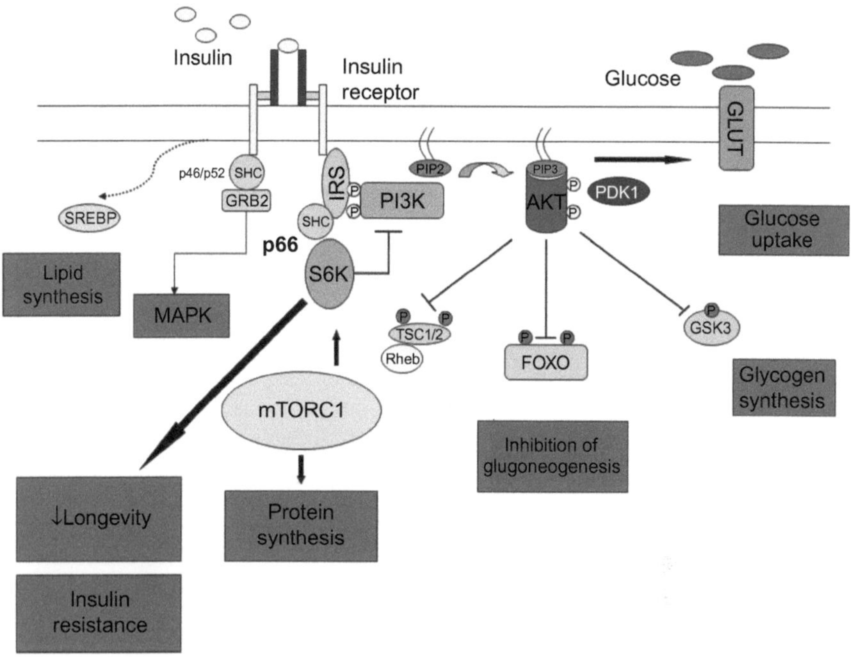

Figure 9.1 Insulin signaling pathways. Activation of the insulin receptor induces transcription of SREBP (Sterol regulatory element-binding protein) and phosphorylation of members of the IRS and SHC families resulting in the activation of a variety of signaling pathways, including MAPK, PI3K, and consequently Akt. These pathways act in a coordinated manner to regulate glucose, lipid, and protein metabolism. Insulin inhibits hepatic gluconeogenesis, promotes glucose uptake and storage into cells by inducing translocation of the glucose transporter and increasing glycogen synthesis, and in addition, stimulates amino acid and fatty acid uptake and protein and lipid synthesis. Chronic abnormal activation of these cascades alters insulin sensibility and accelerates physiological aging of the organism. (See Color Insert.)

PI3K, thereby inhibiting insulin and IGF-1 signaling (Fig. 9.1), a feature frequently associated with longevity (Yamamoto et al., 2005). Thus insulin signaling and susceptibility to oxidative stress appear as two closely related and deeply interconnected determinants of life span extension also in murine models of genetic longevity.

2. ADIPOSE TISSUE AND LONGEVITY IN MAMMALS

There are also several long-living mutant mice made by organ-specific gene mutation: one of the most studied tissues in aging is certainly white fat. Well more than a mere energy reserve, fat is a major target for insulin, exerts important immune and endocrine functions, and lies at the crossroad of multiple mechanisms and pathways involved in longevity, age-related diseases, inflammation, and metabolic dysfunction (Tchkonia, Corkey, & Kirkland, 2006).

Accordingly, obesity leads to reduced life span and increases the risk of clinical pathologies and morbidity similar to those common in aging (Ahima, 2009). However, conditions that inhibit fat tissue accumulation extend life span: these include calorie restriction that reduces visceral fat (Barzilai & Gupta, 1999), mutant mice harboring fat-specific inactivation of the insulin receptor (FIRKO) or of its major target insulin receptor substrate-1 (IRS-1), mice devoid of the downstream insulin effector S6 kinase 1 (S6K) (Bluher, Kahn, & Kahn, 2003; Selman et al., 2008, 2009; Um et al., 2004), or mice treated chronically with the drug rapamycin (a specific inhibitor of the insulin-activated mTOR/S6K cascade, see below), which limits fat tissue development (Chang et al., 2009).

For example, fat insulin receptor knockout (FIRKO) mice exhibit 50% less fat mass in adipose tissue and live 18% longer than wild type although their food intake is normal (Bluher et al., 2003). This suggests that a reduction of fat mass due to reduced insulin activity, rather than calorie restriction *per se*, can increase longevity in mice. Along similar lines of thinking, it has been suggested that obesity may represent an accelerated form of fat tissue aging (Ahima, 2009).

Among the several insulin-activated signaling cascades regulating adipose metabolism and potentially involved in the regulation of life span, attention has recently focused on the pathway triggered by the mammalian target of rapamycin (mTOR). This cascade integrates nutrient, hormone, and GF signals and regulates crucial cell responses ranging from protein synthesis, to energy metabolism, to progression through the cell cycle (Fig. 9.1). Not a surprise, the list of roles for this cascade in both physiology and human disease is rapidly expanding (Zoncu, Efeyan, & Sabatini, 2011). In Drosophila inhibition of mTOR signaling by genetic manipulation of diverse components (S6K1, TSC1, and TSC2) of this nutrient-sensitive cascade causes life span extension (Kapahi et al., 2004). Likewise, mice deficient for ribosomal S6K1, a key downstream effector of mTOR, are protected against obesity and insulin resistance by increase of lipolysis and consequently inhibition of fat accumulation (Um et al., 2004) and show increased life span and resistance to age-related pathologies (Selman et al., 2009). Additionally, as mentioned above, chronic administration of the mTOR inhibitor rapamycin to adult mice results in life span extension (Chang et al., 2009).

3. ADIPOSE TISSUE, INSULIN RESISTANCE, AND LONGEVITY

In view of the compelling genetic evidence for a direct involvement of the insulin/IGF cascade in life span limitation, the fact that many long-lived mouse mutants display increased insulin sensitivity, and that diabetes (no matter whether due to lack of insulin, like in T1D, or a defect in its signaling activity as in T2D) leads to a general acceleration of aging and age-related diseases, may appear paradoxical. However, these seemingly conflicting evidence can be easily reconciled based on a number of considerations. First of all, in the absence of insulin (like in type 1 diabetes), the ensuing chronic hyperglycemia and hyperlipidemia act as direct tissue damaging agents, promoting vascular, nerve, and ocular pathologic changes that recapitulate, in an exacerbated form, those associated with "normal" senescence. On the other hand, a common feature of obesity-associated insulin resistance and type-two diabetes is elevated plasma insulin. Hyperinsulinemia occurs as an adaptive response to chronic hyperglycemia due to reduced glucose clearance by insulin-resistant liver, muscle, and adipose tissue; in the current view, the raise in circulating insulin leads to aberrant signaling in the same and other tissues (such as in endothelial and VSMC cells) through biochemical branches (as the RAS–MAPK cascade) distinct from the canonical metabolic pathways mediated by PI3K (Muniyappa & Quon, 2007). Consequences of this "pathway-selective" insulin unresponsiveness include generation of ROS and increased cell proliferation and are believed to play a role in diabetes-associated vascular aging and atherosclerosis and also, to some extent, in increased susceptibility to cancer. Thus, in this perspective, mutations and treatments that increase systemic insulin responsiveness promote longevity, at least in part, by reducing the concentration of circulating insulin and its deleterious effects on peripheral, nonmetabolic tissues (i.e., vasculature, kidney, brain) involved in age-related diseases (Fig. 9.2).

The mTOR/S6K cascade has been recently recognized a central role in the negative feedback loop that leads to peripheral insulin resistance. In fact, in insulin target tissues, hyperstimulation through the insulin receptor (as it occurs as consequence of excess nutrient availability) leads to the S6K-dependent phosphorylation of the IRS-1 on serine residues, which in turn inhibits phosphotyrosine-dependent recruitment of PI3K and downstream signaling to AkT (Um, D'Alessio, & Thomas, 2006). Accordingly, in several experimental models *in vitro* and *in vivo*, attenuation of mTOR signaling

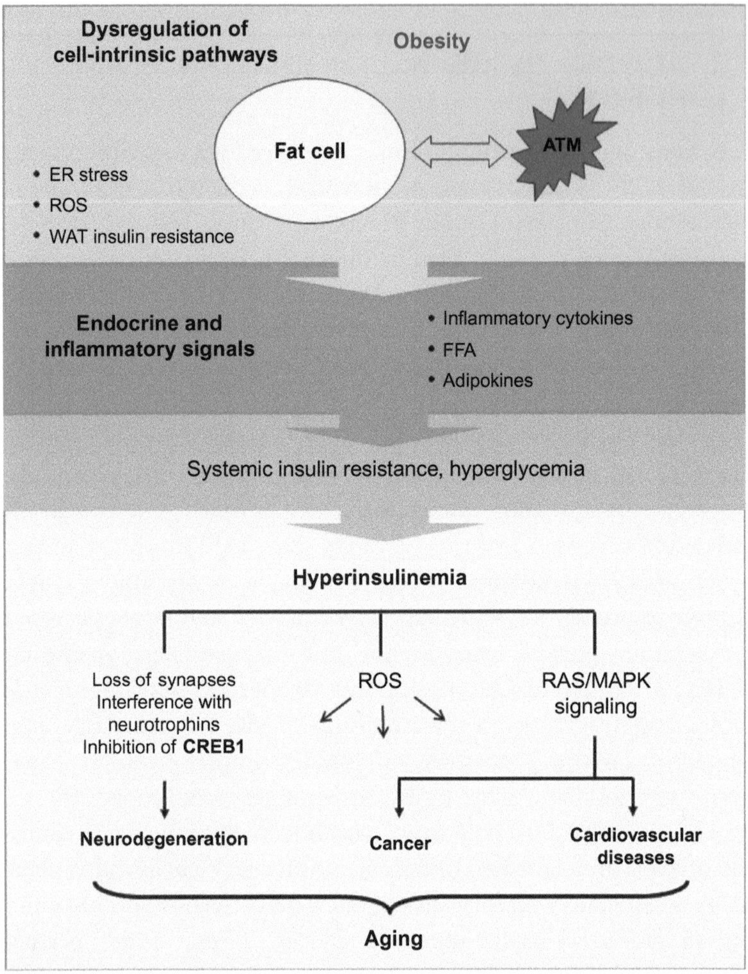

Figure 9.2 Mechanisms linking obesity-associated insulin resistance and aging. Increased adipose mass predisposes to systemic insulin resistance through interplay of cell-extrinsic and cell-intrinsic mechanisms. Adipocytes and adipose tissue macrophages (ATMs) produce adipokines and inflammatory cytokines interfering with insulin signaling in all tissues (in particular, liver, muscle, and adipose tissue); excess of nutrients and fatty acids (FAs) has a negative impact on insulin signaling and increases reactive oxygen species (ROS) generation by mitochondrial β oxidation, leading to mitochondrial dysfunction and induction of cellular endoplasmic reticulum (ER) stress. Hyperinsulinemia plays a central role in obesity-associated pathology, by acting through canonical and noncanonical signaling branches; the latter include hyperstimulation of the ERK cascade in endothelial cells (atherosclerosis) and epithelial cells (cancer), as well as interference with neurotrophin signaling in the brain (neurodegeneration, Alzheimer's Disease). (See Color Insert.)

leads to increased insulin responsiveness through the PI3K/AkT cascade (Haruta et al., 2000; Um et al., 2004).

Adiposity has a main role in the establishment of peripheral insulin resistance and hyperinsulinemia, due to (a) decreased clearance of glucose and (b) increased release of free fatty acids, that cause insulin resistance in other organs, primarily the liver. Thus, in the context of obesity, a decrease of mTOR activity in adipocytes is anticipated to have the double beneficial effect of reducing fat accumulation, and of promoting insulin responsiveness in fat and other insulin-dependent tissues, with a general improvement of glucose homeostasis (Um et al., 2004). Thus, it is conceivable that at least some of the described life span-extending effects of mTOR inhibition in mice indeed occur in fat cells, as in part suggested by the phenotype of fat-specific Raptor KO mice (Polak et al., 2008), and are related to a general improvement of glucose homeostasis and a reduction of circulating insulin. To date, and in spite of accumulating knowledge in the field, the mechanisms whereby the IGF-1/insulin signaling circuit works as a pacemaker of aging are far from being completely clarified, posing a major challenge to research on aging and age-related disorders. Clarification of these networks may help to discover potential drug targets to improve health and extend life span in humans.

4. p66Shc AND LONGEVITY IN MICE

In recent years, another putative longevity gene has attracted major interest in aging research: the one encoding the 66 kDa isoform of Shc ($p66^{ShcA}$). The genetic deletion of this gene in mice surprisingly resulted in about 30% extension of the mean and maximum life span, with wild-type animals having an average life span of 761 days, whereas the heterozygous or homozygous mice having an average life span of 815 or 973 days, respectively (Migliaccio et al., 1999).

Although no metabolic measurements were taken in this initial study, there were no reported differences in body weight or food intake between the groups. Phenotypical and histopathological analysis revealed no obvious abnormalities in the $p66Shc^{-/-}$ mice such as changes in tumor frequency and endocrinological defects (Migliaccio et al., 1999).

These transgenic animals seemed better equipped to handle oxidative stress than their wild-type counterparts: p66Shc(−/−) mice (henceforth indicated as p66 KO) had a 40% increased survival time in response to injection of paraquat when compared to wild-type littermates (p66 WT);

moreover, mouse embryonic fibroblasts (MEFs) derived from p66 KO rodents exposed to H2O2 or UV treatment showed greatly reduced levels of apoptosis when compared to wild-type cells, while overexpression of a wild-type p66Shc transgene in normal MEFs resulted in increased sensitivity to apoptosis under the same conditions (Migliaccio et al., 1999). This enhanced resistance to oxidative stress in p66Shc null mice was proposed as the mechanism underlying increased longevity, to the point that this novel transgenic mouse became one of the strongest argument in support of the Harman's "free radical theory of aging" (Harman, 1956).

Substantial experimental evidence suggested that p66Shc played a crucial role in the regulation of oxidative stress response and apoptosis (Pani, Koch, & Galeotti, 2009; Trinei et al., 2002). Interestingly, it appeared that p66Shc was both a downstream mediator and a primary source of oxidative stress. In fact, in the model of Migliaccio and colleagues, p66Shc responds to a variety of proapoptotic stimuli, such as UV irradiation and oxidant compounds by increasing ROS levels in the mitochondrial intermembrane space, through an inherent ROS producing activity that triggers initiation of the mitochondrial apoptosis pathway. Behaving like an oxidoreductase, p66Shc is able to shuttle electrons from cytochrome *c* to molecular oxygen and generates hydrogen peroxide which in turn induces mitochondrial permeability and apoptosis (Giorgio et al., 2005; Orsini et al., 2004). Along different lines of investigation, the capacity of p66Shc to increase the levels of intracellular ROS has been linked to the reduction of ROS scavenging through the redox-dependent inactivation of FKHRL1, a transcription factor whose homologue, DAF-16, controls life span in *C. elegans* (Nemoto & Finkel, 2002); finally, an increase of membrane oxidases activity was reported as a further mechanism for ROS increase by p66Shc (Khanday et al., 2006).

Importantly, the proapoptotic activity of p66Shc is strictly dependent on phosphorylation of Serine 36 in the CH2 domain by stress kinases such as Jnk-1 and Pkc-B (Pinton et al., 2007), which occurs in the cytoplasm and is necessary for Shc activation and translocation to the mitochondrial intermembrane space.

5. p66Shc AND INSULIN SIGNALING

All the three isoforms of p66Shc have been reported to reside predominantly in the cytoplasm, and some redistribute to the plasma membrane after GF stimulation (Lotti et al., 1996); thus, only a marginal fraction of cellular p66Shc is associated with mitochondria, suggesting that additional mechanisms may contribute to the connection between loss of p66Shc and organismal longevity and health.

The p66Shc protein was first described as a GF "adapter" protein, just like the two other members, p52 and p46, of the ShcA family (Pelicci et al., 1992). These proteins couple the activated epidermal growth factor (EGF) receptor to Ras and the MAP kinase cascade and have a unique, highly conserved domain organization characterized by one carboxy terminal Src homology 2 (SH2) domain, a central region highly enriched in glycine and proline residues named collagen homologous 1(CH1) due to its homology with collagen protein (CH1), and one N-terminal phosphotyrosine-binding domain (PTB) (Luzi, Confalonieri, Di Fiore, & Pelicci, 2000).

P66 differs for the presence of an additional N-terminal proline-rich collagen-homology domain (CH2) (Luzi et al., 2000; Ventura, Luzi, Pacini, Baldari, & Pelicci, 2002), which contains the serine phosphorylation site (Ser36) reportedly crucial for its "proapoptotic" properties (Migliaccio et al., 1999).

ShcA proteins can use the SH2 domain in receptor interactions such as EGF receptor and ErbB-2 (Bonfini, Migliaccio, Pelicci, Lanfrancone, & Pelicci, 1996). The PTB domains, instead, can bind to phospholipids, implying a role of PI3K in the activation of Shc (Bonfini et al., 1996; Okada et al., 1997).

The p46 and p52Shc isoforms have been implicated in mitogenic signaling and oncogenesis, by assisting receptor tyrosine kinase activity. Upon GF stimulation, these proteins are rapidly and efficiently tyrosine phosphorylated by all the tyrosine kinase receptors tested and have been found to recruit the Grb2/Sos complex to a membrane-proximal localization where it activates the Ras and the downstream MAP kinase cascade (Bonfini et al., 1996; Giorgetti, Pelicci, Pelicci, & Van Obberghen, 1994; Okada et al., 1997).

Although p66Shc can be phosphorylated on tyrosine and interacts with GRb2, unlike p46 and p52, it has been shown to inhibit activation of the Ras/MAP kinase pathway triggered by the EGF and IGF-1 receptors, as well as by cytokines in lymphocytes (Migliaccio et al., 1997; Okada et al., 1997; Pacini et al., 2004).

Therefore, p66Shc may compete with p52 for Grb2 binding, thus acting as a dominant negative regulator of the Ras-mediated mitogenic signaling pathway.

Accordingly, Clemmons et al. reported that p66Shc inhibits IGF-1-stimulated p52Shc tyrosine phosphorylation and Grb2 recruitment as well as MAPK activation by competitively binding to Src homology 2 domain-containing protein tyrosine phosphatase substrate-1 (SHPS-1); this prevents the assembly of the SHP-2/Src/p52Shc/Grb2 signaling complex, thus resulting in a decrease in IGF-1-stimulated proliferation and migration (Xi, Shen, & Clemmons, 2008).

In vascular smooth muscle cells, p66Shc also appears to inhibit IGF-dependent PI3K/Akt activation and cell survival through a distinct mechanism that involves the interaction of a polyproline sequence contained in the unique CH2 region of p66Shc with the SH3 domain in Src kinase, that leads to impaired Src activation, and that is, of note, independent of p66Shc redox properties. In fact, ROS are associated with enhancement, and not inhibition of the IGF-1 cascade, and were in absolute terms increased, in these studies, in cells overexpressing p66Shc, in spite of overall attenuation of IGF signaling (Xi, Shen, & Clemmons, 2010; Xi, Shen, Radhakrishnan, Maile, & Clemmons, 2010).

A similar signal inhibitory role for p66Shc has been described by Baldari and colleagues in T and B lymphocytes, where p66Shc acts also as a negative regulator of TCR and BCR signaling becoming a gatekeeper of autoimmunity and leukemia (Finetti et al., 2008). In fact, p66Shc effectively downregulates both TCR and BCR signaling, as shown by the hyperreactivity to antigen receptor engagement of peripheral T and B cells from p66 KO mice both *in vitro* and *in vivo*, and by the increased proliferative response of p66 KO T and B cells when stimulated *in vitro*.

Again, the enhancement in BCR signaling in B cells from p66 KO mice is not limited to Erk, as expected from the known inhibitory activity of p66Shc on Ras activation, but extends to other signaling mediators like PI3K/Akt pathway (Capitani et al., 2010) In addition to their hyperreactivity to antigen receptor triggering, p66 KO lymphocytes are long lived compared to their wild-type counterparts when cultured *ex vivo* and are more resistant to proapoptotic stimuli, including those involved in immune tolerance (Finetti et al., 2008; Pacini et al., 2004). These phenotypes can be accounted for, at least in part, by the enhanced activation of Akt downstream of both the TCR and the BCR.

A common theme emerging from the above and other studies is that, in several cell types including peripheral blood mononuclear cells (Pagnin et al., 2005), kidney cells (Menini et al., 2006), endothelial progenitor cells (Di Stefano et al., 2009), and smooth muscle cells (Xi, Shen, Radhakrishnan, et al., 2010), the expression of p66Shc and its signaling inhibitory effects are increased in response to hyperglycemia. Moreover, in skeletal muscle myoblasts, overexpression of p66Shc results in changes in the glucose transport system, including lower basal transport and decreased protein levels of

GLUT1 and GLUT3 transporters that are opposite to those observed when p66Shc is knocked down and again seem to depend on p66Shc's action on the ERK cascade.

Thus, collectively, the above findings outline a potential homeostatic circuitry whereby cells defend themselves from excess glucose by a p66-mediated inhibition of its transport (Natalicchio et al., 2004). Along similar lines of investigation, studies on the role of p66Shc in cell response to insulin have led to conflicting results. In fact, Berniakovich and colleagues have provided convincing evidence that p66Shc potentiates insulin signaling through its capacity to generate ROS, which in turn facilitate tyrosine kinase signaling by transient and reversible inhibition of tyrosine phosphatases. In particular, p66Shc-generated H_2O_2 appears to increase the intensity of insulin/IGF-1 signaling by transiently inactivating tyrosine and serine/threonine phosphatases such as PP2A and PTEN that regulate PI3K and Akt activation (Lee et al., 2002). Accordingly, these authors demonstrate that p66Shc is highly expressed in primary adipocytes where it is rapidly phosphorylated on Ser-36 after insulin treatment.

Through its redox effect on the insulin-signaling cascade, p66Shc can reduce mitochondrial oxygen consumption and favors TG accumulation inhibiting insulin-induced FA oxidation and basal and insulin-induced UCP1 expression.

So, the authors suggest p66Shc as a genetic determinant of fat development in adult mice and regulator of energetic metabolism and adipogenesis due to its redox properties (Berniakovich et al., 2008).

On the contrary, and in line with above-mentioned studies on IGF and antigen receptor signaling, Cortopassi et al. have shown that p66Shc inhibits the insulin pathway as negative adaptor, and because of this property, more than for its redox activity, p66Shc can modulate longevity and healthy aging in mice.

P66 KO mice have alterations in the expression of PI3K and insulin-related transcripts, and in accordance to our result (see below), they found these mice weigh less, because of significantly reduced fat mass, and are more insulin sensitive and glucose tolerant than wild-type controls. Moreover, unlike reported by Berniakovich et al., p66 KO mice tissues displayed stronger activation of Akt, a reliable indicator of insulin responsiveness. Authors concluded that, at the organismal, tissue, and cellular level, p66Shc opposes insulin signaling through PI3K and its depletion sensitizes to insulin (Tomilov et al., 2011).

In our studies, discussed below, we have demonstrated that p66Shc indeed plays both a positive and a negative role on insulin signaling and identified in its dual action upstream and downstream of the mTOR/S6K pathway a molecular key to solve this apparent contradiction (Ranieri et al., 2010).

6. p66Shc IN OBESITY-INDUCED INSULIN RESISTANCE

We thought to investigate the molecular interaction of p66Shc with signaling cascades triggered by insulin and nutrients in Ob/Ob mice, a genetic model of obesity and diabetes in which mTOR/S6K pathway is hyperfunctioning.

This mouse, homozygous for the LepOb mutation, is hyperphagic due to complete deficiency of leptin and develops overt obesity, hyperinsulinemia, and glucose intolerance, thereby modeling human metabolic syndrome (Westman, 1968).

Double-mutant animals (Ob/Ob-p66 KO and Ob/Ob-p66 WT littermates) were generated and monitored for weight gain, metabolic phenotype, and longevity. These studies led to a number of interesting observations: first of all, we noticed that obese p66 KO mice lived longer compared with p66 WT controls, with a survival median of 72 weeks versus 58 weeks of p66-proficient obese animals. This effect was proportionally comparable to the reported life span-extending effect of p66Shc deletion in lean mice. Second, analysis of metabolic setting in 20-week-old mice revealed that fasting glycemia values were significantly lower in p66 KO than in p66 WT animals. In line with this finding, insulin and glucose tolerance tests confirmed that glucose intolerance was less severe and peripheral response to insulin largely preserved in p66 KO obese mice compared to p66 WT animals, although, at that indicated age, the two groups displayed similar body weight and composition, and weight-matched animals were used for comparisons. Of note, we found no difference in plasma insulin levels between the p66 KO and p66 WT animals.

Overall, these data suggested that p66Shc deletion attenuates insulin resistance of Ob/Ob mice independent of body weight, thus pointing to a direct signaling effect of p66Shc in glucose homeostasis.

A third important difference among the two strains of obese mice emerged from the careful evaluation of weight gain curves: in fact, p66 KO animals displayed a lower maximum body weight and an earlier age-associated weight loss (that normally occurs in the Ob/Ob model) than the p66-proficient animals, with significantly different values between 34th and 54th week, in spite of no differences in food consumption between the two groups. Analysis of body composition of these animals confirmed that weight differences were mostly accounted for by a reduction in fat mass and in particular in visceral fat. Moreover, in keeping with a cell-autonomous role for p66Shc in fat accumulation, primary preadipocyte purified from the fat of obese p66 KO mice displayed much reduced triglyceride accumulation in response to an insulin and dexamethasone mix, compared with precursors derived from WT animals.

Prompted by these evidence and considering the high expression of p66Shc in adipocytes (Kao, Waters, Okada, & Pessin, 1997) and the key role of fat tissue in modulating organismal insulin responsiveness and life span especially in this mouse model of metabolic dysfunction, we analyzed the insulin pathway in this tissue with particular attention at the mTOR/S6K pathway, whose enhanced activity by high fat diet or leptin deficiency promotes fat accumulation, induces insulin resistance, and shortens mouse life span (Um et al., 2006, 2004). Interestingly, we found that the levels of the IRS-1 and its tyrosine phosphorylated form were markedly decreased (down to 40–50%) in this tissue, but not in the liver, in parallel with a constitutive activation of the S6K and its downstream effector S6 in p66 WT obese mice compared with p66 KO littermates. These findings suggested that p66 participates in insulin desensitization selectively in adipose tissue, through the S6K-dependent degradation of IRS-1. To further support this hypothesis, we exposed predipocytes isolated from white fat of p66 KO and WT to excess free fatty acids, in order to model insulin resistance *in vitro*. In these conditions, the presence of p66Shc was associated with a stronger inhibitory phosphorylation of the IRS-1 on serines 307 and 636–639 in response to insulin, a reduced recruitment of the PI3K, and impaired downstream signaling through the Akt/PKB kinase on serine 473. A similar difference, based on phosphorylation of Akt and uptake of 2-deoxy-D-[1-3H] glucose, was also observed in mature adipocytes freshly isolated from p66 WT and p66 KO adipose tissues. Analysis of the mTOR pathway in primary preadipocytes of p66 KO mice revealed a lower basal phosphorylation of S6 on serines 285–286, but neither that of another major mTOR substrate, 4EBP1, nor of the upstream mTOR regulator AMPK. Moreover, when exposed to glucose after nutrient starvation, the p66 KO cells showed a blunted reactivation of the S6 kinase and its substrate S6 compared to WT preadipocytes.

We were also able to reproduce the same results in another cell model, the 3T3L1 preadipocyte cell line, in which p66Shc overexpression or down-regulation by siRNA modifies the level of S6K activation and the ensuing inhibitory phosphorylation of IRS-1 on serine in response to insulin and IGF-1. Importantly, in 3T3L1 cells exposed to excess fatty acids, S6 phosphorylation (Ser-285–286), and phosphorylation of p66Shc on serine 36, occurred concomitantly and in parallel with the establishment of insulin desensitization.

The role of p66Shc in S6K-dependent downregulation of insulin signaling was further corroborated by biochemical evidence that IRS-1, p66Shc, and S6K form a complex *in vivo*, with p66Shc acting as an adapter in keeping with the structural features of this protein shc family protein. In fact, the association between S6K and IRS-1, that underlies IRS-1 phosphorylation/degradation (Haruta et al., 2000; Smith, 2002), was increased by p66Shc upon overexpression in the HEK 293 cell line.

Of note, although p66Shc increases cell oxidative damage in a number of pathophysiological settings, the redox properties of p66 are unlikely responsible for the stimulatory effects on the S6K pathway. Infact, we did not find any difference in the oxidation profile between p66 KO and WT preadipocytes, as investigated by means of a mitochondrially targeted redox-sensitive fluorescent protein (rxYFP). Moreover, overexpression of p66Shc and p66Shcqq, a mutant p66Shc devoid of prooxidant activity (Giorgio et al., 2005), activated S6K to a comparable extent in serum-stimulated 3T3L1 cells, while having no effect on the redox-sensitive stress kinase p38 MAPK. Finally, S6K activation by p66Shc was not inhibited by treatment with high concentrations of oxidant scavengers *N*-acetyl cysteine and catalase.

Thus, in our model, p66Shc operates as an adaptor molecule by participating in the assembly of supramolecular complexes relevant to the regulation of nutrient and insulin signaling (Fig. 9.3). By favoring S6K recruitment to IRS-1, p66Shc not only promotes adipogenic signaling through the mTOR/S6K pathway but also enhances the S6K-dependent retroinhibitory feedback and negatively impacts on downstream signal propagation through the PI3K cascade leading to insulin resistance (Ranieri et al., 2010).

7. p66 AND AGING: THE ROLE OF mTOR/S6K

These evidences offer a better knowledge of the molecular players and biochemical mechanisms linking overnutrition, excess body fat, and glucose intolerance but, above all, support a novel perspective to the linkage between p66Shc and longevity. P66 appears to affect life span not only for its redox properties but also for its adaptor capacity, able to potentiate the damaging effect of mTOR/S6K pathway in the past years of life. It has been proposed that aging may be the consequence of excess growth signals such as those mediated by the mTOR pathway (Blagosklonny, 2008; Pani, 2011).

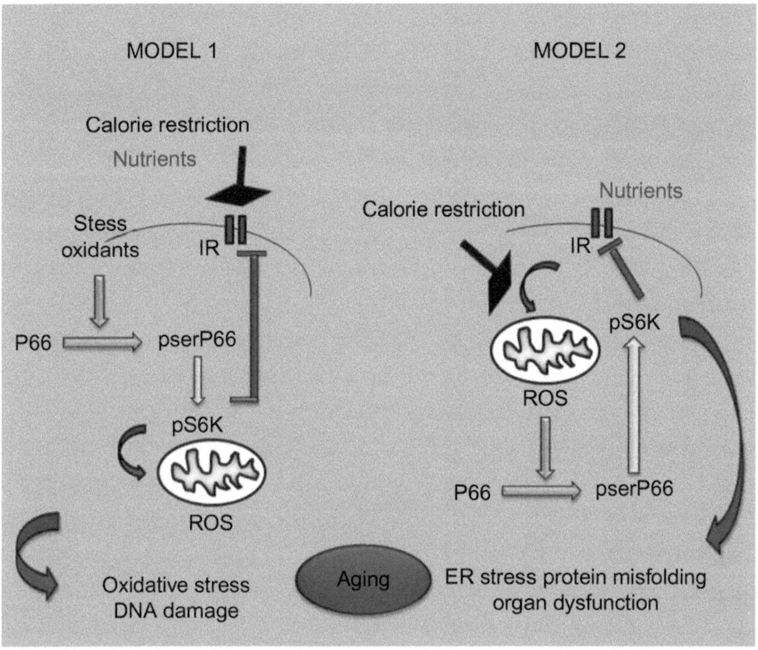

Figure 9.3 P66 Shc:TOR and ROS. Hypothetical redox circuits linking p66Shc and mTOR/S6K to mitochondrial nutrient signaling, calorie restriction, and aging: the two different models are not necessarily alternative. Model 1: p66Shc may rise intracellular ROS by activating mTOR/S6K and promoting the stimulatory effect of this cascade on mitochondrial metabolism (Scheme 1: p66 → S6K → ROS). Model 2: p66Shc as a redox sensor activates S6K in response to nutrient catabolism favoring feedback inibitory effects on insulin responsiveness (Scheme 2: ROS → p66 → S6K). (See Color Insert.)

This serine/threonine protein kinase absolutely necessary early in life is essential in organism development (Gangloff et al., 2004) and regulates cell growth, proliferation, motility, and survival, as well as protein synthesis and transcription. It is able to integrate the input from multiple upstream pathways, including insulin, GFs, and mitogens and functions as a sensor of cellular nutrient, energy levels, and redox status (Sarbassov, Ali, & Sabatini, 2005). For the same properties, it may be responsible for organismal aging. It is well known that natural selection declines with age, so nature does not switch TOR off later in life, and in such a way, mTOR mediates cellular hyperfunction leading to cellular damage and organ failure. Cells become hypertrophic and hyperactive and develop compensatory resistance to GFs signals such as insulin. Evidence emerges that calorie restriction delays aging via the TOR pathway (Bonawitz, Chatenay-Lapointe, Pan, & Shadel,

2007; Panieri et al., 2010; Vellai et al., 2003) and, however, overactivation of this pathway is involved in age-related diseases such as cancer, diabetes and its complication, obesity, atherosclerosis, cardiac hypertrophy, and neurodegeneration (Inoki, Corradetti, & Guan, 2005; Krebs et al., 2007). Traditionally linked to oxidative stress, these diseases are being increasingly related to inhibition of autophagy, stimulation of cell mass growth and hypertrophy and ribosomal synthesis, accumulation of aggregation-prone proteins, increase in GFs secretion and resistance to GFs and insulin, all potential consequences of TOR hyperactivation.

Parallel evidences also involve p66Shc as a determinant of tissue damage in the same degenerative diseases. It has been reported that p66Shc mRNA levels are higher in peripheral blood mononuclear cells of diabetics compared to healthy individuals (Pagnin et al., 2005), additionally, a recent study indicated an association between p66Shc mRNA expression and coronary atherosclerosis in humans and acute coronary syndrome (Franzeck et al., 2012). Moreover, p66 KO mice are protected from hyperglycemia-induced endothelial dysfunction (Camici et al., 2007), glomerulopathy (Menini et al., 2006), cardiomyopathy (Rota et al., 2006), and delayed wound-healing process in the setting of diabetes and ischemia (Fadini et al., 2010).

Diabetes, obesity, and organ disfunction associated with metabolic derangement recapitulate and accelerate many pathologic changes associated with aging. Based on our findings, p66 KO mice appear "calorie restricted" without actual calorie restriction because of reduced activity of insulin signaling through the mTOR pathway. We have proposed that this mechanism could account, at least in part, for the long-lived phenotype of p66-deficient mice.

But how does this relate to the redox activity of p66Shc, and the profound effect of this molecule on cell/tissue susceptibility to oxidative stress?

8. p66Shc: TOR AND ROS?

We know that calorie-restricted animals have reduction in ROS content and enhanced redox scavenging genes expression (Lee, Klopp, Weindruch, & Prolla, 1999); furthermore, ROS are considered messengers in nutrient sensing other than inducer of molecular damage (Bae et al., 1999). p66Shc, besides directly generating ROS in mitochondria, may rise intracellular ROS by activating mTOR/S6K and promoting the stimulatory effect of this cascade on mitochondrial metabolism (Fig. 9.3, model1: p66 → S6K → ROS); on the other hand, as a redox sensor (Nemoto & Finkel, 2002), p66Shc may activate S6K in response to several stress and oxidant signals, including those generated by nutrient catabolism, eventually favoring feedback inhibitory effects on insulin responsiveness (Fig. 9.3, model 2: ROS → p66 → S6K). The latter would represent an intriguing alternative route for nutrients to signal, *via* mitochondria, ROS, and p66Shc to the mTOR/S6K cascade (Fig. 9.3).

Such a role for p66Shc in nutrient sensing may explain why p66Shc, so dangerous for life span and metabolism derangement, has been evolutionarily selected to provide an advantage in energy metabolism under natural condition and famine (Giorgio et al., 2012).

ACKNOWLEDGMENTS

The authors wish to thank colleagues and laboratory members for helpful discussions. Original research described in this chapter was funded by the European Association for the Study of Diabetes (EASD), EFSD/GSK Program for the Study of Metabolic Toxicity in Diabetes (to G. P.), the Italian Ministry of University and Research and Catholic University (MIUR ex 60% linea D1 to G. P.), and the Italian Association for Cancer Research (AIRC, Grant IG8634/2009) to G. P.

REFERENCES

Accili, D., & Arden, K. C. (2004). FoxOs at the crossroads of cellular metabolism, differentiation, and transformation. *Cell, 117*, 421–426.

Ahima, R. S. (2009). Connecting obesity, aging and diabetes. *Nature Medicine, 15*, 996–997.

Bae, G. U., Seo, D. W., Kwon, H. K., Lee, H. Y., Hong, S., Lee, Z. W., et al. (1999). Hydrogen peroxide activates p70(S6k) signaling pathway. *The Journal of Biological Chemistry, 274*, 32596–32602.

Bartke, A. (2006). Life extension in the dwarf mouse. In P. M. Conn (Ed.), *Handbook of models for the study of human aging* (pp. 403–414). Waltham, MA, USA: Elsevier Academic Press.

Barzilai, N., & Gupta, G. (1999). Revisiting the role of fat mass in the life extension induced by caloric restriction. *The Journals of Gerontology. Series A, Biological Sciences and Medical Sciences, 54*, B89–B96.

Berniakovich, I., Trinei, M., Stendardo, M., Migliaccio, E., Minacci, S., Bernardi, P., et al. (2008). p66Shc generated oxidative signal promotes fat accumulation. *The Journal of Biological Chemistry, 283*, 34283–34293.

Berryman, D. E., Christiansen, J. C., Johannsson, G., Thorner, M. O., & Kopchick, J. J. (2008). Role of the GH/IGF-1 axis in lifespan and healthspan: Lessons from animal models. *Growth Hormone & IGF Research, 18*(6), 455–471.

Blagosklonny, M. V. (2008). Aging: ROS or TOR. *Cell Cycle, 7*, 3344–3354.

Bluher, M., Kahn, B. B., & Kahn, C. R. (2003). Extended longevity in mice lacking the insulin receptor in adipose tissue. *Science, 299*, 572–574.

Bonawitz, N. D., Chatenay-Lapointe, M., Pan, Y., & Shadel, G. S. (2007). Reduced TOR signaling extends chronological life span via increased respiration and upregulation of mitochondrial gene expression. *Cell Metabolism, 5*, 265–277.

Bonfini, L., Migliaccio, E., Pelicci, G., Lanfrancone, L., & Pelicci, P. G. (1996). Not all Shc's roads lead to Ras. *Trends in Biochemical Sciences, 21*, 257–261.

Brown-Borg, H. M. (2006). Longevity in mice: Is stress resistance a common factor? *Age*, *28*, 145–162.

Brown-Borg, H. M., Borg, K. E., Meliska, C. J., & Bartke, A. (1996). Dwarf mice and the aging process. *Nature*, *384*, 33.

Camici, G. G., Schiavoni, M., Francia, P., Bachschmid, M., Martin-Padura, I., Hersberger, M., et al. (2007). Genetic deletion of p66(Shc) adaptor protein prevents hyperglycemia-induced endothelial dysfunction and oxidative stress. *Proceedings of the National Academy of Sciences of the United States of America*, *104*, 5217–5222.

Capitani, N., Lucherini, O. M., Sozzi, E., Ferro, M., Giommoni, N., Finetti, F., et al. (2010). Impaired expression of p66Shc, a novel regulator of B cell survival, in chronic lymphocytic leukemia. *Blood*, *115*, 3726–3736.

Chang, G. R., Chiu, Y. S., Wu, Y. Y., Chen, W. Y., Liao, J. W., Chao, T. H., et al. (2009). Rapamycin protects against high fat diet-induced obesity in C57BL/6J mice. *Journal of Pharmacological Sciences*, *109*, 496–503.

Clancy, D. J., Gems, D., Harshman, L. G., Oldham, S., Stocker, H., Hafen, E., et al. (2001). Extension of life-span by loss of CHICO, a Drosophila insulin receptor substrate protein. *Science*, *292*, 104–106.

Coschigano, K. T., Clemmons, D., Bellush, L. L., & Kopchick, J. J. (2000). Assessment of growth parameters and lifespan of GHR/BP gene-disrupted mice. *Endocrinology*, *141*, 2608–2613.

Di Stefano, V., Cencioni, C., Zaccagnini, G., Magenta, A., Capogrossi, M. C., & Martelli, F. (2009). p66ShcA modulates oxidative stress and survival of endothelial progenitor cells in response to high glucose. *Cardiovascular Research*, *82*, 421–429.

Fadini, G. P., Albiero, M., Menegazzo, L., Boscaro, E., Pagnin, E., Iori, E., et al. (2010). The redox enzyme p66Shc contributes to diabetes and ischemia-induced delay in cutaneous wound healing. *Diabetes*, *59*, 2306–2314.

Finetti, F., Pellegrini, M., Ulivieri, C., Savino, M. T., Paccagnini, E., Ginanneschi, C., et al. (2008). The proapoptotic and antimitogenic protein p66SHC acts as a negative regulator of lymphocyte activation and autoimmunity. *Blood*, *111*, 5017–5027.

Flurkey, K., Papaconstantinou, J., & Harrison, D. E. (2002). The Snell dwarf mutation Pit1 (dw) can increase lifespan in mice. *Mechanisms of Ageing and Development*, *123*, 121–130.

Fontana, L. (2009). The scientific basis of caloric restriction leading to longer life. *Current Opinion in Gastroenterology*, *25*, 144–150.

Franzeck, F. C., Hof, D., Spescha, R. D., Hasun, M., Akhmedov, A., Steffel, J., et al. (2012). Expression of the aging gene p66Shc is increased in peripheral blood monocytes of patients with acute coronary syndrome but not with stable coronary artery disease. *Atherosclerosis*, *220*, 282–286.

Gangloff, Y. G., Mueller, M., Dann, S. G., Svoboda, P., Sticker, M., Spetz, J. F., et al. (2004). Disruption of the mouse mTOR gene leads to early postimplantation lethality and prohibits embryonic stem cell development. *Molecular and Cellular Biology*, *24*, 9508–9516.

Giorgetti, S., Pelicci, P. G., Pelicci, G., & Van Obberghen, E. (1994). Involvement of Src homology/collagen (SHC) proteins in signaling through the insulin receptor and the insulin-like growth- factor-I-receptor. *European Journal of Biochemistry*, *223*, 195–202.

Giorgio, M., Berry, A., Berniakovich, I., Poletaeva, I., Trinei, M., Stendardo, M., et al. (2012). The p66Shc knocked out mice are short lived under natural condition. *Aging Cell*, *11*, 162–168.

Giorgio, M., Migliaccio, E., Orsini, F., Paolucci, D., Moroni, M., Contursi, C., et al. (2005). Electron transfer between cytochrome c and p66Shc generates reactive oxygen species that trigger mitochondrial apoptosis. *Cell*, *122*, 221–233.

Harman, D. (1956). Aging: A theory based on free radical and radiation chemistry. *Journal of Gerontology*, *11*, 298–300.

Harper, J. M., Salmon, A. B., Chang, Y., Bonkowski, M., Bartke, A., & Miller, R. A. (2006). Stress resistance and aging: Influence of genes and nutrition. *Mechanisms of Ageing and Development*, *127*, 687–694.

Haruta, T., Uno, T., Kawahara, J., Takano, A., Egawa, K., Sharma, P. M., et al. (2000). A rapamycin-sensitive pathway down-regulates insulin signaling via phosphorylation and proteasomal degradation of insulin receptor substrate-1. *Molecular Endocrinology*, *14*, 783–794.

Holzenberger, M., Dupont, J., Ducos, B., Leneuve, P., Geloen, A., Even, P. C., et al. (2003). IGF-1 receptor regulates lifespan and resistance to oxidative stress in mice. *Nature*, *421*, 182–187.

Imura, A., Iwano, A., Tohtyama, O., Tsuji, Y., Nozaki, K., Hashimoto, N., et al. (2004). Secreted Klotho protein in sera and CSF: Implication for post-translational cleavage in release of Klotho protein from cell membrane. *FEBS Letters*, *565*, 143–147.

Inoki, K., Corradetti, M. N., & Guan, K. L. (2005). Dysregulation of the TSC-mTOR pathway in human disease. *Nature Genetics*, *37*, 19–24.

Kao, A. W., Waters, S. B., Okada, S., & Pessin, J. E. (1997). Insulin stimulates the phosphorlation o the 66- and 52-kilodalton Shc isoforms by distinct pathways. *Endocrinology*, *138*, 2474–2480.

Kapahi, P., Zid, B. M., Harper, T., Koslover, D., Sapin, V., & Benzer, S. (2004). Regulation of lifespan in Drosophila by modulation of genes in the TOR signaling pathway. *Current Biology*, *14*(10), 885–890.

Kenyon, C. (2001). A conserved regulatory system for aging. *Cell*, *105*, 165–168.

Kenyon, C., Chang, J., Gensch, E., Rudner, A., & Tabtiang, R. A. (1993). C. elegans mutant that lives twice as long as wild type. *Nature*, *366*, 461–464.

Khanday, F. A., Yamamori, T., Mattagajasingh, I., Zhang, Z., Bugayenko, A., Naqvi, A., et al. (2006). Rac1 leads to phosphorylation-dependent increase in stability of the p66shc adaptor protein: Role in Rac1-induced oxidative stress. *Molecular Biology of the Cell*, *17*, 122–129.

Krebs, M., Brunmair, B., Brehm, A., Artwohl, M., Szendroedi, J., Nowotny, P., et al. (2007). The mammalian target of rapamycin pathway regulates nutrient-sensitive glucose uptake in man. *Diabetes*, *56*, 1600–1607.

Kurosu, H., Yamamoto, M., Clark, J. D., Pastor, J. V., Nandi, A., Gurnani, P., et al. (2005). Suppression of aging in mice by the hormone Klotho. *Science*, *309*(5742), 1829–1833.

Lee, C. K., Klopp, R., Weindruch, R., & Prolla, T. A. (1999). Gene expression profile of aging and its retardation by caloric restriction. *Science*, *285*, 1390–1393.

Lee, S. R., Yang, K. S., Kwon, J., Lee, C., Jeong, W., & Rhee, S. G. (2002). Reversible inactivation of the tumor suppressor PTEN by H2O2. *The Journal of Biological Chemistry*, *277*, 20336–20342.

Lotti, L. V., Lanfrancone, L., Migliaccio, E., Zompetta, C., Pelicci, G., Salcini, A. E., et al. (1996). Sch proteins are localized on endoplasmic reticulum membranes and are redistributed after tyrosine kinase receptor activation. *Molecular and Cellular Biology*, *16*, 1946–1954.

Luzi, L., Confalonieri, S., Di Fiore, P. P., & Pelicci, P. G. (2000). Evolution of Shc functions from nematode to human. *Current Opinion in Genetics & Development*, *10*, 668–674.

Menini, S., Amadio, L., Oddi, G., Ricci, C., Pesce, C., Pugliese, F., et al. (2006). Deletion of p66Shc longevity gene protects against experimental diabetic glomerulopathy by preventing diabetes-induced oxidative stress. *Diabetes*, *55*, 1642–1650.

Migliaccio, E., Giorgio, M., Mele, S., Pelicci, G., Reboldi, P., Pandolfi, P. P., et al. (1999). The p66shc adaptor protein controls oxidative stress response and life span in mammals. *Nature*, *402*, 309–313.

Migliaccio, E., Mele, S., Salcini, A. E., Pelicci, G., Lai, K. M., Superti- Furga, G., et al. (1997). Opposite effects of the p52shc/p46shc and p66shc splicing isoforms on the EGF receptor-MAP kinase-fos signaling pathway. *The EMBO Journal*, *16*, 706–716.

Muniyappa, R., & Quon, M. J. (2007). Insulin action and insulin resistance in vascular endothelium. *Current Opinion in Clinical Nutrition and Metabolic Care, 10*, 23–30.

Natalicchio, A., Laviola, L., De Tullio, C., Renna, L. A., Montrone, C., Perrini, S., et al. (2004). Role of the p66Shc isoform in insulin-like growth factor I receptor signaling through MEK/Erk and regulation of actin cytoskeleton in rat myoblasts. *The Journal of Biological Chemistry, 279*, 43900–43909.

Nemoto, S., & Finkel, T. (2002). Redox regulation of forkhead proteins through a p66shc-dependent signaling pathway. *Science, 295*(5564), 2450–2452.

Okada, S., Kao, A. W., Ceresa, B. P., Blaikie, P., Margolis, B., & Pessin, J. E. (1997). The 66-kDa Shc isoform is a negative regulator of the epidermal growth factor-stimulated mitogen-activated protein kinase pathway. *The Journal of Biological Chemistry, 272*, 28042–28049.

Orsini, F., Migliaccio, E., Moroni, M., Contursi, C., Raker, V. A., Piccini, D., et al. (2004). The life span determinant p66Shc localizes to mitochondria where it associates with mitochondrial heat shock protein 70 and regulates trans-membrane potential. *The Journal of Biological Chemistry, 27*, 25689–25695.

Pacini, S., Pellegrini, M., Migliaccio, E., Patrussi, L., Ulivieri, C., Ventura, A., et al. (2004). P66SHC promotes apoptosis and antagonizes mitogenic signaling in T cells. *Molecular and Cellular Biology, 24*, 1747–1757.

Pagnin, E., Ladini, G., De Toni, R., Tiengo, A., Calo, L., & Avogaro, A. (2005). Diabetes induces p66shc gene expression in human peripheral blood mononuclear cells: Relationship to oxidative stress. *The Journal of Clinical Endocrinology and Metabolism, 90*, 1130–1136.

Pani, G. (2011). From growing to secreting: New roles for mTOR in aging cells. *Cell Cycle, 10*, 2450–2453.

Pani, G., Koch, O. R., & Galeotti, T. (2009). The p53-shc-Manganese Superoxide Dismutase (MnSOD) network: A mitochondrial intrigue to generate reactive oxygen species. *The International Journal of Biochemistry & Cell Biology, 41*, 1002–1005.

Panieri, E., Toletta, G., Mele, M., Labate, V., Ranieri, S. C., Fusco, S., et al. (2010). Nutrient withdrawal rescues growth factor-deprived cells from mTOR-dependent damage. *Aging (Albany NY), 2*, 487–503.

Pelicci, G., Lanfrancone, L., Grignani, F., McGlade, J., Cavallo, F., Forni, G., et al. (1992). A novel transforming protein (SHC) with an SH2 domain is implicated in mitogenic signal transduction. *Cell, 70*, 93–104.

Pinton, P., Rimessi, A., Marchi, S., Orsini, F., Migliaccio, E., Giorgio, M., et al. (2007). Protein kinase C beta and prolyl isomerase 1 regulate mitochondrial effects of the life-span determinant p66Shc. *Science, 315*, 659–663.

Polak, P., Cybulski, N., Feige, J. N., Auwerx, J., Rüegg, M. A., & Hall, M. N. (2008). Adipose-specific knockout of raptor results in lean mice with enhanced mitochondrial respiration. *Cell Metabolism, 8*(5), 99–410.

Ranieri, S. C., Fusco, S., Panieri, E., Labate, V., Mele, M., Tesori, V., et al. (2010). Mammalian life-span determinant p66shcA mediates obesity induced insulin resistance. *Proceedings of the National Academy of Sciences of the United States of America, 107*, 13420–13425.

Rota, M., LeCapitaine, N., Hosoda, T., Boni, A., De Angelis, A., Padin-Iruegas, M. E., et al. (2006). Diabetes promotes cardiac stem cell aging and heart failure, which are prevented by deletion of the p66shc gene. *Circulation Research, 99*, 42–52.

Sarbassov, D. D., Ali, S. M., & Sabatini, D. M. (2005). Growing roles for the mTOR pathway. *Current Opinion in Cell Biology, 17*, 596–603.

Selman, C., Lingard, S., Choudhury, A. I., Batterham, R. L., Claret, M., Clements, M., et al. (2008). Evidence for lifespan extension and delayed age-related biomarkers in insulin receptor substrate 1 null mice. *The FASEB Journal, 22*, 807–818.

Selman, C., Tullet, J. M., Wieser, D., Irvine, E., Lingard, S. J., Choudhury, A. I., et al. (2009). Ribosomal protein S6 kinase 1 signaling regulates mammalian life span. *Science*, *326*, 140–144.

Smith, U. (2002). Impaired ('diabetic') insulin signaling and action occur in fat cells long before glucose intolerance—Is insulin resistance initiated in the adipose tissue? *International Journal of Obesity and Related Metabolic Disorders*, *26*, 897–904.

Tatar, M., Kopelman, A., Epstein, D., Tu, M. P., Yin, C. M., & Garofalo, R. S. (2001). A mutant Drosophila insulin receptor homolog that extends life-span and impairs neuroendocrine function. *Science*, *292*, 107–110.

Tchkonia, T., Corkey, B. E., & Kirkland, J. L. (2006). Current views of the fat cell as an endocrine cell: Lipotoxicity. *Endocrine Updates*, *26*, 105–118.

Tomilov, A. A., Ramsey, J. J., Hagopian, K., Giorgio, M., Kim, K. M., Lam, A., et al. (2011). The Shc locus regulates insulin signaling and adiposity in mammals. *Aging Cell*, *10*, 55–65.

Trinei, M., Giorgio, M., Cicalese, A., Barozzi, S., Ventura, A., Migliaccio, E., et al. (2002). A p53-Shc signalling pathway controls intracellular redox status, levels of oxidation-damaged DNA and oxidative stress-induced apoptosis. *Oncogene*, *21*, 3872–3878.

Um, S. H., D'Alessio, D., & Thomas, G. (2006). Nutrient overload, insulin resistance, and ribosomal protein S6 kinase 1, S6K1. *Cell Metabolism*, *3*, 393–402.

Um, S. H., Frigerio, F., Watanabe, M., Picard, F., Joaquin, M., Sticker, M., et al. (2004). Absence of S6K1 protects against age- and diet-induced obesity while enhancing insulin sensitivity. *Nature*, *431*, 200–205.

Vellai, T., Takacs-Vellai, K., Zhang, Y., Kovacs, A. L., Orosz, L., & Muller, F. (2003). Genetics: Influence of TOR kinase on lifespan in C. elegans. *Nature*, *426*, 620.

Ventura, A., Luzi, L., Pacini, S., Baldari, C. T., & Pelicci, P. G. (2002). The p66Shc longevity gene is silenced through epigenetic modifications of an alternative promoter. *The Journal of Biological Chemistry*, *277*, 22370–22376.

Westman, S. (1968). Development of the obese-hyperglycaemic syndrome in mice. *Diabetologia*, *4*, 141–149.

Xi, G., Shen, X., & Clemmons, D. R. (2008). p66shc negatively regulates insulin-like growth factor I signal transduction via inhibition of p52shc binding to Src homology 2 domain-containing protein tyrosine phosphatase substrate-1 leading to impaired growth factor receptor-bound protein-2 membrane recruitment. *Molecular Endocrinology*, *22*, 2162–2175.

Xi, G., Shen, X., & Clemmons, D. R. (2010). p66shc inhibits IGF-I signaling via direct binding to Src through its polyproline and SH2 domains resulting in impairment of Src kinase activation. *The Journal of Biological Chemistry*, *285*, 6937–6951.

Xi, G., Shen, X., Radhakrishnan, Y., Maile, L., & Clemmons, D. (2010). Hyperglycemia-induced p66shc inhibits insulin-like growth factor I-dependent cell survival via impairment of Src kinase-mediated phosphoinositide-3 kinase/AKT activation in vascular smooth muscle cells. *Endocrinology*, *151*, 3611–3623.

Yamamoto, M., Clark, J. D., Pastor, J. V., Gurnani, P., Nandi, A., Kurosu, H., et al. (2005). Regulation of oxidative stress by the anti-aging hormone Klotho. *The Journal of Biological Chemistry*, *280*, 38029–38034.

Zhou, Y., Xu, B. C., Maheshwari, H. G., He, L., Reed, M., Lozykowski, M., et al. (1997). A mammalian model for Laron syndrome produced by targeted disruption of the mouse growth hormone receptor/binding protein gene (the Laron mouse). *Proceedings of the National Academy of Sciences of the United States of America*, *94*, 13215–13220.

Zoncu, R., Efeyan, A., & Sabatini, D. M. (2011). mTOR: From growth signal integration to cancer, diabetes and ageing. *Nature Reviews Molecular Cell Biology*, *12*, 21–35.

CHAPTER TEN

The Emerging Role of Constitutive Androstane Receptor and Its Cross talk with Liver X Receptors and Peroxisome Proliferator-Activated Receptor A in Lipid Metabolism

Lei Xiao[*,†], **Jing Wang**[‡], **Mengxi Jiang**[§], **Wen Xie**[§], **Yonggong Zhai**[*,‡,1]

[*]Key Laboratory for Cell Proliferation and Regulation Biology, Ministry of Education, Biomedicine Research Institute and College of Life Sciences, Beijing Normal University, Beijing, China
[†]Cardiovascular Research Center, Xi'an Jiaotong University School of Medicine, Xi'an, China
[‡]Beijing Key Laboratory of Gene Resource and Molecular Development, College of Life Sciences, Beijing Normal University, Beijing, China
[§]Center for Pharmacogenetics, University of Pittsburgh, Pittsburgh, Pennsylvania, USA
[1]Corresponding author: e-mail address: ygzhai@bnu.edu.cn

Contents

Abstract

The regulation of lipid metabolism is central to energy homeostasis in higher multicellular organisms. Lipid homeostasis depends on factors that are able to transduce metabolic parameters into regulatory events representing the fundamental components of the general control system. Nuclear receptors form a superfamily of ligand-activated transcription factors implicated in various physiological functions including energy metabolism. The constitutive androstane receptor (CAR, NR1I3), initially identified as a xenobiotic-sensing receptor, may also have roles in lipid homeostasis. The nuclear receptors liver X receptors (LXRs, NR1H2/3) and peroxisome proliferator-activated

Vitamins and Hormones, Volume 91
ISSN 0083-6729
http://dx.doi.org/10.1016/B978-0-12-407766-9.00010-9

receptors (PPARs, NR1C) have been known for their roles in lipid metabolism. LXR is a sterol sensor that promotes lipogenesis, whereas PPARα controls a variety of genes in several pathways of lipid metabolism. This chapter focuses primarily on the role of CAR in lipid metabolism directly or through its cross talk with LXRs and PPARα.

1. INTRODUCTION

Lipids are essential for energy homeostasis, reproductive and organ physiology, and numerous aspects of cellular biology. Disruption of lipid metabolism might trigger various pathological metabolic processes, such as atherosclerosis, diabetes, and obesity (Lee, Olson, & Evans, 2003). The human body has evolved a sophisticated lipoprotein transport system to deliver cholesterol and fatty acids (FAs) to the periphery to regulate lipid metabolism (Lee et al., 2003). Lipoproteins are composed of triglycerides (TGs), cholesterol esters, phospholipids, and apolipoproteins, which modulate lipoprotein catabolism (Reaven, 1994). The liver is an essential organ in metabolic homeostasis (Xiao, Xie, & Zhai, 2010). In the forward transport system, TG-rich very low-density lipoprotein (VLDL) released by the liver delivers FAs to adipocytes for storage and to cardiac and skeletal muscle for energy consumption. Lipoprotein lipase (LPL), secreted by the adipocyte, muscle, and macrophage, plays an important role in VLDL FA release and its subsequent conversion to low-density lipoprotein (LDL). Cholesterol ester-rich LDL delivers cholesterol to peripheral tissues for steroidogenesis and maintaining cell membrane integrity. Conversely, in the reverse transport system, high-density lipoprotein (HDL) transports excess cholesterol from extrahepatic cells, such as macrophages at the vessel wall, to the liver, where it can be recycled or catabolized to bile acids (Russell, 1992).

Nuclear receptors were first cloned in the middle 1980s, and they have since been established as ligand-dependent transcriptional factors that play important roles in metabolic homeostasis (Wada, Kang, Jetten, & Xie, 2008). Genomic studies have indicated that there are 48 and 50 nuclear receptor genes in rodents and humans, respectively. Most, if not all, nuclear receptors contain an N-terminal DNA-binding domain and a C-terminal ligand-binding domain (Fig. 10.1). Nuclear receptors often regulate gene expression by binding, as homodimers, retinoid X receptor (RXR) heterodimers or as monomers, to their responsive elements present in target gene promoters.

One of the nuclear receptors which can function as an important metabolic sensor in the liver is constitutive androstane receptor (CAR). CAR is active in the absence of ligand with the unique capability to be further

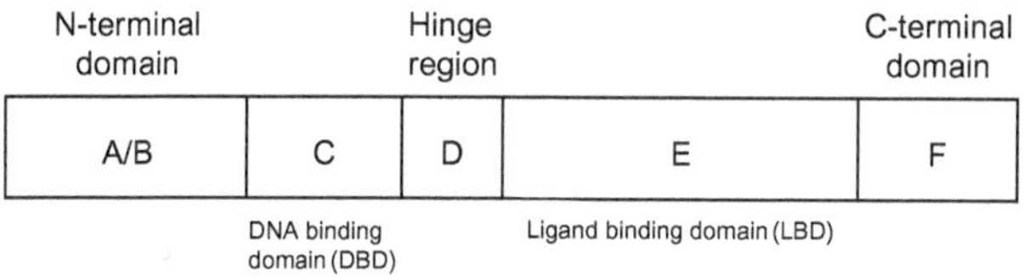

Figure 10.1 General domain structure of a nuclear receptor.

regulated by agonists or antagonists. CAR is most frequently identified as a xenobiotic-sensing receptor because it regulates enzymes that metabolize xenobiotics. However, more recent studies show that CAR may also have a role in lipid homeostasis. CAR has the potential roles to impact numerous signaling pathways via the genes it modulates, directly and or cross talk with other nuclear receptor signaling pathways (Swales & Negishi, 2004).

Two nuclear receptors that play an important role in lipid homeostasis are liver X receptors (LXRs) and peroxisome proliferator-activated receptors (PPARs). LXRs, both LXRα and LXRβ whose ligands are oxidized derivatives of cholesterol (oxysterols), promote lipogenesis and are known as sterol sensors. PPARs contain a signature type II zinc finger DNA-binding motif and a hydrophobic ligand-binding pocket (Chawla, Repa, Evans, & Mangelsdorf, 2001). PPARα regulates the expression of a number of genes critical for lipid and lipoprotein metabolism (Yoon, 2009). In this chapter, we describe the physiological aspects related to LXRs and PPARα expression and functional consequences of their activation. We also discuss the mechanism of CAR in the regulation of lipid metabolism and its cross talk with LXRs and PPARα in the regulation of lipid metabolism.

2. INITIAL CHARACTERIZATION OF CAR AND ITS EMERGING ROLES IN LIPID METABOLISM

2.1. The xenobiotic receptor CAR

The nuclear receptor CAR is a key regulator of such xenobiotic-metabolizing enzymes as cytochrome P450 (CYP), UDP-glucuronosyltransferase (Honkakoski, Moore, Washburn, & Negishi, 1998; Honkakoski, Zelko, Sueyoshi, & Negishi, 1998; Sueyoshi, Kawamoto, Zelko, Honkakoski, & Negishi, 1999), and multidrug resistance-associated protein (Cherrington, Hartley, Li, Johnson, & Klaassen, 2002; Kast et al., 2002). CAR forms heterodimers with the RXR and then bind to its target gene promoters and transactivate target genes in an apparently constitutive manner (Baes et al., 1994; Choi et al., 1997). In general, transcriptional activation by the nuclear receptors involves a ligand-dependent conformational change that promotes the recruitment of transcriptional regulatory proteins, termed coactivators (McKenna et al., 1999; Xu, Glass, & Rosenfeld, 1999). CAR is the opposite of the classical model of ligand activation (Tzameli & Moore, 2001). The constitutive activity of CAR could reflect a ligand-independent recruitment of coactivators, whereas binding of CAR inverse agonist androstanol could cause coactivator release (Tzameli & Moore, 2001). To protect the body against foreign chemicals (xenobiotics) and the buildup of toxic endogenous chemicals, CAR functions in a metabolic cascade to regulate detoxification and elimination (Chawla et al., 2001). CAR is expressed primarily in the liver, intestine, and kidney, also with some expression being detected in the heart and brain (Baes et al., 1994; Choi et al., 1997), which are all consistent with its role as a xenobiotic sensor. The CAR activity can be modulated by a variety of compounds that include phenobarbital (PB), androstane steroidal compounds (i.e., 3α, 5α-androstanol), 5α-pregnane-3, 20-dione, retinoic acids, clotrimazole, chlorpromazine (CPZ), o, p′-DDT, methoxychlor, and to hydrocarbons such as 2, 3, 3′, 4′, 5′, 6-hexachlorobiphenyl, 6-(4-chlorophenyl) imidazo [2,1-b][1,3]thiazole-5-carbaldehyde O-(3,4-dichlorobenzyl) oxime (CITCO), and 1,4-bis[2-(3,5-dichloropyridyloxy)]benzene (TCPOBOP) (Kobayashi, Yamanaka, & Iwazaki, 2005; Moore, Parks, & Jones, 2000; Poland & Rubin, 1981; Tontonoz & Mangelsdorf, 2003).

As mentioned earlier, CAR is a central regulator of xenobiotic metabolism. CAR was initially proposed to be constitutively active, ligands with negative (androstanes) and positive PB effects were soon found. CAR regulates the expression of genes that are responsible for drug metabolism and elimination including CYP2B6 and CYP3A4 (Sueyoshi et al., 1999). CAR binds to and activates the CYP2B promoter in response to PB-like molecules, certain androgens, and the muscle relaxant zoxazolamine (Chawla et al., 2001). The CYP3A enzyme is responsible for metabolizing and clearing over 60% of clinically prescribed drugs, and its induction plays a pivotal role in the clearance of hepatotoxic bile salts. Moreover, recent experiments demonstrate that CAR-mediated hepatomegaly is a transient, adaptive response to acute xenobiotic stress. In contrast, chronic CAR activation results in hepatocarcinogenesis (Huang, Zhang, & Washington, 2005). In the acute response, direct upregulation of *Mdm2* expression by CAR contributes to both increased DNA replication and inhibition of p53-mediated apoptosis (Huang et al., 2005). Thus, the demonstration of a key role for CAR in tumor promotion provides a novel and potentially important link between environmental stress and tumorigenesis. In addition to its xenobiotic function, an increasing body of evidence suggests that CAR also has an endobiotic function that impacts energy homeostasis through the regulation of lipid metabolism.

2.2. Effect of CAR on lipid metabolism

The liver is the major tissue responsible for lipid homeostasis and drug metabolism (Finn, Henderson, Scott, & Wolf, 2009). Besides its role as a xenobiotic receptor, it is now evident that CAR also has important endobiotic roles in energy metabolism by affecting the metabolism of FAs and lipids (Wada, Gao, & Xie, 2009).

The first endogenous modifiers of CAR activity identified were the steroids androstanol and androstenol, which inhibit the constitutive activity of the receptor by directly dissociating the interaction between CAR and the steroid receptor coactivator-1 (Forman et al., 1998). CAR activity is required for a normal decrease in thyroid hormone (TH) levels during fasting in mice (Maglich et al., 2004). One possible mechanism by which CAR affects fasting responses is via its regulation of TH metabolism, through the regulation of Phase II TH-metabolizing enzymes, such as *Ugt1a1*, *Sultn*, *Sult1a1*, and *Sult2a1* (Maglich et al., 2004; Qatanani, Zhang, & Moore, 2005). The induction of CAR under fasting conditions might also involve the coactivator PGC-1α because PGC-1α increases CAR activity in a ligand-independent manner and potentiates the transcriptional activity of CAR in the presence of CAR agonists (Shiraki, Sakai, Kanaya, & Jingami, 2003). The activation of CAR can suppress lipogenesis and lower serum TG levels by reducing the protein levels of the active form of sterol regulatory element-binding protein 1 (SREBP-1) and stearoyl-CoA desaturase 1 (SCD-1) (Dong et al., 2009; Gao, He, Zhai, Wada, & Xie, 2009; Roth, Looser, Kaufmann, & Meyer, 2008). Activation of CAR in mouse liver results in activation of Insig-1, a protein with anti-lipogenic properties along with reduced protein levels of the active form of SREBP-1 (Roth et al., 2008). Insig-1 is also a target of AMP-activated protein kinase (AMPK), the hepatic activity of which is increased by activators of CAR and is known to cause a reduction of TGs (Roth et al., 2008). Therefore, CAR can cross talk with AMPK, an indicator of cellular energy status, and function as an energy sensor to influence lipid metabolism (Wada et al., 2009). The CAR activator PB can increase AMPK activity (Rencurel et al., 2005), which is essential for the induction of CAR

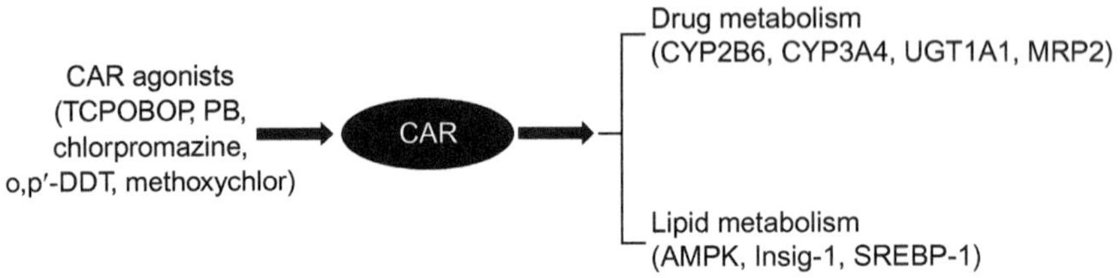

Figure 10.2 Proposed model of CAR-mediated regulation of drug and lipid metabolism.

target genes (Rencurel et al., 2006). A pharmacological inhibitor of AMPK or a dominant-negative form of AMPK abolished the response to PB, whereas an AMPK agonist enhanced PB-responsive activation of CAR target genes and nuclear translocation of CAR in the mouse liver (Rencurel et al., 2006; Wada et al., 2009). Furthermore, it has been shown that CAR deficient mice were defective in fasting adaptation and lost more weight in response to calorie restriction (Qatanani et al., 2005).

In addition, PB decreases both the mitochondrial carnitine palmitoyltransferase 1 (CPT-1) and the enoyl CoA isomerase mRNA content in wild-type mice but not in CAR −/− mice (Ueda et al., 2002) and the increases in serum TGs observed after a high-fat diet regime were not observed in CAR −/− animals. Other studies indicate that CAR might be involved in the pathogenesis of nonalcoholic steatohepatitis (Yamazaki et al., 2007), regulation of serum TG levels under metabolic stresses (Maglich, Lobe, & Moore, 2009), and regulation of HDL in wild-type and human apoA-I transgenic mice (Masson et al., 2008). The proposed roles of CAR in lipid metabolism are summarized in Fig. 10.2.

3. LXRs AND PPARα IN LIPID METABOLISM

LXRs and PPARs play important roles in endobiotic homeostasis, which is the balanced production and elimination of endobiotics, such as lipids (Zhai, Wada, & Zhang, 2010). Lipid homeostasis in higher organisms is under the control of an integrated system that has the capacity to rapidly respond to metabolic changes (Lee et al., 2003). Here, we specifically discuss the roles of LXRs and PPARα in the regulation of genes involved in lipid homeostasis.

3.1. The sterol sensors LXRs

The LXRs, LXRα and LXRβ are now recognized to be the central regulators of cholesterol and lipid metabolism in mammals (Repa & Mangelsdorf, 2000). LXRα is highly expressed in liver tissue and with lower levels present in adipose tissue, intestine, kidney, and splenic tissues (Willy et al., 1995), whereas LXRβ is expressed in all tissues examined (Repa & Mangelsdorf, 2000). Both LXRα and LXRβ isoforms preferentially bind with their heterodimeric partner RXR, LXR/RXR heterodimers then bind to LXR response elements in the promoters of their target genes and activate target gene expression (Apfel et al., 1994; Song, Kokontis, Hiipakka, & Liao, 1994; Teboul et al., 1995; Willy et al., 1995). Some of the most potent endogenous activators of LXRs include 22(*R*)-hydroxycholesterol, an intermediate metabolite of steroid hormone, and 24(*S*)-hydroxycholesterol, an abundant cholesterol metabolite in brain tissue, as well as 24(*S*), 25-epoxycholesterol which is produced in hepatocytes and macrophages (Bjorkhem, Meaney, & Diczfalusy, 2002; Janowski, Willy, Devi, Falck, & Mangelsdorf, 1996, Janowski et al., 1999; Lehmann et al., 1997). 27-Hydroxycholesterol, which is produced in macrophages in response to cholesterol loading, also activates LXRs and has been proposed to be an important physiological ligand of LXRs (Fu et al., 2001). In addition to natural ligands, a number of potent synthetic LXR agonists such as T0901317 and GW3965 have also been developed (Collins et al., 2002; Schultz et al., 2000).

Considerable evidence suggests that the targets for the two LXRs are highly overlapping and that the receptors functionally compensate for one another in most physiological contexts (Zelcer & Tontonoz, 2006). LXRs respond to elevated cholesterol levels via transactivation of genes involved in sterol transport (*ABCA1*, *ABCG1*, *ABCG5*, and *ABCG8*), cholesterol efflux, HDL metabolism (apoE, cholesterol ester transfer protein (CETP), and PLTP), and sterol catabolism (CYP7A1) (Peters et al., 1997).

In the liver, bile acid synthesis is initiated by the enzyme cholesterol 7α-hydroxylase (CYP7A1), which catalyzes the rate-limiting step in the biosynthetic pathway. Expression of CYP7A1 in rodents is positively regulated by oxysterol-activated LXRα (Peet et al., 1998). Recent studies have demonstrated that ABCA1 participates in the efflux of cholesterol from the basolateral membrane, but not the apical membrane, of the enterocyte (Mulligan et al., 2003). In macrophages, LXRα activation contributes to sterol homeostasis through upregulation of ABCA1, ABCG1, and apoE expression. Both ABCG5 and ABCG8 are LXRs target genes function in cholesterol absorption (Berge et al., 2000; Lee et al., 2001; Repa et al., 2000). Additionally, LXRα affects remodeling of apoB-containing lipoproteins via induction of hepatic expression of CETP (Luo & Tall, 2000; Zhang, Repa, Gauthier, & Mangelsdorf, 2001), which facilitates the transfer of CE from HDL to apoB-containing lipoproteins. LXRα activation also induces the

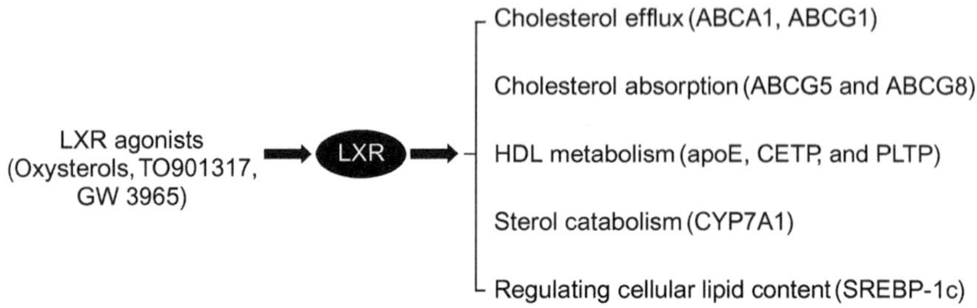

Figure 10.3 Proposed model of LXR-mediated regulation of lipid metabolism.

expression of LPL, which is responsible for the hydrolysis of postprandial TG-rich particles, and coordinately upregulates expression of the apoE/apoC-I/apoC-IV/apoC-II gene cluster (Mak et al., 2002). Recent study has revealed a novel lithogenic role of LXR as well as a functional interplay between LXR and LDLR in gallbladder cholesterol crystallization and cholesterol gallstone disease (Uppal et al., 2008). Moreover, LXRs play a central role in regulating cellular lipid content through activation of the master regulator of *de novo* lipogenesis named SREBP-1c (Shimano, Yahagi, Amemiya-Kudo, et al., 1999) and its target genes including acetyl-CoA carboxylase-1 (ACC-1), fatty acid synthase (FAS), and SCD-1 (Kim & Spiegelman, 1996; Shimano et al., 1999). The roles of LXR in lipid metabolism are summarized in Fig. 10.3.

3.2. The FA receptor PPARα

The PPARs are FA receptors that have been implicated to play an important role in obesity-related metabolic diseases such as hyperlipidemia, insulin resistance, and coronary artery disease. Once activated, PPARs affect FA transport and metabolism, adipogenesis, peroxisome biogenesis, cholesterol and bile acid biosynthesis, and proteasome activation (Lau, Abbott, Corton, & Cunningham, 2010). Among the three PPRA isoforms, PPARα has been shown to mediate the hypolipidemic effects of fibrates on lipid and lipoprotein metabolism (Schoonjans et al., 1996; Staels et al., 1998). PPARα can be activated by polyunsaturated FAs and leukotriene B4 (Luquet et al., 2005). The expression of PPARα can be detected in a variety of tissues and cells involved in FA oxidation, such as liver, kidney, heart, skeletal muscle, brown fat, monocytes, vascular endothelial, and smooth muscle cells (Auboeuf et al., 1997; Braissant, Foufelle, Scotto, Dauca, & Wahli, 1996; Chinetti et al., 1998; Inoue, Shino, Noji, Awata, & Katayama, 1998; Staels et al., 1998). Many lines of evidence have demonstrated that PPARα directly regulates

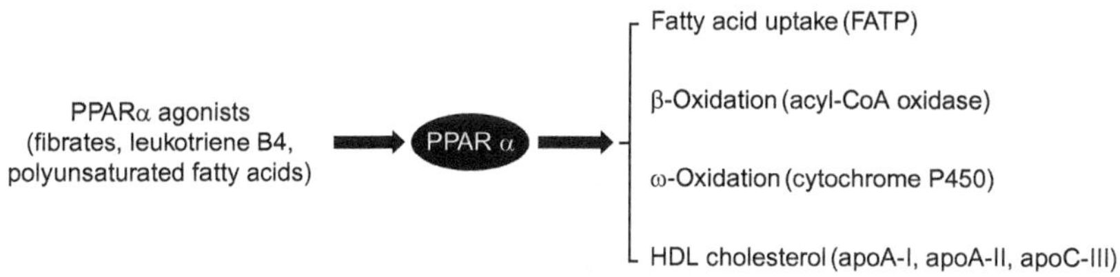

Figure 10.4 Proposed model of PPARα-mediated regulation of lipid metabolism.

genes involved in FA uptake (Motojima, Passilly, Peters, Gonzalez, & Latruffe, 1998; Schoonjans et al., 1995), β-oxidation (acyl-CoA oxidase), and ω-oxidation (cytochrome P450) (Yoon, 2009). PPARα participates in FA uptake across the membranes, FA binding in cells, their activation to acyl-CoA esters and degradation via FA oxidation pathways, and ketone body synthesis. After the passage across the plasma membrane, PPARα-stimulated acyl-CoA synthase esterifies the FAs to acyl-CoA esters, thereby preventing their efflux from the cell and activating the FA for further metabolism (Schoonjans et al., 1995). Furthermore, PPARα induces the entry of FA into the mitochondria by stimulating mitochondrial CPT-1 and CPT-II (Louet et al., 2001; Mascaró et al., 1998). PPARα has also been reported to have antiobesity effects and relieve obesity-related disorders (Serrano et al., 2008; Zhang, Lavan, & Gregoire, 2004). PPARα ligand fibrates have been used for the treatment of dyslipidemia due to their ability to lower plasma TG levels and elevate HDL cholesterol levels (Barter & Rye, 2008; Berger, Akiyama, & Meinke, 2005; Jay & Ren, 2007; Staels, 2007). Activation of PPARα has also been shown to increase apoA-I and apoA-II and decrease apoC-III, an inhibitor of LPL (Berthou et al., 1996; Vu-Dac et al., 1998). Recent work also demonstrates that the fibroblast growth factor 21 is a direct target gene of PPARα and is responsible for many of the downstream effects that permit the body to adapt to fasting and starvation and to a ketogenic diet (Mangelsdorf, 2010). The roles of PPARα in lipid metabolism are summarized in Fig. 10.4.

4. FUNCTIONAL CROSSTALK BETWEEN CAR AND LXR OR CAR AND PPAR IN THE REGULATION OF LIPID METABOLISM

As discussed earlier, LXRs, PPARα, and CAR can individually and directly regulate lipid metabolism. More recent studies suggest that CAR can also affect lipid metabolism through its functional cross talk with LXRs and PPARα.

We have recently shown that LXR and CAR are functionally related *in vivo*. Specifically, loss of CAR increased the expression of lipogenic LXR target genes, leading to increased hepatic TG accumulation; whereas activation of CAR inhibited the expression of LXR target genes and LXR ligand-induced lipogenesis (Zhai et al., 2010). CAR agonist treatment decreased the expression of the lipogenic transcription factor SREBP-1c and its downstream genes. These responses are consistent with a recent report that TCPOBOP treatment decreased nuclear levels of the mature SREBP-1c protein (Roth et al., 2008). CAR can induce the phase II enzyme *SULT2B1b*, which sulfates and inactivates oxysterol agonists for LXR, a well-known positive regulator of SREBP-1c expression. The suppressive effect of CAR on lipogenic gene expression is substantially impaired in SULT2B1b knock-out mice (Dong et al., 2009).

The effect of CAR on lipid homeostasis can also be achieved through its functional cross talk with PPARα, a major regulator of hepatic fatty oxidation (Wada et al., 2009). It has been reported that the PPARα agonist WY14643 induces CAR gene expression in a PPARα-dependent manner and also potentiates PB-induced transcription of CYP2B, a prototypical target gene of CAR (Wieneke, Hirsch-Ernst, Kuna, Kersten, & Pueschel, 2007). In addition, in wild-type mice, the CAR agonist PB induces serum ketone bodies and decreases the expression of mitochondrial CPT-1, the rate-limiting enzyme of β-oxidation, an effect not observed in CAR knock-out mice (Baskin-Bey, Anan, Isomoto, Bronk, & Gores, 2007; Kiyosawa et al., 2004). CAR activation significantly represses the expression of both ACC-1 and ACC-2, strongly suggesting that CAR induces β-oxidation by decreasing malonyl CoA levels. Recent studies have shown that SCD-1 ablation protects against hepatic steatosis by the combined effects of decreased lipogenesis and increased β-oxidation (Dobrzyn et al., 2004). Hence, reduced SCD-1 expression by CAR activation may have also contributed to the induction of β-oxidation.

5. CONCLUSION REMARKS

Recent findings from many laboratories have clearly suggested that nuclear receptors CAR not only has its "traditional" function as a xenobiotic receptor to regulate the transcription of drug-metabolizing enzymes and transporters, which are essential in protecting our bodies from the accumulation of harmful chemicals, but also has interesting functions in lipid metabolism directly or through its cross talk with LXRs and PPARs. The cross talk

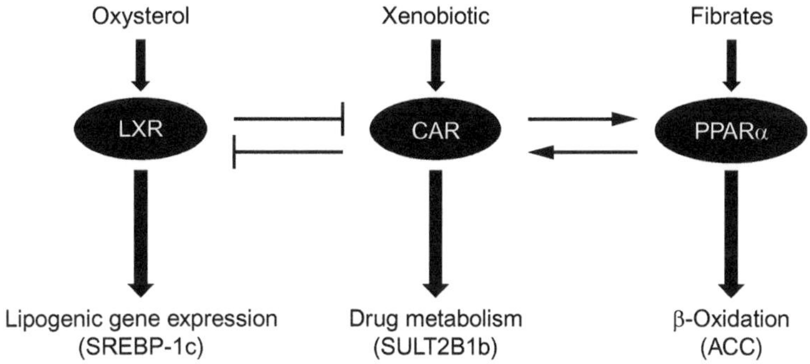

Figure 10.5 Proposed model of the cross talks of CAR–LXR and CAR–PPAR and their implications in lipid metabolism.

of CAR–LXR and CAR–PPAR and their implications in lipid metabolism are summarized in Fig. 10.5. These pioneering studies advocate continued exploration of the mechanisms by which nuclear receptors execute control of lipid metabolism and integrate these pathways into complex biological systems. It is hoped that these studies may establish CAR as a therapeutic target and ultimately lead to the development of new therapeutics for metabolic diseases.

ACKNOWLEDGMENTS

This work was supported by grants from the NSFC (NO. 30870926, NO. 31071027 to Y. Z.), the National Natural Science Foundation of China, and 973 Project (NO. 2011CB915504).

REFERENCES

Apfel, R., Benbrook, D., Lernhardt, E., Ortiz, M. A., Salbert, G., & Pfahl, M. (1994). A novel orphan receptor-specific for a subset of thyroid hormone-responsive elements and its interaction with the retinoid/thyroid hormone-receptor subfamily. *Molecular and Cellular Biology*, *14*, 7025–7035.

Auboeuf, D., Rieusset, J., Fajas, L., Vallier, P., Frering, V., & Riou, J. P. (1997). Tissue distribution and quantification of the expression of mRNAs of peroxisome proliferator-activated receptors and liver X receptor-alpha in humans—No alteration in adipose tissue of obese and NIDDM patients. *Diabetes*, *46*, 1319–1327.

Baes, M., Gulick, T., Choi, H. S., Martinoli, M. G., Simha, D., & Moore, D. D. (1994). A new orphan member of the nuclear hormone-receptor superfamily that interacts with a subset of retinoic acid response elements. *Molecular and Cellular Biology*, *14*, 1544–1552.

Barter, P. J., & Rye, K. A. (2008). Is there a role for fibrates in the management of dyslipidemia in the metabolic syndrome? *Arteriosclerosis, Thrombosis, and Vascular Biology*, *28*, 39–46.

Baskin-Bey, E. S., Anan, A., Isomoto, H., Bronk, S. F., & Gores, G. J. (2007). Constitutive androstane receptor agonist, TCPOBOP, attenuates steatohepatitis in the methionine choline-deficient diet-fed mouse. *World Journal of Gastroenterology*, *13*, 5635–5641.

Berge, K. E., Tian, H., Graf, G. A., Yu, L., Grishin, N. V., & Schultz, J. (2000). Accumulation of dietary cholesterol in sitosterolemia caused by mutations in adjacent ABC transporters. *Science, 290*, 1771–1775.

Berger, J. P., Akiyama, T. E., & Meinke, P. T. (2005). PPARs: Therapeutic targets for metabolic disease. *Trends in Pharmacological Sciences, 26*, 244–251.

Berthou, L., Duverger, N., Emmanuel, F., Langouët, S., Auwerx, J., & Guillouzo, A. (1996). Opposite regulation of human versus mouse apolipoprotein A-I by fibrates in human apolipoprotein A-I transgenic mice. *The Journal of Clinical Investigation, 97*, 2408–2416.

Bjorkhem, I., Meaney, S., & Diczfalusy, U. (2002). Oxysterols in human circulation: Which role do they have? *Current Opinion in Lipidology, 13*, 247–253.

Braissant, O., Foufelle, F., Scotto, C., Dauca, M., & Wahli, W. (1996). Differential expression of peroxisome proliferator-activated receptors (PPARs): Tissue distribution of PPAR-alpha, -beta, and -gamma in the adult rat. *Endocrinology, 137*, 354–366.

Chawla, A., Repa, J. J., Evans, R. M., & Mangelsdorf, D. J. (2001). Nuclear receptors and lipid physiology: Opening the X-files. *Science, 294*, 1866–1870.

Cherrington, N. J., Hartley, D. P., Li, N., Johnson, D. R., & Klaassen, C. D. (2002). Organ distribution of multidrug resistance proteins 1, 2, and 3 (Mrp1, 2, and 3) mRNA and hepatic induction of mrp3 by constitutive androstane receptor activators in rats. *The Journal of Pharmacology and Experimental Therapeutics, 300*, 97–104.

Chinetti, G., Griglio, S., Antonucci, M., Torra, I. P., Delerive, P., & Majd, Z. (1998). Activation of proliferator-activated receptors alpha and gamma induces apoptosis of human monocyte-derived macrophages. *The Journal of Biological Chemistry, 273*, 25573–25580.

Choi, H. S., Chung, M. R., Tzameli, I., Simha, D., Lee, Y. K., Seol, W., et al. (1997). Differential transactivation by two isoforms of the orphan nuclear hormone receptor CAR. *The Journal of Biological Chemistry, 272*, 23565–23571.

Collins, J. L., Fivush, A. M., Watson, M. A., Galardi, C. M., Lewis, M. C., & Moore, L. B. (2002). Identification of a nonsteroidal liver X receptor agonist through parallel array synthesis of tertiary amines. *Journal of Medicinal Chemistry, 45*, 1963–1966.

Dobrzyn, P., Dobrzyn, A., Miyazaki, M., Cohen, P., Asilmaz, E., & Hardie, D. G. (2004). Stearoyl-CoA desaturase 1 deficiency increases fatty acid oxidation by activating AMP-activated protein kinase in liver. *Proceedings of the National Academy of Sciences of the United States of America, 101*, 6409–6414.

Dong, B., Saha, P. K., Huang, W. D., Chend, W., Abu-Elheigae, L. A., & Wakil, S. J. (2009). Activation of nuclear receptor CAR ameliorates diabetes and fatty liver disease. *Proceedings of the National Academy of Sciences of the United States of America, 106*, 18831–18836.

Finn, R. D., Henderson, C. J., Scott, C. L., & Wolf, C. R. (2009). Unsaturated fatty acid regulation of cytochrome P450 expression via a CAR-dependent pathway. *The Biochemical Journal, 417*, 43–54.

Forman, B. M., Tzameli, I., Choi, H. S., Chen, J., Simha, D., & Seol, W. (1998). Androstane metabolites bind to and deactivate the nuclear receptor CAR-beta. *Nature, 395*, 612–615.

Fu, X., Menke, J. G., Chen, Y. L., Zhou, G. C., MacNaul, K. L., & Wright, S. D. (2001). 27-hydroxycholesterol is an endogenous ligand for liver X receptor in cholesterol-loaded cells. *The Journal of Biological Chemistry, 276*, 38378–38387.

Gao, J., He, J., Zhai, Y., Wada, T., & Xie, W. (2009). The constitutive androstane receptor is an anti-obesity nuclear receptor that improves insulin sensitivity. *The Journal of Biological Chemistry, 284*, 25984–25992.

Honkakoski, P., Moore, R., Washburn, K. A., & Negishi, M. (1998). Activation by diverse xenochemicals of the 51-base pair phenobarbital-responsive enhancer module in the CYP2B10 gene. *Molecular Pharmacology, 53*, 597–601.

Honkakoski, P., Zelko, I., Sueyoshi, T., & Negishi, M. (1998). The nuclear orphan receptor CAR-retinoid X receptor heterodimer activates the phenobarbital-responsive enhancer module of the CYP2B gene. *Molecular and Cellular Biology, 18*, 5652–5658.

Huang, W., Zhang, J., & Washington, M. (2005). Xenobiotic stress induces hepatomegaly and liver tumors via the nuclear receptor constitutive androstane receptor. *Molecular Endocrinology, 19*, 1646–1653.

Inoue, I., Shino, K., Noji, S., Awata, T., & Katayama, S. (1998). Expression of peroxisome proliferator-activated receptor alpha (PPAR alpha) in primary cultures of human vascular endothelial cells. *Biochemical and Biophysical Research Communications, 246*, 370–374.

Janowski, B. A., Grogan, M. J., Jones, S. A., Wisely, G. B., Kliewer, S. A., Corey, E. J., & Mangelsdorf, D. J. (1999). Structural requirements of ligands for the oxysterol liver X receptors LXR alpha and LXR beta. *Proceedings of the National Academy of Sciences of the United States of America, 96*, 266–271.

Janowski, B. A., Willy, P. J., Devi, T. R., Falck, J. R., & Mangelsdorf, D. J. (1996). An oxysterol signalling pathway mediated by the nuclear receptor LXR alpha. *Nature, 383*, 728–731.

Jay, M. A., & Ren, J. (2007). Peroxisome proliferator-activated receptor (PPAR) in metabolic syndrome and type 2 diabetes mellitus. *Current Diabetes Reviews, 3*, 33–39.

Kast, H. R., Goodwin, B., Tarr, P. T., Jones, S. A., Anisfeld, A. M., & Stoltz, C. M. (2002). Regulation of multidrug resistance-associated protein 2 (ABCC2) by nuclear receptors pregnane X receptor, farnesoid X-activated receptor, and constitutive androstane receptor. *The Journal of Biological Chemistry, 277*, 2908–2915.

Kim, J. B., & Spiegelman, B. M. (1996). ADD1/SREBP1 promotes adipocyte differentiation and gene expression linked to fatty acid metabolism. *Genes & Development, 10*, 1096–1107.

Kiyosawa, N., Tanaka, K., Hirao, J., Ito, K., Niino, N., & Sakuma, K. (2004). Molecular mechanism investigation of phenobarbital-induced serum cholesterol elevation in rat livers by microarray analysis. *Archives of Toxicology, 78*, 435–442.

Kobayashi, K., Yamanaka, Y., & Iwazaki, N. (2005). Identification of HMG-CoA reductase inhibitors as activators for human, mouse and rat constitutive androstane receptor. *Drug Metabolism and Disposition, 33*, 924–929.

Lau, C., Abbott, B. D., Corton, J. C., & Cunningham, M. L. (2010). PPARs and xenobiotic induced adverse effects: Relevance to human health. *PPAR Research, 2010*, 954639. [displayed 7 June 2011]. Available at http://www.ncbi.nlm.nih.gov/pmc/articles/PMC3124005/?tool=pubmed.

Lee, M. H., Lu, K., Hazard, S., Yu, H. W., Shulenin, S., & Hidaka, H. (2001). Identification of a gene, ABCG5, important in the regulation of dietary cholesterol absorption. *Nature Genetics, 27*, 79–83.

Lee, C. H., Olson, P., & Evans, R. M. (2003). Minireview: Lipid metabolism, metabolic diseases, and peroxisome proliferator-activated receptors. *Endocrinology, 144*, 2201–2207.

Lehmann, J. M., Kliewer, S. A., Moore, L. B., Smith-Oliver, T. A., Oliver, B. B., & Su, J. L. (1997). Activation of the nuclear receptor LXR by oxysterols defines a new hormone response pathway. *The Journal of Biological Chemistry, 272*, 3137–3140.

Louet, J. F., Chatelain, F., Decaux, J. F., Park, E. A., Kohl, C., & Pineau, T. (2001). Long chain fatty acids regulate liver carnitine palmitoyltransferase I gene (L-CPTI) expression through a peroxisome-proliferator-activated receptor alpha (PPARalpha)-independent pathway. *The Biochemical Journal, 354*, 189–197.

Luo, Y., & Tall, A. R. (2000). Sterol upregulation of human CETP expression in vitro and in transgenic mice by an LXR element. *The Journal of Clinical Investigation, 105*, 513–520.

Luquet, S., Gaudel, C., Holst, D., Lopez-Soriano, J., Jehl-Pietri, C., Fredenrich, A., & Grimaldi, P. A. (2005). Roles of PPAR delta in lipid absorption and metabolism: A new target for the treatment of type 2 diabetes. *Biochimica Et Biophysica Acta-Molecular Basis of Disease, 1740*, 313–317.

Maglich, J. M., Lobe, D. C., & Moore, J. T. (2009). The nuclear receptor CAR (NR1I3) regulates serum triglyceride levels under conditions of metabolic stress. *Journal of Lipid Research, 50*, 439–445.

Maglich, J. M., Watson, J., McMillen, P. J., Goodwin, B., Willson, T. M., & Moore, J. T. (2004). The nuclear receptor CAR is a regulator of thyroid hormone metabolism during caloric restriction. *The Journal of Biological Chemistry*, *279*, 19832–19838.

Mak, P. A., Laffitte, B. A., Desrumaux, C., Joseph, S. B., Curtiss, L. K., & Mangelsdorf, D. J. (2002). Regulated expression of the apolipoprotein E/C-I/C-IV/C-II gene cluster in murine and human macrophages—A critical role for nuclear liver X receptors alpha and beta. *The Journal of Biological Chemistry*, *277*, 31900–31908.

Mangelsdorf, D. J. (2010). Commentary: The year in nuclear receptor control of metabolism. *Molecular Endocrinology*, *24*, 2075–2080.

Mascaró, C., Acosta, E., Ortiz, J. A., Marrero, P. F., Hegardt, F. G., & Haro, D. (1998). Control of human muscle-type carnitine palmitoyltransferase I gene transcription by peroxisome proliferator-activated receptor. *The Journal of Biological Chemistry*, *273*, 8560–8563.

Masson, D., Qatanani, M., Sberna, A. L., Xiao, R., Barros, J. P. P., & Grober, J. (2008). Activation of the constitutive androstane receptor decreases HDL in wild-type and human apoA-I transgenic mice. *Journal of Lipid Research*, *49*, 1682–1691.

McKenna, N. J., Xu, J. M., Nawaz, Z., Tsai, S. Y., Tsai, M. J., & O'Malley, B. W. (1999). Nuclear receptor coactivators: Multiple enzymes, multiple complexes, multiple functions. *The Journal of Steroid Biochemistry and Molecular Biology*, *69*, 3–12.

Moore, L. B., Parks, D. J., & Jones, S. A. (2000). Orphan nuclear receptors constitutive androstane receptor and pregnane X receptor share xenobiotic and steroid ligands. *The Journal of Biological Chemistry*, *275*, 15122–15127.

Motojima, K., Passilly, P., Peters, J. M., Gonzalez, F. J., & Latruffe, N. (1998). Expression of putative fatty acid transporter genes are regulated by peroxisome proliferator-activated receptor alpha and gamma activators in a tissue- and inducer-specific manner. *The Journal of Biological Chemistry*, *273*, 16710–16714.

Mulligan, J. D., Flowers, M. T., Tebon, A., Bitgood, J. J., Wellington, C., Hayden, M. R., & Attie, A. D. (2003). ABCA1 is essential for efficient basolateral cholesterol efflux during the absorption of dietary cholesterol in chickens. *The Journal of Biological Chemistry*, *278*, 13356–13366.

Peet, D. J., Turley, S. D., Ma, W. Z., Janowski, B. A., Lobaccaro, J. M. A., Hammer, R. E., & Mangelsdorf, D. J. (1998). Cholesterol and bile acid metabolism are impaired in mice lacking the nuclear oxysterol receptor LXR alpha. *Cell*, *93*, 693–704.

Peters, J. M., Hennuyer, N., Staels, B., Fruchart, J. C., Fievet, C., Gonzalez, F. J., & Auwerx, J. (1997). Alterations in lipoprotein metabolism in peroxisome proliferator-activated receptor alpha-deficient mice. *The Journal of Biological Chemistry*, *272*, 27307–27312.

Poland, R. E., & Rubin, R. T. (1981). Radioimmunoassay of haloperidol in human-serum correlation of serum haloperidol with serum prolactin. *Life Sciences*, *29*, 1837–1845.

Qatanani, M., Zhang, J., & Moore, D. D. (2005). Role of the constitutive androstane receptor in xenobiotic-induced thyroid hormone metabolism. *Endocrinology*, *146*, 995–1002.

Reaven, G. (1994). Syndrome X: 6 years later. *Journal of Internal Medicine. Supplement*, *736*, 13–22.

Rencurel, F., Foretz, M., Kaufmann, M. R., Stroka, D., Looser, R., & Leclerc, I. (2006). Stimulation of AMP-activated protein kinase is essential for the induction of drug metabolizing enzymes by phenobarbital in human and mouse liver. *Molecular Pharmacology*, *70*, 1925–1934.

Rencurel, F., Stenhouse, A., Hawley, S. A., Friedberg, T., Hardie, D. G., Sutherland, C., & Wolf, C. R. (2005). AMP-activated protein kinase mediates phenobarbital induction of CYP2B gene expression in hepatocytes and a newly derived human hepatoma cell line. *The Journal of Biological Chemistry*, *280*, 4367–4373.

Repa, J. J., & Mangelsdorf, D. J. (2000). The role of orphan nuclear receptors in the regulation of cholesterol homeostasis. *Annual Review of Cell and Developmental Biology*, *16*, 459–481.

Repa, J. J., Turley, S. D., Lobaccaro, J. M. A., Medina, J., Li, L., & Lustig, K. (2000). Regulation of absorption and ABC1-mediated efflux of cholesterol by RXR heterodimers. *Science*, *289*, 1524–1529.

Roth, A., Looser, R., Kaufmann, M., & Meyer, U. A. (2008). Sterol regulatory element binding protein 1 interacts with pregnane X receptor and constitutive androstane receptor and represses their target genes. *Pharmacogenetics and Genomics*, *18*, 325–337.

Roth, A., Looser, R., Kaufmann, M., Blättler, S. M., Rencurel, F., & Huang, W. D. (2008). Regulatory cross-talk between drug metabolism and lipid homeostasis: Constitutive androstane receptor and pregnane X receptor increase Insig-1 expression. *Molecular Pharmacology*, *73*, 1282–1289.

Russell, D. W. (1992). Cholesterol biosynthesis and metabolism. *Cardiovascular Drugs and Therapy*, *6*, 103–110.

Schoonjans, K., PeinadoOnsurbe, J., Lefebvre, A. M., Heyman, R. A., Briggs, M., & Deeb, S. (1996). PPAR alpha and PPAR gamma activators direct a distinct tissue-specific transcriptional response via a PPRE in the lipoprotein lipase gene. *The EMBO Journal*, *15*, 5336–5348.

Schoonjans, K., Watanabe, M., Suzuki, H., Mahfoudi, A., Krey, G., & Wahli, W. (1995). Induction of the acyl-coenzyme-A synthetase gene by fibrates and fatty-acids is mediated by a peroxisome proliferator response element in the C-promoter. *The Journal of Biological Chemistry*, *270*, 19269–19276.

Schultz, J. R., Tu, H., Luk, A., Repa, J. J., Medina, J. C., & Li, L. (2000). Role of LXRs in control of lipogenesis. *Genes & Development*, *14*, 2831–2838.

Serrano, A., Del Arco, I., Javier Pavón, F., Macías, M., Perez-Valero, V., & Rodríguez de Fonseca, F. (2008). The cannabinoid CB1 receptor antagonist SR141716A (Rimonabant) enhances the metabolic benefits of long-term treatment with oleoylethanolamide in Zucker rats. *Neuropharmacology*, *54*, 226–234.

Shimano, H., Yahagi, N., Amemiya-Kudo, M., et al. (1999). Sterol regulatory element-binding protein-1 as a key transcription factor for nutritional induction of lipogenic enzyme genes. *The Journal of Biological Chemistry*, *274*, 35832–35839.

Shiraki, T., Sakai, N., Kanaya, E., & Jingami, H. (2003). Activation of orphan nuclear constitutive androstane receptor requires subnuclear targeting by peroxisome proliferator-activated receptor gamma coactivator-1 alpha—Possible link between xenobiotic response and nutritional state. *The Journal of Biological Chemistry*, *278*, 11344–11350.

Song, C., Kokontis, J. M., Hiipakka, R. A., & Liao, S. S. (1994). Ubiquitous Receptor—A receptor that modulates gene activation by retinoic acid and thyroid-hormone receptors. *Proceedings of the National Academy of Sciences of the United States of America*, *91*, 10809–10813.

Staels, B. (2007). PPAR agonists and the metabolic syndrome. *Therapie*, *62*, 319–326.

Staels, B., Koenig, W., Habib, A., Merval, R., Lebret, M., & Torra, I. P. (1998). Activation of human aortic smooth-muscle cells is inhibited by PPAR alpha but not by PPAR gamma activators. *Nature*, *393*, 790–793.

Sueyoshi, T., Kawamoto, T., Zelko, I., Honkakoski, P., & Negishi, M. (1999). The repressed nuclear receptor CAR responds to phenobarbital in activating the human CYP2B6 gene. *The Journal of Biological Chemistry*, *274*, 6043–6046.

Swales, K., & Negishi, M. (2004). CAR, driving into the future. *Molecular Endocrinology*, *18*, 1589–1598.

Teboul, M., Enmark, E., Li, Q., Wikstrom, A. C., Peltohuikko, M., & Gustafsson, J. A. (1995). OR-1, a member of the nuclear receptor superfamily that interacts with the 9-cis-retinoic acid receptor. *Proceedings of the National Academy of Sciences of the United States of America*, *92*, 2096–2100.

Tontonoz, P., & Mangelsdorf, D. J. (2003). Liver X receptor signaling pathways in cardiovascular disease. *Molecular Endocrinology*, *17*, 985–993.

Tzameli, I., & Moore, D. D. (2001). Role reversal: New insights from new ligands for the xenobiotic receptor CAR. *Trends in Endocrinology and Metabolism*, *12*, 7–10.

Ueda, A., Hamadeh, H. K., Webb, H. K., Yamamoto, Y., Sueyoshi, T., & Afshari, C. A. (2002). Diverse roles of the nuclear orphan receptor CAR in regulating hepatic genes in response to phenobarbital. *Molecular Pharmacology*, *61*, 1–6.

Uppal, H., Zhai, Y. G., Gangopadhyay, A., Khadem, S., Ren, S. R., Moser, J. A., & Xie, W. (2008). Activation of liver X receptor sensitizes mice to gallbladder cholesterol crystallization. *Hepatology*, *47*, 1331–1342.

Vu-Dac, N., Chopin-Delannoy, S., Gervois, P., Bonnelye, E., Martin, G., & Fruchart, J. C. (1998). The nuclear receptors peroxisome proliferator-activated receptor alpha and Rev-erb alpha mediate the species-specific regulation of apolipoprotein A-I expression by fibrates. *The Journal of Biological Chemistry*, *273*, 25713–25720.

Wada, T., Gao, J., & Xie, W. (2009). PXR and CAR in energy metabolism. *Trends in Endocrinology and Metabolism*, *20*, 273–279.

Wada, T., Kang, H. S., Jetten, A. M., & Xie, W. (2008). The emerging role of nuclear receptor ROR alpha and its crosstalk with LXR in xeno- and endobiotic gene regulation. *Experimental Biology and Medicine*, *233*, 1191–1201.

Wieneke, N., Hirsch-Ernst, K. I., Kuna, M., Kersten, S., & Pueschel, G. P. (2007). PPAR alpha-dependent induction of the energy homeostasis-regulating nuclear receptor NR1i3 (CAR) in rat hepatocytes: Potential role in starvation adaptation. *FEBS Letters*, *581*, 5617–5626.

Willy, P. J., Umesono, K., Ong, E. S., Evans, R. M., Heyman, R. A., & Mangelsdorf, D. J. (1995). LXR, a nuclear receptor that defines a distinct retinoid response pathway. *Genes & Development*, *9*, 1033–1045.

Xiao, L., Xie, X. N., & Zhai, Y. G. (2010). Functional crosstalk of CAR-LXR and ROR-LXR in drug metabolism and lipid metabolism. *Advanced Drug Delivery Reviews*, *62*, 1316–1321.

Xu, L., Glass, C. K., & Rosenfeld, M. G. (1999). Coactivator and corepressor complexes in nuclear receptor function. *Current Opinion in Genetics & Development*, *9*, 140–147.

Yamazaki, Y., Kakizaki, S., Horiguchi, N., Sohara, N., Sato, K., & Takagi, H. (2007). The role of the nuclear receptor constitutive androstane receptor in the pathogenesis of non-alcoholic steatohepatitis. *Gut*, *56*, 565–574.

Yoon, M. (2009). The role of PPAR alpha in lipid metabolism and obesity: Focusing on the effects of estrogen on PPAR alpha actions. *Pharmacological Research*, *60*, 151–159.

Zelcer, N., & Tontonoz, P. (2006). Liver X receptors as integrators of metabolic and inflammatory signaling. *The Journal of Clinical Investigation*, *116*, 607–614.

Zhai, Y., Wada, T., & Zhang, B. (2010). A functional crosstalk between LXR and CAR links lipogenesis and xenobiotic responses. *Molecular Pharmacology*, *78*, 666–674.

Zhang, F., Lavan, B., & Gregoire, F. M. (2004). Peroxisome proliferator-activated receptors as attractive antiobesity targets. *Drug News & Perspectives*, *17*, 661–669.

Zhang, Y., Repa, J. J., Gauthier, K., & Mangelsdorf, D. J. (2001). Regulation of lipoprotein lipase by the oxysterol receptors, LXR alpha and LXR beta. *The Journal of Biological Chemistry*, *276*, 43018–43024.

CHAPTER ELEVEN

Lecithin Cholesterol Acyltransferase Deficiency Protects from Diet-Induced Insulin Resistance and Obesity—Novel Insights from Mouse Models

Dominic S. Ng[*,†,1]
[*]Keenan Research Center, Li Ka Shing Knowledge Institute, St Michael's Hospital, Toronto, Ontario, Canada
[†]Department of Medicine, University of Toronto, Toronto, Ontario, Canada
[1]Corresponding author: e-mail address: ngd@smh.ca

Contents

Abstract

Reduced plasma level of high-density lipoprotein cholesterol is an independent risk factor for atherosclerotic heart disease and is also a major diagnostic feature for the metabolic syndrome. Lecithin cholesterol acyltransferase (LCAT), an enzyme mediating the

Vitamins and Hormones, Volume 91
ISSN 0083-6729
http://dx.doi.org/10.1016/B978-0-12-407766-9.00011-0

esterification of cholesterol in circulating lipoproteins, is one of the major modulators of high-density lipoprotein levels and composition. Loss-of-function mutations of LCAT invariably results in profound HDL deficiency and also modest hypertriglyceridemia (HTG). While intense effort has been devoted to investigate the role of LCAT in atherogenesis, which remains controversial, much less is known about whether LCAT also modulates glucose and energy homeostasis. In recent years, findings from studying the LCAT knockout mice began to suggest that LCAT deficiency, in spite of its unfavorable high triglyceride/low HDL lipid phenotypes, may confer protection from the development of insulin resistance and obesity. To date, alterations in specific metabolic pathways in liver, white adipose tissue, and skeletal muscle have been implicated. A better mechanistic understanding in the metabolic linkage between the primary biochemical action of LCAT and the downstream protective phenotypes will greatly facilitate the identification of potential novel pathways and targets in the treatment of obesity and diabetes.

1. INTRODUCTION

1.1. The role of lecithin cholesterol acyltransferase in lipid metabolism and atherosclerosis

Lecithin cholesterol acyltransferase (LCAT) is a lipoprotein-associated enzyme and its enzymatic action provides the major source of esterified cholesterol in circulating lipoproteins through hydrolysis of phosphatidylcholine at the *sn-2* position and transfer of the fatty acid to free cholesterol (FC) to form cholesterol ester (CE) (Glomset, 1962). LCAT activity is detected both in high-density lipoproteins (HDLs), termed α-activity, and apolipoprotein (apo) B-containing lipoproteins, β-activity (O et al., 1993). LCAT plays a key role in sustaining the circulating level of HDL as loss-of-function mutations in LCAT results in profound HDL deficiency transmitted in an autosomal co-dominant manner (Kuivenhoven, Pritchard, et al., 1997). Homozygotes or compound heterozygotes for a number of reported mutations that result in loss of both α- and β-activities develop complete LCAT deficiency (CLD). In addition to severe HDL deficiency, CLD subjects also develop elevated plasma FC/CE ratio, modest hypertriglyceridemia (HTG); Frohlich, McLeod, et al., 1988; Nishiwaki, Ikewaki, et al., 2006; Ng et al., 2004), and in some cases, accumulation of lipoprotein-X (LpX), the latter being implicated in the development of LCAT deficiency-associated nephropathy (Ng, 2012). Clinically, mild form of anemia and corneal opacity are commonly observed, but progressive glomerulopathy is the major cause of morbidity, including renal failure (Borysiewicz, Soutar, et al., 1982; Zhu, Herzenberg, et al., 2004). A number of mutations that result in only partial loss of LCAT activity with lone HDL deficiency but severe corneal opacity are coined the term fish eye disease

(FED). In spite of the unfavorable HTG/low HDL dyslipidemic phenotypes, the impact of CLD in atherosclerosis continues to be clouded with controversy. Many observational studies suggest a lack of accelerated atherosclerosis in spite of marked reduction in high-density lipoprotein cholesterol in LCAT-deficient subjects (Calabresi, Baldassarre, et al., 2011). Recent studies suggest that subjects with FED subjects might be more prone to atherosclerosis than those with CLD (Calabresi, Simonelli, et al., 2012; Holleboom, Kuivenhoven, et al., 2011; Ng, 2012), but this hypothesis needs to be thoroughly tested.

1.2. LCAT knockout mouse models: Understanding the LCAT–atherosclerosis–nephropathy links

Two lines of LCAT knockout mice were generated and each had been shown to recapitulate the key lipoprotein phenotypes seen in humans with CLD, namely, a profound HDL deficiency, moderate HTG, increased plasma FC/CE ratio, and abnormal morphological changes in apoB-containing particles (Ng, Francone, et al., 1997; Sakai, Vaisman, et al., 1997). These two laboratories independently showed that breeding of LCAT null mice into the apolipoprotein (apo) E knockout background, the latter being a model for hyperlipidemia-induced atherosclerosis, resulted in less atherosclerosis than their apoE knockout controls (Lambert, Sakai, et al., 2001; Ng, Maguire, et al., 2002). This paradoxical atheroprotective phenotype was in part linked to the rescue of the serum paraoxonase 1 (PON1) activity and reduced oxidative stress in LCAT deficiency in spite of the profound HDL deficiency (Ng et al., 2002). However, conflicting findings were reported when the effect of LCAT deficiency on atherosclerosis was tested in the low-density lipoprotein receptor (LDLR) knockout background (Furbee, Sawyer, et al., 2002; Lambert et al., 2001). The reason for the discordant findings is not known. The possibility of an induced reduction in PON1 expression and activity, a putative atheroprotective factor in LCAT deficiency, in response to the atherogenic diet in the study by Furbee et al. was not explored. A unique model through cross-breeding the LCAT knockout mice into the sterol-responsive element-binding protein 1a transgenic mice, the latter being a model for unregulated hepatic lipogenesis, was generated and provided compelling evidence to support that LpX being the key inciting lipid fraction for the development of nephropathy (Zhu et al., 2004). By cross-breeding the LCAT knockout mice into appropriate dyslipidemic strains, studies to date suggest that this mouse model faithfully recapitulates the major classic phenotypes and provides important mechanistic insights.

2. NOVEL METABOLIC PHENOTYPES IN MURINE MODELS OF LCAT DEFICIENCY

2.1. Exploring the role of LCAT in glucose and energy homeostasis: Novel linkages in LCAT null mice

While a great deal of attention has been devoted to address the role of LCAT in atherogenesis, little is known whether LCAT plays a role in the modulation of glucose or energy homeostasis, including humans with LCAT deficiency. The identification of a link between CLD and altered glucose metabolism was first reported by Ng et al. (2004). In the background of LDLR knockout mice, deletion of the LCAT gene resulted in an enhanced insulin sensitivity. The LDLR/LCAT double knockout mice fed a regular chow diet were found to develop lower fasting insulin, fasting glucose, and improvement in glucose tolerance and insulin tolerance, when compared to the LDLR knockout control mice. When fed a high fat high sucrose (HFHS) diet, or frequently referred to as a high fat diet, the LDLR/LCAT double knockout mice also demonstrated marked attenuation in the development of impaired glucose tolerance (Li, Hossain, et al., 2011). In the absence of the LDLR knockout background, the chow-fed LCAT knockout mice also showed a trend in improved glucose tolerance when compared to the wild-type control thus suggesting that insulin hypersensitivity is inherently attributable to LCAT deficiency. To date, a number of cellular and molecular pathways have been identified as putative mechanisms for this protective metabolic phenotype. In chow-fed LCAT/LDLR double knockout mice, Li et al. demonstrated insulin signaling, namely, the phosphorylation of insulin receptor, IRS, and Akt in the liver is upregulated. Meanwhile, nuclear abundance of the transcription factor FoxO1 in the fasting state is elevated in the chow-fed LCAT/LDLR double knockout mice in conjunction with a reduced mRNA level of its target gene PEPCK (Li, Naples, et al., 2007). These changes are expected to cause reduced gluconeogenesis and is consistent with the observed reduced fasting glucose previously reported (Ng et al., 2004). Concomitantly, hepatic expression of a number of genes were shown altered in favor of enhanced insulin sensitivity in a coordinated manner and they include SOCS1 and TCFE3 (Li et al., 2007).

2.2. LCAT deficiency modulates hepatic unfolded protein response in response to nutritional excess

In recent years, emerging evidence continue to implicate ER stress in playing a central role in the development of obesity, type 2 diabetes (T2DM), and a variety of other chronic metabolic diseases (Hotamisligil,

2010; Ozcan & Tabas, 2012). Excess burden of unfolded proteins in the ER triggers a set of well-defined transduction signals, namely the unfolded protein response (UPR). As an adaptive response, the canonical UPR invokes three distinct pathways that cooperatively attenuate global gene expression and protein translation. In the liver, a direct linkage between UPR induction and activation of JNK signaling, a pathway known to cause impairment of insulin signaling through serine phosphorylation of insulin receptor substrate, provides convincing mechanistic role of ER stress in the development of insulin resistance (IR) and T2DM (Hotamisligil, 2008; Ozcan, Cao, et al., 2004; Ozcan, Yilmaz, et al., 2006). Activation of the UPR in ER stress may also activate other pathologic processes including oxidative stress, autophagy, inflammation and, under specific conditions, apoptosis. It is well established that nutrient excess in high fat diet-induced obesity activates ER stress in the liver and other tissues (Ozcan et al., 2004; Ozcan & Tabas, 2012). Treatment with chemical chaperone led to a rapid amelioration of ER stress and its glucose intolerance independent of weight loss (Ozcan et al., 2006). In humans, disruption of chronic ER stress with 4-phenylbutyrate also resulted in acute reversal of IR and pancreatic beta cell dysfunction (Xiao, Giacca, et al., 2011), supportive of the pathophysiologic role of ER stress in nutrient excess-induced IR and as a potential therapeutic target (Engin & Hotamisligil, 2010).

2.3. LDL receptor deficiency augments the protective phenotypes of LCAT deficiency in mice

LDLR knockout mice develop accentuated high fat diet-induced obesity and IR when compared to the C57Bl/6 wild-type mice, but the underlying mechanism remains poorly understood (Schreyer, Vick, et al., 2002). Mechanistically, Li et al. recently demonstrated that LDLR knockout mice developed elevated hepatic ER stress even under chow-fed condition and an accentuated induction of ER stress in response to a HFHS diet (or high fat diet) when compared to the C57Bl/6 wild-type mice. In this LDL receptor null metabolic background, Li et al. reported that rendering the LDLR knockout mice also LCAT deficient led to a normalization of baseline hepatic ER stress under chow-fed condition. Further, the LDLR knockout mice made LCAT deficient also showed marked resistance to the HFHS diet induction of ER stress in conjunction with a dramatic protection from HFHS diet-induced obesity and glucose intolerance (Li et al., 2011). The underlying mechanism of the protection from ER stress in LCAT deficiency is not yet known. It is conceivable that LCAT deficiency may, through yet-to-be defined pathways, directly modulate hepatic ER stress. In the case of protection from HFHS diet-induced ER stress, it is also possible that this is a consequence of the LDLR/LCAT double knockout mice being protected from diet-induced obesity.

Regarding the question whether LCAT deficiency directly modulates hepatic ER stress, my laboratory has recently provided preliminary evidence in support of this notion. First, we showed that the LDLR knockout mice developed elevated hepatic total and FC when compared to the wild-type control, which correlate with the elevated basal expression of the ER stress markers. In the concurrent absence of LCAT, namely, the LDLR/LCAT double knockout, the ER stress marker expression becomes normalized to the wild-type mice level, in association with a reversal of the whole hepatic tissue total and FC levels to those of the wild-type mice. We also measured FC in the ER fraction from chow-fed wild type, LDLR knockout, and LDLR/LCAT double knockout mice and observed changes in ER cholesterol being parallel to the tissue cholesterol changes. Upon feeding these three genotypic groups of mice with a 2% high cholesterol diet (HCD), we observed significant elevation of hepatic cholesterol and a parallel increase in hepatic ER stress in the LDLR knockout mice in response to the HCD. Unexpectedly, the LDLR/LCAT double knockout mice ER cholesterol and ER stress became lower than those of the wild-type mice in spite of a significant rise in the tissue cholesterol level. Combining the findings from these six groups of animals, we observed a strong correlation between hepatic ER stress and ER cholesterol, a finding further substantiating the direct role of LCAT in the modulation of hepatic ER stress, in part through regulation of ER cholesterol (Hager, Li, et al., 2012).

2.4. Antiobesity phenotypes in LDLR/LCAT double knockout mouse adipose tissue and skeletal muscle

In the report by Li et al., (2011) in addition to the effect of LCAT deficiency on hepatic ER stress, the authors also reported additional antiobesity phenotypes in white adipose tissue (WAT) and in the skeletal muscle. The chow-fed LDLR/LCAT double knockout mice showed reduced fat mass and smaller adipocyte sizes. Further, the WAT expression of the two key adipogenic genes, PPARγ and CEBPα are reduced in association with increased expression of a panel of target genes of the canonical Wnt-signaling pathway when compared to the LDLR knockout controls. This gene profile resembles those observed in transgenic mice selectively overexpressing the Wnt ligand Wnt10b in the adipose tissue, which were found to strongly resist high fat diet-induced obesity through inhibition of adipogenesis

(Wright, Longo, et al., 2007). Further, Li et al. also reported an unexpected presence of ectopic brown adipose tissue (BAT) in skeletal muscle of the LDLR/LCAT double knockout mice, consistent with the observation of increased energy expenditure and resistance to diet-induced obesity (Li et al., 2011). Although a similar phenotype was previously reported in an inbred mouse strain (Almind, Manieri, et al., 2007), the detection of ectopic BAT in the interfiber regions in skeletal muscle of the LCAT-deficient mice represents the first such findings in monogenic dyslipidemia of a known gene. Full exploration of the development origin of such ectopic BAT is of great interest in the context of identifying novel targets for the treatment of obesity.

2.5. Potential developmental origins of ectopic BAT in LCAT null mice

Recent studies in embryonic development of skeletal muscle, BAT, and WAT revealed that the BAT and skeletal muscle share the same progenitor cell lineage, namely, the myf5+ myoblasts, whereas the origin of the WAT is distinctly different (Seale, Kajimura, et al., 2009). Further, PRD1-BF1-R1Z1 homologous domain containing 16 (PRDM16), one of the positive transcriptional regulators of brown fat development, has been shown to determine the fate of brown fat cells in a cell-autonomous manner. Ectopic expression of PRDM16 in the C2C12 myoblast cell line, an immortalized cell line derived from the skeletal muscle adult stem cell—the satellite cells, enriched with a promyogenic culture condition prevented the differentiation into terminal myocytes. Enrichment of the PRDM16-transfected cells with adipogenic inducers efficiently led to lipid-filled adipocytes with high levels of expression of brown fat cell-selective genes, including Ucp1, Pgc1α, Elovl3, and Cidea. The brown fat/skeletal muscle switch is bidirectional as knockdown of PRDM16 in brown preadipocyte cell lines led to myogenic differentiation (Seale, Bjork, et al., 2008). Identity of the signaling molecules that control the timing and specificity of PRDM16 expression in myogenic cells is not known but is of great interest for development of novel BAT-based strategies in combating obesity and diabetes.

2.6. Embryonic myoblast to BAT switch as possible origin of ectopic BAT in LCAT-deficient mice

During embryonic development, progression of specific mesodermal precursor cells to the myogenic lineage requires the expression of MyoD and myf5, two basic helix-loop-helix transcriptional factors, to transactivate the myogenic regulatory factor family. Proliferating myoD and/or myf5-positive myogenic cells are termed *myoblasts*. Committed myoblasts, or myogenic precursor cells (mpcs), migrate laterally to form the myotome. In case of limb muscles, mpc migrates into the limb bud by embryonic days E9.5–E10 (Bladt, Riethmacher, et al., 1995). Fetal brown adipogenesis likely begins when PRDM16 complexes with C/EBPβ to initiate the myoblast-to-brown fat switch (Kajimura, Bjork, et al., 2009) as expression of PPARγ, a crucial adipogenic factor, become detectable around E14.5 (Kliewer, Forman, et al., 1994). In the case of LCAT deficiency, it is therefore quite plausible that some of the laterally migrated myoblasts, otherwise destined to differentiate into mature myocytes, are triggered to express PRDM16 and differentiate into colonies of BAT during embryonic development. In this scenario, candidate endogenous factors that could induce expression of PRDM16 include PPARα (Hondares, Rosell, et al., 2011; Sun, Xie, et al., 2011), bone morphogenic protein (BMP7) (Townsend, Suzuki, et al., 2012; Tseng, Kokkotou, et al., 2008), and possibly PPARγ (Ohno, Shinoda, et al., 2012). Preliminary data in our lab revealed that PPARγ gene expression is markedly upregulated in the skeletal muscle of the LCAT-deficient mice and this upregulation is markedly reversed when the mice were fed a 2% HLD, raising a possible role for cellular cholesterol in the modulation of the PRDM16 switch and brown fat adipogenesis via PPARγ (Li L & Ng DS, 2012, unpublished data).

2.7. Could the ectopic BAT be due to browning of the WAT?

Although WAT does not share the same progenitor cells as BAT and skeletal muscle as described, mature WAT may be induced to develop a brown fat-like phenotype. The WAT will express a gene program highly characteristic of BAT, including expression of UCP1, PRDM16 through trans-differentiation, most notably under cold exposure or direct β3 adrenergic stimulation (Barbatelli, Murano, et al., 2010). A study by Seale, Conroe, et al. (2011) showed that forced expression of PRDM16 in selective WAT depots can generate a brown adipogenic program, expressing PRDM16 and UCP1. More recently, Bostrom et al. reported that increased expression of a FNDC5, which encodes for a membrane protein with the cleavage product being a circulating hormone irisin, will result in the induction of a brown fat gene program including UCP1 in WAT (Bostrom, Wu, et al., 2012). A recent study by Ohno et al. (2012) reported a novel role of PPARγ in WAT biology. In addition to being crucial gene for the final step in adipogenesis, PPARγ has very recently been shown to play a major role in the "browning" of the white fat. Activation of PPARγ by its agonist is crucial in the induction of the brown adipocyte gene program in WAT through stabilization of the WAT-derived PRDM16 protein. We tested the possible presence of browning of the WAT in LCAT-deficient mice. In spite of our preliminary finding of a 1.6-fold upregulation of FNDC5 mRNA level in skeletal muscle of the LDLR/LCAT double knockout mice, we did not observe any significant increase in the protein level of UCP1 in various WAT depots (Li et al., 2011). In addition, the PPARγ mRNA level in WAT is lower in the LDLR/LCAT double knockout mice as compared to their LDLR knockout control, effectively diminishing the possible role of PPARγ in the browning of WAT. Taken together, our current data are suggestive of the ectopic brown fat seen in the skeletal muscle being myoblastic in origin. Western blot analysis led to an initial estimate that the UCP1 protein mass in skeletal muscle of LCAT-deficient mice is approximately 20% of that of the whole body BAT, a level of abundance sufficient to confer energy expenditure to prevent diet-induced obesity.

3. CONCLUSIONS

LCAT plays a crucial role in the regulation of HDL metabolism as loss-of-function mutations invariably result in profound HDL deficiency. However, susceptibility of LCAT deficiency to accelerated atherosclerosis is highly variable and is likely dependent on the nature of the mutations. LCAT knockout mice as a model for LCAT deficiency have been informative in providing mechanistic insight into the complex LCAT–atherosclerosis link. Recent studies with this model, particularly when bred into the LDL receptor knockout background, have provided novel findings to suggest a potentially protective milieu against IR and obesity, involving at least three insulin-sensitive tissues, namely, the liver, WAT, and skeletal muscle. The identification of ectopic brown fat in skeletal muscle of LDLR/LCAT double knockout mice is most striking, opening the possibility of novel BAT-based target for combating the duel obesity and diabetes epidemics. Future studies include an in-depth mechanistic understanding of how LCAT deficiency leads to the induction of ectopic brown fat, the modulation of hepatic ER stress, and adipose tissue adipogenesis. Expanding the studies to examine the effect of LCAT deficiency on the pancreatic β cell insulin-secretory function and the central nervous system function in relation to metabolic regulations is also warranted.

REFERENCES

Almind, K., Manieri, M., Sivitz, W. I., Cinti, S., & Kahn, C. R. (2007). Ectopic brown adipose tissue in muscle provides a mechanism for differences in risk of metabolic syndrome in mice. *Proceedings of the National Academy of Sciences of the United States of America*, *104*(7), 2366–2371.

Barbatelli, G., Murano, I., Madsen, L., Hao, Q., Jimenez, M., Kristiansen, K., et al. (2010). The emergence of cold-induced brown adipocytes in mouse white fat depots is determined predominantly by white to brown adipocyte transdifferentiation. *American Journal of Physiology. Endocrinology and Metabolism*, *298*(6), E1244–E1253.

Bladt, F., Riethmacher, D., Isenmann, S., Aguzzi, A., & Birchmeier, C. (1995). Essential role for the c-met receptor in the migration of myogenic precursor cells into the limb bud. *Nature*, *376*(6543), 768–771.

Borysiewicz, L. K., Soutar, A. K., Evans, D. J., Thompson, G. R., & Rees, A. J. (1982). Renal failure in familial lecithin:cholesterol acyltransferase deficiency. *The Quarterly Journal of Medicine*, *51*, 411–426.

Boström, P., Wu, J., Jedrychowski, M. P., Korde, A., Ye, L., Lo, J. C., et al. (2012). A PGC1-a-dependent myokine that drives brown-fat-like development of white fat and thermogenesis. *Nature*, *481*(7382), 463–468.

Calabresi, L., Baldassarre, D., Simonelli, S., Gomaraschi, M., Amato, M., Castelnuovo, S., et al. (2011). Plasma lecithin:cholesterol acyltransferase and carotid intima-media thickness in European individuals at high cardiovascular risk. *Journal of Lipid Research*, *52*(8), 1569–1574.

Calabresi, L., Simonelli, S., Gomaraschi, M., & Franceschini, G. (2012). Genetic lecithin: cholesterol acyltransferase deficiency and cardiovascular disease. *Atherosclerosis*, *222*, 299–306.

Engin, F., & Hotamisligil, G. S. (2010). Restoring endoplasmic reticulum function by chemical chaperones: An emerging therapeutic approach for metabolic diseases. *Diabetes, Obesity & Metabolism*, *12*(Suppl. 2), 108–115.

Frohlich, J., McLeod, R., Pritchard, P. H., Fesmire, J., & McConathy, W. (1988). Plasma lipoprotein abnormalities in heterozygotes for familial lecithin:cholesterol acyltransferase deficiency. *Metabolism*, *37*(1), 3–8.

Furbee, J. W., Jr., Sawyer, J. K., & Parks, J. S. (2002). Lecithin:cholesterol acyltransferase deficiency increases atherosclerosis in the low density lipoprotein receptor and apolipoprotein E knockout mice. *The Journal of Biological Chemistry*, *277*, 3511–3519.

Glomset, J. A. (1962). The mechanism of the plasma cholesterol esterification reaction: Plasma fatty acid transferase. *Biochimica et Biophysica Acta*, *65*, 128–135.

Hager, L., Li, L., Pun, H., Liu, L., Hossain, M. A., Maguire, G. F., et al. (2012). Lecithin: cholesterol acyltransferase deficiency protects against cholesterol-induced hepatic endoplasmic reticulum stress in mice. *The Journal of Biological Chemistry*, *287*(24), 20755–22768.

Holleboom, A. G., Kuivenhoven, J. A., Peelman, F., Schimmel, A. W., Peter, J., Defesche, J. C., et al. (2011). High prevalence of mutations in LCAT in patients with low HDL cholesterol levels in The Netherlands: Identification and characterization of eight novel mutations. *Human Mutation*, *32*(11), 1290–1298.

Hondares, E., Rosell, M., Díaz-Delfín, J., Olmos, Y., Monsalve, M., Iglesias, R., et al. (2011). Peroxisome proliferator-activated receptor a (PPARa) induces PPARg coactivator 1a (PGC-1a) gene expression and contributes to thermogenic activation of brown fat: Involvement of PRDM16. *The Journal of Biological Chemistry*, *286*(50), 43112–43122.

Hotamisligil, G. S. (2008). Inflammation and endoplasmic reticulum stress in obesity and diabetes. *International Journal of Obesity*, *32*(Suppl. 7), S52–S54.

Hotamisligil, G. S. (2010). Endoplasmic reticulum stress and the inflammatory basis of metabolic disease. *Cell, 140*(6), 900–917.

Kajimura, S., Seale, P., Kubota, K., Lunsford, E., Frangioni, J. V., Gygi, S. P., et al. (2009). Initiation of myoblast to brown fat switch by a PRDM16-C/EBP-beta transcriptional complex. *Nature, 460*(7259), 1154–1158.

Kliewer, S. A., Forman, B. M., Blumberg, B., Ong, E. S., Borgmeyer, U., Mangelsdorf, D. J., et al. (1994). Differential expression and activation of a family of murine peroxisome proliferator-activated receptors. *Proceedings of the National Academy of Sciences of the United States of America, 91*(15), 7355–7359.

Kuivenhoven, J. A., Pritchard, H., Hill, J., Frohlich, J., Assmann, G., & Kastelein, J. (1997). The molecular pathology of lecithin:cholesterol acyltransferase (LCAT) deficiency syndromes. *Journal of Lipid Research, 38*(2), 191–205.

Lambert, G., Sakai, N., Vaisman, B. L., Neufeld, E. B., Marteyn, B., Chan, C. C., et al. (2001). Analysis of glomerulosclerosis and atherosclerosis in lecithin cholesterol acyltransferase-deficient mice. *The Journal of Biological Chemistry, 276*(18), 15090–15098.

Li, L., Hossain, M. A., Sadat, S., Hager, L., Liu, L., Tam, L., et al. (2011). Lecithin cholesterol acyltransferase null mice are protected from diet-induced obesity and insulin resistance in a gender-specific manner through multiple pathways. *The Journal of Biological Chemistry, 286*(20), 17809–17820.

Li, L., Naples, M., Song, H., Yuan, R., Ye, F., Shafi, S., et al. (2007). LCAT-null mice develop improved hepatic insulin sensitivity through altered regulation of transcription factors and suppressors of cytokine signaling. *American Journal of Physiology. Endocrinology and Metabolism, 293*(2), E587–E594.

Ng, D. S. (2012). The role of lecithin:cholesterol acyltransferase in the modulation of cardiometabolic risks - a clinical update and emerging insights from animal models. *Biochimica et Biophysica Acta, 1821*(4), 654–659.

Ng, D. S., Francone, O. L., Forte, T. M., Zhang, J., Haghpassand, M., & Rubin, E. M. (1997). Disruption of the murine lecithin:cholesterol acyltransferase gene causes impairment of adrenal lipid delivery and up-regulation of scavenger receptor class B type I. *The Journal of Biological Chemistry, 272*(25), 15777–15781.

Ng, D. S., Maguire, G. F., Wylie, J., Ravandi, A., Xuan, W., Ahmed, Z., et al. (2002). Oxidative stress is markedly elevated in lecithin:cholesterol acyltransferase-deficient mice and is paradoxically reversed in the apolipoprotein E knockout background in association with a reduction in atherosclerosis. *The Journal of Biological Chemistry, 277*(14), 11715–11720.

Ng, D. S., Xie, C., Maguire, G. F., Zhu, X., Ugwu, F., Lam, E., et al. (2004). Hypertriglyceridemia in lecithin-cholesterol acyltransferase-deficient mice is associated with hepatic overproduction of triglycerides, increased lipogenesis, and improved glucose tolerance. *The Journal of Biological Chemistry, 279*(9), 7636–7642.

Nishiwaki, M., Ikewaki, K., Bader, G., Nazih, H., Hannuksela, M., Remaley, A. T., et al. (2006). Human lecithin:cholesterol acyltransferase deficiency: in vivo kinetics of low-density lipoprotein and lipoprotein-X. *Arteriosclerosis, Thrombosis, and Vascular Biology, 26*(6), 1370–1375.

O, K., Hill, J. S., Wang, X., & Pritchard, P. H. (1993). Recombinant lecithin:cholesterol acyltransferase containing a Thr123->Ile mutation esterifies cholesterol in low density lipoprotein but not in high density lipoprotein. *Journal of Lipid Research, 34*(1), 81–88.

Ohno, H., Shinoda, K., Spiegelman, B. M., & Kajimura, S. (2012). PPARg agonists induce a white-to-brown fat conversion through stabilization of PRDM16 protein. *Cell Metabolism, 15*(3), 395–404.

Ozcan, U., Cao, Q., Yilmaz, E., Lee, A. H., Iwakoshi, N. N., Ozdelen, E., et al. (2004). Endoplasmic reticulum stress links obesity, insulin action, and type 2 diabetes. *Science, 306*(5695), 457–461.

Ozcan, L., & Tabas, I. (2012). Role of endoplasmic reticulum stress in metabolic disease and other disorders. *Annual Review of Medicine, 63*, 317–328.

Ozcan, U., Yilmaz, E., Ozcan, L., Furuhashi, M., Vaillancourt, E., Smith, R. O., et al. (2006). Chemical chaperones reduce ER stress and restore glucose homeostasis in a mouse model of type 2 diabetes. *Science, 313*(5790), 1137–1140.

Sakai, N., Vaisman, B. L., Koch, C. A., Hoyt, R. F., Jr., Meyn, S. M., Talley, G. D., et al. (1997). Targeted disruption of the mouse lecithin:cholesterol acyltransferase (LCAT) gene. Generation of a new animal model for human LCAT deficiency. *The Journal of Biological Chemistry, 272*(11), 7506–7510.

Schreyer, S. A., Vick, C., Lystig, T. C., Mystkowski, P., & LeBoeuf, R. C. (2002). LDL receptor but not apolipoprotein E deficiency increases diet-induced obesity and diabetes in mice. *American Journal of Physiology. Endocrinology and Metabolism, 282*(1), E207–E214.

Seale, P., Bjork, B., Yang, W., Kajimura, S., Chin, S., Kuang, S., et al. (2008). PRDM16 controls a brown fat/skeletal muscle switch. *Nature, 454*(7207), 961–967.

Seale, P., Conroe, H. M., Estall, J., Kajimura, S., Frontini, A., Ishibashi, J., et al. (2011). Prdm16 determines the thermogenic program of subcutaneous white adipose tissue in mice. *The Journal of Clinical Investigation, 121*(1), 96–105.

Seale, P., Kajimura, S., & Spiegelman, B. M. (2009). Transcriptional control of brown adipocyte development and physiological function—Of mice and men. *Genes & Development, 23*(7), 788–797.

Sun, L., Xie, H., Mori, M. A., Alexander, R., Yuan, B., Hattangadi, S. M., et al. (2011). Mir193b-365 is essential for brown fat differentiation. *Nature Cell Biology, 13*(8), 958–965.

Townsend, K. L., Suzuki, R., Huang, T. L., Jing, E., Schulz, T. J., Lee, K., et al. (2012). Bone morphogenetic protein 7 (BMP7) reverses obesity and regulates appetite through a central mTOR pathway. *The FASEB Journal, 26*, 2187–2196.

Tseng, Y. H., Kokkotou, E., Schulz, T. J., Huang, T. L., Winnay, J. N., Taniguchi, C. M., et al. (2008). New role of bone morphogenetic protein 7 in brown adipogenesis and energy expenditure. *Nature, 454*(7207), 1000–1004.

Wright, W. S., Longo, K. A., et al. (2007). Wnt10b inhibits obesity in ob/ob and agouti mice. *Diabetes, 56*(2), 295–303.

Xiao, C., Giacca, A., & Lewis, G. F. (2011). Sodium phenylbutyrate, a drug with known capacity to reduce endoplasmic reticulum stress, partially alleviates lipid-induced insulin resistance and beta-cell dysfunction in humans. *Diabetes, 60*(3), 918–924.

Zhu, X., Herzenberg, A. M., Eskandarian, M., Maguire, G. F., Scholey, J. W., Connelly, P. W., et al. (2004). A novel in vivo lecithin-cholesterol acyltransferase (LCAT)-deficient mouse expressing predominantly LpX is associated with spontaneous glomerulopathy. *The American Journal of Pathology, 165*(4), 1269–1278.

CHAPTER TWELVE

Sphingolipids in Lipid Microdomains and Obesity

Susumu Mitsutake[1], Yasuyuki Igarashi
Laboratory of Biomembrane and Biofunctional Chemistry, Faculty of Advanced Life Sciences, Hokkaido University, Sapporo, Japan
[1]Corresponding author: e-mail address: susumu-m@pharm.hokudai.ac.jp

Contents

Abstract

Sphingolipids are major constituents of the plasma membrane, where they are known to form lipid microdomains with cholesterol. Lipid microdomains are thought to be important not only for cellular signal transduction but also for the absorption of extracellular lipids or nutrients. Inhibition of sphingolipid biosynthesis suggested an importance for sphingolipids in fatty acid uptake via lipid microdomains. Additionally, we recently reported that the function of lipid microdomains was dynamically regulated by the sphingomyelin synthase SMS2 on the plasma membrane and that SMS2-deficient mice exhibit resistance against high-fat diet-induced increases in body weight, glucose intolerance, and fatty liver. Now, biosynthesis or metabolism of sphingolipids is thought to be involved in obesity, diabetes, and cardiovascular diseases. In this review, I focus on the functions of sphingolipids in lipid microdomains and describe their contributions to obesity and diabetes.

Vitamins and Hormones, Volume 91
ISSN 0083-6729
http://dx.doi.org/10.1016/B978-0-12-407766-9.00012-2

1. INTRODUCTION

Sphingolipids, a major class of lipids, are ubiquitous constituents of the eukaryotic cells membrane. They comprise in part a sphingoid base (sphingosine, sphinganine, or phytosphingosine), which is synthesized in several steps from serine and palmitoyl-coenzyme A. Many types of fatty acyl-coenzyme A could be condensed to the sphingoid base forming amide linkage to be ceramide (Cer), which can be further converted into more complex compounds (sphingomyelin (SM) and glycosphingolipids (GSLs)) via modification of its C-1 hydroxyl group (Fig. 12.1; Hirabayashi, Igarashi, & Merrill, 2006). Cer can also be produced via degradation of SM, the most abundant membrane sphingolipid.

Cer is known to act as a second messenger involved in the regulation of apoptosis and differentiation of cells (Hannun & Obeid, 1995). Sphingosine, the metabolic product of Cer, can itself be converted to sphingosine 1-phosphate (S1P; Igarashi, 1997), which is known to be a ligand of the G-protein-coupled receptors, $S1P_1$–$S1P_5$, and to be involved in apoptosis inhibition, angiogenesis, and cell migration (Kihara, Mitsutake, Mizutani, & Igarashi, 2007). Since Cer and S1P often perform opposite roles in cell processes, the balance of these two lipids is proposed to be a rheostatic switch that determines whether a cell will undergo apoptosis or proliferation (Hannun & Obeid, 2008).

Over the past two decades, Cer, S1P, and other sphingolipid messengers have been attentively studied for their signaling roles. Since SM and GSLs are abundant in the plasma membrane, these lipids would also likely have important functions at that site. In 1972, Singer and Nicolson proposed the "fluid mosaic model" of the plasma membrane (Singer & Nicolson, 1972),

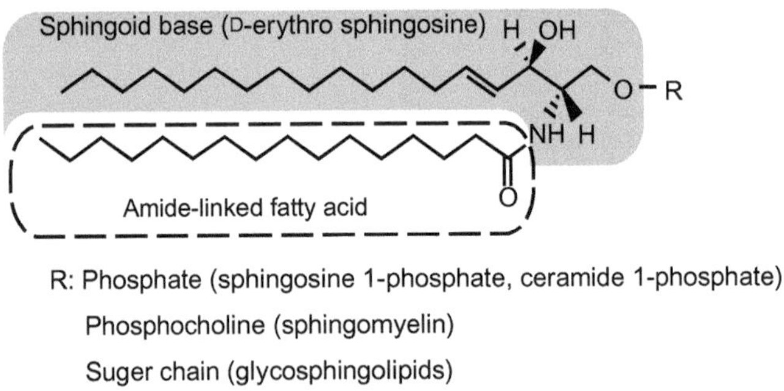

Figure 12.1 General structure of sphingolipids.

which enabled us to understand the composition of the membrane and its resident polar lipids and proteins. However, the model did not completely address the fact that many types of polar lipids exist in the plasma membrane, including glycerolipids, SM, and GSLs. In these lipids, the fatty acid moieties of Cer and diacylglycerol exhibit much variation. Each variation would change the local physical properties of the plasma membrane. In light of this information, Dr Hakomori and Simons proposed a new concept for the model, named "lipid microdomains" or "lipid rafts" (Hakomori, 2010; Simons & Gerl, 2010), in which membranes are embedded with detergent-resistant subdomains having characteristic lipid compositions. Lipid microdomains are enriched with SM, GSLs, and cholesterol, and the acyl chains of the lipids in these subdomains tend to be more rigid and in a less fluid state. Lipid microdomains also contain many proteins, including receptors, signaling molecules, and adhesion molecules, organized into functional complexes and cytoskeletal proteins aligned to form platforms for communication between extracellular and intracellular environments. Lipid microdomains are also thought to be important not only for signal transduction but also in the absorption of nutrients or lipids from the extracellular space (Cheng, Singh, Marks, & Pagano, 2006). In this review, I focus on the functions of sphingolipids in lipid microdomains and their contributions to obesity and diabetes.

2. SPHINGOLIPIDS FUNCTION IN LIPID MICRODOMAINS

2.1. Caveolae and lipid droplets

"Caveolae" are important structures in the function of lipid microdomains. These flask-shaped, small invaginations of the plasma membrane are characterized by specific scaffolding proteins and contain many lipid microdomains (Sonnino & Prinetti, 2009). Caveolae are involved in vesicle-mediated uptake of lipids or nutrients into the cell interior as well as in organizing signal transduction. In a 2005 study by Dr Straalfors' group in intact primary adipocytes, exogenous oleic acid was taken up and directly converted to triglyceride (TG) in the plasma membrane, in a novel subclass of caveolae that specifically contain the protein perilipin (Ortegren, Aboulaich, Ost, & Stralfors, 2007). Surprisingly, biochemically isolated caveolae catalyzed *de novo* TG synthesis from oleic acid and glycerol 3-phosphate (Ost, Ortegren, Gustavsson, Nystrom, & Stralfors, 2005). Caveolae scaffolding proteins, caveolin-1, -2, and -3, were also discovered to be localized in lipid droplets under some circumstances (Fujimoto, Kogo, Ishiguro, Tauchi, &

Nomura, 2001; Martin & Parton, 2005; Ostermeyer et al., 2001). Caveolin-1-deficient mice were lean, resistant to diet-induced obesity, and exhibited hypertriglyceridemia with adipocyte abnormalities (Razani et al., 2002). In this context, caveolae appear to be important metabolic platforms able to control uptake and storage of neutral lipids, with implications of roles in obesity and diabetes.

2.2. Lipid microdomains and the fatty acid transporter CD36/FAT

Dr Stremmel and Ehehalt investigated caveolae and lipid microdomain functions on fatty acid transport across the plasma membrane, especially focusing on the fatty acid transporter CD36/FAT (Ehehalt et al., 2008). They found that CD36/FAT is localized in lipid microdomains and that this localization is important for its activity as a fatty acid transporter (Ehehalt et al., 2006). Interestingly, they also found that a deficiency in caveolin-1 decreased fatty acid transport by CD36/FAT (Ring, Le Lay, Pohl, Verkade, & Stremmel, 2006). Their findings would seem to indicate that caveolae and lipid microdomains are a center for fatty acid uptake by CD36/FAT. However, in their experiments, the authors used methyl-β-cyclodextrin (MβCD) to disrupt lipid microdomains. MβCD is known to remove cholesterol from the plasma membrane, so any contributions by other constituents of lipid microdomains, such as SM or GSLs, would be difficult to investigate. Other experiments, though, using an inhibitor for serine-palmitoyltransferase, myriocin (ISP-1), also suggest the importance of sphingolipids on fatty acid uptake of CD36/FAT in lipid microdomains.

2.3. SM and lipid microdomains

SM is synthesized from phosphatidylcholine and Cer by catalytic action of SM synthase (SMS). The gene encoding SMS was identified in 2004 (Huitema, van den Dikkenberg, Brouwers, & Holthuis, 2004; Yamaoka, Miyaji, Kitano, Umehara, & Okazaki, 2004). SMS has three isoforms, SMS1, SMS2, and SMSr. SMS1 is responsible for bulk of SM synthesis in the Golgi apparatus, whereas SMS2 is located in the plasma membrane. SMSr synthesizes ceramide phosphoethanolamine in the endoplasmic reticulum, controlling Cer homeostasis (Vacaru et al., 2009). To date, most of the studies regarding the cellular functions of lipid microdomains have been performed using MβCD to disrupt lipid microdomains and thereby eliminate cellular cholesterol. The function of SM on lipid microdomains has only been examined using the inhibitors of sphingolipid biosynthesis myriocin (ISP-1) and fumonisin B1. The identification of the SMSs and the responsible gene opened the door to investigating SM function in lipid microdomains on a genetic level.

Dr Umehara's group established a SM-knockdown cell, which is deficient in the membrane expression of SM, and demonstrated that translocation of the T-cell receptor (TCR), ZAP-70, and PKCθ into lipid microdomains was markedly decreased in this cell. They concluded that membrane SM is crucial for TCR signal transduction, and subsequent T-cell activation, through lipid raft function (Jin et al., 2008). They also reported that cisplatin combined with an anti-FAS antibody was not able to induce apoptosis in SM-deficient cells, but was able to do in SMS-restored cells, indicating that SM plays an important role in the formation of FAS-associated signaling complexes (Lafont et al., 2010). In addition, Dr Okazaki's group demonstrated that SM synthesis directs the transferrin-receptor complex to undergo clathrin-dependent endocytosis and recycling, thereby promoting the proliferation of lymphoma cells (Shakor et al., 2011). SMS2 is expressed less in leukocytic cells, so deletion of SMS1 would lead to almost complete depletion of SM from lipid microdomains.

The experiments discussed earlier demonstrate that SM synthesis in the Golgi apparatus or the certain amount of SM is necessary for the formation and function of lipid microdomains. Further, we found that SMS2 is important for the regulation of lipid microdomains (Mitsutake et al., 2011). SMS2 has been identified as a homologue of SMS1 and was found to be localized in the plasma membrane and Golgi apparatus. However, whether SMS2 has any activity on the plasma membrane or any function in lipid microdomains remains to be determined (Milhas, Clarke, & Hannun, 2010). Dr Tafesse et al. investigated the differences between SMS1 and SMS2 expression patterns and their location of activity using a combination of overexpression and siRNA experiments in HeLa cells. They determined that SMS2 was able to catalyze the conversion of the substrate NBD-Cer into NBD-SM on the cell surface (Tafesse et al., 2007). They also demonstrated that siRNA-mediated depletion of either SMS1 or SMS2 reduced SM content and blocked cell growth, yet the addition of exogenous SM did not restore cell growth. The authors concluded that the biological role of SMS would beyond formation of SM. Their results strongly indicate that SMS is not only involved in SM synthesis but also in the regulation of lipid microdomains that contribute to cell growth. Recently, we uncovered additional information regarding SMS2 function (Mitsutake et al., 2011). To completely distinguish SMS2 function from SMS1 function, we established SM-null cells from an embryonic fibroblast of SMS1 and SMS2 double knockout mice, naming the clone “ZS2.” We stably expressed SMS1 or SMS2 in ZS2 cells (ZS2/SMS1 or ZS2/SMS2, respectively) (Fig. 12.2). ZS2 cells

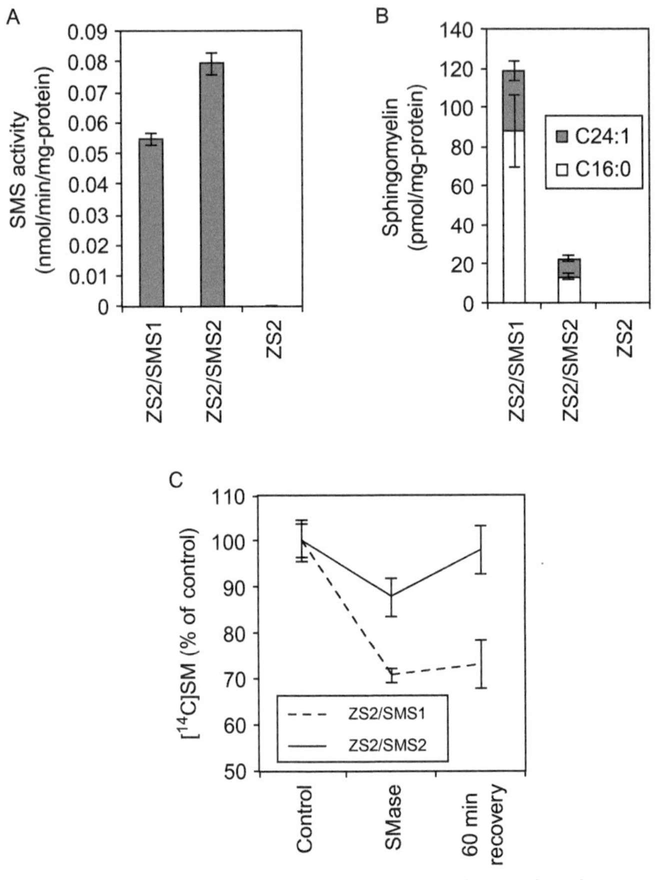

Figure 12.2 SMS2 can convert ceramide to sphingomyelin on the plasma membrane. We isolated SM-null fibroblasts from the embryos of SMS1 and SMS2 double knockout mice and named the cell line “ZS2.” We stably expressed SMS1 and SMS2 in ZS2 cells to generate reconstituted cells, ZS2/SMS1 and ZS2/SMS2, as described previously (Mitsutake et al., 2011). *In vitro* sphingomyelin synthase activity (A) and SM levels (B) were determinant using ZS2, ZS2/SMS1, and ZS2/SMS2 cells. (C) ZS2/SMS1 and ZS2/SMS2 cells were prelabeled with [^{14}C]choline, and bacterial sphingomyelinase was added to generate ceramide on the plasma membrane, then recovery of SM was monitored. The experiments contained dynasore, a dynamin II inhibitor, to inhibit ceramide recycling. The details of all experiments were described in our previous paper (Mitsutake et al., 2011). *Part of this figure was originally published in Mitsutake et al. (2011) © 2011. The American Society for Biochemistry and Molecular Biology.*

have no SMS activity or SM indicating that only SMS1 and SMS2 are synthases for SM, at least in mouse fibroblasts (Fig. 12.2A). Interestingly, ZS2/SMS2 cells exhibited significant SMS activity *in vitro*, whereas ZS2/SMS2 cells had only one-sixth the amount of SM comparatively (Fig. 12.2A and B). We examined whether SMS2 could convert extracellular NBD-Cer in these cells and confirmed that SMS2 acts on the outer leaflet of the plasma membrane; the results agreed with Tafesse et al. (2007). Additionally, we examined whether the Cer produced on the outer leaflet of the plasma membrane could be converted into SM by SMS2, by employing bacterial sphingomyelinase (bSMase) and inhibiting the endocytosis for recycling the Cer produced using the dynamin II inhibitor, dynasore. SM levels in the ZS2/SMS2 cells had completely recovered at 60 min after removing the bSMase, but the SM levels in the ZS2/SMS1 cells showed only a moderate recovery (Fig. 12.2C). Interestingly, the reduction in SM levels following bSMase treatment was milder in ZS2/SMS2 cells than in ZS2/SMS1 cells. Thus, the Cer produced was quickly converted to SM by SMS2, which strongly suggests that SMS2 would be able to convert Cer produced at the outer leaflets of the plasma membrane into SM. Further, we found that SMS2 is localized in lipid microdomains, where it interacts with the fatty acid transporter CD36/FAT and caveolin-1 to regulate "caveolae-dependent endocytosis." Thus, SMS2 is able to regulate the dynamic structure of SM-rich lipid microdomains on the plasma membrane.

In the experiments described earlier, we used bSMase to produce Cer on the plasma membrane. The question remains whether Cer is produced on the plasma membrane in physiological conditions. Recent studies have revealed that wounded cells secrete lysosomal acid SMase (ASMase) to promote endocytosis and plasma membrane repair (Tam et al., 2010). It is also known that exogenously added bSMase induces ATP-independent endocytosis (Zha et al., 1998). These studies imply that conversion of SM to Cer might cause a structural change in the plasma membrane and thereby facilitates endocytosis. Both the conversion of SM into Cer, and of Cer into SM, would be important processes for endocytosis, and a change in lipid composition might facilitate these processes. ASMase reportedly is localized in lipid microdomains (Opreanu et al., 2011), so both ASMase and SMS2 might be regulators of structure and function in lipid microdomains. There is growing evidences that impaired structure and function of lipid microdomains might cause obesity, diabetes, and metabolic syndrome (MetS) as described in Section 3.

3. SPHINGOLIPID AND OBESITY

3.1. *In vivo* experiments for atherosclerosis using an inhibitor of sphingolipid biosynthesis

MetS is associated with a clustering of risk factors that include high blood pressure, obesity, glucose intolerance, and dyslipidemia; it is also associated with an increased risk of type 2 diabetes and cardiovascular disease. MetS is known to increase with the level of obesity, which is reaching epidemic proportions in some countries. Thus, it is becoming increasingly important to understand the relationships among MetS, obesity, type 2 diabetes, and cardiovascular disease.

Initial *in vivo* studies of sphingolipids in cardiovascular disease were reported by the laboratory of Dr Jiang, who employed the serine-palmitoyltransferase inhibitor myriocin (ISP-1) in an atherosclerosis animal model of apoE-deficient mice (Fig. 12.3). The myriocin treatment caused significant decreases in plasma levels of SM, Cer, and S1P to 54%, 32%,

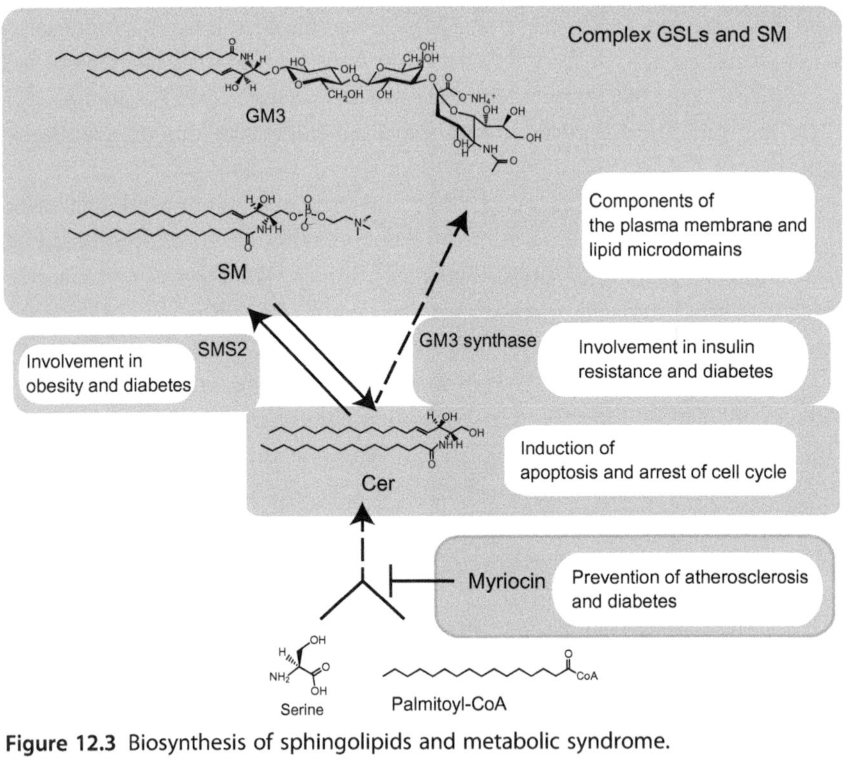

Figure 12.3 Biosynthesis of sphingolipids and metabolic syndrome.

and 73%, respectively, and also caused a 42% decrease in the average area of atherosclerotic lesions (Hojjati et al., 2005). However, the myriocin treatment induced no change in plasma cholesterol or TGs. This is notable, as many inhibitors of cholesterol biosynthesis are in fact used as drugs for hyperlipidemia and MetS. The findings of Jiang et al. suggest that the inhibitor of sphingolipid biosynthesis myriocin might be a novel drug acting through a completely new mode of action. Their study garnered much attentions and inspired studies by several other groups (Glaros et al., 2007; Park, Rosebury, Kindt, Kowala, & Panek, 2008). However, since myriocin exerts broad effects on sphingolipid biosynthesis, exactly which sphingolipids are important for obesity or diabetes could not be determined.

As described earlier, SM and GSLs contribute to the formation of lipid microdomains, together with cholesterol. Caveolae, specialized lipid microdomains, are known to be important for insulin signaling and for controlling plasma glucose levels (Sonnino & Prinetti, 2009). The function of lipid microdomains in obesity and diabetes appear to be important.

3.2. The *in vivo* experiments for obesity and diabetes using a mouse deficient in a sphingolipid biosynthesis-related gene

A recent study using gene-targeted mice revealed a correlation between sphingolipids and diabetes. In 2003, the laboratory of Dr Proia described that mice deficient in the gene encoding ganglioside GM3-synthase exhibited resistance to high-fat diet-induced insulin resistance (Fig. 12.3; Yamashita et al., 2003). Interestingly, higher levels of phosphorylation on the insulin receptor were found in the skeletal muscle of the knockout mice. Dr Inokuchi's group performed more detail analysis of the correlation between GM3 and the insulin receptor on the plasma membrane (Kabayama et al., 2007). They found that excessive amounts of GM3 were bound to the insulin receptor at "noncaveolae lipid microdomains," so they removed the insulin receptors from caveolae and found that this loss impaired insulin signaling. Their study added new insight into the field of diabetes.

As described in Section 1, SM is the most abundant sphingolipid found in the plasma membrane, and, together with cholesterol and GSLs, it is important for the formation of lipid microdomains. SM is essential for the growth of cells (Hanada et al., 1992), and a deficiency in the enzyme responsible for the production of the majority of cellular SM, SMS1, results in a severe phenotype that includes reduced body weight, loss of fat tissue, moderate male infertility, and impaired pancreatic β-cells function (Yano et al., 2011). However, SMS2-deficient mice are seemingly healthy and display almost no apparent abnormalities, and so these mice are used in experiments for obesity and atherosclerosis. The group of Dr Jiang transplanted SMS2-knockout mouse bone marrow into LDL-receptor knockout mice to create a mouse model exhibiting SMS2-deficiency in macrophages. After 3 months on a western-type diet, the mice showed a 57% decrease in the average area of atherosclerotic lesions in the aortic arch root (Liu et al., 2009). In addition, we performed diet-induced obesity tests using SMS2-deficient mice, and demonstrated that SMS-deficiency prevents high-fat diet-induced body weight gain, obesity, and insulin resistance (Fig. 12.4; Mitsutake et al., 2011). Interestingly, in the liver of SMS2-knockout mice, almost no large and mature lipid droplets were observed. As described in the previous section, we found that SMS2 binds to caveolin-1 and CD36/FAT in lipid microdomains and that caveolae and lipid microdomains are important for fatty acid uptake and lipid droplet formation. Indeed, treatment with

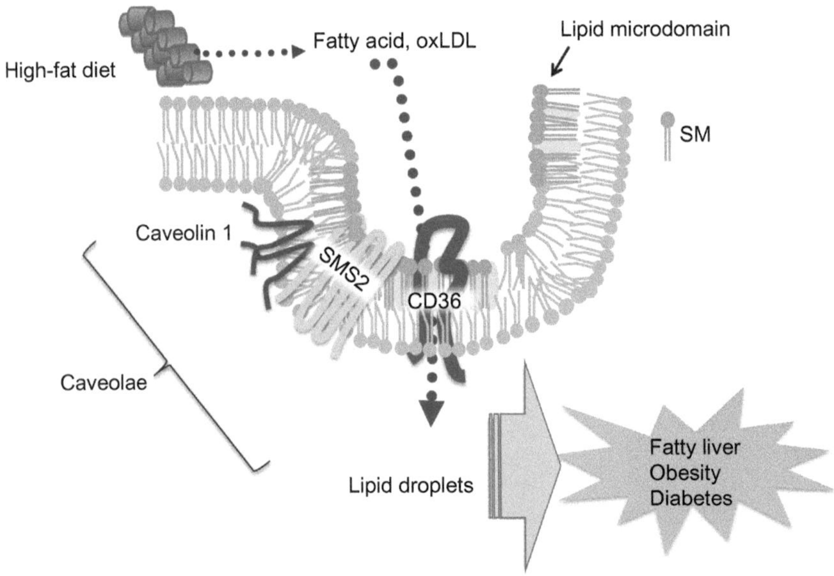

Figure 12.4 SMS2 could be a novel regulator of lipid microdomain structure and function. After a high-fat diet, increased fatty acid or oxidized low density lipoprotein (oxLDL) would be incorporated into the cells via CD36/FAT, resulting in triglyceride (TG) accumulation in lipid droplets. SMS2 is localized in lipid microdomains, where it associates with CD36/FAT and caveolin-1. Thus, SMS2 could be a novel regulator of lipid microdomain structure and function; such regulation would be crucial for obesity and diabetes. *Part of this figure was originally published in Mitsutake et al. (2011) © 2011. The American Society for Biochemistry and Molecular Biology.* (See Color Insert.)

siRNA for SMS2 also decreased the number of large lipid droplets in HepG2 cells. Further, the siRNA for SMS2 decreased the accumulation of TGs in the livers of leptin-deficient (*ob/ob*) mice. Recently, the group of Dr Jiang also performed DIO-tests and obtained results similar to ours (Li et al., 2011).

4. CONCLUSIONS AND FUTURE DIRECTIONS

At this time, lipid microdomains are thought to be formed at the Golgi apparatus and to have function as sorting platforms for targeted transport of transmembrane and glycosylphosphatidylinositol-anchored proteins. Thus, the functional regulation of these microdomains is thought to exclusively account for the modification of proteins such as their phosphorylation, lipidation, or ubiquitinations. In the case of GSL-rich lipid microdomains, since most of the GSL-synthase and GLS-degrading enzymes are located in Golgi apparatus and lysosomes, respectively, lipid microdomains would be stable on plasma membrane. However, SM-rich lipid microdomains could be dynamically modulated by the combination of ASMase and SMS2. The conversion of SM into Cer, and its reverse reaction, would drive conformational changes in the plasma membrane, facilitating the uptake of extracellular fatty acids or nutrient factors. Such regulation in lipid microdomains was shown to be involved in lipid droplet formation, obesity, diabetes, and atherosclerosis, *in vivo*. Recent studies demonstrated that the biosynthesis or dynamic regulation of sphingolipids has important roles in the functions of lipid microdomains and caveolae. The caveolin-1 knockout mouse exhibited impaired caveolae structure and "lean" and high levels of fatty acid in plasma (Razani et al., 2002). Caveolae and lipid microdomains are emerging as important regulators of cell metabolism. Thus, complete regulation of sphingolipid function might lead to novel types of pharmaceuticals for obesity and type 2 diabetes, which would act to modulate lipid microdomain function and regulate protein functions in lipid microdomains.

REFERENCES

Cheng, Z. J., Singh, R. D., Marks, D. L., & Pagano, R. E. (2006). Membrane microdomains, caveolae, and caveolar endocytosis of sphingolipids. *Molecular Membrane Biology, 23*, 101–110.

Ehehalt, R., Fullekrug, J., Pohl, J., Ring, A., Herrmann, T., & Stremmel, W. (2006). Translocation of long chain fatty acids across the plasma membrane—Lipid rafts and fatty acid transport proteins. *Molecular and Cellular Biochemistry, 284*, 135–140.

Ehehalt, R., Sparla, R., Kulaksiz, H., Herrmann, T., Fullekrug, J., & Stremmel, W. (2008). Uptake of long chain fatty acids is regulated by dynamic interaction of FAT/CD36 with cholesterol/sphingolipid enriched microdomains (lipid rafts). *BMC Cell Biology, 9*, 45.

Fujimoto, T., Kogo, H., Ishiguro, K., Tauchi, K., & Nomura, R. (2001). Caveolin-2 is targeted to lipid droplets, a new "membrane domain" in the cell. *The Journal of Cell Biology, 152*, 1079–1085.

Glaros, E. N., Kim, W. S., Wu, B. J., Suarna, C., Quinn, C. M., Rye, K. A., et al. (2007). Inhibition of atherosclerosis by the serine palmitoyl transferase inhibitor myriocin is associated with reduced plasma glycosphingolipid concentration. *Biochemical Pharmacology, 73*, 1340–1346.

Hakomori, S. I. (2010). Glycosynaptic microdomains controlling tumor cell phenotype through alteration of cell growth, adhesion, and motility. *FEBS Letters, 584*, 1901–1906.

Hanada, K., Nishijima, M., Kiso, M., Hasegawa, A., Fujita, S., Ogawa, T., et al. (1992). Sphingolipids are essential for the growth of Chinese hamster ovary cells. Restoration of the growth of a mutant defective in sphingoid base biosynthesis by exogenous sphingolipids. *The Journal of Biological Chemistry, 267*, 23527–23533.

Hannun, Y. A., & Obeid, L. M. (1995). Ceramide: An intracellular signal for apoptosis. *Trends in Biochemical Sciences, 20*, 73–77.

Hannun, Y. A., & Obeid, L. M. (2008). Principles of bioactive lipid signalling: Lessons from sphingolipids. *Nature Reviews. Molecular Cell Biology, 9*, 139–150.

Hirabayashi, Y., Igarashi, Y., & Merrill, A. H. (2006). *Sphingolipid biology*. Tokyo: Springer.

Hojjati, M. R., Li, Z., Zhou, H., Tang, S., Huan, C., Ooi, E., et al. (2005). Effect of myriocin on plasma sphingolipid metabolism and atherosclerosis in apoE-deficient mice. *The Journal of Biological Chemistry, 280*, 10284–10289.

Huitema, K., van den Dikkenberg, J., Brouwers, J. F., & Holthuis, J. C. (2004). Identification of a family of animal sphingomyelin synthases. *The EMBO Journal, 23*, 33–44.

Igarashi, Y. (1997). Functional roles of sphingosine, sphingosine 1-phosphate, and methylsphingosines: In regard to membrane sphingolipid signaling pathways. *Journal of Biochemistry, 122*, 1080–1087.

Jin, Z. X., Huang, C. R., Dong, L., Goda, S., Kawanami, T., Sawaki, T., et al. (2008). Impaired TCR signaling through dysfunction of lipid rafts in sphingomyelin synthase 1 (SMS1)-knockdown T cells. *International Immunology, 20*, 1427–1437.

Kabayama, K., Sato, T., Saito, K., Loberto, N., Prinetti, A., Sonnino, S., et al. (2007). Dissociation of the insulin receptor and caveolin-1 complex by ganglioside GM3 in the state of insulin resistance. *Proceedings of the National Academy of Sciences of the United States of America, 104*, 13678–13683.

Kihara, A., Mitsutake, S., Mizutani, Y., & Igarashi, Y. (2007). Metabolism and biological functions of two phosphorylated sphingolipids, sphingosine 1-phosphate and ceramide 1-phosphate. *Progress in Lipid Research, 46*, 126–144.

Lafont, E., Milhas, D., Carpentier, S., Garcia, V., Jin, Z. X., Umehara, H., et al. (2010). Caspase-mediated inhibition of sphingomyelin synthesis is involved in FasL-triggered cell death. *Cell Death and Differentiation, 17*, 642–654.

Li, Z., Zhang, H., Liu, J., Liang, C. P., Li, Y., Teitelman, G., et al. (2011). Reducing plasma membrane sphingomyelin increases insulin sensitivity. *Molecular and Cellular Biology, 31*, 4205–4218.

Liu, J., Huan, C., Chakraborty, M., Zhang, H., Lu, D., Kuo, M. S., et al. (2009). Macrophage sphingomyelin synthase 2 deficiency decreases atherosclerosis in mice. *Circulation Research, 105*, 295–303.

Martin, S., & Parton, R. G. (2005). Caveolin, cholesterol, and lipid bodies. *Seminars in Cell & Developmental Biology, 16*, 163–174.

Milhas, D., Clarke, C. J., & Hannun, Y. A. (2010). Sphingomyelin metabolism at the plasma membrane: Implications for bioactive sphingolipids. *FEBS Letters, 584*, 1887–1894.

Mitsutake, S., Zama, K., Yokota, H., Yoshida, T., Tanaka, M., Mitsui, M., et al. (2011). Dynamic modification of sphingomyelin in lipid microdomains controls development of obesity, fatty liver, and type 2 diabetes. *The Journal of Biological Chemistry, 286*, 28544–28555.

Opreanu, M., Tikhonenko, M., Bozack, S., Lydic, T. A., Reid, G. E., McSorley, K. M., et al. (2011). The unconventional role of acid sphingomyelinase in regulation of retinal microangiopathy in diabetic human and animal models. *Diabetes, 60*, 2370–2378.

Ortegren, U., Aboulaich, N., Ost, A., & Stralfors, P. (2007). A new role for caveolae as metabolic platforms. *Trends in Endocrinology and Metabolism, 18*, 344–349.

Ost, A., Ortegren, U., Gustavsson, J., Nystrom, F. H., & Stralfors, P. (2005). Triacylglycerol is synthesized in a specific subclass of caveolae in primary adipocytes. *The Journal of Biological Chemistry, 280*, 5–8.

Ostermeyer, A. G., Paci, J. M., Zeng, Y., Lublin, D. M., Munro, S., & Brown, D. A. (2001). Accumulation of caveolin in the endoplasmic reticulum redirects the protein to lipid storage droplets. *The Journal of Cell Biology, 152*, 1071–1078.

Park, T. S., Rosebury, W., Kindt, E. K., Kowala, M. C., & Panek, R. L. (2008). Serine palmitoyltransferase inhibitor myriocin induces the regression of atherosclerotic plaques in hyperlipidemic ApoE-deficient mice. *Pharmacological Research, 58*, 45–51.

Razani, B., Combs, T. P., Wang, X. B., Frank, P. G., Park, D. S., Russell, R. G., et al. (2002). Caveolin-1-deficient mice are lean, resistant to diet-induced obesity, and show hypertriglyceridemia with adipocyte abnormalities. *The Journal of Biological Chemistry, 277*, 8635–8647.

Ring, A., Le Lay, S., Pohl, J., Verkade, P., & Stremmel, W. (2006). Caveolin-1 is required for fatty acid translocase (FAT/CD36) localization and function at the plasma membrane of mouse embryonic fibroblasts. *Biochimica et Biophysica Acta, 1761*, 416–423.

Shakor, A. B., Taniguchi, M., Kitatani, K., Hashimoto, M., Asano, S., Hayashi, A., et al. (2011). Sphingomyelin synthase 1-generated sphingomyelin plays an important role in transferrin trafficking and cell proliferation. *The Journal of Biological Chemistry, 286*, 36053–36062.

Simons, K., & Gerl, M. J. (2010). Revitalizing membrane rafts: New tools and insights. *Nature Reviews. Molecular Cell Biology, 11*, 688–699.

Singer, S. J., & Nicolson, G. L. (1972). The fluid mosaic model of the structure of cell membranes. *Science, 175*, 720–731.

Sonnino, S., & Prinetti, A. (2009). Sphingolipids and membrane environments for caveolin. *FEBS Letters, 583*, 597–606.

Tafesse, F. G., Huitema, K., Hermansson, M., van der Poel, S., van den Dikkenberg, J., Uphoff, A., et al. (2007). Both sphingomyelin synthases SMS1 and SMS2 are required for sphingomyelin homeostasis and growth in human HeLa cells. *The Journal of Biological Chemistry, 282*, 17537–17547.

Tam, C., Idone, V., Devlin, C., Fernandes, M. C., Flannery, A., He, X., et al. (2010). Exocytosis of acid sphingomyelinase by wounded cells promotes endocytosis and plasma membrane repair. *The Journal of Cell Biology, 189*, 1027–1038.

Vacaru, A. M., Tafesse, F. G., Ternes, P., Kondylis, V., Hermansson, M., Brouwers, J. F., et al. (2009). Sphingomyelin synthase-related protein SMSr controls ceramide homeostasis in the ER. *The Journal of Cell Biology, 185*, 1013–1027.

Yamaoka, S., Miyaji, M., Kitano, T., Umehara, H., & Okazaki, T. (2004). Expression cloning of a human cDNA restoring sphingomyelin synthesis and cell growth in sphingomyelin synthase-defective lymphoid cells. *The Journal of Biological Chemistry, 279*, 18688–18693.

Yamashita, T., Hashiramoto, A., Haluzik, M., Mizukami, H., Beck, S., Norton, A., et al. (2003). Enhanced insulin sensitivity in mice lacking ganglioside GM3. *Proceedings of the National Academy of Sciences of the United States of America, 100*, 3445–3449.

Yano, M., Watanabe, K., Yamamoto, T., Ikeda, K., Senokuchi, T., Lu, M., et al. (2011). Mitochondrial dysfunction and increased reactive oxygen species impair insulin secretion in sphingomyelin synthase 1-null mice. *The Journal of Biological Chemistry*, *286*, 3992–4002.

Zha, X., Pierini, L. M., Leopold, P. L., Skiba, P. J., Tabas, I., & Maxfield, F. R. (1998). Sphingomyelinase treatment induces ATP-independent endocytosis. *The Journal of Cell Biology*, *140*, 39–47.

CHAPTER THIRTEEN

Ghrelin At the Interface of Obesity and Reward

Harriët Schellekens[*,¶], **Timothy G. Dinan**[*,§,†], **John F. Cryan**[*,§,‡,1]
[*]Food for Health Ireland, University College Cork, Cork, Ireland
[§]Laboratory of Neurogastroenterology, Alimentary Pharmabiotic Centre, University College Cork, Cork, Ireland
[†]Department of Psychiatry, University College Cork, Cork, Ireland
[‡]Department of Anatomy and Neuroscience, University College Cork, Cork, Ireland
[¶]School of Pharmacy, University College Cork, Cork, Ireland
[1]Corresponding author: e-mail address: j.cryan@ucc.ie

Contents

Abstract

The prevalence of obesity continues to increase and has reached epidemic proportions. Accumulating data over the past few decades have given us key insights and broadened our understanding of the peripheral and central regulation of energy homeostasis. Despite this, the currently available pharmacological treatments, reducing body weight, remain limited due to poor efficacy and side effects. The gastric peptide ghrelin has been identified as the only orexigenic hormone from the periphery to act in the hypothalamus to stimulate food intake. Recently, a role for ghrelin and its receptor at the interface between homeostatic control of appetite and reward circuitries modulating the hedonic aspects of food has also emerged. Nonhomeostatic factors such as the rewarding and motivational value of food, which increase with food palatability and caloric content, can override homeostatic control of food intake. This nonhomeostatic decision to eat leads to overconsumption beyond nutritional needs and is being

Vitamins and Hormones, Volume 91
ISSN 0083-6729
http://dx.doi.org/10.1016/B978-0-12-407766-9.00013-4

recognized as a key component in the underlying causes for the increase in obesity incidence worldwide. In addition, the hedonic feeding behavior has been linked to food addiction and an important role for ghrelin in the development of addiction has been suggested. Moreover, plasma ghrelin levels are responsive to conditions of stress, and recent evidence has implicated ghrelin in stress-induced food-reward behavior. The prominent role of the ghrelinergic system in the regulation of feeding gives rise to it as an effective target for the development of successful antiobesity pharmacotherapies that not only affect satiety but also selectively modulate the rewarding properties of food and reduce the desire to eat.

1. INTRODUCTION

The obesity epidemic is gaining momentum worldwide and poses an increasingly heavy burden on health and associated health care costs (Bloom et al., 2008). Obesity or being overweight both arise from an excess accumulation of body fat or adiposity, brought about following excess consumption of calorie dense and palatable foods (i.e., hyperphagia), more often than not combined with a decrease in energy expenditure due to physical inactivity, leading to an imbalance in overall energy homeostasis (Chakrabarti, 2009). Obesity is often associated with comorbid diseases within the framework of what is called the metabolic syndrome, including atherosclerosis, hypertension, insulin resistance or diabetes mellitus type 2, glucose intolerance, dyslipidemia, and a general proinflammatory phenotype. Together, obesity and the additional metabolic syndrome-associated complications have a significant adverse effect on health and decrease life expectancy (Cheng & Leiter, 2006; Mikhail, 2009). A body mass index (BMI), comparing body weight and height, of between 25 and 30 kg/m^2 defines being overweight, while obesity is associated with a BMI of greater than 30 kg/m^2 (Eknoyan, 2008). Even though, obesity can originate following physical inactivity and the subsequent decreases in energy expenditure, obese individuals generally tend to display an increased metabolic rate compared to their lean counterparts due to the energy required to maintain a greater body mass. In addition, genetic predisposition can increase the likelihood for the development of obesity but genetic factors cannot account for all variation in BMI. Interestingly, obese gut microbiota was shown to have an increased capacity to harvest energy from dietary nutrients leading to increased body fat mass when transferred to germ-free mice and the gut microbiome therefore may represents an additional factor contributing to the pathophysiology of obesity (Turnbaugh et al., 2006). However, the abnormal high accumulation of adiposity in most obese individuals is mainly lifestyle dependent and due to dietary habits (i.e., calorie overconsumption), low physical activity, and excessive storage of energy (Marti, Moreno-Aliaga, Hebebrand, & Martinez, 2004). This increase in food intake is not driven by a nutritional need but mediated by the individual's natural sensitivity to food stimuli and the pleasurable feelings associated with eating of calorie-dense foods. This nonhomeostatic feeding component is increasingly being recognized as an important underlying cause for the increase in the incidence of obesity worldwide (Berthoud, 2006).

The ghrelinergic system targets both the peripheral and central nervous system regulating a plethora of biological activities, including appetite and food intake. Primarily, ghrelin has been strongly implicated in the regulation of homeostatic energy metabolism due to its orexigenic effects and has thus received much attention as an antiobesity drug target. However, it is now becoming clear that the orexigenic ghrelin and its receptor, the growth hormone secretagogue (GHS-R1a) receptor, play a pivotal role in a wide range of food intake behaviors, including behavioral reward aspects of food intake and in the neurobiology of addiction. In addition, accumulating data suggest involvement of the ghrelin system in stress-induced food intake (Chuang et al., 2011; Diz-Chaves, 2011). Thus, ghrelin not only regulates the homeostatic regulation of energy metabolism but also has an impact on the hedonic rewarding and motivational pathways modulating food intake (for review, see Dickson et al., 2011; Egecioglu et al., 2011; Skibicka & Dickson, 2011). This chapter will describe the homeostatic ghrelin-mediated appetite signaling as well as recent advances in the understanding of ghrelin's role in the nonhomeostatic rewarding aspect of feeding and the potential obligatory role for ghrelin in stress-induced eating behavior. In addition, the concept of food addiction has been postulated to contribute to the obesity epidemic and will be discussed. Ultimately, this field of research may contribute to the future development of novel therapeutics, specifically targeting the ghrelin-mediated reinforcing effects of palatable foods and drugs of abuse. Ghrelin-specific drug, selectively reducing the rewarding properties of food may be identified through behavioral analysis and may maximize the antiobesity potential of pharmaceutical therapies in reducing the desire to eat.

2. GHRELIN IN OBESITY

2.1. Ghrelin and ghrelin receptor expression and signaling

The ghrelin peptide, discovered in 1999, is a 28-amino acid peptide, primarily synthesized by gastric endocrine cells (Kojima et al., 1999). Human ghrelin is encoded by the ghrelin gene (*ghrl*) on chromosome 3p25–26 (Fig. 13.1A),

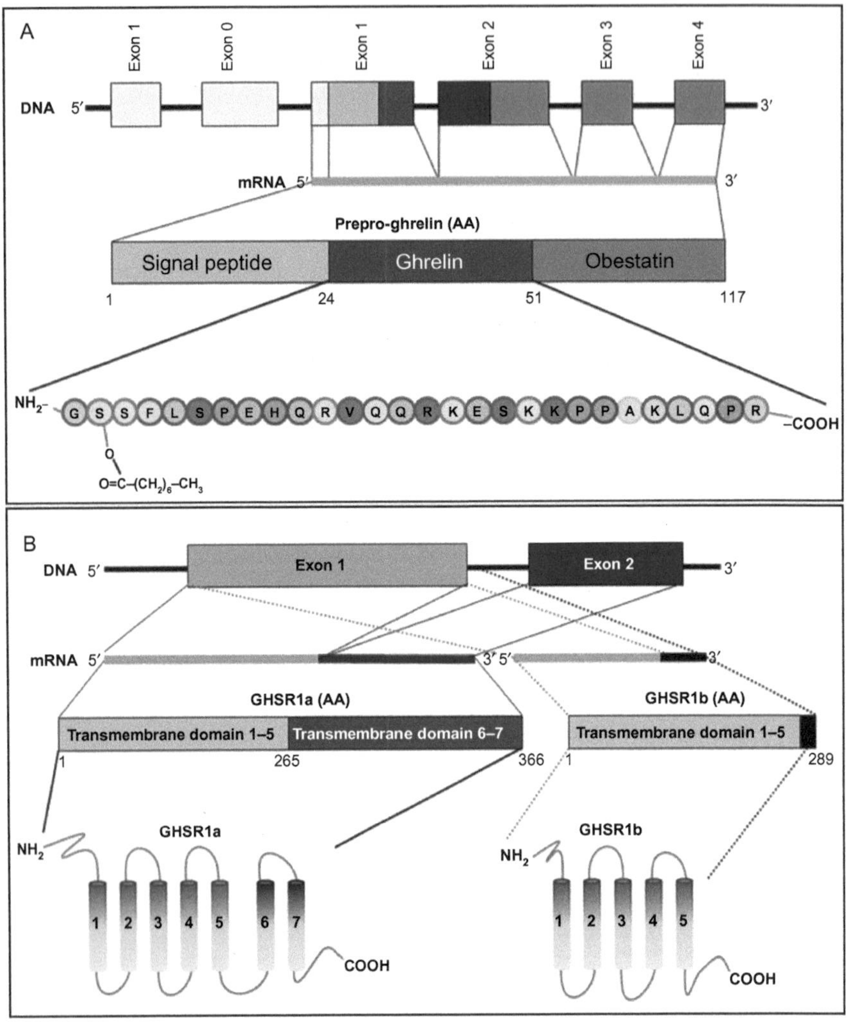

Figure 13.1 Human ghrelin and growth secretagogue receptor gene structure. (A) The human ghrelin gene, located on chromosome 3p25–26, is composed of six exons (boxes) and five introns (lines), including a 20-bp noncoding first exon (exon 0). After transcription, the mRNA is translated into the 117 amino acid preproghrelin, which is enzymatically cleaved into proghrelin and a C-terminal polypeptide. The latter C-peptide is processed into obestatin, a peptide with anorexigenic biological activities opposing those of ghrelin (Zhang et al., 2005). Proghrelin is finally processed into mature ghrelin following cleavage of the signal peptide and posttranslational addition of the unique octanoyl group on Ser 3, crucial for biological activity. (B) The human full-length GHS-R1a is located on chromosome 3p26.2 and processed from mRNA encoded from exons 1 and 2 after splicing out intronic sequence. Alternative splicing yields GHS-R1b mRNA encoded by exon 1 and the intron and an alternative polyadenylation site. *Figure taken from Schellekens, Dinan, and Cryan (2009).*

spanning 7.2 kb of genomic DNA consisting of six exons (Seim, Collet, Herington, & Chopin, 2007). Recent developments have shown that alternative splicing and posttranslational modification can yield a variety of other bioactive molecules besides ghrelin, des-acyl ghrelin, and obestatin (Gahete et al., 2011; Hosoda, Kojima, Matsuo, & Kangawa, 2000; Jeffery et al., 2005; Kineman, Gahete, & Luque, 2007; Seim et al., 2007; Seim, Josh, Cunningham, Herington, & Chopin, 2011; Soares & Leite-Moreira, 2008; Tanaka, Hayashida, Nakao, Nakai, & Nakashima, 2001). The mature ghrelin peptide is derived following enzymatic cleavage from preproghrelin after which the enzyme, *O*-acetyltransferase (GOAT), activates the peptide via *n*-octanoylation on the serine 3 residue yielding acyl-ghrelin (Gualillo, Lago, & Dieguez, 2008; Gutierrez et al., 2008; Yang, Brown, Liang, Grishin, & Goldstein, 2008; Yang, Zhao, Goldstein, & Brown, 2008). However, other types of esterification have also been observed to yield active ghrelin (Hosoda, Kojima, Mizushima, Shimizu, & Kangawa, 2003; Ohgusu et al., 2009). Both the acylated and unacylated ghrelin isoforms are found in the blood circulation but only the activated ghrelin was shown to function as the endogenous ligand for the growth hormone secretagogue (GHS-R) receptor (Bednarek et al., 2000). However, unacylated ghrelin is the most abundant form in the circulation and has been suggested to be the active ligand for additional, yet unidentified, GHS-R subtypes. The GHS-R was initially designated as an orphan receptor specific for the peptidyl synthetic growth hormone secretagogues (GHS), herexalin, GHRP6, GHRP1, GHRP2, and the nonpeptidyl ligand, MK-0677 (Howard et al., 1996; Muccioli, Baragli, Granata, Papotti, & Ghigo, 2007; Muccioli et al., 2002), which was shown to stimulate the release of growth hormone (GH) from the pituitary. The endogenous ligand, ghrelin in its acylated form, was identified following reversed pharmacology (Kojima et al., 1999) and the human GHS was designated as the ghrelin receptor (Davenport et al., 2005). The ghrelin receptor (Fig. 13.1B) is expressed as two variants from the same gene (*ghsr*) on chromosome 3p26.2, the biological functional GHS-R1a, and its splice variant GHS-R1b (for review, see Schellekens et al., 2009). The fully functional GHS-R1a is a G-protein-coupled receptor (GPCR) spanning 366 amino acids of approximately 41 kDa with the seven-transmembrane domain structure, typical for GPCRs. The GHS-R1b receptor is a truncated polypeptide of 289 amino acids with only five-transmembrane domains and does not bind ghrelin. The GHS-R1b receptor alone displays no functional activity but attenuates GHS-R1a signaling when expressed as a heterodimer (Chan & Cheng, 2004).

GPCRs, like the GHS-R1a receptor, signal via coupling to trimeric G-proteins which mediates their subsequent downstream signaling following receptor activation (Luttrell, 2008). The ghrelin receptor couples mainly to Gαq/11 following ligand binding and activates phospholipase C, which triggers the inositol phosphate (IP) cascade, generating inositol 1,4,5-triphosphate (IP3) and diacylglycerol (DAG). Next, IP3 triggers the release of Ca^{2+} from intracellular stores and DAG activates PKC (Eglen, Bosse, & Reisine, 2007; Luttrell, 2008). The initial influx of intracellular Ca^{2+} activates voltage-independent K^+ channels, leading to an increase in K^+ and a subsequent hyperpolarization of the cells (Camina, 2006). In addition, the ghrelin-mediated initial Ca^{2+} influx is followed by a sustained Ca^{2+} influx through L-type voltage-sensitive Ca^{2+} channels during ghrelin-induced GH release from somatotropes (Malagon et al., 2003). Moreover, in neurons of the arcuate (ARC) nucleus of the hypothalamus, ghrelin induces neuropeptide Y (NPY) release, which is mediated through N-type voltage-gated Ca^{2+} channels (Kohno, Gao, Muroya, Kikuyama, & Yada, 2003). In addition, ghrelin activates voltage-dependent K^+ channels in pancreatic island cells (Dezaki, Kakei, & Yada, 2007). It is hypothesized that ghrelin signaling can also be mediated via differential tissue-specific G-protein coupling (Fig. 13.2). In pancreatic islet β-cells, the ghrelin receptor-mediated suppression of glucose-induced Ca^{2+} signaling is mediated via coupling to Gαi2, which results in attenuated insulin release (Dezaki et al., 2007). Moreover, the GHS-R1a-mediated Ca^{2+} mobilization in NPY neurons is proposed to be achieved via the Gs-cAMP-protein kinase A (PKA) signaling pathway (Kohno et al., 2003). The PKA dependency was demonstrated by the attenuation of the ghrelin-mediated calcium mobilization following pharmacological PKA blockade (Kohno et al., 2003).

The ghrelinergic system is ubiquitously expressed both centrally and peripherally. The ghrelin hormone is primarily produced in the X/A cells of the gastric fundus mucosa (Date et al., 2000; Kojima et al., 1999) and in much lower amounts in human umbilical vein endothelial cells (Conconi et al., 2004), chondrocytes (Caminos et al., 2005), cardiomyocytes (Iglesias et al., 2004), and T-cells (Dixit et al., 2004) and in other tissues, including pituitary, thyroid, lung, pancreas, kidney, adrenal gland, intestine, placenta, and gonads (Caminos et al., 2003; Korbonits, Goldstone, Gueorguiev, & Grossman, 2004; Korbonits & Grossman, 2004; Korbonits, Kojima, Kangawa, & Grossman, 2001; Rindi, Torsello, Locatelli, & Solcia, 2004). The ghrelin receptor, GHS-R1a, is abundantly expressed in peripheral tissue, with high levels in the stomach and pituitary gland, and lower levels in

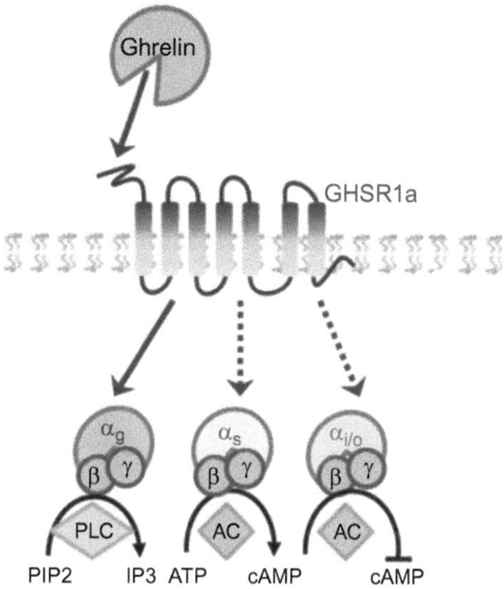

Figure 13.2 Hypothesized model of G-protein coupling following GHS-R1a activation. It is hypothesized that ghrelin signaling can be mediated via differential G-protein coupling which is tissue-specific or cell-type dependent. In somatotropic pituitary cells, the growth hormone secretagogue receptor type 1a (GHS-R1a) mainly signals via $G\alpha q/11$ following ghrelin binding, which triggers the release of Ca^{2+} from intracellular stores. However, in neurons of the arcuate nucleus, ghrelin induces coupling to Gs, while in pancreatic islet β-cells, the ghrelin receptor preferentially couples to $G\alpha i$, leading, respectively, to an increase or attenuation of cyclic adenosine monophosphate (cAMP).

pancreas, spleen, intestine, heart, kidney, thyroid, gonads, adrenal, adipose tissue and vasculature, as well as in several endocrine and endocrine tumors. In addition, the GHS-R receptor is expressed in the brain with the highest levels in the hypothalamus (Zigman, Jones, Lee, Saper, & Elmquist, 2006). The abundant expression of ghrelin and its receptor implicates its involvement in a plethora of biological functions (for review, see Schellekens et al., 2009). The primary functions of ghrelin are the secretion of GH from the anterior pituitary cells (Howard et al., 1996; Kamegai et al., 2004) and the central regulation of energy homeostasis through the modulation of appetite and food intake (for review, see Andrews, 2011; Castaneda, Tong, Datta, Culler, & Tschop, 2010; Depoortere, 2009; Kola & Korbonits, 2009; Schellekens et al., 2009). The appetite-inducing effects of ghrelin are mainly linked to peripherally produced ghrelin, which exerts its effects centrally after passing through the blood–brain barrier (Banks, Burney, & Robinson, 2008; Banks, Tschop, Robinson, & Heiman, 2002). However, some evidence suggests ghrelin synthesis in the brain, albeit at a much lower levels, in specific neuronal cells of the hypothalamus (Cowley et al., 2003; Kojima, Hosoda, & Kangawa, 2001; Lu et al., 2002; Mondal et al., 2005; Sato, Fukue, Teranishi, Yoshida, & Kojima, 2005). The existence of ghrelin-immunoreactive neurons was demonstrated in neurons adjacent to the third ventricle and between the dorsal medial nucleus and the ventromedial nucleus, as well as the paraventricular nucleus (PVN) and the ARC nucleus of the hypothalamus (Cowley et al., 2003). In addition, these ghrelin neurons project to hypothalamic NPY neurons and proopiomelanocortin (POMC) and cocaine- and amphetamine-related transcript (CART) neurons, as well as corticotrophin-releasing hormone-containing neurons and have terminals on orexin fibers in the lateral hypothalamic area (LHA) (Cowley et al., 2003; Toshinai et al., 2003). Moreover, a transgenic ghrelin-EGFP mouse model confirmed the existence of ghrelin-producing neurons in the ARC (Kageyama et al., 2008) and ghrelin expression was also detected in human hypothalamic areas (Menyhert et al., 2006). However, many discrepancies exist between studies and the expression of ghrelin in the brain has not been consistently shown and therefore remains highly controversial (Furness et al., 2011). Finally, biological functions of ghrelin other than GH secretion and appetite regulation, have been described, including glucose and lipid metabolism, gastrointestinal motility, cardiovascular, pulmonary and immune functions, sleep duration, learning, memory and behavior, cellular proliferation, reproduction, and bone physiology (Carlini et al., 2008; Charoenthongtrakul et al., 2009; Diano, 2008; Ghigo et al., 2001; Kojima & Kangawa, 2006; Leite-Moreira & Soares, 2007; Lutter et al., 2008; Muccioli et al., 2002; Ochi et al., 2008; Soares & Leite-Moreira, 2008; Taheri, Lin, Austin, Young, & Mignot, 2004; van der Lely, Tschop, Heiman, & Ghigo, 2004).

2.2. Physiological relevance of ghrelin in obesity

The likelihood for the development of obesity is enhanced through genetic predisposition (Bouchard, 2008; Dahlman & Arner, 2007; Loos & Bouchard, 2003; Willer et al., 2009). Typical genetic determinants, negatively influencing the body's metabolism, include variants and mutations of the fat mass and obesity-associated gene and the melanocortin 4 receptor (MC_4R) (Li & Loos, 2008; Loos et al., 2008; Willer et al., 2009). Ghrelin gene polymorphisms and mutations in humans have been associated with an increased risk to develop obesity (Hinney et al., 2002; Ukkola et al., 2001). Mutations in the ghrelin receptor involved in an altered metabolism, include four different point mutations, I134T, V160M, A204E, and F279L, which mainly affect ligand-independent or constitutive receptor activity (Liu, Fortin, Beinborn, & Kopin, 2007). In addition, polymorphisms in the GHS-R promoter may modify changes in body weight via alteration of GHS-R gene expression (Mager et al., 2008).

The important role of ghrelin peptide in the regulation of appetite and food intake has been well established and the orexigenic effects of peripheral or central ghrelin administration have been widely documented (Naleid, Grace, Cummings, & Levine, 2005; Tschop, Smiley, & Heiman, 2000; Wren et al., 2001). Ghrelin plays a role in the sensation of hunger and meal initiation, which correlates with high plasma ghrelin levels just before a meal and decrease post food ingestion (Cummings et al., 2001; Tschop et al., 2000). Aberrant circulating levels of plasma ghrelin suggest a role of ghrelin and its receptor in the metabolic disorder of obesity. Plasma ghrelin levels are negatively correlated with adiposity, leading to reduced plasma ghrelin levels in general obesity, but the nocturnal plasma ghrelin increase is also blunted in obese individuals. However, high ghrelin levels exist in Prader-Willi syndrome-associated obesity, following a blunted age-related decline in ghrelin (Cummings et al., 2002; DelParigi et al., 2002).

In addition, several rodent studies have investigated the importance of ghrelin in obesity (Shrestha, Wickwire, & Giraudo, 2009; Wortley et al., 2005; Zigman et al., 2005). The ghrelin-mediated increase in food intake is independent of its GH-secreting potential, as demonstrated by a marked increase in food intake in a GH-deficient spontaneous dwarf rat model (Nakazato et al., 2001). In addition, chemical ablation of the ARC and antisense GHS-R1a mRNA blocked food intake following central administration of ghrelin in this rat model (Shuto et al., 2002; Tamura et al., 2002). Moreover, ghrelin exposure was shown to increase neuronal NPY and agouti-related peptide (AgRP) mRNA expression and induce the neuronal activity markers c-Fos and Egr-1. Ghrelin's orexigenic effect was attenuated following pretreatment with a nonpeptide NPY Y1 receptor antagonist and specific antiserum against NPY and AgRP. In addition, intranuclear injection of ghrelin into neurons of the PVN similarly increased appetite. This reinforces the ghrelin-induced orexigenic effects to be mediated via NPY/AgRP neurons in the ARC. Moreover, hypothalamic GHS-R expression has been shown to be increased following fasting or chronic food restriction (Kurose et al., 2005).

Despite ghrelin's potent orexigenic effects, several studies using knockout mouse models, whereby either ghrelin or its receptor is deficient, suggest that ghrelin is not the dominant and critical regulator of appetite

(Sun, Ahmed, & Smith, 2003; Sun, Butte, Garcia, & Smith, 2008; Wortley et al., 2004). Specifically, neither knockdown affected normal food intake behavior, growth rate, or body composition, suggesting that the ghrelin system is not an essential regulator of energy balance. In contrast, the ghrelin-induced secretion of GH is clearly mediated through the GHS-R1a receptor, since the effect was absent in GHS-R null mice following acute ghrelin treatment (Sun, Wang, Zheng, & Smith, 2004). However, appetite and food intake are mediated by several interconnected molecular signaling pathways and the lack of ghrelinergic signaling in knockdown models may have been blunted by compensatory mechanisms controlling appetite and food intake. In line with this finding, is the study by Zigman et al., demonstrating normal sized mice with normal food intake behavior in NPY and AgRP knockout mice (Zigman et al., 2005). Nevertheless, it has been demonstrated that ghrelin-deficient mice are resistant to a high-fat diet (HFD)-induced obesity. Indeed, GHS-R1a null mice on a HFD were shown to accumulate less body weight and adiposity compared to control animals, presumably through the preference of utilizing fat as an energy substrate (Zigman et al., 2005). In a different study, ghrelin knockout mice similarly increased fat utilization when fed a HFD (Wortley et al., 2004). In more recent double knockout studies, in which both the ghrelin peptide and the ghrelin receptor are deleted, body weight was decreased independent of food intake when mice were fed a standard chow diet. At the same time, energy expenditure and motor activity were increased (Pfluger et al., 2008). These data reinforce the evolutionary determined role of ghrelin signaling in accumulating and preserving fat reserves in times of abundance to survive poor nutritional availability, such as famine. Thus, a large body of evidence demonstrates that inhibition of the ghrelin/GHS-R pathway results in reduced food intake and this may consequently lead to reductions in body weight and adiposity. This confirms the ghrelin system as a biologically relevant signaling pathway regulating appetite and food intake and therapeutic interventions attenuating or blocking ghrelin signaling may be used in the fight against obesity and its complications (Soares, Roncon-Albuquerque, & Leite-Moreira, 2008).

Interestingly, while in a study by Sun et al., single ghrelin or GHS-R knockout mice did not demonstrate resistance to diet-induced obesity an impaired glucose homeostasis (i.e., lower blood glucose levels) was observed under conditions of negative energy balance (i.e., caloric restriction) (Sun et al., 2008). Indeed, pancreatic islet-derived ghrelin has been identified as a physiological regulator of glucose-induced insulin release (Dezaki et al., 2006). Both ghrelin and its receptor are expressed in the pancreas where ghrelin release and receptor activation were shown to attenuate glucose-induced insulin release. Glucose-induced insulin release was enhanced following endogenous ghrelin immunoneutralization and GHS-R1a receptor antagonism and suppressed following exogenous ghrelin exposure. In addition, ghrelin knockout mice displayed enhanced glucose-induced insulin release from isolated islets, whereas islet density, size, insulin content, and insulin mRNA levels were unaltered. In addition, the HFD-mediated glucose intolerance was largely prevented in ghrelin knockout mice. Thus, antagonism of the ghrelinergic system may enhance insulin release and normalize glycemic control in HFD-induced obesity. Therefore, ghrelin and its receptor may also represent appropriate therapeutic targets to counteract the metabolic syndrome by curbing the progression of type 2 diabetes associated with obesity.

3. GHRELIN IN THE MULTICIRCUIT CONTROL OF APPETITE

3.1. Homeostatic regulation of energy balance

Food intake and ingested nutrients modulate the peripheral release of a plethora of hormones and gut peptides, which all act through the brain to coordinate appetite and food intake and affect satiety. Interaction between the components within this complex network maintains a homeostatic energy balance between energy intake and expenditure. The peripheral peptides influence the brain either via the general blood circulation or via the vagus nerve (Venkova & Greenwood-Van Meerveld, 2008). The integration and processing of peripheral peptides with the brain and brain-gut feedback signaling can lead to meal termination and satiety. In addition, the time between subsequent meals may be extended or decreased. Cholecystokinin (CCK), amylin, and pancreatic glucagons are examples of peptides involved in meal termination. These peptides are released rapidly from the intestine and pancreas with the onset of feeding and have short durations of action. The postprandial released CCK was the first gut hormone demonstrating to have an effect on food intake (Gibbs, Young, & Smith, 1973). Amylin is secreted from the pancreatic A-cell secretory vesicles in response to food and has a glucose-regulatory and an anorexigenic action. Amylin reduces food intake and mediates satiety via action on the hindbrain area postrema and central nucleus of the amygdala (Amyg) (Lutz, 2006). Peptide tyrosine-tyrosine (PYY), glucagon-like peptide 1 (GLP-1), and oxyntomodulin are peptides for which longer term feeding inhibitory actions have been detected. They are released from intestinal L cells in the distal intestine and rise slowly during a meal and peak after meal termination, after which they remain elevated for several hours. Pancreatic polypeptide is also released postprandial into the circulation in proportion to calories ingested to reduce food intake. Moreover, GLP-1, derived from proglucagon, together with glucose-dependent insulinotropic polypeptide (GIP) belong to the secretin family of gastrointestinal hormones referred to as incretins which stimulate insulin release from the beta cells of the Islets of Langerhans in response to food, independent of blood glucose levels. They reduce gastric emptying, glucagon release from alpha cells of the Islets of Langerhans and the absorption rate of nutrients into the blood stream. GIP (also named gastric inhibitory polypeptide) is released from K cells in the mucosa of the duodenum and the jejunum of the gastrointestinal tract. In addition, adipokines including leptin, adiponectin, and resistin are secreted, in proportion to fat mass, by the adipose tissue and mediate their effects centrally via the hypothalamus to affect food intake as well as energy expenditure (Ahima & Lazar, 2008; Kadowaki, Yamauchi, & Kubota, 2008; Tovar et al., 2005; Zhang et al., 1994). Leptin counteracts the effect of NPY and the endogenous cannabinoid neurotransmitter, anandamide, in the hypothalamus and promotes the synthesis of α-melanocyte-stimulating hormone (α-MSH), an appetite suppressant. Leptin-mediated inhibition of food intake is long term, in contrast to the rapid inhibition by CCK and the slower suppression of hunger between meals mediated by PYY. Insulin, secreted by the pancreas upon increase in glucose load is, like leptin, proportional to fat mass and has equally profound effects on appetite reaching the brain via receptor-mediated transport across the blood–brain barrier. Furthermore, the gastric peptide ghrelin increases appetite and subsequent food intake and shortens meal intervals.

3.2. Ghrelin and homeostatic control of food intake

The stomach-derived ghrelin peptide is the first and only known peripheral hormone identified exerting an orexigenic effect and is therefore aptly named the "hunger hormone" (Nakazato et al., 2001). Considerable evidence supports the role for the hormone ghrelin in mealtime hunger and meal initiation, increasing food intake, and body weight (Cummings et al., 2001; Tschop et al., 2000). Plasma ghrelin levels have been shown to fall and rise following food intake and conditions of food deprivation, respectively. However, circulating levels of the ghrelin hormone often display aberrant signaling in obesity and the role of ghrelin and its receptor in the homeostatic imbalance leading to obesity have been extensively reviewed (Andrews, 2011; Naleid et al., 2005; Schellekens et al., 2009; Tschop et al., 2000; Wren et al., 2001). The orexigenic effects of ghrelin are mediated via activation of the both peripherally and centrally expressed GHS-R1a receptors. Neuronal excitation following activation of centrally expressed GHS-R1a in the ARC of the hypothalamus, leads to downstream signaling via centrally expressed melanocortin receptors (MC_3 and MC_4) expressed on hypothalamic neurons in the PVN and the LHA (Schellekens et al., 2009). Ghrelin increases appetite and food intake via direct GHS-R1a-mediated expression of NPY from arcuate NPY/AgRP neurons and inhibition of neurons expressing POMC/CART (Andrews et al., 2008; Chen et al., 2004). Ghrelin-mediated GABA release from the NPY/AgRP neurons leads to an indirect inhibition of the POMC/CART neurons, which do not express the GHS-R1a receptor (Cowley et al., 2003). Moreover, ghrelin-mediated expression of AgRP contributes to the inhibition of satiety through antagonism of the α-MSH and inverse agonism of the constitutively active MC_3 and MC_4 receptors (Tolle & Low, 2008). Thus, ghrelin acts in an opposite manner to that of adipose-derived leptin, expressed from the *ob* gene (Zhang et al., 1994). Leptin expression is increased in proportion to fat mass and causes an upregulation of POMC expression, the precursor of α-MSH, while decreasing NPY and AgRP (Obici, 2009).

3.3. Ghrelin and food-reward behavior

Classically, homeostatic control of food intake has been attributed to the neuronal circuits in both the hypothalamus and brainstem. However, food also provides strong pleasurable and rewarding signals that can override satiety and stimulate appetite independently of metabolic needs (Kenny, 2011b). Recently, it has become apparent that control of feeding is an integrated multiprocess neurocircuitry, involving cortical areas as well as areas within the mesolimbic pathway, including the ventral tegmental area (VTA), nucleus accumbens (NAcc), hippocampus (Hipp), and Amyg, which are all involved in the modulation of the hedonic aspects of food intake (Skibicka & Dickson, 2011). The majority of the classical feeding peptides regulating energy homeostasis, including the gastric-derived orexigenic ghrelin peptide, also interact with these cortical and mesolimbic circuits. The ghrelin receptor (GHS-R1a) is mainly expressed in the ARC of the hypothalamus, which reinforces its crucial role in neuroendocrine and appetite-stimulating activities. However, recently, a role for the GHS-R1a receptor at the interface between homeostatic control and neurobiological circuits involved in reward and motivational aspects of food has emerged (for review, see Dickson et al., 2011; Egecioglu et al., 2011; Skibicka & Dickson, 2011). Expression of the ghrelin receptor in hypothalamic nuclei other than the ARC is in line with this hypothesis. Indeed, the GHS-R1a receptor has been found to be expressed in the suprachiasmatic, anterior hypothalamic, paraventricular, anteroventral preoptic, and tuberomammillary nuclei (Zigman et al., 2006). Moreover, the GHS-R1a receptor was found to be expressed in extrahypothalamic brain areas, such as the substantia nigra, the dorsal and median raphe nuclei, the VTA, and the Hipp (Zigman et al., 2006). These discoveries sparked the interest in the possibility of a role for ghrelin in the hedonic aspects of food and suggested the involvement of ghrelin in choice and selection of rewarding foods. Several studies have investigated the effects of ghrelin on the rewarding value of food and reward-based eating behavior (Table 13.1). Ghrelin was shown to increase preference for rewarding foods, such as diets rich in fat (Shimbara et al., 2004) as well as saccharin solutions (Disse et al., 2010). The involvement of the ghrelin in food preference was demonstrated following central ghrelin injection in rats, which was shown to preferentially increase fat intake over carbohydrate intake (Shimbara et al., 2004). Furthermore, increased saccharin consumption in mice was observed after peripherally administered ghrelin, while this was not observed in GHS-R1a knockout mice (Disse et al., 2010). Moreover, the ghrelin-mediated shift in preference to sweet tasting foods was independently of caloric content (Disse et al., 2010). In addition, a decreased consumption in palatable food (not chow) was reported in a free-choice paradigm (chow vs. rewarding food) following ghrelin administration in mice deficient for the GHS-R1a receptor or upon GHS-R1a antagonist treatment in rats (Egecioglu et al., 2010). The importance of ghrelin in the enhancement of the rewarding value of food was further demonstrated by recent studies using conditioned place preference (CPP), which measures the time rodents spend in an environment previously paired with a palatable food reward (Disse et al., 2011; Egecioglu et al., 2010; Perello et al., 2010). Increases in plasma ghrelin by peripheral ghrelin administration or caloric restriction were shown to enhance the CPP response for HFD (Disse et al., 2011; Perello et al., 2010), thus increasing the motivation to obtain rewards in the form of palatable food (Skibicka, Hansson, Alvarez-Crespo, Friberg, & Dickson, 2011; Skibicka, Hansson, Egecioglu, & Dickson, 2012).

Table 13.1 Ghrelin and reward-motivated feeding

Route of ghrelin administration	Food reward	Animal model	Test paradigm	Antagonist	Conclusion	References
ICV (lateral cerebral ventricle)	HF or HC diet	Male Wistar rats, HF, or HC preferring	Preference test	n/a	Centrally administered ghrelin enhanced fat ingestion.	Shimbara et al. (2004)
Direct VTA or NAcc microinjection; caloric restriction	Regular chow	Sprague–Dawley rats	Cumulative food intake	Opioid antagonist, NTX at alternate site before injecting ghrelin	Increased chow intake with VTA and NAcc injection of ghrelin demonstrating hedonic responses to be modulated by homeostatic factors. NTX injected into NAcc or VTA did not affect feeding elicited by ghrelin injection in alternate site.	Naleid et al. (2005)
Sc; caloric restriction	High-fat diet	C57BL6/J mice; GHS-R1a (−/−); orexin (−/−) mice	CPP; operant conditioning	GHS-R1a antagonist compound 26 (oral gavage); orexin 1 receptor antagonist SB-334867 (IP)	Ghrelin is obligatory in rewarding aspects of food and requires intact orexin signaling, but separate from homeostasis-related compensatory hyperphagia under caloric restriction.	Perello et al. (2010)

Continued

Table 13.1 Ghrelin and reward-motivated feeding—cont'd

Route of ghrelin administration	Food reward	Animal model	Test paradigm	Antagonist	Conclusion	References
IP; caloric restriction	High-fat diet	C57BL6/J mice; GHS-R1a (−/−)	CPP	Cholinergic blockade by mecamylamine	The ghrelin receptor is obligatory to acquire a food CPP. Strength of food-induced CPP response correlates to plasma ghrelin level and is partly mediated by the cholinergic pathway.	Disse et al. (2011)
IP	Saccharin 0.3% solution, noncaloric	C57BL6/J mice; GHS-R1a (−/−)	Single-bottle tests; free-choice preference test	n/a	Peripheral ghrelin enhances intake and preference for a sweet food, regardless of whether the food has caloric content and may serve as additional enhancers of energy intake.	Disse et al. (2010)
Direct bilateral intra-VTA (mice)	Peanut butter	NMRI mice; GHS-R1a KO	Locomoter activity; free-choice paradigm	n/a	The GHS-R1a is necessary for accumbal dopamine release (microdialysis) and intake of rewarding food in free-choice paradigm and CPP-food response. This supports the hypothesis that central	Egecioglu et al. (2010)
ICV (VTA lesion study in rats)	Ensure and chocolate pellets	Sprague–Dawley rats	Food preference free-choice paradigm; CPP	GHS-R1a antagonist JMV2959 IP		

					ghrelin signaling at the level of the VTA is important for the incentive value of rewarding food but not standard chow.	
Sc	HFD	C57BL6/J mice; GHS-R1a (−/−); GHS-R1a-TH mice, expressing only in TH-expressing neurons	CPP; CSDS stress	n/a	GHS-R1a-expressing catecholaminergic neurons are responsible for stress-induced eating.	Chuang et al. (2011)
IP or ICV (third ventricle)	Sucrose	Sprague–Dawley rats	Operant conditioning	GHS-R1a antagonist JMV2959 IP or ICV (third ventricle)	Ghrelin plays a key role in motivation and reinforcement for sucrose and impacts on the expression of the major reward neurotransmitter receptors dopamine and acetylcholine encoding genes in the mesolimbic reward circuitry.	Skibicka, Hansson, Egecioglu, et al. (2011)

Continued

Table 13.1 Ghrelin and reward-motivated feeding—cont'd

Route of ghrelin administration	Food reward	Animal model	Test paradigm	Antagonist	Conclusion	References
Direct VTA or NAcc microinjection; caloric restriction	Sucrose	Sprague–Dawley rats	Operant conditioning	GHS-R1a antagonist JMV2959 direct in VTA or NAcc	Blockade of the GHS-R1a in the VTA is sufficient to decrease motivation to work for a sugar reward but not to reduce fasting-induced chow hyperphagia in VTA and NAcc. The VTA but not the NAcc is sufficient as a direct but necessary target site for ghrelin's action on food motivation.	Skibicka, Hansson, Alvarez-Crespo, et al. (2011)

CPP, conditioned place preference; HC, high carbohydrate; HF, high fat; ICV, intracerebroventricular; NTX, naltrexone; IP, intraperitoneal; Sc, subcutane; VTA, ventral tegmental area; NAcc, nucleus accumbens.

The CPP response to chocolate was shown to be blocked using GHS-R1a antagonists in satiated rats (Egecioglu et al., 2010). Moreover, peripherally and centrally administered ghrelin elevated the incentive motivation for sucrose rewards in a satiated rat in progressive ratio operant conditioning and blockade of ghrelin signaling reduced the operant responding for sugar in hungry rats to the level of a satiated rat (Skibicka, Hansson, Egecioglu, et al., 2012).

In addition, the impact of ghrelin on the motivational aspects of food-associated reward has been investigated using operant conditioning (Finger, Dinan, & Cryan, 2012; Perello et al., 2010; Skibicka, Hansson, Alvarez-Crespo, et al., 2011; Skibicka, Hansson, Egecioglu, et al., 2012). Operant lever pressing or nose poking for palatable rewards, including sucrose and HFD, was increased following ghrelin exposure in rodents. Furthermore, a resistance to ghrelin-induced motivation to obtain a food reward (i.e., reduced operant responding) was observed in diet-induced obese mice, in parallel to the documented blunted orexigenic effect of ghrelin in this mouse model of obesity (Briggs, Enriori, Lemus, Cowley, & Andrews, 2010; Finger et al., 2012).

3.4. Neuronal circuitry underlying ghrelin-mediated food reward

The rewarding effects of food intake are mediated via dopamine (DA) neurons in the mesolimbic circuitry system (Hoebel, 1985; Parylak, Koob, & Zorrilla, 2011; Volkow, Wang, & Baler, 2010; Volkow, Wang, Fowler, Tomasi, & Baler, 2012; Wise, 2006; Wise & Rompre, 1989). DA plays a key role in the regulation of normal brain functioning including locomotion, behavior, learning, emotion, reward, and motivated behavior. Deficits in the dopaminergic system have been linked to many pathophysiologies, such as schizophrenia, Parkinson's disease, attention deficit hyperactivity disorder, depression, and drug addiction. The VTA is the main region involved both in the selection of palatable foods and in regulating the rewarding properties of food and the motivational drive to obtain these. Neuronal populations in the VTA have mesolimbic dopaminergic projections, which terminate in the ventral striatum and the prefrontal cortex and are important for anticipatory food-reward and food-seeking behavior (Bassareo & Di Chiara, 1999; Richardson & Gratton, 1998). The high level of ghrelin receptors expressed in the VTA of the mesolimbic system support a role for ghrelin in the dopaminergic VTA-mediated reward signaling (Guan et al., 1997; Zigman et al., 2006). Indeed, ingestion of palatable foods elicits rewarding effects via neuronal DA secretion from VTA originating projections, which is suggested to mediate ghrelin's ability to alter food reward (Dickson et al., 2011). Reward is mediated following a ghrelin-induced overflow of DA within the NAcc (Jerlhag et al., 2007). Thus, while direct effects of ghrelin on the hypothalamic nuclei regulate energy homeostasis, direct effects of ghrelin on VTA and NAcc have been shown to affect reward and incentive motivational aspects of food intake (Skibicka & Dickson, 2011).

Moreover, the increase in DA release in the NAcc following direct ghrelin administration in the VTA is in support of this hypothesis (Jerlhag et al., 2007). In addition, DA release following rewarding food was absent in GHS-R1a knockout mice (Egecioglu et al., 2010). Furthermore, chronic central ghrelin exposure increased gene expression of the DA receptor D5 and the acetylcholine receptor nAChRβ2 in the VTA and decreased expression of D1, D3, D5, and nAChRα3 in the NAcc (Skibicka, Hansson, Egecioglu, et al., 2012).

Interestingly, direct injection of ghrelin in VTA or NAcc was shown to increase feeding behavior (Naleid et al., 2005) and it was suggested that intra-VTA administration of ghrelin is responsible for the increased consumption of rewarding foods in mice (Egecioglu et al., 2010). The observed increase of intake of rewarding foods (but not standard chow) after central ghrelin administration and the subsequent attenuation of both intake and CPP response after chemical VTA lesion strongly support the involvement of the VTA on ghrelin-mediated choice and selection of rewarding (high-calorie) foods (Egecioglu et al., 2010). Moreover, an additional study by the Dickson group demonstrated that direct microinjection of ghrelin into the VTA not only increased free feeding of chow but also enhanced motivated behavior for a sucrose reward, assessed in an operant conditioning paradigm in rats (Skibicka, Hansson, Alvarez-Crespo, et al., 2011). Interestingly, neither stimulation nor blockade of the GHS-R1a had an effect on the NAcc-mediated motivated behavior, while regular chow intake was still intact (Skibicka, Hansson, Alvarez-Crespo, et al., 2011). These results identify the VTA but not the NAcc as a direct target site for ghrelin's action on food-motivated behavior. These findings lead to the hypothesis that ghrelin signaling impacts the NAcc to increase hunger and decrease satiety and that the VTA also affects ghrelin-mediated rewarding properties of food, the selection of rewarding foods as well as the food-motivated behavior (Dickson et al., 2011; Skibicka & Dickson, 2011). Alternate networks originating from the VTA, including those projecting to the LHA, Amyg, Hipp, and medial prefrontal cortex (mPFC) have also been implicated in ghrelin's

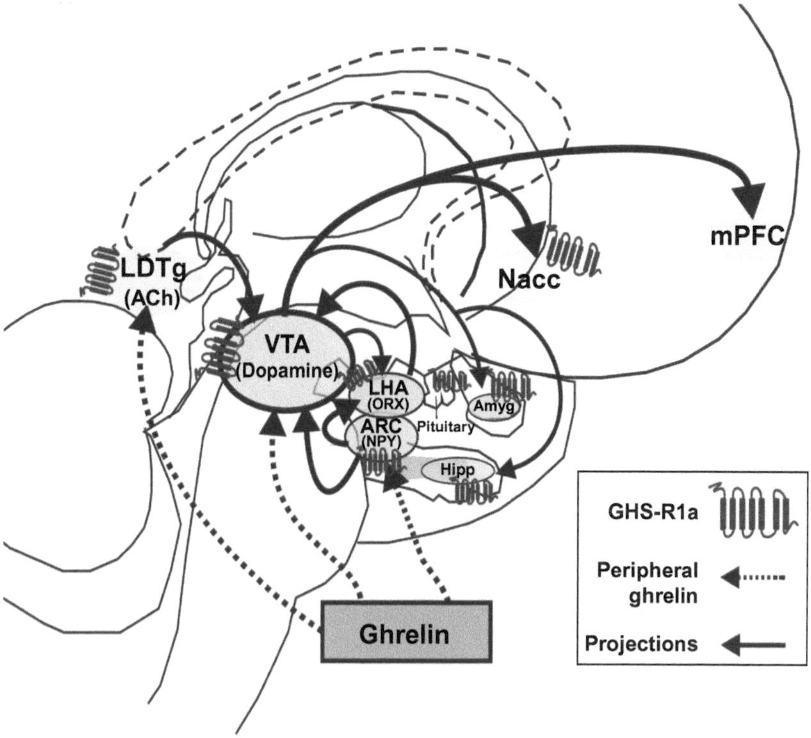

Figure 13.3 Ghrelin signaling in the mesolimbic reward circuitry in the human brain. Schematic representation of ghrelin-mediated regulation of homeostatic food intake and food reward in the human brain. Key pathways and regions implicated in the regulation of food intake and reward-seeking behavior are depicted. Peripheral ghrelin increases food intake via homeostatic circuits in the ARC including direct activation of NPY-expressing neurons. Moreover, ghrelin directly activates VTA dopaminergic neurons which project to the NAcc. The involvement of ghrelin in signaling in projections from VTA to other brain regions, including the LHA, amygdala, hippocampus, and mPFC, has also been suggested. A cholinergic-dopaminergic reward link has been suggested via GHS-R1a signaling in the LDTg. Expression of the GHS-R1a is indicated. Amyg, amygdala; ARC, arcuate nucleus; LDTg, laterodorsal tegmental area; LHA, lateral hypothalamic area; Hipp, hippocampus; mPFC, medial prefrontal cortex; NAcc, nucleus accumbens; NPY, neuropeptide Y; ORX, orexin; VTA, ventral tegmental area. (See Color Insert.)

effect on food reward but remain unclear. The neuroanatomical interactions of ghrelin with central brain regions involved in both the homeostatic and hedonic aspects of food intake are summarized in Fig. 13.3. Together, these data suggest not only that ghrelin increases the motivation to eat which leads to consumption of food in the absence of hunger but also that ghrelin signaling is a requirement for the food-motivated response. In addition, it seems likely that the orexigenic effect of ghrelin on palatable foods is mediated via the VTA but that the VTA is not required for ghrelin-induced chow intake.

Additional pathways and signaling molecules have also been suggested to mediate ghrelin's effect on food-reward behavior. To this effect, the involvement of a cholinergic-dopaminergic reward circuitry in ghrelin-mediated reward signaling upon palatable food was recently demonstrated (Disse et al., 2011). In this study, the food-reward seeking behavior, assessed using classical CPP, was directly correlated to the level of plasma ghrelin and the further enhancement of food reward following pharmacological increase in ghrelin was demonstrated to involve the cholinergic pathway after cholinergic blockade by mecamylamine (Disse et al., 2011). Moreover, in line with these findings are the blunted mesolimbic effects of ghrelin administration upon cholinergic blockade shown by others (Dickson et al., 2010; Jerlhag et al., 2006). Thus, ghrelin may activate cholinergic neurons emanating from the laterodorsal tegmental area (LDTg) which may similarly impact on food-induced reward. Moreover, the GHS-R1a receptor is also expressed on the orexin neurons of the lateral hypothalamus and GHS-R1a-expressing NPY neurons project to orexin neurons where ghrelin can directly activate orexin neurons and increases food intake when microinjected into the LHA (Lawrence, Snape, Baudoin, & Luckman, 2002; Olszewski et al., 2003; Yamanaka et al., 2003). Furthermore, reduced expression of orexins in the hypothalamus of *ob/ob* mice, in contrast to upregulated NPY levels, have been suggested to potentially act as a counter regulatory system necessary to limit the adverse effects of NPY on food intake and body weight in this mouse model of obesity (Stricker-Krongrad, Richy, & Beck, 2002). It has been hypothesized, that the ghrelin-mediated increase in high-fat food reward occurs in an orexin-dependent manner, since ghrelin's effects on HFD reward response in CPP and operant conditioning were absent following pharmacologic or genetic blockade of the orexin receptor (Perello et al., 2010). Moreover, hyperphagia after chronic calorie restriction was not inhibited following ghrelin receptor antagonism or in GHS-R1a receptor deficient mice suggesting a separation in signaling pathways mediating the homeostatic function of ghrelin and the regulation of hedonic values of food in conditions of caloric restriction. Ghrelin has also been shown to directly act on areas of the Hipp and Amyg, affecting learning and memory as well as areas of the prefrontal cortex which indicates a role for ghrelin in the control and decision making aspects (Dickson et al., 2011).

Finally, the impact of ghrelin on hedonic appetite signaling independent of the homeostatic regulation has been suggested to involve dimerization of the GHS-R1a receptor. Recent investigations demonstrated evidence for a dimer between the dopamine D1 receptor and the growth hormone secretagogue (GHS-R1) receptor leading to enhanced DA signaling (Jiang, Betancourt, & Smith, 2006; Schellekens et al., 2012a). In addition, the involvement of a dimer between the GHS-R1a and the dopamine D2 receptor in the regulation of appetite has recently been shown (Kern, Albarran-Zeckler, Walsh, & Smith, 2012). This implicates a potential role for GHS-R1a receptor dimerization in the ghrelin-mediated effect on the rewarding properties of food and warrants further investigation.

In conclusion, the ghrelin system has the ability to enhance preference for palatable foods rich in sugar or fat as well as to alter the rewarding value of food through modulation of the DA system and reward behavior. However, the precise integration of ghrelin-mediated DA signaling with additional appetite-mediated molecular signaling pathways warrants further investigation.

4. STRESS-INDUCED GHRELIN-MEDIATED FOOD REWARD

Communication between the brain and the gut, the brain-gut axis, regulates the body's energy homeostasis and its deregulation has been linked to psychological conditions such as stress, anxiety, and depression. An increased ingestion of caloric-dense "comfort foods" often correlates with chronic stress or atypical depression as well as being a major contributor to the increased prevalence of obesity. Indeed, an increase in food intake (hyperphagia) and body weight have been associated with major depressive disorders according to criteria of the diagnostic and statistical manual of mental disorders (DSM-IV-TR) (American Psychiatric Association, 2000). In addition, chronic stress was shown to inhibit the metabolic effects of a HFD using an unpredictable chronic social defeat stress (CSDS) paradigm, a model of prolonged psychosocial stress featuring aspects of post-traumatic stress disorder as well as major depression (Finger et al., 2012). Diet-induced increases in weight gain were attenuated following the CSDS paradigm and decreased the caloric intake as well as lowered the insulin resistance and plasma leptin and insulin levels, highlighting the importance of considering the chronic aspects of both stress and diet and their time-dependent interplay. Moreover, in most humans, stress was shown to affect feeding behavior in both directions, with some increasing their food intake while others eat less (Dallman, 2010; Gibson, 2006; Oliver & Wardle, 1999). However, an overall increase in consumption of highly palatable or caloric-dense foods has been reported, independent of hyperphagia or hypophagia following stress. Considering ghrelin's big impact on homeostatic and hedonic aspects of food intake, a role for ghrelin-mediated signaling in stress-induced food-reward behavior has also been suggested (Schellekens et al., 2012b; Dickson et al., 2011; Perello et al., 2010; Skibicka & Dickson, 2011). Plasma ghrelin levels and preproghrelin mRNA levels have been shown to be increased in conditions of stress in rodents and humans (Asakawa et al., 2001; Chuang & Zigman, 2010; Kristenssson et al., 2006; Lutter et al., 2008; Raspopow, Abizaid, Matheson, & Anisman, 2010). The enhanced levels of ghrelin during stress have been suggested as a potential defense mechanism against symptoms of stress-induced depression and anxiety (Lutter et al., 2008). This may potentially explain the phenomenon of "comfort eating" observed in conditions of stress. It is tempting to speculate that the ghrelin system has evolved as an energy deficit signal to act in times of energy insufficiency by favoring consumption of calorie-dense palatable foods and protect the storage of fat (Wells, 2009). However, circulating levels of ghrelin, as well as GOAT levels, have been observed to respond differentially depending on the type of stressor (Stengel, Wang, & Tache, 2011). Indeed, ghrelin levels were shown to be elevated by metabolic stressors, including cold exposure, acute fasting, and caloric restriction, as well as psychological stressors, including chronic social defeat or unpredictable stress. Interestingly, plasma ghrelin levels were reduced following physical stressors, such as abdominal surgery, immunological/endotoxin injection, and exercise (Stengel et al., 2011). The pathways underlying the alterations in ghrelin plasma concentration under differential stress conditions are still largely unknown but may be mediated via sympathoadrenal response following activation of the sympathetic nervous system and catecholamine release (Mundinger, Cummings, & Taborsky, 2006; Zhao et al., 2010).

Recently, evidence has suggested a potential role for ghrelin in the mediation of stress-induced food reward (Chuang et al., 2011). Mice increased CPP responding and intake of HFD following stress exposure using the chronic psychosocial stress model of CSDS. The study showed that the stress-induced food-reward behavior was ghrelin dependent and that both ghrelin's orexigenic and antidepressant-like effect as well as the food-reward behavioral effects were mediated by catecholaminergic neurons (Chuang et al., 2011). Finally, ghrelin plays a role in stress-induced gastric motility and ghrelin release from the stomach is increased in the chronic phase of stress and mediates an accelerated gastric emptying. In contrast, continuous stress delays gastric emptying at the acute phase via sympathetic pathway

(Ochi et al., 2008). The potential role of ghrelin in stress-induced gastric motility is reinforced by the enhanced GHS-R agonist-mediated gastric emptying (Charoenthongtrakul et al., 2009).

4.1. Ghrelin in food addiction

Accumulating evidence supports the premise of overlapping neurocircuits between drug reward and food addiction, in particular, the mesolimbic dopaminergic system, which is heavily involved in the hedonic and motivational aspects of food intake and drugs of abuse (Hoebel, 1985; Kenny, 2011a, 2011b; Parylak et al., 2011; Volkow et al., 2010, 2012; Wang, Volkow, Thanos, & Fowler, 2009; Wise, 2006). Drugs of abuse as well as highly palatable, calorie-dense foods are consumed by humans or self-administered by laboratory animals because they are inherently rewarding, an effect that is mediated through their DA-enhancing properties in the mesolimbic circuitry system (Hoebel, 1985; Volkow et al., 2010, 2011; Wise, 2006). The repeated overconsumption of palatable (i.e., pleasurable) foods shares many characteristics with addictive behaviors, including low impulse control and continued use despite negative consequences. Both drug dependence and repeated excessive palatable food consumption are accompanied with neuroadaptive changes within the central reward circuitries, such as altered gene expression and altered brain responsivity to food, similar to those occurring upon drug dependence (Avena, Bocarsly, Hoebel, & Gold, 2011; Avena, Rada, & Hoebel, 2008; Leggio et al., 2011; Parylak et al., 2011; Volkow et al., 2012). In addition, repeated overeating also increases the vulnerability toward overeating (Burger & Stice, 2011). Heightened food-reward sensitivity and excess food consumption result from a hyperactivation of the hedonic system (Stice, Yokum, Blum, & Bohon, 2010). Indeed, obese individuals display enhanced reward-related responsivity to palatable food cues (Stoeckel et al., 2008). Moreover, neuroimaging in humans has demonstrated that palatable food dependence activates similar brain regions implicated with substance addiction (Gearhardt, Corbin, & Brownell, 2009a, 2009b; Gearhardt, Yokum, et al., 2011). Moreover, altered reward responsivity and brain alterations similar to those occurring in addiction have been revealed in overweight individuals using fMRI and PET (Wang et al., 2001). The combination of low impulse control and high motivation to eat reinforce the excessive feeding behavior and can contribute to the development of obesity (Carr, Daniel, Lin, & Epstein, 2011). In particular, the periodic intermittent bouts of overeating seen in binge-eating disorder, which can occur in some obesity types, has been associated with the development of food addiction (Corwin, 2011; Gearhardt, White, & Potenza, 2011). Moreover, repeated exposure to drugs and palatable foods is associated with a transition away from the acute pleasurable reinforcing effects toward dependence and concurrent enhancement of negative reinforcement (or antireward circuitry), as the rewarding properties wear off and in response to stress upon withdrawal. This is followed by a stimulus anticipation phase that precedes renewed drug intake. The initial rewarding effects of addiction, including food addiction, are mediated via DA neuron firing and are diminished at dependence, which can be correlated to a downregulation of the dopaminergic system in the brain reward circuitry. It has been hypothesized that the continuing overfeeding compensates this reward deficit and a role for the orexigenic ghrelin has been proposed in this vicious circle.

Central ghrelin-mediated signaling has been shown to be required for the reinforcing effect of drug of abuse, such as alcohol, as well as the rewarding aspects of food. In particular, human studies have demonstrated that the ghrelin system plays a role in alcohol dependence (Leggio et al., 2011). Alcohol consumption was shown to significantly reduces ghrelin levels in healthy controls (Calissendorff, Danielsson, Brismar, & Rojdmark, 2005, 2006; Zimmermann, Buchmann, Steffin, Dieterle, & Uhr, 2007) and reduced ghrelin levels were reported in alcohol-dependent patients compared to control (Addolorato et al., 2006; Badaoui et al., 2008). Moreover, significant increases in plasma ghrelin levels were observed in abstinent alcoholics (Kim et al., 2005; Kraus et al., 2005). In addition, plasma ghrelin levels have been shown to correlate with alcohol craving in alcoholics, which suggest a potential role for the brain-gut ghrelin peptide in alcohol-seeking behavior. Furthermore, the ghrelin system was shown to be involved in the consumption of rewarding substances such as sweets (Landgren et al., 2011), which reinforces the hypothesis of the common reward mechanism activated by both alcohol and sweet substances. Indeed, higher sweet cravings have been reported in alcohol-dependent individuals compared to healthy controls. Moreover, a link between glucose levels and alcohol-seeking behavior has been suggested (Connelly, Unwin, & Taberner, 1983; Leggio, Ray, Kenna, & Swift, 2009; Overstreet et al., 1993; Zito, Vickers, Telford, & Roberts, 1984). Alcohol is metabolized to sugar and rats bred for high sucrose preference have increased alcohol consumption compared to nonsucrose preferring rats (Landgren et al., 2011), reinforcing sweet liking as a possible endophenotype for alcohol dependence (for review, see Kampov-Polevoy,

Garbutt, & Khalitov, 2003). In addition, certain ghrelin gene haplotypes have been associated with increased sucrose or alcohol consumption (Landgren et al., 2011). Moreover, self-administration of sucrose in rats as well as saccharin intake in mice was increased after ghrelin administration and subsequently reduced following GHS-R1a antagonism in a sucrose (5%) versus water two-bottle-choice drinking protocol as well as in the operant sucrose self-administration paradigm (Landgren et al., 2011). This clearly indicates ghrelin to play an important role in the reward pathways leading to addiction and the ghrelinergic system may well have important consequences for the development of addictive behavior toward certain types of food.

Up until recently, the behavioral impact of abused drugs and compulsive food intake has been somewhat studied independently, despite the known overlap in neuronal signaling mechanisms in both scenarios. However, now it is becoming clear that certain foods have strong addictive properties in certain individuals (Avena et al., 2011; Parylak et al., 2011). The concept of food addiction has been reinforced by empirical studies and the impact of compulsive behaviors related to food and their role in the obesity epidemic are now actively being investigated. Understanding the contributions of the ghrelinergic system on the dopaminergic output following maladaptive food intake and the reward aspects associated with food intake will increase our knowledge of the pathology of obesity and other eating disorders and its relation to food addiction and dependence.

5. CONCLUSION AND FUTURE PERSPECTIVES

With obesity being one of the leading causes of illness and mortality the development of novel strategies to reduce body weight as well as alleviate associated diseases within the metabolic syndrome are urgently sought after. Ghrelin is the only peripheral derived orexigenic hormone, which acts centrally to regulate food intake and body weight homeostasis (Kojima, Hosoda, & Kangawa, 2004; Nakazato et al., 2001; Tschop et al., 2000). Therefore, it is not surprising that the obesity epidemic has led to the attention and effort of the pharmaceutical industry to the development of antiobesity drugs targeting the ghrelinergic system (Chollet, Meyer, & Beck-Sickinger, 2009; Leite-Moreira & Soares, 2007; Lu et al., 2009; Moulin, Ryan, Martinez, & Fehrentz, 2007; Soares et al., 2008; Zorrilla et al., 2006). The ubiquitous expression of the ghrelinergic system strongly implicates that the biological activities go far beyond GH secretion and energy homeostasis (Guan et al., 1997; Howard et al., 1996; Zigman et al., 2006). Recently, the importance ghrelin in hedonic reward-based eating (Perello et al., 2010) and in stress-induced food-reward behavior (Chuang et al., 2011) have been demonstrated. Food palatability is the biggest contributor to nonhomeostatic feeding and strongly influences the decision to eat as it can override satiation, leading to excess food consumption, and hence, obesity. Ghrelin was shown to enhance the rewarding values of palatable food and increase feeding behavior via the direct targeting of the VTA, a key brain region mediating rewarding effects of food and the motivation to eat. Ghrelin enhances hedonic eating behavior via the modulation of the mesolimbic dopaminergic reward pathways. In addition, a role for heterodimerization of the GHS-R1a receptor with DA receptors has been suggested to impact the neurocircuitry of reward driven hedonic eating behavior and warrants further investigation (Jiang et al., 2006).

These findings suggest that the central ghrelin signaling system provides therapeutic potential for the development of novel drug strategies to combat obesity through the multicircuitry regulation of food intake behaviors, including both the ghrelin-mediated homeostatic and the hedonic control of food intake. In addition, novel drugs targeting ghrelin in the mesolimbic dopaminergic circuitries may also be useful in the treatment of drug dependence and addictive behaviors other than food addiction. Conversely, individuals with anorexia nervosa or cachexia may benefit from ghrelin mimetics to help stimulate food intake via increased appetite and enhancement of the desire to eat. In addition, the inherent rewarding properties of ghrelin may contribute to a better sense of well being to minimize the frequent comorbid condition of depression. However, despite the increased understanding of the ghrelin-mediated multifaceted control on food intake, these multiple actions of the ghrelinergic system may actually also result in limitations to its potential as an effective antiobesity drug target, as many central acting antiobesity agents have been withdrawn due to side effects, including rimonabant, sibutramine, and the sympathomimetic weight loss drugs fenfluramine and dexfenfluramine. Nevertheless, deciphering and dissecting the mechanisms by which the ghrelinergic signaling modulates rewarding effects of palatable foods versus its homeostatic effect, would contribute significantly to the knowledge needed by the pharmaceutical industry to develop appropriate therapeutic strategies to either reduce ghrelin's appetite-inducing effect alone or combined with a reduction in incentive salience and rewarding properties of food. In addition, the role of ghrelin in the concept of "food addiction," describing an unhealthy dependence on calorically dense, palatable foods, is certainly poised to have a key role in the present-day obesity crisis, as well as in certain eating disorders.

ACKNOWLEDGMENTS

The work was supported by Enterprise Ireland under Grant Number CC20080001. J. F. C. and T. G. D. are also supported in part by Science Foundation Ireland (SFI) in the form of a centre grant (Alimentary Pharmabiotic Centre) through the Irish Government's National Development Plan. The authors and their work were supported by SFI (Grant Numbers 02/CE/B124 and 07/CE/B1368). J. F. C. is funded by European Community's Seventh Framework Programme (Grant Number FP7/2007-2013, Grant Agreement 201714).

REFERENCES

Addolorato, G., Capristo, E., Leggio, L., Ferrulli, A., Abenavoli, L., Malandrino, N., et al. (2006). Relationship between ghrelin levels, alcohol craving, and nutritional status in current alcoholic patients. *Alcoholism, Clinical and Experimental Research, 30*, 1933–1937.

Ahima, R. S., & Lazar, M. A. (2008). Adipokines and the peripheral and neural control of energy balance. *Molecular Endocrinology, 22*, 1023–1031.

American Psychiatric Association. (2000). *Diagnostic and statistical manual of mental disorders*, (4th ed., text revision). Washington, DC: American Psychiatric Association.

Andrews, Z. B. (2011). Central mechanisms involved in the orexigenic actions of ghrelin. *Peptides, 32*, 2248–2255.

Andrews, Z. B., Liu, Z. W., Walllingford, N., Erion, D. M., Borok, E., Friedman, J. M., et al. (2008). UCP2 mediates ghrelin's action on NPY/AgRP neurons by lowering free radicals. *Nature, 454*, 846–851.

Asakawa, A., Inui, A., Kaga, T., Yuzuriha, H., Nagata, T., Fujimiya, M., et al. (2001). A role of ghrelin in neuroendocrine and behavioral responses to stress in mice. *Neuroendocrinology, 74*, 143–147.

Avena, N. M., Bocarsly, M. E., Hoebel, B. G., & Gold, M. S. (2011). Overlaps in the nosology of substance abuse and overeating: The translational implications of "food addiction" *Current Drug Abuse Reviews, 4*, 133–139.

Avena, N. M., Rada, P., & Hoebel, B. G. (2008). Evidence for sugar addiction: Behavioral and neurochemical effects of intermittent, excessive sugar intake. *Neuroscience and Biobehavioral Reviews, 32*, 20–39.

Badaoui, A., De Saeger, C., Duchemin, J., Gihousse, D., de Timary, P., & Starkel, P. (2008). Alcohol dependence is associated with reduced plasma and fundic ghrelin levels. *European Journal of Clinical Investigation, 38*, 397–403.

Banks, W. A., Burney, B. O., & Robinson, S. M. (2008). Effects of triglycerides, obesity, and starvation on ghrelin transport across the blood-brain barrier. *Peptides, 29*, 2061–2065.

Banks, W. A., Tschop, M., Robinson, S. M., & Heiman, M. L. (2002). Extent and direction of ghrelin transport across the blood-brain barrier is determined by its unique primary structure. *The Journal of Pharmacology and Experimental Therapeutics, 302*, 822–827.

Bassareo, V., & Di Chiara, G. (1999). Modulation of feeding-induced activation of mesolimbic dopamine transmission by appetitive stimuli and its relation to motivational state. *The European Journal of Neuroscience, 11*, 4389–4397.

Bednarek, M. A., Feighner, S. D., Pong, S. S., McKee, K. K., Hreniuk, D. L., Silva, M. V., et al. (2000). Structure-function studies on the new growth hormone-releasing peptide, ghrelin: Minimal sequence of ghrelin necessary for activation of growth hormone secretagogue receptor 1a. *Journal of Medicinal Chemistry, 43*, 4370–4376.

Berthoud, H. R. (2006). Homeostatic and non-homeostatic pathways involved in the control of food intake and energy balance. *Obesity (Silver Spring)*, *14*(Suppl. 5), 197S–200S.

Bloom, S. R., Kuhajda, F. P., Laher, I., Pi-Sunyer, X., Ronnett, G. V., Tan, T. M., et al. (2008). The obesity epidemic: Pharmacological challenges. *Molecular Interventions*, *8*, 82–98.

Bouchard, C. (2008). Gene-environment interactions in the etiology of obesity: Defining the fundamentals. *Obesity (Silver Spring)*, *16*(Suppl. 3), S5–S10.

Briggs, D. I., Enriori, P. J., Lemus, M. B., Cowley, M. A., & Andrews, Z. B. (2010). Diet-induced obesity causes ghrelin resistance in arcuate npy/agrp neurons. *Endocrinology*, *151*, 4745–4755.

Burger, K. S., & Stice, E. (2011). Variability in reward responsivity and obesity: Evidence from brain imaging studies. *Current Drug Abuse Reviews*, *4*, 182–189.

Calissendorff, J., Danielsson, O., Brismar, K., & Rojdmark, S. (2005). Inhibitory effect of alcohol on ghrelin secretion in normal man. *European Journal of Endocrinology*, *152*, 743–747.

Calissendorff, J., Danielsson, O., Brismar, K., & Rojdmark, S. (2006). Alcohol ingestion does not affect serum levels of peptide YY but decreases both total and octanoylated ghrelin levels in healthy subjects. *Metabolism*, *55*, 1625–1629.

Camina, J. P. (2006). Cell biology of the ghrelin receptor. *Journal of Neuroendocrinology*, *18*, 65–76.

Caminos, J. E., Gualillo, O., Lago, F., Otero, M., Blanco, M., Gallego, R., et al. (2005). The endogenous growth hormone secretagogue (ghrelin) is synthesized and secreted by chondrocytes. *Endocrinology*, *146*, 1285–1292.

Caminos, J. E., Nogueiras, R., Blanco, M., Seoane, L. M., Bravo, S., Alvarez, C. V., et al. (2003). Cellular distribution and regulation of ghrelin messenger ribonucleic acid in the rat pituitary gland. *Endocrinology*, *144*, 5089–5097.

Carlini, V. P., Martini, A. C., Schioth, H. B., Ruiz, R. D., Fiol de Cuneo, M., & de Barioglio, S. R. (2008). Decreased memory for novel object recognition in chronically food-restricted mice is reversed by acute ghrelin administration. *Neuroscience*, *153*, 929–934.

Carr, K. A., Daniel, T. O., Lin, H., & Epstein, L. H. (2011). Reinforcement pathology and obesity. *Current Drug Abuse Reviews*, *4*, 190–196.

Castaneda, T. R., Tong, J., Datta, R., Culler, M., & Tschop, M. H. (2010). Ghrelin in the regulation of body weight and metabolism. *Frontiers in Neuroendocrinology*, *31*, 44–60.

Chakrabarti, R. (2009). Pharmacotherapy of obesity: Emerging drugs and targets. *Expert Opinion on Therapeutic Targets*, *13*, 195–207.

Chan, C. B., & Cheng, C. H. (2004). Identification and functional characterization of two alternatively spliced growth hormone secretagogue receptor transcripts from the pituitary of black seabream *Acanthopagrus schlegeli*. *Molecular and Cellular Endocrinology*, *214*, 81–95.

Charoenthongtrakul, S., Giuliana, D., Longo, K. A., Govek, E., Nolan, A., Gagne, S., et al. (2009). Enhanced gastrointestinal motility with orally active ghrelin receptor agonists. *The Journal of Pharmacology and Experimental Therapeutics*, *329*, 1178–1186.

Chen, H. Y., Trumbauer, M. E., Chen, A. S., Weingarth, D. T., Adams, J. R., Frazier, E. G., et al. (2004). Orexigenic action of peripheral ghrelin is mediated by neuropeptide Y and agouti-related protein. *Endocrinology*, *145*, 2607–2612.

Cheng, A. Y., & Leiter, L. A. (2006). Metabolic syndrome under fire: Weighing in on the truth. *The Canadian Journal of Cardiology*, *22*, 379–382.

Chollet, C., Meyer, K., & Beck-Sickinger, A. G. (2009). Ghrelin—A novel generation of anti-obesity drug: Design, pharmacomodulation and biological activity of ghrelin analogues. *Journal of Peptide Science*, *15*, 711–730.

Chuang, J. C., Perello, M., Sakata, I., Osborne-Lawrence, S., Savitt, J. M., Lutter, M., et al. (2011). Ghrelin mediates stress-induced food-reward behavior in mice. *The Journal of Clinical Investigation, 121*, 2684–2692.

Conconi, M. T., Nico, B., Guidolin, D., Baiguera, S., Spinazzi, R., Rebuffat, P., et al. (2004). Ghrelin inhibits FGF-2-mediated angiogenesis in vitro and in vivo. *Peptides, 25*, 2179–2185.

Connelly, D. M., Unwin, J. W., & Taberner, P. V. (1983). The role of the blood glucose level in determining voluntary ethanol consumption in the LACG and diabetogenic C57BL strains of mice. *Biochemical Pharmacology, 32*, 221–226.

Corwin, R. L. (2011). The face of uncertainty eats. *Current Drug Abuse Reviews, 4*, 174–181.

Cowley, M. A., Smith, R. G., Diano, S., Tschop, M., Pronchuk, N., Grove, K. L., et al. (2003). The distribution and mechanism of action of ghrelin in the CNS demonstrates a novel hypothalamic circuit regulating energy homeostasis. *Neuron, 37*, 649–661.

Cummings, D. E., Clement, K., Purnell, J. Q., Vaisse, C., Foster, K. E., Frayo, R. S., et al. (2002). Elevated plasma ghrelin levels in Prader Willi syndrome. *Nature Medicine, 8*, 643–644.

Cummings, D. E., Purnell, J. Q., Frayo, R. S., Schmidova, K., Wisse, B. E., & Weigle, D. S. (2001). A preprandial rise in plasma ghrelin levels suggests a role in meal initiation in humans. *Diabetes, 50*, 1714–1719.

Dahlman, I., & Arner, P. (2007). Obesity and polymorphisms in genes regulating human adipose tissue. *International Journal of Obesity, 31*, 1629–1641.

Dallman, M. F. (2010). Stress-induced obesity and the emotional nervous system. *Trends in Endocrinology and Metabolism, 21*, 159–165.

Date, Y., Kojima, M., Hosoda, H., Sawaguchi, A., Mondal, M. S., Suganuma, T., et al. (2000). Ghrelin, a novel growth hormone-releasing acylated peptide, is synthesized in a distinct endocrine cell type in the gastrointestinal tracts of rats and humans. *Endocrinology, 141*, 4255–4261.

Davenport, A. P., Bonner, T. I., Foord, S. M., Harmar, A. J., Neubig, R. R., Pin, J. P., et al. (2005). International Union of Pharmacology. LVI. Ghrelin receptor nomenclature, distribution, and function. *Pharmacological Reviews, 57*, 541–546.

DelParigi, A., Tschop, M., Heiman, M. L., Salbe, A. D., Vozarova, B., Sell, S. M., et al. (2002). High circulating ghrelin: A potential cause for hyperphagia and obesity in Prader-Willi syndrome. *The Journal of Clinical Endocrinology and Metabolism, 87*, 5461–5464.

Depoortere, I. (2009). Targeting the ghrelin receptor to regulate food intake. *Regulatory Peptides, 156*, 13–23.

Dezaki, K., Kakei, M., & Yada, T. (2007). Ghrelin uses Galphai2 and activates voltage-dependent K+ channels to attenuate glucose-induced Ca2+ signaling and insulin release in islet beta-cells: Novel signal transduction of ghrelin. *Diabetes, 56*, 2319–2327.

Dezaki, K., Sone, H., Koizumi, M., Nakata, M., Kakei, M., Nagai, H., et al. (2006). Blockade of pancreatic islet-derived ghrelin enhances insulin secretion to prevent high-fat diet-induced glucose intolerance. *Diabetes, 55*, 3486–3493.

Diano, S. (2008). Ghrelin's role in feeding behavior and memory performance. *Appetite, 51*, 363.

Dickson, S. L., Egecioglu, E., Landgren, S., Skibicka, K. P., Engel, J. A., & Jerlhag, E. (2011). The role of the central ghrelin system in reward from food and chemical drugs. *Molecular and Cellular Endocrinology, 340*, 80–87.

Dickson, S. L., Hrabovszky, E., Hansson, C., Jerlhag, E., Alvarez-Crespo, M., Skibicka, K. P., et al. (2010). Blockade of central nicotine acetylcholine receptor signaling attenuate ghrelin-induced food intake in rodents. *Neuroscience, 171*, 1180–1186.

Disse, E., Bussier, A. L., Deblon, N., Pfluger, P. T., Tschop, M. H., Laville, M., et al. (2011). Systemic ghrelin and reward: Effect of cholinergic blockade. *Physiology and Behavior, 102*, 481–484.

Disse, E., Bussier, A. L., Veyrat-Durebex, C., Deblon, N., Pfluger, P. T., Tschop, M. H., et al. (2010). Peripheral ghrelin enhances sweet taste food consumption and preference, regardless of its caloric content. *Physiology and Behavior*, *101*, 277–281.

Dixit, V. D., Schaffer, E. M., Pyle, R. S., Collins, G. D., Sakthivel, S. K., Palaniappan, R., et al. (2004). Ghrelin inhibits leptin- and activation-induced proinflammatory cytokine expression by human monocytes and T cells. *The Journal of Clinical Investigation*, *114*, 57–66.

Diz-Chaves, Y. (2011). Ghrelin, appetite regulation, and food reward: Interaction with chronic stress. *International Journal of Peptides*, *2011*, 898450.

Egecioglu, E., Jerlhag, E., Salome, N., Skibicka, K. P., Haage, D., Bohlooly, Y. M., et al. (2010). Ghrelin increases intake of rewarding food in rodents. *Addiction Biology*, *15*, 304–311.

Egecioglu, E., Skibicka, K. P., Hansson, C., Alvarez-Crespo, M., Friberg, P. A., Jerlhag, E., et al. (2011). Hedonic and incentive signals for body weight control. *Reviews in Endocrine & Metabolic Disorders*, *12*, 141–151.

Eglen, R. M., Bosse, R., & Reisine, T. (2007). Emerging concepts of guanine nucleotide-binding protein-coupled receptor (GPCR) function and implications for high throughput screening. *Assay and Drug Development Technologies*, *5*, 425–451.

Eknoyan, G. (2008). Adolphe Quetelet (1796–1874)—The average man and indices of obesity. *Nephrology, Dialysis, Transplantation*, *23*, 47–51.

Finger, B. C., Dinan, T. G., & Cryan, J. F. (2012). Diet-induced obesity blunts the behavioural effects of ghrelin: Studies in a mouse-progressive ratio task. *Psychopharmacology*, *220*, 173–181.

Furness, J. B., Hunne, B., Matsuda, N., Yin, L., Russo, D., Kato, I., et al. (2011). Investigation of the presence of ghrelin in the central nervous system of the rat and mouse. *Neuroscience*, *193*, 1–9.

Gahete, M. D., Cordoba-Chacon, J., Hergueta-Redondo, M., Martinez-Fuentes, A. J., Kineman, R. D., Moreno-Bueno, G., et al. (2011). A novel human ghrelin variant (In1-ghrelin) and ghrelin-O-acyltransferase are overexpressed in breast cancer: Potential pathophysiological relevance. *PLoS One*, *6*, e23302.

Gearhardt, A. N., Corbin, W. R., & Brownell, K. D. (2009a). Food addiction: An examination of the diagnostic criteria for dependence. *Journal of Addiction Medicine*, *3*, 1–7.

Gearhardt, A. N., Corbin, W. R., & Brownell, K. D. (2009b). Preliminary validation of the Yale Food Addiction Scale. *Appetite*, *52*, 430–436.

Gearhardt, A. N., White, M. A., & Potenza, M. N. (2011). Binge eating disorder and food addiction. *Current Drug Abuse Reviews*, *4*, 201–207.

Gearhardt, A. N., Yokum, S., Orr, P. T., Stice, E., Corbin, W. R., & Brownell, K. D. (2011). Neural correlates of food addiction. *Archives of General Psychiatry*, *68*, 808–816.

Ghigo, E., Arvat, E., Giordano, R., Broglio, F., Gianotti, L., Maccario, M., et al. (2001). Biologic activities of growth hormone secretagogues in humans. *Endocrine*, *14*, 87–93.

Gibbs, J., Young, R. C., & Smith, G. P. (1973). Cholecystokinin elicits satiety in rats with open gastric fistulas. *Nature*, *245*, 323–325.

Gibson, E. L. (2006). Emotional influences on food choice: Sensory, physiological and psychological pathways. *Physiology and Behavior*, *89*, 53–61.

Gualillo, O., Lago, F., & Dieguez, C. (2008). Introducing GOAT: A target for obesity and anti-diabetic drugs? *Trends in Pharmacological Sciences*, *29*, 398–401.

Guan, X. M., Yu, H., Palyha, O. C., McKee, K. K., Feighner, S. D., Sirinathsinghji, D. J., et al. (1997). Distribution of mRNA encoding the growth hormone secretagogue receptor in brain and peripheral tissues. *Brain Research. Molecular Brain Research*, *48*, 23–29.

Gutierrez, J. A., Solenberg, P. J., Perkins, D. R., Willency, J. A., Knierman, M. D., Jin, Z., et al. (2008). Ghrelin octanoylation mediated by an orphan lipid transferase. *Proceedings of the National Academy of Sciences of the United States of America*, *105*, 6320–6325.

Hinney, A., Hoch, A., Geller, F., Schafer, H., Siegfried, W., Goldschmidt, H., et al. (2002). Ghrelin gene: Identification of missense variants and a frame shift mutation in extremely obese children and adolescents and healthy normal weight students. *The Journal of Clinical Endocrinology and Metabolism, 87*, 2716.

Hoebel, B. G. (1985). Brain neurotransmitters in food and drug reward. *The American Journal of Clinical Nutrition, 42*, 1133–1150.

Hosoda, H., Kojima, M., Matsuo, H., & Kangawa, K. (2000). Purification and characterization of rat des-Gln14-Ghrelin, a second endogenous ligand for the growth hormone secretagogue receptor. *The Journal of Biological Chemistry, 275*, 21995–22000.

Hosoda, H., Kojima, M., Mizushima, T., Shimizu, S., & Kangawa, K. (2003). Structural divergence of human ghrelin. Identification of multiple ghrelin-derived molecules produced by post-translational processing. *The Journal of Biological Chemistry, 278*, 64–70.

Howard, A. D., Feighner, S. D., Cully, D. F., Arena, J. P., Liberator, P. A., Rosenblum, C. I., et al. (1996). A receptor in pituitary and hypothalamus that functions in growth hormone release. *Science, 273*, 974–977.

Iglesias, M. J., Pineiro, R., Blanco, M., Gallego, R., Dieguez, C., Gualillo, O., et al. (2004). Growth hormone releasing peptide (ghrelin) is synthesized and secreted by cardiomyocytes. *Cardiovascular Research, 62*, 481–488.

Jeffery, P. L., Duncan, R. P., Yeh, A. H., Jaskolski, R. A., Hammond, D. S., Herington, A. C., et al. (2005). Expression of the ghrelin axis in the mouse: An exon 4-deleted mouse proghrelin variant encodes a novel C terminal peptide. *Endocrinology, 146*, 432–440.

Jerlhag, E., Egecioglu, E., Dickson, S. L., Andersson, M., Svensson, L., & Engel, J. A. (2006). Ghrelin stimulates locomotor activity and accumbal dopamine-overflow via central cholinergic systems in mice: Implications for its involvement in brain reward. *Addiction Biology, 11*, 45–54.

Jerlhag, E., Egecioglu, E., Dickson, S. L., Douhan, A., Svensson, L., & Engel, J. A. (2007). Ghrelin administration into tegmental areas stimulates locomotor activity and increases extracellular concentration of dopamine in the nucleus accumbens. *Addiction Biology, 12*, 6–16.

Jiang, H., Betancourt, L., & Smith, R. G. (2006). Ghrelin amplifies dopamine signaling by cross talk involving formation of growth hormone secretagogue receptor/dopamine receptor subtype 1 heterodimers. *Molecular Endocrinology, 20*, 1772–1785.

Kadowaki, T., Yamauchi, T., & Kubota, N. (2008). The physiological and pathophysiological role of adiponectin and adiponectin receptors in the peripheral tissues and CNS. *FEBS Letters, 582*, 74–80.

Kageyama, H., Kitamura, Y., Hosono, T., Kintaka, Y., Seki, M., Takenoya, F., et al. (2008). Visualization of ghrelin-producing neurons in the hypothalamic arcuate nucleus using ghrelin-EGFP transgenic mice. *Regulatory Peptides, 145*, 116–121.

Kamegai, J., Tamura, H., Shimizu, T., Ishii, S., Tatsuguchi, A., Sugihara, H., et al. (2004). The role of pituitary ghrelin in growth hormone (GH) secretion: GH-releasing hormone-dependent regulation of pituitary ghrelin gene expression and peptide content. *Endocrinology, 145*, 3731–3738.

Kampov-Polevoy, A. B., Garbutt, J. C., & Khalitov, E. (2003). Family history of alcoholism and response to sweets. *Alcoholism, Clinical and Experimental Research, 27*, 1743–1749.

Kenny, P. J. (2011a). Common cellular and molecular mechanisms in obesity and drug addiction. *Nature Reviews. Neuroscience, 12*, 638–651.

Kenny, P. J. (2011b). Reward mechanisms in obesity: New insights and future directions. *Neuron, 69*, 664–679.

Kern, A., Albarran-Zeckler, R., Walsh, Heidi E., & Smith, Roy G. (2012). Apo-ghrelin receptor forms heteromers with DRD2 in hypothalamic neurons and is essential for anorexigenic effects of DRD2 agonism. *Neuron, 73*, 317–332.

Kim, D. J., Yoon, S. J., Choi, B., Kim, T. S., Woo, Y. S., Kim, W., et al. (2005). Increased fasting plasma ghrelin levels during alcohol abstinence. *Alcohol and Alcoholism*, *40*, 76–79.

Kineman, R. D., Gahete, M. D., & Luque, R. M. (2007). Identification of a mouse ghrelin gene transcript that contains intron 2 and is regulated in the pituitary and hypothalamus in response to metabolic stress. *Journal of Molecular Endocrinology*, *38*, 511–521.

Kohno, D., Gao, H. Z., Muroya, S., Kikuyama, S., & Yada, T. (2003). Ghrelin directly interacts with neuropeptide-Y-containing neurons in the rat arcuate nucleus: Ca2+ signaling via protein kinase A and N-type channel-dependent mechanisms and cross-talk with leptin and orexin. *Diabetes*, *52*, 948–956.

Kojima, M., Hosoda, H., Date, Y., Nakazato, M., Matsuo, H., & Kangawa, K. (1999). Ghrelin is a growth-hormone-releasing acylated peptide from stomach. *Nature*, *402*, 656–660.

Kojima, M., Hosoda, H., & Kangawa, K. (2001). Purification and distribution of ghrelin: The natural endogenous ligand for the growth hormone secretagogue receptor. *Hormone Research*, *56*(Suppl. 1), 93–97.

Kojima, M., Hosoda, H., & Kangawa, K. (2004). Clinical endocrinology and metabolism. Ghrelin, a novel growth-hormone-releasing and appetite-stimulating peptide from stomach. *Best Practice & Research. Clinical Endocrinology & Metabolism*, *18*, 517–530.

Kojima, M., & Kangawa, K. (2006). Drug insight: The functions of ghrelin and its potential as a multitherapeutic hormone. *Nature Clinical Practice. Endocrinology & Metabolism*, *2*, 80–88.

Kola, B., & Korbonits, M. (2009). Shedding light on the intricate puzzle of ghrelin's effects on appetite regulation. *The Journal of Endocrinology*, *202*, 191–198.

Korbonits, M., Goldstone, A. P., Gueorguiev, M., & Grossman, A. B. (2004). Ghrelin—A hormone with multiple functions. *Frontiers in Neuroendocrinology*, *25*, 27–68.

Korbonits, M., & Grossman, A. B. (2004). Ghrelin: Update on a novel hormonal system. *European Journal of Endocrinology*, *151*(Suppl. 1), S67–S70.

Korbonits, M., Kojima, M., Kangawa, K., & Grossman, A. B. (2001). Presence of ghrelin in normal and adenomatous human pituitary. *Endocrine*, *14*, 101–104.

Kraus, T., Schanze, A., Groschl, M., Bayerlein, K., Hillemacher, T., Reulbach, U., et al. (2005). Ghrelin levels are increased in alcoholism. *Alcoholism, Clinical and Experimental Research*, *29*, 2154–2157.

Kristenssson, E., Sundqvist, M., Astin, M., Kjerling, M., Mattsson, H., Dornonville de la Cour, C., et al. (2006). Acute psychological stress raises plasma ghrelin in the rat. *Regulatory Peptides*, *134*, 114–117.

Kurose, Y., Iqbal, J., Rao, A., Murata, Y., Hasegawa, Y., Terashima, Y., et al. (2005). Changes in expression of the genes for the leptin receptor and the growth hormone-releasing peptide/ghrelin receptor in the hypothalamic arcuate nucleus with long-term manipulation of adiposity by dietary means. *Journal of Neuroendocrinology*, *17*, 331–340.

Landgren, S., Simms, J. A., Thelle, D. S., Strandhagen, E., Bartlett, S. E., Engel, J. A., et al. (2011). The ghrelin signalling system is involved in the consumption of sweets. *PLoS One*, *6*, e18170.

Lawrence, C. B., Snape, A. C., Baudoin, F. M., & Luckman, S. M. (2002). Acute central ghrelin and GH secretagogues induce feeding and activate brain appetite centers. *Endocrinology*, *143*, 155–162.

Leggio, L., Addolorato, G., Cippitelli, A., Jerlhag, E., Kampov-Polevoy, A. B., & Swift, R. M. (2011). Role of feeding-related pathways in alcohol dependence: A focus on sweet preference, NPY, and ghrelin. *Alcoholism, Clinical and Experimental Research*, *35*, 194–202.

Leggio, L., Ray, L. A., Kenna, G. A., & Swift, R. M. (2009). Blood glucose level, alcohol heavy drinking, and alcohol craving during treatment for alcohol dependence: Results from the Combined Pharmacotherapies and Behavioral Interventions for Alcohol

Dependence (COMBINE) Study. *Alcoholism, Clinical and Experimental Research, 33*, 1539–1544.

Leite-Moreira, A. F., & Soares, J. B. (2007). Physiological, pathological and potential therapeutic roles of ghrelin. *Drug Discovery Today, 12*, 276–288.

Li, S., & Loos, R. J. (2008). Progress in the genetics of common obesity: Size matters. *Current Opinion in Lipidology, 19*, 113–121.

Liu, G., Fortin, J. P., Beinborn, M., & Kopin, A. S. (2007). Four missense mutations in the ghrelin receptor result in distinct pharmacological abnormalities. *The Journal of Pharmacology and Experimental Therapeutics, 322*, 1036–1043.

Loos, R. J., & Bouchard, C. (2003). Obesity—Is it a genetic disorder? *Journal of Internal Medicine, 254*, 401–425.

Loos, R. J., Lindgren, C. M., Li, S., Wheeler, E., Zhao, J. H., Prokopenko, I., et al. (2008). Common variants near MC4R are associated with fat mass, weight and risk of obesity. *Nature Genetics, 40*, 768–775.

Lu, S., Guan, J. L., Wang, Q. P., Uehara, K., Yamada, S., Goto, N., et al. (2002). Immunocytochemical observation of ghrelin-containing neurons in the rat arcuate nucleus. *Neuroscience Letters, 321*, 157–160.

Lu, S. C., Xu, J., Chinookoswong, N., Liu, S., Steavenson, S., Gegg, C., et al. (2009). An acyl-ghrelin-specific neutralizing antibody inhibits the acute ghrelin-mediated orexigenic effects in mice. *Molecular Pharmacology, 75*, 901–907.

Lutter, M., Sakata, I., Osborne-Lawrence, S., Rovinsky, S. A., Anderson, J. G., Jung, S., et al. (2008). The orexigenic hormone ghrelin defends against depressive symptoms of chronic stress. *Nature Neuroscience, 11*, 752–753.

Luttrell, L. M. (2008). Reviews in molecular biology and biotechnology: Transmembrane signaling by G protein-coupled receptors. *Molecular Biotechnology, 39*, 239–264.

Lutz, T. A. (2006). Amylinergic control of food intake. *Physiology and Behavior, 89*, 465–471.

Mager, U., Degenhardt, T., Pulkkinen, L., Kolehmainen, M., Tolppanen, A. M., Lindstrom, J., et al. (2008). Variations in the ghrelin receptor gene associate with obesity and glucose metabolism in individuals with impaired glucose tolerance. *PLoS One, 3*, e2941.

Malagon, M. M., Luque, R. M., Ruiz-Guerrero, E., Rodriguez-Pacheco, F., Garcia-Navarro, S., Casanueva, F. F., et al. (2003). Intracellular signaling mechanisms mediating ghrelin-stimulated growth hormone release in somatotropes. *Endocrinology, 144*, 5372–5380.

Marti, A., Moreno-Aliaga, M. J., Hebebrand, J., & Martinez, J. A. (2004). Genes, lifestyles and obesity. *International Journal of Obesity and Related Metabolic Disorders, 28*(Suppl. 3), S29–S36.

Menyhert, J., Wittmann, G., Hrabovszky, E., Szlavik, N., Keller, E., Tschop, M., et al. (2006). Distribution of ghrelin-immunoreactive neuronal networks in the human hypothalamus. *Brain Research, 1125*, 31–36.

Mikhail, N. (2009). The metabolic syndrome: Insulin resistance. *Current Hypertension Reports, 11*, 156–158.

Mondal, M. S., Date, Y., Yamaguchi, H., Toshinai, K., Tsuruta, T., Kangawa, K., et al. (2005). Identification of ghrelin and its receptor in neurons of the rat arcuate nucleus. *Regulatory Peptides, 126*, 55–59.

Moulin, A., Ryan, J., Martinez, J., & Fehrentz, J. A. (2007). Recent developments in ghrelin receptor ligands. *ChemMedChem, 2*, 1242–1259.

Muccioli, G., Baragli, A., Granata, R., Papotti, M., & Ghigo, E. (2007). Heterogeneity of ghrelin/growth hormone secretagogue receptors. Toward the understanding of the molecular identity of novel ghrelin/GHS receptors. *Neuroendocrinology, 86*, 147–164.

Muccioli, G., Tschop, M., Papotti, M., Deghenghi, R., Heiman, M., & Ghigo, E. (2002). Neuroendocrine and peripheral activities of ghrelin: Implications in metabolism and obesity. *European Journal of Pharmacology, 440*, 235–254.

Mundinger, T. O., Cummings, D. E., & Taborsky, G. J., Jr. (2006). Direct stimulation of ghrelin secretion by sympathetic nerves. *Endocrinology, 147*, 2893–2901.

Nakazato, M., Murakami, N., Date, Y., Kojima, M., Matsuo, H., Kangawa, K., et al. (2001). A role for ghrelin in the central regulation of feeding. *Nature, 409*, 194–198.

Naleid, A. M., Grace, M. K., Cummings, D. E., & Levine, A. S. (2005). Ghrelin induces feeding in the mesolimbic reward pathway between the ventral tegmental area and the nucleus accumbens. *Peptides, 26*, 2274–2279.

Obici, S. (2009). Minireview: Molecular targets for obesity therapy in the brain. *Endocrinology, 150*(6), 2512–2517.

Ochi, M., Tominaga, K., Tanaka, F., Tanigawa, T., Shiba, M., Watanabe, T., et al. (2008). Effect of chronic stress on gastric emptying and plasma ghrelin levels in rats. *Life Sciences, 82*, 862–868.

Ohgusu, H., Shirouzu, K., Nakamura, Y., Nakashima, Y., Ida, T., Sato, T., et al. (2009). Ghrelin O-acyltransferase (GOAT) has a preference for n-hexanoyl-CoA over n-octanoyl-CoA as an acyl donor. *Biochemical and Biophysical Research Communications, 386*, 153–158.

Oliver, G., & Wardle, J. (1999). Perceived effects of stress on food choice. *Physiology and Behavior, 66*, 511–515.

Olszewski, P. K., Li, D., Grace, M. K., Billington, C. J., Kotz, C. M., & Levine, A. S. (2003). Neural basis of orexigenic effects of ghrelin acting within lateral hypothalamus. *Peptides, 24*, 597–602.

Overstreet, D. H., Kampov-Polevoy, A. B., Rezvani, A. H., Murrelle, L., Halikas, J. A., & Janowsky, D. S. (1993). Saccharin intake predicts ethanol intake in genetically heterogeneous rats as well as different rat strains. *Alcoholism, Clinical and Experimental Research, 17*, 366–369.

Parylak, S. L., Koob, G. F., & Zorrilla, E. P. (2011). The dark side of food addiction. *Physiology and Behavior, 104*, 149–156.

Perello, M., Sakata, I., Birnbaum, S., Chuang, J. C., Osborne-Lawrence, S., Rovinsky, S. A., et al. (2010). Ghrelin increases the rewarding value of high-fat diet in an orexin-dependent manner. *Biological Psychiatry, 67*, 880–886.

Pfluger, P. T., Kirchner, H., Gunnel, S., Schrott, B., Perez-Tilve, D., Fu, S., et al. (2008). Simultaneous deletion of ghrelin and its receptor increases motor activity and energy expenditure. *American Journal of Physiology. Gastrointestinal and Liver Physiology, 294*, G610–G618.

Raspopow, K., Abizaid, A., Matheson, K., & Anisman, H. (2010). Psychosocial stressor effects on cortisol and ghrelin in emotional and non-emotional eaters: Influence of anger and shame. *Hormones and Behavior, 58*, 677–684.

Richardson, N. R., & Gratton, A. (1998). Changes in medial prefrontal cortical dopamine levels associated with response-contingent food reward: An electrochemical study in rat. *The Journal of Neuroscience, 18*, 9130–9138.

Rindi, G., Torsello, A., Locatelli, V., & Solcia, E. (2004). Ghrelin expression and actions: A novel peptide for an old cell type of the diffuse endocrine system. *Experimental Biology and Medicine (Maywood, N.J.), 229*, 1007–1016.

Sato, T., Fukue, Y., Teranishi, H., Yoshida, Y., & Kojima, M. (2005). Molecular forms of hypothalamic ghrelin and its regulation by fasting and 2-deoxy-d-glucose administration. *Endocrinology, 146*, 2510–2516.

Schellekens, H., Dinan, T. G., & Cryan, J. F. (2009). Lean mean fat reducing "ghrelin" machine: Hypothalamic ghrelin and ghrelin receptors as therapeutic targets in obesity. *Neuropharmacology, 58*, 2–16.

Schellekens, H., et al. (2012a). Promiscuous dimerization of the growth hormone secretagogue receptor (GHS-R1a) attenuates ghrelin-mediated signalling. *J Biol Chem*.

Schellekens, H., et al. (2012b). Ghrelin signalling and obesity: At the interface of stress, mood and food reward. *Pharmacology and therapeutics, 135*(3), 316–326.

Seim, I., Collet, C., Herington, A. C., & Chopin, L. K. (2007). Revised genomic structure of the human ghrelin gene and identification of novel exons, alternative splice variants and natural antisense transcripts. *BMC Genomics*, *8*, 298.

Seim, I., Josh, P., Cunningham, P., Herington, A., & Chopin, L. (2011). Ghrelin axis genes, peptides and receptors: Recent findings and future challenges. *Molecular and Cellular Endocrinology*, *340*, 3–9.

Shimbara, T., Mondal, M. S., Kawagoe, T., Toshinai, K., Koda, S., Yamaguchi, H., et al. (2004). Central administration of ghrelin preferentially enhances fat ingestion. *Neuroscience Letters*, *369*, 75–79.

Shrestha, Y. B., Wickwire, K., & Giraudo, S. (2009). Effect of reducing hypothalamic ghrelin receptor gene expression on energy balance. *Peptides*, *30*, 1336–1341.

Shuto, Y., Shibasaki, T., Otagiri, A., Kuriyama, H., Ohata, H., Tamura, H., et al. (2002). Hypothalamic growth hormone secretagogue receptor regulates growth hormone secretion, feeding, and adiposity. *The Journal of Clinical Investigation*, *109*, 1429–1436.

Skibicka, K. P., & Dickson, S. L. (2011). Ghrelin and food reward: The story of potential underlying substrates. *Peptides*, *32*, 2265–2273.

Skibicka, K. P., Hansson, C., Alvarez-Crespo, M., Friberg, P. A., & Dickson, S. L. (2011). Ghrelin directly targets the ventral tegmental area to increase food motivation. *Neuroscience*, *180*, 129–137.

Skibicka, K. P., Hansson, C., Egecioglu, E., & Dickson, S. L. (2012). Role of ghrelin in food reward: Impact of ghrelin on sucrose self-administration and mesolimbic dopamine and acetylcholine receptor gene expression. *Addiction Biology*, *17*(1), 95–107.

Soares, J. B., & Leite-Moreira, A. F. (2008). Ghrelin, des-acyl ghrelin and obestatin: Three pieces of the same puzzle. *Peptides*, *29*, 1255–1270.

Soares, J. B., Roncon-Albuquerque, R., Jr., & Leite-Moreira, A. (2008). Ghrelin and ghrelin receptor inhibitors: Agents in the treatment of obesity. *Expert Opinion on Therapeutic Targets*, *12*, 1177–1189.

Stengel, A., Wang, L., & Tache, Y. (2011). Stress-related alterations of acyl and desacyl ghrelin circulating levels: Mechanisms and functional implications. *Peptides*, *32*, 2208–2217.

Stice, E., Yokum, S., Blum, K., & Bohon, C. (2010). Weight gain is associated with reduced striatal response to palatable food. *The Journal of Neuroscience*, *30*, 13105–13109.

Stoeckel, L. E., Weller, R. E., Cook, E. W., 3rd, Twieg, D. B., Knowlton, R. C., & Cox, J. E. (2008). Widespread reward-system activation in obese women in response to pictures of high-calorie foods. *NeuroImage*, *41*, 636–647.

Stricker-Krongrad, A., Richy, S., & Beck, B. (2002). Orexins/hypocretins in the ob/ob mouse: Hypothalamic gene expression, peptide content and metabolic effects. *Regulatory Peptides*, *104*, 11–20.

Sun, Y., Ahmed, S., & Smith, R. G. (2003). Deletion of ghrelin impairs neither growth nor appetite. *Molecular and Cellular Biology*, *23*, 7973–7981.

Sun, Y., Butte, N. F., Garcia, J. M., & Smith, R. G. (2008). Characterization of adult ghrelin and ghrelin receptor knockout mice under positive and negative energy balance. *Endocrinology*, *149*, 843–850.

Sun, Y., Wang, P., Zheng, H., & Smith, R. G. (2004). Ghrelin stimulation of growth hormone release and appetite is mediated through the growth hormone secretagogue receptor. *Proceedings of the National Academy of Sciences of the United States of America*, *101*, 4679–4684.

Taheri, S., Lin, L., Austin, D., Young, T., & Mignot, E. (2004). Short sleep duration is associated with reduced leptin, elevated ghrelin, and increased body mass index. *PLoS Medicine*, *1*, e62.

Tanaka, M., Hayashida, Y., Nakao, N., Nakai, N., & Nakashima, K. (2001). Testis-specific and developmentally induced expression of a ghrelin gene-derived transcript that encodes a novel polypeptide in the mouse. *Biochimica et Biophysica Acta*, *1522*, 62–65.

Tolle, V., & Low, M. J. (2008). In vivo evidence for inverse agonism of Agouti-related peptide in the central nervous system of proopiomelanocortin-deficient mice. *Diabetes, 57*, 86–94.

Toshinai, K., Date, Y., Murakami, N., Shimada, M., Mondal, M. S., Shimbara, T., et al. (2003). Ghrelin-induced food intake is mediated via the orexin pathway. *Endocrinology, 144*, 1506–1512.

Tovar, S., Nogueiras, R., Tung, L. Y., Castaneda, T. R., Vazquez, M. J., Morris, A., et al. (2005). Central administration of resistin promotes short-term satiety in rats. *European Journal of Endocrinology, 153*, R1–R5.

Tschop, M., Smiley, D. L., & Heiman, M. L. (2000). Ghrelin induces adiposity in rodents. *Nature, 407*, 908–913.

Turnbaugh, P. J., Ley, R. E., Mahowald, M. A., Magrini, V., Mardis, E. R., & Gordon, J. I. (2006). An obesity-associated gut microbiome with increased capacity for energy harvest. *Nature, 444*, 1027–1031.

Ukkola, O., Ravussin, E., Jacobson, P., Snyder, E. E., Chagnon, M., Sjostrom, L., et al. (2001). Mutations in the preproghrelin/ghrelin gene associated with obesity in humans. *The Journal of Clinical Endocrinology and Metabolism, 86*, 3996–3999.

van der Lely, A. J., Tschop, M., Heiman, M. L., & Ghigo, E. (2004). Biological, physiological, pathophysiological, and pharmacological aspects of ghrelin. *Endocrine Reviews, 25*, 426–457.

Venkova, K., & Greenwood-Van Meerveld, B. (2008). Application of ghrelin to gastrointestinal diseases. *Current Opinion in Investigational Drugs, 9*, 1103–1107.

Volkow, N. D., Wang, G. J., & Baler, R. D. (2010). Reward, dopamine and the control of food intake: Implications for obesity. *Trends in Cognitive Sciences, 15*, 37–46.

Volkow, N. D., Wang, G. J., Fowler, J. S., Tomasi, D., & Baler, R. (2012). Food and drug reward: Overlapping circuits in human obesity and addiction. *Current Topics in Behavioral Neurosciences, 11*, 1–24.

Wang, G. J., Volkow, N. D., Logan, J., Pappas, N. R., Wong, C. T., Zhu, W., et al. (2001). Brain dopamine and obesity. *The Lancet, 357*, 354–357.

Wang, G. J., Volkow, N. D., Thanos, P. K., & Fowler, J. S. (2009). Imaging of brain dopamine pathways: Implications for understanding obesity. *Journal of Addiction Medicine, 3*, 8–18.

Wells, T. (2009). Ghrelin—Defender of fat. *Progress in Lipid Research, 48*, 257–274.

Willer, C. J., Speliotes, E. K., Loos, R. J., Li, S., Lindgren, C. M., Heid, I. M., et al. (2009). Six new loci associated with body mass index highlight a neuronal influence on body weight regulation. *Nature Genetics, 41*, 25–34.

Wise, R. A. (2006). Role of brain dopamine in food reward and reinforcement. *Philosophical Transactions of the Royal Society of London. Series B, Biological Sciences, 361*, 1149–1158.

Wise, R. A., & Rompre, P. P. (1989). Brain dopamine and reward. *Annual Review of Psychology, 40*, 191–225.

Wortley, K. E., Anderson, K. D., Garcia, K., Murray, J. D., Malinova, L., Liu, R., et al. (2004). Genetic deletion of ghrelin does not decrease food intake but influences metabolic fuel preference. *Proceedings of the National Academy of Sciences of the United States of America, 101*, 8227–8232.

Wortley, K. E., del Rincon, J. P., Murray, J. D., Garcia, K., Iida, K., Thorner, M. O., et al. (2005). Absence of ghrelin protects against early-onset obesity. *The Journal of Clinical Investigation, 115*, 3573–3578.

Wren, A. M., Seal, L. J., Cohen, M. A., Brynes, A. E., Frost, G. S., Murphy, K. G., et al. (2001). Ghrelin enhances appetite and increases food intake in humans. *The Journal of Clinical Endocrinology and Metabolism, 86*, 5992.

Yamanaka, A., Beuckmann, C. T., Willie, J. T., Hara, J., Tsujino, N., Mieda, M., et al. (2003). Hypothalamic orexin neurons regulate arousal according to energy balance in mice. *Neuron, 38*, 701–713.

Yang, J., Brown, M. S., Liang, G., Grishin, N. V., & Goldstein, J. L. (2008). Identification of the acyltransferase that octanoylates ghrelin, an appetite-stimulating peptide hormone. *Cell*, *132*, 387–396.

Yang, J., Zhao, T. J., Goldstein, J. L., & Brown, M. S. (2008). Inhibition of ghrelin O-acyltransferase (GOAT) by octanoylated pentapeptides. *Proceedings of the National Academy of Sciences of the United States of America*, *105*, 10750–10755.

Zhang, Y., Proenca, R., Maffei, M., Barone, M., Leopold, L., & Friedman, J. M. (1994). Positional cloning of the mouse obese gene and its human homologue. *Nature*, *372*, 425–432.

Zhang, J. V., Ren, P. G., Avsian-Kretchmer, O., Luo, C. W., Rauch, R., Klein, C., et al. (2005). Obestatin, a peptide encoded by the ghrelin gene, opposes ghrelin's effects on food intake. *Science*, *310*, 996–999.

Zhao, T. J., Sakata, I., Li, R. L., Liang, G., Richardson, J. A., Brown, M. S., et al. (2010). Ghrelin secretion stimulated by {beta}1-adrenergic receptors in cultured ghrelinoma cells and in fasted mice. *Proceedings of the National Academy of Sciences of the United States of America*, *107*, 15868–15873.

Zigman, J. M., Jones, J. E., Lee, C. E., Saper, C. B., & Elmquist, J. K. (2006). Expression of ghrelin receptor mRNA in the rat and the mouse brain. *The Journal of Comparative Neurology*, *494*, 528–548.

Zigman, J. M., Nakano, Y., Coppari, R., Balthasar, N., Marcus, J. N., Lee, C. E., et al. (2005). Mice lacking ghrelin receptors resist the development of diet-induced obesity. *The Journal of Clinical Investigation*, *115*, 3564–3572.

Zimmermann, U. S., Buchmann, A., Steffin, B., Dieterle, C., & Uhr, M. (2007). Alcohol administration acutely inhibits ghrelin secretion in an experiment involving psychosocial stress. *Addiction Biology*, *12*, 17–21.

Zito, K. A., Vickers, G., Telford, L., & Roberts, D. C. (1984). Experimentally induced glucose intolerance increases oral ethanol intake in rats. *Alcohol*, *1*, 257–261.

Zorrilla, E. P., Iwasaki, S., Moss, J. A., Chang, J., Otsuji, J., Inoue, K., et al. (2006). Vaccination against weight gain. *Proceedings of the National Academy of Sciences of the United States of America*, *103*, 13226–13231.

CHAPTER FOURTEEN

Endocannabinoids and Obesity

Guoxun Chen[*,1], **Zhen Pang**[†]
[*]Department of Nutrition, University of Tennessee at Knoxville, Knoxville, Tennessee, USA
[†]Department of Immuno-inflammation, Sanofi US, Bridgewater, New Jersey, USA
[1]Corresponding author: e-mail address: gchen6@utk.edu

Contents

Vitamins and Hormones, Volume 91
ISSN 0083-6729
http://dx.doi.org/10.1016/B978-0-12-407766-9.00014-6

Abstract

A safe and effective antiobesity drug is needed to combat the global obesity epidemic. The discovery of cannabinoids from medicinal herbs has revealed the endocannabinoid system (ECS) in animals and humans, which regulates various physiological activities such as feeding, thermogenesis, and body weight (BW). Although cannabinoid receptors 1 (CB1) antagonists have shown antiobesity efficacies in animal models and in the clinic, they failed to establish as a treatment due to their psychological side effects. Recent studies indicate that CB1 in various peripheral tissues may mediate some of the therapeutic effects of CB1 antagonists, such as improved lipid and glucose homeostasis. It rationalizes the development of compounds with limited brain penetration, for minimizing the side effects while retaining the therapeutic efficacies. A survey of the literature has revealed some controversies about how the ECS affects obesity. This review summarizes the research progresses and discusses some future perspectives.

1. INTRODUCTION

It is essential for cells in a multicellular organism to receive environmental stimuli and respond accordingly. The long evolutionary process has equipped human and animal bodies with a variety of resilient and efficient systems for this purpose. The rise of obese population in the United States has become a concern of public health (Haslam & James, 2005). The interactions of genes and environments/nutritional status play a role in the current epidemic of obesity (Popkin & Gordon-Larsen, 2004; Wells, 2009). In the search for the obesity treatments, a group of cell membrane receptors for mediating the actions of cannabinoids, the active ingredients of medicinal herb *Cannabis sativa*, have come to play (Di Marzo, 2008b; Felder & Glass, 1998; Mechoulam & Hanuš, 2000). The withdrawal of drugs targeting cannabinoid receptor 1 (CB1) for obesity treatment (Jones, 2008) not only shows the challenges facing the development of antiobesity drugs but also demonstrates opportunities for exploring novel ideas based on the lessons learned (Di Marzo & Després, 2009). This review tries to summarize the connection between cannabinoid receptors and obesity, and highlight the recent progresses toward understanding the mechanisms of actions of CB1 antagonists.

2. DISCOVERY OF THE ENDOCANNABINOID SYSTEM

2.1. The cannabinoids

The *C. sativa* is a herbaceous flowering plant, which can grow in a variety of soil and climate conditions. The use of cannabis as a medicine can be dated back to thousand years ago in many parts of the world (Grinspoon & Bakalar, 1997).

Physical evidence of medicinal use of cannabis in the Middle East was discovered in a tomb dating to the fourth century AD (Zlas et al., 1993). Hashish (marijuana), the psychoactive preparation of *C. sativa*, is one of the most widely used illicit drugs. Medicinal uses include treatment for nausea associated with chemotherapy, arthritic pain, epilepsy, hypertension, and cachexia. The first active component of hashish was isolated from its hexane extract and indentified as Δ^9-tetrahydrocannabinol (Δ^9-THC, originally named Δ^1-THC), which caused ataxia in experimental dogs (Gaoni & Mechoulam, 1964). Subsequently, Δ^8-THC (originally named $\Delta^{1(6)}$-THC) as the second psychoactive component was reported (Hively, Mosher, & Hoffmann, 1966). Later on, the psychopharmacological activities of some naturally occurring and synthetic cannabinoids were tested in a variety of animals (Grunfeld & Edery, 1969).

The term of cannabinoids was introduced to mean the C21 compounds typically present in *C. sativa*, a definition that was focused more on the origin of the molecules regardless of their functions (Mechoulam, 1970). With increased understanding of the physiological functions of natural and synthetic compounds, the term is frequently used for those that are pharmacologically active (Pacher, Btkai, & Kunos, 2006; van Diepen, Schlicker, & Michel, 2008).

2.2. Identification of cannabinoid receptor

The analysis of the synthetic cannabinoids indicated that there are structural requirements for their activities (Edery, Grunfeld, Ben-Zvi, & Mechoulam, 1971). The potencies of THC stereoisomers were compared in different experimental animals and were found to be dependent on the animal species or pharmacological effects tested (Martin, Balster, Razdan, Harris, & Dewey, 1981; Wilson, May, Martin, & Dewey, 1976). The hydrophobic nature of THC makes it difficult to determine specific binding (Harris, Carchman, & Martin, 1978). The development of methods to efficiently synthesize optically active cannabinoids and their metabolites helped to understand the relationships of structural features and activities (Mechoulam, Braun, & Gaoni, 1972; Mechoulam, Varconi, Ben-Zvi, Edery, & Grunfeld, 1972). The modifications of Δ^9-THC resulted in stereo-specific derivatives which are more potent analgetics than Δ^9-THC and suitable for receptor binding and distribution studies (Johnson et al., 1981). The discovery of a highly potent analgesic cannabinoid compound, CP-55,940 (Howlett, Johnson, Melvin, & Milne, 1988), allowed the determination, characterization, and distribution of cannabinoid receptors in rat brain (Devane, Dysarz, Johnson, Melvin, & Howlett, 1988; Herkenham et al., 1991). Subsequent analysis of cellular effects of these compounds showed that they inhibited cAMP production in cells, such as the N18TG2 neuroblastoma cell line (Allyn, 1984; Howlett & Fleming, 1984; Martin, 1986). These characteristics of cannabinoid effects suggested that their activities are probably mediated by a membrane receptor.

The first cannabinoid receptor (CB1) was cloned from a rat cerebral cortex cDNA library. Its identification was supported by the similarity of its mRNA distribution and the cannabinoid binding patterns in the brain, its predicted activity on cAMP, and the confirmation of CB1 as a G-protein-coupled receptor (GPCR) (Matsuda, Lolait, Brownstein, Young, & Bonner, 1990). Subsequently, the CB1 receptors were cloned from human (Gérard, Mollereau, Vassart, & Parmentier, 1991), mouse (Abood, Ditto, Noel, Showalter, & Tao, 1997), fish (Soderstrom & Johnson, 2001; Yamaguchi, Macrae, & Brenner, 1996), newt (Soderstrom, Leid, Moore, & Murray, 2000), frog (Cottone, Salio, Conrath, & Franzoni, 2003), and chicken. Two more splice variants of CB1 receptors have been identified (Ryberg et al., 2005; Shire et al., 1995). In addition to their expression in the brain, human CB1a (with an altered amino-terminus) and CB1b (with an in-frame deletion of 33 amino acids at the amino-terminus) isoforms are detected in a variety of human tissues (Ryberg et al., 2005; Shire et al., 1995). Moreover, a polymorphism of silent mutation in the coding sequence of human CB1 receptor gene (*CNR1*) was also observed (Gadzicki, Müller-Vahl, & Stuhrmann, 1999).

The attempt to hunt for novel GPCRs in human myeloid cells led to the identification of a cDNA with homology to the CB1 and high level of expression in spleen, but not brain, which was designated CB2 (Munro, Thomas, & Abu-Shaar, 1993). After that, the CB2 cDNAs for mouse (Shire et al., 1996), rat (Griffin, Tao, & Abood, 2000), dog (Ndong, O'Donnell, Ahmad, & Groblewski, 2011), and zebra fish (Rodriguez-Martin et al., 2007) were cloned.

Since there was binding of agonist to the central nerve system (CNS) regions of CB1 knockout (KO) mice, and anandamide had residual effect on the brain membranes of the KO mice in the presence of CB1 and CB2 antagonists, the existence of additional cannabinoid receptor, other than CB1 and CB2, was suggested (Breivogel, Griffin, Di Marzo, & Martin, 2001; Di Marzo et al., 2000). Recently, GPR55, an orphan GPCR, which is considered as lysophosphatidylinositol receptor (Oka, Nakajima, Yamashita, Kishimoto, & Sugiura, 2007), has been suggested as an atypical cannabinoid receptor (Hiley & Kaup, 2007; Kapur et al., 2009; Ryberg et al., 2007).

The cloning of CB1 and CB2 allowed the discovery for receptor-specific compounds, which led to the identification of the first high affinity and orally active CB1 antagonist, SR141716A, also known as rimonabant

(Rinaldi-Carmona et al., 1994). SR141716A has a 1000-fold higher affinity for the CB1 than for CB2. SR141716A blocks the inhibitory effects of cannabinoids on the foskolin-induced adenylyl cyclase activity in CHO cells transfected with CB1, but not CB2 cDNA. It is able to antagonize the pharmacological effects of cannabinoid receptor agonists *in vitro* and *in vivo* (Rinaldi-Carmona et al., 1994).

2.3. Endocannabinoids

After the cloning of cannabinoid receptors, attempts were made to identify the endogenous molecules that bind to CB1. Organic porcine brain extract and its fractions were analyzed for their abilities to displace the binding of $[^3H]$HU-243 in rat synaptosomal membranes. From the fractions that had more effects on the binding to the membrane, but less effects on the binding to the wall of the test tubes, the first endogenous agonist of CB1, arachidonoylethanolamide, was isolated (0.4 pmol/mg protein) and named anandamide, after the Indian Sanskrit word "ananda," meaning bliss (Devane et al., 1992). The second identified endogenous molecule with the functions of cannabinoids is 2-arachidonoyl glycerol (2-AG). It was isolated from canine intestine due to its binding to CB1 or CB2 and inhibition of cAMP production in isolated mouse spleen cells (Mechoulam et al., 1995). It was also tested directly for its ability to compete with the specific binding of $[^3H]$CP55940, an agonist for CB1 and CB2 (Sugiura et al., 1995). Additionally, 2-arachidonyl glycerol ether (noladin ether), which is an ether-type endocannabinoid, was isolated from the water–methanol layer of porcine brain extracted with chloroform/methanol, due to its ability to displace the binding of $[^3H]$HU-243 to CB2. In the process to measure anandamide in brain and peripheral tissues, *O*-arachidonoyl ethanolamine (virodhamine) was found to act as an antagonist to CB1 and agonist to CB2 (Porter et al., 2002). Virodhamine has the same molecular mass as anandamide and is arachidonic acid (AA) and ethanolamine linked by an ester bond in the opposite orientation of the amide linkage in anandamide. It is named from the Indian Sanskrit word *virodha*, meaning opposition (Porter et al., 2002).

2.4. Metabolism of endocannabinoids

Endocannabinoids are synthesized and degraded in the body through enymatic reactions as shown in Fig. 14.1. Anandamide can be synthesized through two routes. The first pathway involves a specific phospholipase D, which produces anandamide after the cleavage of *N*-arachidonoyl phosphatidylethanolamine

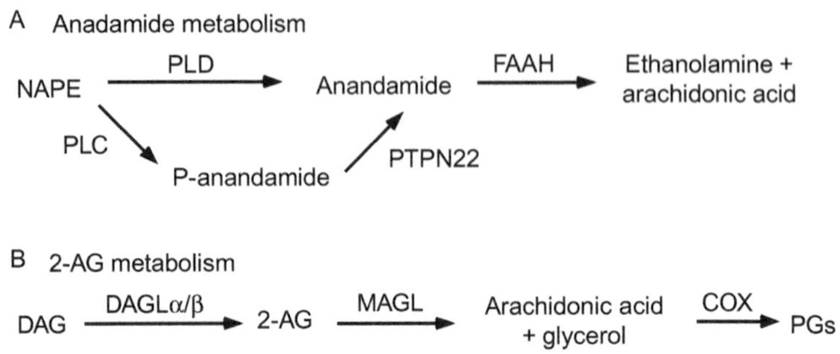

Figure 14.1 Enzymatic reactions for the synthesis and degradation of anandamide (A) and 2-AG (B). COX, cyclo-oxygenase; DAG, diacylglycerol; DAGL, diacylglycerol lipase; FAAH, fatty acid amide hydrolase; MAGL, monoacylglycerol lipase; NAPE, *N*-arachidonoyl phosphatidylethanolamine; PLC, phospholipase C; PLD, phospholipase D; P-anandamide, phosphoanandamide; PTPN22, protein tyrosine phosphatase N22.

(NAPE) (Di Marzo et al., 1994; Okamoto, Morishita, Tsuboi, Tonai, & Ueda, 2004). The second pathway mediates anandamide synthesis in mouse brain and RAW264.7 mouse macrophages, in which phospholipase C catalyzes the cleavage of NAPE to produce phosphoanandamide (Liu et al., 2006), which is then dephosphorylated into anandamide by a protein tyrosine phosphatase PTPN22 (Cohen, Dadi, Shaoul, Sharfe, & Roifman, 1999). This pathway is upregulated by endotoxins in macrophages and may be responsible for the hypotension of septic shock and advanced liver cirrhosis (Liu et al., 2006). Anandamide is thought to be degraded by fatty acid amide hydrolase (FAAH), a serine hydrolase which also has high activity against oleamide, a natural sleep-inducing substance (Cravatt et al., 1996).

2-AG can be generated from hydrolysis of 2-arachidonoyl diacylglycerol, which can be derived from phospholipids or triacylglycerols (Di Marzo, 2008a). The hydrolysis is catalyzed by specific *sn*-1 diacylglycerol lipases (DAGL). Two DAGLs, designated DAGLα and DAGLβ have been identified in human genome. Both of them are membrane-associated proteins. It seems that DAGLβ prefers *sn*-1-oleoyl-2-acyl-*sn*-glycerols with linoleic ≥ oleic > arachidonic > stearic acid on the two position, whereas DAGLα works equally well to all of them. In addition to high expression levels in the brain, they are also expressed in a variety of human tissues (Bisogno et al., 2003). The degradation of 2-AG is thought to be mediated by monoacylglycerol lipase (MAGL), a serine hydrolase which is expressed in a variety of tissues with high expression levels in fat (Karlsson, Contreras, Hellman, Tornqvist, & Holm, 1997; Tornqvist & Belfrage, 1976).

3. OBESITY AND ENDOCANNABINOID SYSTEM

3.1. The obesity epidemic

Obesity is defined as the excessive accumulation of body fat, which is typically determined by the body mass index. The high obesity prevalence in the U.S. population (Yanovski & Yanovski, 2011) indicates that there will be an increase of the number of patients with noninsulin-dependent diabetes mellitus (NIDDM) and other metabolic diseases (Schulze & Hu, 2005). This has become a concern of public health (Haslam & James, 2005). Reduction of BW in obese and diabetic patients improves insulin sensitivity (Reaven, 2005) and corrects abnormalities of NIDDM (Dixon et al., 2008). Both genetic and dietary factors contribute to the development of obesity. Overconsumption of nutrients, such as fructose in sweetened beverages, has also been suggested to play a role in the rise of obese population (Bray, Nielsen, & Popkin, 2004). Genetic mutations, such as those in leptin and its receptor, cause the development of obesity (Friedman, 2009). The association of genes with the development of human obesity or NIDDM has been reported (Gaulton et al., 2008; O'Rahilly & Farooqi, 2008). Genes regulating food intake (FI) often control energy expenditure as well, demonstrating their strong influence on BW (Friedman, 2004; O'Rahilly & Farooqi, 2008). The link between overfeeding and obesity is obvious. Therefore, any system with the potential of controlling FI and energy expenditure is likely to regulate the obesity development.

3.2. The relationship of cannabinoids with feeding behavior and body weight

The use of cannabis has been associated with increased appetite (Abel, 1971, 1975; Hollister, 1971; Mattes, Engelman, Shaw, & Elsohly, 1994), which has been attributed to the weight gain of those users (Greenberg, Kuehnle, Mendelson, & Bernstein, 1976). Smoking of cannabis in patients with a history of HIV infection resulted in elevated ghrelin and leptin, but not insulin levels, demonstrating the effects of cannabinoids on the hormones that control appetite in humans (Riggs et al., 2012). Cannabinoids have been prescribed as an appetite-enhancing medicine for patients with AIDS and cancer experiencing chemotherapy (Beal et al., 1995; Berry & Mechoulam, 2002; Hanus et al., 2003).

2-AG, but not anandamide, has been detected in human and animal milk (Di Marzo et al., 1998; Fride et al., 2001), suggesting a role in suckling.

In addition, both anandamide and 2-AG were detected in the brain of neonates and adult rats, with the increase of anandamide and decrease of 2-AG during brain development (Berrendero, Sepe, Ramos, Di Marzo, & Fernàndez-Ruiz, 1999). CB1 can be detected in a variety of brain structures of new born rats, in which the locations and levels of CB1 mRNA expression change with age (Berrendero et al., 1999). The administration of CB1 antagonist to the mouse pups reduced milk ingestion and growth, and caused the increase of mortality, which can be attenuated by co-administration of THC or 2-AG (Fride et al., 2001).

It has been shown that Δ^9-THC-stimulated FI in rats (Gluck & Ferraro, 1974). However, early studies indicated that Δ^9-THC treatment suppressed FI and growth in rats (Manning, McDonough, Elsmore, Saller, & Sodetz, 1971; Sjödén, Järbe, & Henriksson, 1973; Sofia & Barry, 1974), depending on the route of administration and dosages (Abel, 1975; Hollister, 1971). Later on, kinetic studies demonstrated that oral administration of Δ^9-THC caused substantial hyperphagia in prefed rats (Brown, Kassouny, & Cross, 1977; Williams, Rogers, & Kirkham, 1998). When it was injected into the ventromedial hypothalamus of rats, anandamide induced hyperphagia, which can be attenuated by CB1 antagonist (Jamshidi & Taylor, 2001). Both exogenous (Δ^9-THC) and endogenous (anandamide) cannabinoids stimulate feeding behavior in rats (Williams & Kirkham, 2002a). Anandamide administrated subcutaneously in rats (Williams & Kirkham, 1999) or intraperitoneally in mice (Hao, Avraham, Mechoulam, & Berry, 2000) stimulate FI. Direct administration of 2-AG into the limbic forebrain area induced feeding, which was attenuated by CB1 antagonist (Kirkham, Williams, Fezza, & Marzo, 2002).

3.3. Altered endocannabinoid levels with energy intake and obesity development

In mice, a moderate food restriction (40–60% of food) for 12 days lowered 2-AG level in the hypothalamus, but not hippocampus. However, a 24-h fasting significantly increased the 2-AG level in the mouse brain (Hanus et al., 2003). In rats, fasting increased anandamide and 2-AG levels in the limbic forebrain, and 2-AG in the hypothalamus, whereas refeeding reduced the 2-AG levels in the hypothalamus (Kirkham et al., 2002). Direct manipulation of the endocannabinoid system (ECS) in rat hypothalamus by injection of anandamide, inhibition of FAAH activity or inhibition of anandamide uptake, stimulated FI, which was blocked by CB1 antagonist (Soria-Gómez et al., 2007). It has been shown that the disruption of MAGL activity raised the 2-AG levels dramatically in the brain and liver (Taschler et al., 2011). However, no change of feeding behaviors and hepatic fatty acid biosynthesis was observed, demonstrating tolerance to the high endocannabinoid levels in those tissues (Taschler et al., 2011).

In obese *ob/ob* and *db/db* mice and Zucker fatty rats, the anandamide and 2-AG contents in hypothalamus, but not cerebellum, are elevated, a phenomenon probably contributing to hyperphagia (Di Marzo, Melck, et al., 2001). Leptin treatment reduced the hypothalamic content of these endocannabinoids in normal rats and *ob/ob* mice, suggesting that the ECS may mediate part of the leptin-regulated FI (Di Marzo, Melck, et al., 2001). In the hippocampal regions of diet-induced obese (DIO) mice, 2-AG and anandamide levels were increased in association with increase of 2-AG synthesizing enzyme and CB1 immunoreactivity, indicating that high-fat diet (HFD) induced changes of ECS beyond the areas directly involved in the metabolic regulation (Massa et al., 2010).

Plasma 2-AG, but not anandamide, levels are higher in obese humans than in the lean ones and are positively correlated with visceral fat mass and negatively correlated with insulin sensitivity (Blüher et al., 2006). Plasma levels of anandamide, but not 2-AG, level was significantly higher in women with anorexic and binge-eating disorders than normal and healthy ones (Monteleone et al., 2005). In obese men, BW reduction after lifestyle modification resulted in reduction of plasma 2-AG and anandamide levels. The decrease in the 2-AG level correlates with the reduction of visceral fat (Di Marzo et al., 2009). Analysis of anandamide and 2-AG contents in adipose tissues showed that obese subjects have lower 2-AG content than the lean ones. The weight reduction in obese subjects caused an increase of 2-AG content and CB1 expression levels in abdominal adipose tissue (Bennetzen et al., 2011). All these observations suggest that an overactive ECS may contribute to obesity development.

4. STUDIES OF CANNABINOID RECEPTORS AND ENDOCANNABINOID METABOLIC ENZYMES

4.1. Tissue distribution of CB1 and CB2

Consistent with the broad psychological and physiological effects of cannabinoids, CB1 is widely distributed in both the central and the peripheral nervous systems. According to quantitative reverse transcription PCR analysis of human tissue samples, CB1 is mainly expressed in the CNS and, to a lower extent, in several peripheral tissues such as adrenal gland, heart, lung, prostate, uterus, ovary, testis, bone marrow, thymus, and tonsils, whereas CB1 is undetectable in liver, kidney, and skeletal muscle (Galiègue et al., 1995). In contrast, CB2, which is not expressed in the brain, is particularly abundant in immune cells, with an expression level 10–100-fold higher than that of CB1 (Galiègue et al., 1995). In developing rat embryos, both CB1 and CB2 mRNA can be detected in a variety of tissues (Buckley, Hansson, Harta, & Mezey, 1997). In cultured mouse neural stem cells, both CB1 and CB2 are expressed and can be activated to stimulate cell proliferation by their respective agonists via a mechanism dependent on phosphoinositide-3 kinase/Akt signaling (Molina-Holgado et al., 2007).

Expression of CB1 in the brain has been studied in great details due to the effects of cannabinoids on brain functions. Many neuronal subpopulations in the brain, including GABAergic, glutamatergic, and dopaminergic neurons, express CB1 at various levels, as revealed by receptor autoradiography (Herkenham et al., 1991), *in situ* hybridization (ISH; Marsicano & Lutz, 1999), and immunohistochemistry (IHC; Tsou, Brown, Sañudo-Peña, Mackie, & Walker, 1997). Using double ISH, it was found that CB1 mRNA is colocalized with dopamine and serotonin receptors, and with corticotropin-releasing hormone receptor type 1 (Hermann & Lutz, 2005) in distinct neuronal subpopulations in different regions of mouse brain controlling various functions, such as cognition, motivation, addiction, and feeding. These data provide an anatomical basis for potential interactions of ECS system with other receptor systems. In general, activation of CB1 results in suppression of neural transmission (Freund, Katona, & Piomelli, 2003).

To date, specific cells expressing CB1 proteins have been identified in a variety of tissues by IHC using antibodies against CB1 (see Section 5). This includes tissues with low or undetectable CB1 mRNA levels (Galiègue et al., 1995; Gérard et al., 1991). According to a recent study, several commercially available antibodies failed to specifically detect CB1 proteins in a stringent lab setting (Grimsey et al., 2008). This demonstrates the challenge of interpreting the IHC data in the literature. Therefore, cautions should be taken when the antibodies used for IHC are not tested extensively. Using a CB1-specific antibody validated by comparing wild type (WT) with CB1 KO mouse tissues, it has been shown that CB1 protein is mainly detected in neuronal processes in various brain regions as well as neuronal cells residing in the intestines and the pancreas, but not in liver and brown fat (Fig. 14.2; Pang et al., 2011). In this case, IHC produced positive staining in the brain, small intestines, and pancreas from WT mice, in contrast to a complete absence of staining in the CB1 KO tissues. The use of CB1 KO samples as negative control is the single most important criterion of antibody validation.

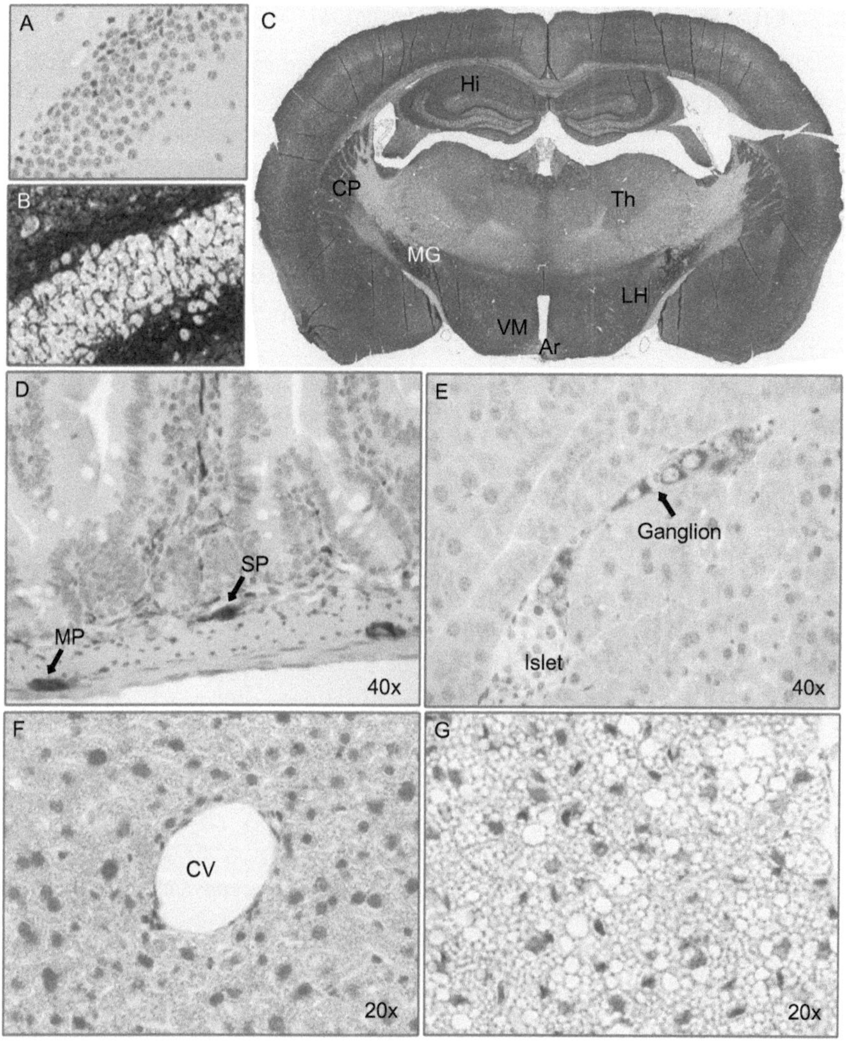

Figure 14.2 Detection of CB1 in the brain and peripheral tissues by immunohistochemistry. Fixed and paraffin-embedded tissues are sectioned and stained using an anti-CB1 polyclonal antibody, as described previously (Pang et al., 2011). Brown color indicates CB1 immunoreactivity. Specificity of the antibody is demonstrated by the complete absence of staining in CB1 knockout tissues, for example, the hippocampal dentate gyrus (A), compared to the wild-type mouse (B). CB1 is detected mainly in neuronal processes in various brain regions (C), as well as in neuronal cells residing in the small intestines (D) and the pancreas (E). CB1 immunoreactivity is undetectable in the liver (F), WAT (not shown) and BAT (G). Hi, hippocampus; CP, caudate putamen; MG, medial globus pallidus; Th, thalamus; Ar, arcuate nucleus; LH, lateral hypothalamus area; VM, ventromedial hypothalamic nucleus; MP, myenteric (Auerbach's) plexus; SP, submucosal (Meissner's) plexus; CV, central vein. (See Color Insert.)

4.2. Effects of CB1 and CB2 KO in mice

The biological functions of ECS have been elucidated by genetic inactivation (KO) of specific genes in mice. Originally, two lines of mice with disruption of CB1 were created by deleting parts of its single exon (Ledent et al., 1999; Zimmer, Zimmer, Hohmann, Herkenham, & Bonner, 1999). Mice missing the first 233 amino acids of CB1 have grossly normal brain and other organs, normal BW over a 6-month period, and normal plasma parameters, according to the initial observations (Ledent et al., 1999). A moderate increase in locomotor activity was observed (Ledent et al., 1999). Zimmer et al. deleted the amino acids 33–447 of CB1, which resulted in CB1 KO mice that were healthy and fertile with normal BW and gross appearance, but increased mortality with about 70% survival rate at 24 weeks of age (Zimmer et al., 1999). Additionally, this line of CB1 KO mice showed reduction of locomotor activity (Zimmer et al., 1999), an unexpected phenotype that is not reported in the first KO line (Ledent et al., 1999) or study using CB1 antagonist (Compton, Aceto, Lowe, & Martin, 1996). The reduction of the locomotor activity in this second line of CB1 KO mice was attributed to the altered gene expression in neurons responsible for locomotion (Steiner, Bonner, Zimmer, Kitai, & Zimmer, 1999).

The effects of CB1 KO on metabolism, which were not discovered by the first two publications (Ledent et al., 1999; Zimmer et al., 1999), were later independently reported by two research groups (Cota et al., 2003; Ravinet Trillou, Delgorge, Menet, Arnone, & Soubrié, 2004). Both groups observed that CB1 KO mice have reduced BW and fat mass, and are resistant to DIO (Cota et al., 2003; Ravinet Trillou et al., 2004). The metabolic phenotypes of CB1 KO mice were investigated in other studies, which also confirmed the decreased BW and resistance to DIO (Fong et al., 2007; Osei-Hyiaman et al., 2005). These phenotypes are accompanied by reduced feeding efficiency (Ravinet Trillou et al., 2004) and increased energy expenditure (Osei-Hyiaman et al., 2005, 2008). In addition, inactivation of CB1 reduced plasma levels of insulin, leptin, cholesterol, and triglyceride (TG), and increased adiponectin levels, suggesting improved hormone sensitivities and lipid metabolism (Cota et al., 2003; Osei-Hyiaman et al., 2008; Quarta et al., 2010). However, the CB1 deletion did not significantly affect fasting glycemia (Cota et al., 2003; Ravinet Trillou et al., 2004).

The single exon of CB2 was deleted to generate CB2 KO mice, which were healthy and fertile with similar size and BW as their WT littermates (Buckley et al., 2000). The CB2 KO mice lack the cannabinoid-mediated modulation of macrophage immune functions (Buckley et al., 2000). The CB1/CB2 double KO mice are also healthy with no gross defects and have the same size and weight as their WT controls (Járai et al., 1999). Interestingly, anandamide and metabolically stable (*R*)-methanandamide still induced mesenteric vasodilation in the CB1/CB2 double KO mice, suggesting the presence of other cannabinoid receptors in the mesenteric vasculature (Járai et al., 1999).

4.3. Antiobesity effects of CB1 antagonists

Given the role of CB1 in FI, CB1 antagonists have been developed as antiobesity drugs. Discovered in 1994, rimonabant (SR141716A) is the first potent and selective CB1 antagonist and a powerful tool for studying the biological functions of CB1 (Rinaldi-Carmona et al., 1994). Rimonabant treatment significantly reduces FI and BW in DIO and genetic obese mice (Di Marzo, Melck, et al., 2001; Jbilo et al., 2005; Poirier et al., 2005; Ravinet Trillou et al., 2003). The effects of rimonabant on FI and BW are lost in CB1 KO mice, demonstrating that they are mediated specifically by CB1 (Ravinet Trillou et al., 2003, 2004; Williams & Kirkham, 2002b). The reduction of BW and FI is accompanied by an increase of oxygen consumption and insulin-stimulated muscle glucose uptake in *ob/ob* mice treated with rimonabant for 7 days (Liu, Connoley, Wilson, & Stock, 2004). A 10-week oral administration of rimonabant (10 mg/kg/day) in DIO mice causes reduction of BW and improvement of plasma lipid profile (Poirier et al., 2005). In DIO mice, rimonabant prevents the increase of both white and brown fat masses, which is accompanied by dramatic changes of expression of genes responsible for lipid oxidation and storage in the fat depots (Jbilo et al., 2005). It is worth noting that rimonabant treatment only reduces plasma leptin levels in mice fed HFD, but has no significant effects in mice fed the control diet (Jbilo et al., 2005). The rimonabant treatment also reduces BW gain in obese mice induced by high-sucrose/fat diet, and changes the expression levels of genes involved in glucose and fatty acid metabolism in liver and adipose tissues (Jourdan et al., 2010). The effects of this treatment on adipose tissues depended on the fat depots (Jourdan et al., 2010). Moreover, rimonabant treatment also decreases the level of monocyte chemoattractant protein-1, a marker of inflammation, in the adipose tissues, liver, and blood in DIO mice (Wang, Perrard, et al., 2011).

Similar effects have been observed in rats. Oral administration of rimonabant (10 mg/kg, daily) for 14 days causes transient reduction of FI and extended reduction of BW in normal (Arnone et al., 1997; Colombo et al., 1998) and

Zucker fatty rats (Di Marzo, Melck, et al., 2001; Vickers, Webster, Wyatt, Dourish, & Kennett, 2003). The treatment also increases basal circulating corticosterone levels and enhances the hypothalamic–pituitary–adrenal axis response to food deprivation in obese rats. The plasma levels of insulin and TG are also reduced in both types of rats (Doyon et al., 2006).

In the clinical trials, rimonabant treatment improves the outcomes of obese, overweight, and diabetic patients. Table 14.1 summarizes the representative clinical trials conducted to examine the effects of rimonabant in these patients. Generally, rimonabant treatment for more than 12 months results in reduction of BW, waist circumference, and plasma TG level in obese and diabetic patients (Després et al., 2005, 2009; Hollander et al., 2010; Nissen et al., 2008; Pi-Sunyer et al., 2006; Scheen et al., 2006; Van Gaal et al., 2008). It also raises the plasma high-density lipoprotein levels (Després et al., 2005, 2009; Hollander et al., 2010; Nissen et al., 2008; Pi-Sunyer et al., 2006; Scheen et al., 2006; Van Gaal et al., 2008). In 2006, rimonabant was approved as a treatment for obesity and for overweight patients with metabolic comorbidities, such as NIDDM and dyslipidemia, in the European Union. However, due to reported side effects (Després et al., 2005, 2009; Hollander et al., 2010; Nissen et al., 2008; Pi-Sunyer et al., 2006; Scheen et al., 2006; Van Gaal et al., 2008), such as depression, anxiety, dizziness, and nausea, rimonabant was withdrawn from the market in 2008 (Jones, 2008).

Another CB1 antagonist, taranabant (MK-0364), inhibits BW gain in male DIO rats (0.3 mg/kg/day administered orally for 14 days) with accompanied loss of fat mass (Fong et al., 2007). Clinical trial of taranabant in obese and overweight patients has demonstrated that the weight loss mediated by 2 mg taranabant in 52 weeks is comparable to that caused by 20 mg rimonabant with similar side effects as well (Proietto et al., 2010). In overweight and obese patients with NIDDM, taranabant (1 and 2 mg/day) for a year, significantly reduced BW and improved the plasma HbA1C level with similar adverse effects as rimonabant (Kipnes et al., 2010). Taranabant used in combination with lifestyle modification improves the maintenance of weight loss despite some adverse effects, which does not include suicidal attempts (Wadden et al., 2010). Nevertheless, its development was also stopped due to safety concerns (Jones, 2008).

An important conclusion derived from the animal and human studies is that the increased energy expenditure induced by CB1 antagonists plays a more important role than the reduction of FI in the reduction of BW. It has been reported that CB1 antagonists stimulate thermogenesis and

Table 14.1 Representative clinical trials of rimonabant

Study	Subject #	Dosages (mg/day)	Duration	BW	Waist cir.	TG	HDL	Others	Side effects	References
Overweight with dyslipidemia	1036	5 or 20	12 months	Down	Down	Down	Up	Diet	Depression, anxiety, and nausea	Després, Golay, and Sjöström (2005)
Overweight or obese	3045	5 or 20	24 months	Down	Down	Down	Up	Diet, exercise	Nausea	Pi-Sunyer et al. (2006)
Overweight or obese with type 2 diabetes	1047	5 or 20	12 months	Down	Down	Down	Up	Diet, exercise	Depression, dizziness, and nausea	Scheen, Finer, Hollander, Jensen, and Van Gaal (2006)
Obese and overweight with hypertension and/or dyslipidemia	1507	5 or 20	24 months	Down	Down	Down	Up	Diet	Depression, dizziness, and nausea	Van Gaal et al. (2008)
Abdominal obesity and coronary artery disease	839	20	18 months	Down	Down	Down	Up	Diet	Depression and anxiety	Nissen et al. (2008)
Type 2 diabetes receiving insulin monotherapy	368	20	48 weeks	Down	Down	Down	Up	Insulin therapy	Depression, hypoglycemia, dizziness, and nausea	Hollander, Amod, Litwak, Chaudhari, and for the ARPEGGIO Study Group (2010)
Obese with atherogenic dyslipidemia	803	20	12 months	Down	Down	Down	Up	Insulin therapy	Depression, hypoglycemia, dizziness, and nausea	Després et al. (2009)

increase energy expenditure (Herling et al., 2007; Herling, Kilp, Elvert, Haschke, & Kramer, 2008; Liu et al., 2004; Osei-Hyiaman et al., 2005, 2008). A 6-week oral administration of rimonabant (10 mg/kg) in Wistar rats revealed that the elevated energy expenditure contributes more to the rimonabant-mediated reduction of BW than reduction of FI (Herling et al., 2008). Data from the rimonabant in obesity trials suggest that 45–57% of the effects on lipid profile, adiponectin, and hemoglobin A1c could not be explained by the observed weight loss (Van Gaal et al., 2008).

As shown in Fig. 14.3 (a schematic view), after the administration of rimonabant to experimental animals an initial drop of FI and BW is observed. Whereas FI (normalized by BW) rebounds quickly and eventually returns to the basal level after 2 weeks, the loss of BW is maintained with the use of the drug. This was observed in the DIO mice (Jbilo et al., 2005; Poirier et al., 2005; Ravinet Trillou et al., 2003), Zucker fatty rats (Doyon et al., 2006; Vickers et al., 2003), and candy-fed rats (Herling et al., 2008). Furthermore, treatment with rimonabant results in more weight loss than pair-fed mice (Ravinet Trillou et al., 2003) or food-restricted rats (Mølhøj et al., 2010). In adult WT mice, caloric restriction (through pair-feeding) does not reduce BW to the levels observed in the KO mice, suggesting that reduced FI alone cannot explain the weight loss and that CB1 receptors in adipose tissue may contribute to the metabolic phenotype (Cota et al., 2003). Together, these results have led to the peripheral CB1 hypothesis which in essence postulates that CB1 receptors expressed in peripheral tissues are responsible for some of the metabolic

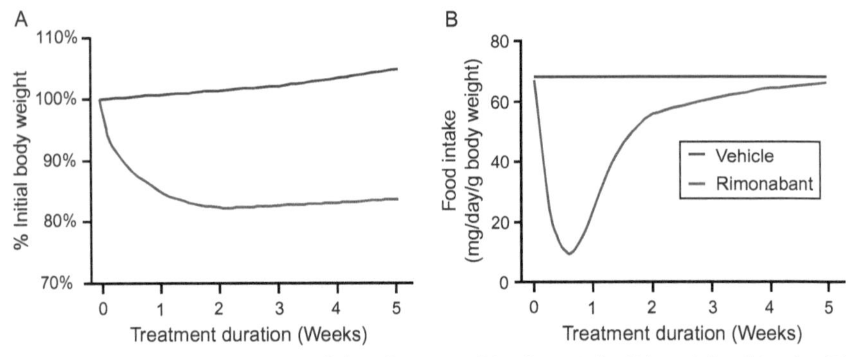

Figure 14.3 A schematic view of the changes of body weight (A) and food intake (B) upon the administration of rimonabant. Treatment with rimonabant results in transient reduction of food intake, but sustained decrease in body weight in mice fed high-fat diet. (See Color Insert.)

effects of CB1 antagonism (Cota et al., 2003; Gómez et al., 2002; Osei-Hyiaman et al., 2005).

Although a growing body of evidence suggests that CB1 is expressed in peripheral tissues that are important for fat and glucose metabolism, such as adipose (Bensaid et al., 2003; Cota et al., 2003), liver (Osei-Hyiaman et al., 2005, 2008), and pancreas (Matias et al., 2006), the existence of these peripheral receptors and their precise roles in mediating metabolic effects are still controversial (Fong & Heymsfield, 2009). Studies pertaining to this important issue will be discussed in Section5.

4.4. Effects of FAAH and MAGL KO in mice

FAAH was originally purified from rat liver membrane preparations with the activity to hydrolyze oleamide into oleic acid and ammonia (Cravatt et al., 1996). This enzyme has activity against fatty acid amides including high activity to anandamide and is highly expressed in brain and liver (Cravatt et al., 1996). The FAAH KO mice have elevated brain anandamide levels, heightened sensitivity to anandamide, but normal BW (Cravatt et al., 2001). It will be very interesting to elucidate the role of FAAH in liver which does not seem to express enzymes for anandamide synthesis (Cohen & Grahame Hardie, 1991; Okamoto et al., 2004).

Since plasma 2-AG level is higher in the obese patients than in the normal subjects (Blüher et al., 2006), its production has been manipulated in animals to determine its effects on obesity. Recombinant adenovirus-mediated overexpression of MAGL in rat brain causes reduction of 2-AG, but not anandamide content (Dinh et al., 2002). Conversely, administration of a specific MAGL inhibitor (JZL184) dramatically increases the brain 2-AG, but not anandamide level, and causes CB1-dependent behavior changes (Long et al., 2009). In agreement with these results, the MAGL KO mice have elevated monoacylglycerol, including 2-AG, levels in their adipose tissue, brain, and liver. However, the MAGL KO mice have normal fat mass, locomotor activity, energy expenditure, and food consumption, probably due to the development of tolerance to the high 2-AG concentrations, suggesting desensitization of the cannabinoid receptors. Interestingly, the KO mice on HFD have improved glucose tolerance and insulin sensitivity comparing to the WT DIO mice (Taschler et al., 2011).

In addition to having agonist activities on CB1 and CB2, 2-AG is also a substrate for the generation of AA for the cyclo-oxygenase (COX)-mediated prostaglandin production (Fig. 14.1) in the brain, liver, and lung, but not gut and spleen (Nomura et al., 2011). The disruption of MAGL activity either by genetic inactivation or by specific inhibitors results in the accumulation of 2-AG in the brain and liver (Nomura et al., 2011; Taschler et al., 2011) and reduction of several prostaglandins and other eicosanoids, such as prostaglandin E2 (Nomura et al., 2011). It also blocks liposaccharide (LPS)-induced inflammatory responses in the brain and provides protection against chemically induced Parkinsonism in mice (Nomura et al., 2011). Traditionally, the supply of AA for eicosanoid production has been considered the role of phospholipase A2 (Buczynski, Dumlao, & Dennis, 2009). However, the reduction of AA contents in the brain and liver was not observed in phospholipase A2 KO mice, but rather in MAGL KO (Nomura et al., 2011). The link between AA generation for prostaglandin production and the rise of cAMP levels in adipocytes has been indicated (Wolf, 2009). The cannabinoid system can also regulate the production of cAMP in adipocytes (Di Marzo et al., 2001; Roche et al., 2006). Further investigations are needed to understand relationships of these two pathways.

5. PHYSIOLOGICAL ROLES OF CB1 IN DIFFERENT TISSUES

5.1. Central nervous system

There is a large body of evidence suggesting that the effects of cannabinoids on FI are mediated by the hypothalamus and the forebrain limbic system. The ECS probably regulates feeding behavior through the release of the orexigenic neuropeptide Y (NPY). In the rat hypothalamus, cannabinoids enhance both the resting and KCl-evoked NPY release. CB1 inverse agonist AM251 inhibits NPY release in the absence of cannabinoids (Gamber, Macarthur, & Westfall, 2005). However, additional study found that CB1 agonist WIN55,212-2 increased β-endorphin, but not NPY immunoreactivity in the arcuate nucleus, whereas AM251 reduced both β-endorphin and NPY in the same locations (Bakkali-Kassemi et al., 2011). The medial shell of nucleus accumbens has been identified as a site of action for endocannabinoids to generate pleasure associated with food consumption (Mahler, Smith, & Berridge, 2007). Injection of FAAH inhibitor, as well as inhibitor of anandamide uptake, into the nucleus accumbens stimulated FI and c-fos expression in the hypothalamus, suggesting that the endocannabinoid degrading enzymes are potential therapeutic targets for eating disorders (Soria-Gómez et al., 2007). Electrophysiological data suggest that hypothalamic melanin-concentrating hormone (MCH) and hypocretin neurons are involved in cannabinoids' action, with cannabinoids having opposite effects on MCH and hypocretin neurons (Huang et al., 2007). Cannabinoids, as well as CB1 KO, affect the expression of hypothalamic neuropeptides, demonstrating hypothalamus as a site of action (Bakkali-Kassemi et al., 2011; Cota et al., 2003). Interestingly, the brain stem has been suggested as a potential site of action for cannabinoids to control feeding, based on data from injection of CP55940 into the 4th ventricle (Miller, Murray, Freeman, & Edwards, 2004), and IHC detection of c-fos induction by rimonabant in certain brain stem nuclei (Jelsing et al., 2009). Furthermore, the dopamine system in the nucleus accumbens may be a substrate for CB1 antagonists, whereas hypothalamic melanocortin neurons are dispensable (Sinnayah et al., 2008). At neurochemical level, data suggest that dopamine D1 signaling is necessary for cannabinoids to stimulate FI (Verty, McGregor, & Mallet, 2004).

Inhibition or stimulation of the AMP-activated protein kinase (AMPK) in hypothalamic regions has anorexigenic or orexigenic effects in rodents, respectively (Andersson et al., 2004; Minokoshi et al., 2004). Both cannabinoids and ghrelin, an orexigenic hormone, moderately stimulate the activation of AMPK in the hypothalamus and the heart, inhibit it in liver and adipose tissue, and have no effect in skeletal muscle in rats (Kola et al., 2005). In CB1 KO mice, ghrelin no longer stimulates FI and AMPK activation (Kola et al., 2008).

Tissue-specific CB1 KO animal models, generated using the Cre/LoxP technology, are instrumental for understanding the relative contributions of CB1 in mediating rimonabant's effects in various tissues. Data derived from conditional KO mouse lines, which lack CB1 in different neuronal subpopulations, have demonstrated that the classic tetrad effects (hypolocomotion, hypothermia, analgesia, and catalepsy) of Δ^9-THC, administrated intraperitoneally, are mediated by the brain principal neurons (Monory et al., 2007), whereas CB1 in GABAergic interneurons is important for the amnesic effects of cannabinoids (Puighermanal et al., 2009).

Recently, the metabolic phenotype of CaMK–CB1 KO mouse model, which is characterized by nearly complete loss of CB1 in the forebrain principal neurons and partial CB1 reduction in the superior cervical ganglia (SCG) has been reported (Quarta et al., 2010). Like the global CB1 KO mice, the CaMK–CB1 KO mice exhibit decreased BW and feeding efficiency, resistance to DIO, and a complete loss of responsiveness to rimonabant. Additionally, increased lipid oxidation and thermogenesis were observed and attributed to increased sympathetic tone in this line of mutant mice (Quarta et al., 2010). These results suggest that the antiobesity effects of rimonabant are neuronally mediated, at the forebrain and/or SCG levels.

This notion has been substantiated by data from a CNS-specific CB1 knockdown mouse model, which was generated by transgenic expression of an artificial microRNA under the control of the neuronal Thy1.2 promoter (Pang et al., 2011). In this model, CB1 expression is reduced in the brain and spinal cord, whereas no change of its expression was observed in the SCG, sympathetic trunk, enteric nervous system, and pancreatic ganglia. Compared to WT littermates, the mutant mice display reduced BW, adiposity, and feeding efficiency, as well as improved plasma insulin, leptin, cholesterol, TG levels, and adiponectin profiles, when animals are challenged with a HFD. Furthermore, the therapeutic effects of rimonabant on FI, BW, and serum parameters are markedly reduced and correlated with the degree of CB1 knockdown. Thus, data from two different transgenic models suggest that blockade of central CB1 is required for rimonabant's antiobesity actions.

5.2. Liver

The expression of CB1 in mouse hepatocytes has been reported and can be induced by HFD feeding (Osei-Hyiaman et al., 2008). It is shown that the CB1/CB2 agonist HU-210 increases the expression of CB1 along with lipogenic genes in the liver and brain of WT mice. This treatment also stimulates fatty acid biosynthesis in the liver and in hepatocytes isolated from WT, but not CB1 KO mice (Osei-Hyiaman et al., 2005). In the conditional hepatocyte-specific CB1 KO mice, HFD-induced liver steatosis, dyslipidemia, hyperglycemia, insulin, and leptin resistance, are reduced or prevented, indicating that these effects are mediated by hepatic CB1 (Osei-Hyiaman et al., 2008). Unlike the CB1 KO mice, the liver-specific CB1 KO mice still gain BW and adiposity in response to HFD feeding (Osei-Hyiaman et al., 2008). In the WT, but not CB1 and apoE double KO mice, treatment with MAGL inhibitor isopropyl dodecylfluorophosphonate results in dramatically elevated 2-AG content in liver, muscle and brown adipose tissues (BATs), and increased plasma levels of TG, cholesterol, and apoE-depleted TG-rich lipoproteins. It seems that these changes are caused by the reduction of plasma TG clearance, but not the altered expression of hepatic lipogenic genes and lipoprotein lipase (LPL) activity (Ruby et al., 2008).

However, analysis of a panel of human tissues by RT-PCR indicates that hepatic CB1 is too low to be quantified (Galiègue et al., 1995). Similarly, CB1 protein is difficult to be detected by IHC in the mouse liver (Fig. 14.2). Further investigations are required to resolve the discrepancies.

CB1 activation is implicated in certain pathological conditions of the liver. In human liver, CB1 is mainly expressed in the myofibroblasts and increased with liver fibrosis (Teixeira-Clerc et al., 2006). Chemically induced liver injury is attenuated in the CB1 KO mice or mice treated with CB1 antagonists (Teixeira-Clerc et al., 2006). In addition, the mRNA levels of CB1 and CB2 in the livers of patients with primary biliary cirrhosis are positively correlated with the severity of the disease (Floreani et al., 2010). In the liver samples of human patients with alcohol liver diseases, CB1 is expressed in areas with advanced fibrosis; acetaldehyde and reactive oxygen species induce its expression in hepatic stellate cells (HSC; Patsenker et al., 2011). The hepatic CB1 has been shown to contribute to ethanol-induced lipogenesis, which is correlated with induction of *Srebp-1c* expression (Trebicka et al., 2011). Ethanol treatment in mice also increases 2-AG production from HSC through the induction of DAGL. The HSC-derived 2-AG acts in a paracrine and CB1 dependent fashion to induce lipogenesis and suppress lipid oxidation in hepatocytes, resulting in the development of alcoholic fatty liver (Jeong et al., 2008). The induction of CB1 expression was also observed in fibrotic livers of patients with chronic hepatitis C, but not hepatitis B viral infection, and this induction can be attributed to the structural proteins of hepatitis C virus (van der Poorten et al., 2010). CB1 expression can also be induced by the treatment of retinoic acid via activation of retinoic acid receptor-gamma (Mukhopadhyay et al., 2010).

It seems that CB1 and CB2 have opposite effects on chemically induced liver damage, since both CB1 antagonist and CB2 agonist protect against liver injury (Trebicka et al., 2011). It is worth noting that co-expression of CB1 and CB2 has been observed in adipocytes (Roche et al., 2006). Whether there is a functional antagonism between CB1 and CB2 during the development of liver diseases deserves further investigation.

5.3. Gastrointestinal tract

2-AG was first identified from methanol extract of canine small intestine as an activity similar to Δ^9-THC (Mechoulam et al., 1995). Anandamide and 2-AG are detected in mouse small intestine and colon (Izzo et al., 2001; Pinto et al., 2002). The anandamide hydrolase and synthase activities are detected in the small intestine (Katayama et al., 1997). The mRNA level of FAAH in the small intestine is similar to that in the liver, the organ with the highest activity (Katayama et al., 1997). CB1 mRNA can be detected along the gastrointestinal (GI) tract of rat embryo at day 15 (Buckley et al., 1997) and the protein in ganglia and smooth muscle cells throughout the GI tract (Casu et al., 2003; Izzo et al., 2001; Pang et al., 2011). As shown in Fig. 14.2, CB1 is expressed at high levels in the enteric nervous system.

It is known that cannabinoids inhibit gastric emptying, gastric acid secretion, and intestinal motility in animals and humans (Pertwee, 2001). Conversely, CB1 antagonist accelerates GI transit (Casu et al., 2003; Izzo et al., 2001; Pang et al., 2011; Pertwee, 2001) and may reduce the absorption of nutrients, explaining in part the reduced feeding efficiency observed in CB1 deficient mouse strains (Pang et al., 2011; Quarta et al., 2010; Ravinet Trillou et al., 2003). These actions seem to be mediated by both central and peripheral CB1 receptors (Pertwee, 2001). For instance, GI transit is inhibited by systemically or centrally administrated CB1 agonists, and the cannabinoid-induced delay of gastric emptying and GI transit in rats are blocked by rimonabant (Landi et al., 2002). Docosatetraenoylethanolamide, an endocannabinoid with selectivity to CB1, differentially affects the discharge of afferent and efferent neural fibers of jejunum depending on the feeding status of the rats (Donovan & Grundy, 2011).

In an attempt to understand the role of endocannabinoids in the intake of palatable foods, the efferent and afferent food-related signals between rat brain and gut were analyzed in a sham feeding model, in which liquid diets were drained from the stomach through a chronically implanted gastric cannula (DiPatrizio, Astarita, Schwartz, Li, & Piomelli, 2011). Sham feeding with fat, but not protein or carbohydrate, specifically increased 2-AG and anandamide levels in the jejunal section of the small intestine, but not in other peripheral tissues or brain regions (DiPatrizio et al., 2011). This induction was caused by elevation of biosynthesis and reduction of degradation of endocannabinoids in jejunum. This positive feedback mechanism can be inhibited if the CB1 signaling is blocked, suggesting a potential way of reducing the intake of high-fat food (DiPatrizio et al., 2011).

In *ob/ob* mice, colonic expression of CB1 was reduced by prebiotics, whereas the activation of CB1 in colon increased the permeability of the gut and caused the elevation of LPS in plasma, which may attenuate adipogenesis (Muccioli et al., 2010). This may be a counter mechanism to suppress the development of obesity in *ob/ob* mice. In addition, it has been shown that HU-308, a specific agonist for CB2, also inhibits intestinal motility at high concentration (20 mg/kg) in mice (Hanuš et al., 1999). CB2 mRNA and protein have been detected in neurons of rat ileum and the CB2 agonist JWH133 blocked LPS-enhanced GI transit (Duncan et al., 2008). Whether the same effect can be observed in CB1 KO mice remains to be seen.

5.4. Pancreas

The expression of CB1 and CB2 in the pancreas and the effects of cannabinoids on the release of pancreatic hormones have been investigated by several groups. However, the results are in disagreement regarding the identities of cells expressing the receptors. One study showed that both CB1 and CB2 were expressed in the mouse pancreas, in which CB2 was expressed in both β- and non-β-cells, and CB1 was expressed in non-β-cells, while the treatment of 2-AG decreased insulin secretion (Juan-Picó et al., 2006). In another study, only CB1 was detected in isolated mouse islets, while anandamide and arachidonylcyclopropylamide inhibited glucose-stimulated insulin secretion (GSIS) via inhibition of calcium oscillation in β-cells (Nakata & Yada, 2008). In isolated human islets analyzed with immunofluorescence, CB1 was colocalized strongly with α-cells and weakly with β-cells, whereas CB2 was detected in δ-cells (Bermúdez-Silva et al., 2008). The presence of enzymes for the production (DAGLα and β) and degradation (MAGL and FAAH) of endocannabinoids are also detected using immunoblot, suggesting that human islets can potentially respond to both circulating and locally produced endocannabinoids (Bermúdez-Silva et al., 2008). Incubation with 11 mM glucose resulted in elevation of 2-AG content in islets along with insulin secretion (Bermúdez-Silva et al., 2008). Activation of CB1-stimulated insulin and glucagon secretion, whereas activation of CB2 suppressed insulin secretion (Bermúdez-Silva et al., 2008). However, analysis of human, mouse, and rat pancreatic islets using confocal immunofluorescence microscopy showed that CB1 was expressed in δ-cells, but not in α-, β-, or pp-cells (Tharp, Lee, Maple, & Pratley, 2008). The enzyme for 2-AG degradation, MAGL, was colocalized with δ-cells, whereas that for anandamide, FAAH, was found to be weakly expressed in β-cells and strongly in α-cells of Zucker fatty rats (Tharp et al., 2008).

A caveat of these earlier studies is that the islets from CB1 KO mice were not used as a control to verify the specificity of CB1 antibodies in immunodetection of the proteins. Recently, IHC using highly validated CB1 antibody has revealed strong staining in neuronal cell bodies of pancreatic ganglia without obvious staining of the islets in mouse pancreas, and this is confirmed by the absence of staining in CB1 KO mice (Fig. 14.2) (Pang et al., 2011). Under certain IHC conditions, strong CB1 immunoreactivity is observed in α-cells of WT mice, as previously reported (Bermúdez-Silva et al., 2008), but the staining patterns are similar in CB1 KO mice, suggesting that the staining of α-cells is an artifact (personal observation, submitted). Recently, a cross talk between CB1 and insulin receptor signaling has been observed. It is shown that CB1 is expressed in β-cells using CB1 antibody validated on CB1 KO samples. Furthermore, inhibition of CB1 signaling increases β-cell proliferation through the insulin receptor substrate 2/AKT pathway (Kim et al., 2011).

Making the matter more complicated, both CB1 and CB2 proteins are detected in exocrine pancreatic lobules and acinar cells of rats, indicating roles in exocrine functions (Linari et al., 2009). In addition, the expression of GPR55, the potential CB3, has been shown in the rat pancreatic β-cells. The activation of GPR55 with low concentrations of its specific agonist O-1602 enhances GSIS, a phenomenon that was not observed in the islets of GPR55 KO mice (Romero-Zerbo et al., 2011). A more definitive picture regarding CB1 expression in the pancreas should be generated in future studies in order to define the roles of these receptors and cannabinoids in the regulation of pancreatic functions.

5.5. Muscle

CB1 was found to be expressed at mRNA and protein levels in differentiated L6 muscle cells at comparable levels as in adipocytes (Esposito et al., 2008). It has been shown that insulin-stimulated glucose uptake in the isolated soleus muscle of *ob/ob* mice treated with rimonabant for 7 days is significantly increased (Liu et al., 2004). Rimonabant treatment results in significant increase of glucose uptake in the L6 muscle cells and mouse primary myocytes. This is attributed to rimonabant-induced post-transcriptional expression of the p85 subunit of phosphatidylinositol-3-kinase, a process mediated by activation of protein kinase A (PKA) (Esposito et al., 2008).

Insulin plays an important role in controlling muscle glucose uptake (Petersen & Shulman, 2002). Available data indicate that endocannabinoids may directly affect the insulin action on muscle cells. In differentiated primary human skeletal muscle cells and mouse muscle cells, the insulin-stimulated phosphorylation of Akt (pSer473) is significantly inhibited by endocannabinoids (Eckardt et al., 2009; Song et al., 2011). This inhibition can be prevented by CB1 antagonists, rimonabant and AM251. In the same cells, anandamide activates MAP kinases, ERK1/2, and P38 (Eckardt et al., 2009). In L6 muscle cells, which express CB1, chronic WIN55,212-2 treatment attenuates insulin-induced phosphorylation of the ERK1/2, but not that of Akt/PKB. This effect of CB1 activation can be blocked by rimonabant (Lipina et al., 2010). In euglycemic–hyperinsulinemic clamp studies, CB1 agonist HU-210 administrated peripherally, but not centrally, decreased muscle glucose uptake in WT, but not CB1 KO mice (Song et al., 2011).

5.6. Adipose tissues

The detection of CB1 mRNA or protein has been reported in mouse (Cota et al., 2003), rat (Bensaid et al., 2003), human (Blüher et al., 2006; Roche et al., 2006), cultured 3T3 F442A (Bensaid et al., 2003; Gary-Bobo et al., 2006), and 3T3-L1 (Bouaboula et al., 2005; Gasperi et al., 2007) adipocytes. CB2 expression was also observed in 3T3-L1 (Bouaboula et al., 2005; Gasperi et al., 2007) and human (Roche et al., 2006) adipocytes. The CB1 expression level is higher in Zucker fatty rats than lean rats (Bensaid et al., 2003). However, the protein results should be read with cautions because of uncertainty about the antibody specificity (Grimsey et al., 2008). The differentiated mouse 3T3 F442A and 3T3-L1 adipocytes have higher expression level of CB1 than the undifferentiated cells (Bensaid et al., 2003; Gasperi et al., 2007). However, the CB2 protein level is reduced during the differentiation of mouse 3T3-L1 cells (Gasperi et al., 2007).

Anandamide and 2-AG increase insulin-stimulated glucose uptake in the differentiated adipocytes, which can be attenuated by rimonabant, suggesting that CB1 agonist and antagonists have the potential to affect insulin action in adipocytes (Gasperi et al., 2007; Motaghedi & McGraw, 2008). Down-regulation of CB1 expression or antagonizing CB1 activity causes trans-differentiation of SV40 immortalized murine white adipocytes into brown adipocytes (Perwitz et al., 2010). Anandamide can activate PPARγ and potentiate differentiation of 3T3-L1 cells (Bouaboula et al., 2005). The CB1 agonist WIN-55,212 induces heparin-releasable LPL activity in cultured primary mouse adipocytes (Cota et al., 2003). Rimonabant treatment induces adiponectin expression in the adipocytes of Zucker fatty rats and 3T3 F442A adipocytes, and reduces hyperinsulinemia in fatty rats (Bensaid et al., 2003).

CB1 expression has been detected in human pre- and mature adipocytes. Its level is lower in obese than in lean subjects, is unchanged with 5% BW reduction, and is negatively associated with the plasma 2-AG and anandamide levels (Engeli et al., 2005). The CB1 and FAAH mRNA levels in the visceral and subcutaneous adipose depots of lean subjects are significantly higher than those in the respective fat depots of obese subjects (Blüher et al., 2006). When the CB1 and CB2 expression levels were analyzed in human adipose tissue samples obtained from subjects undergoing liposuction (subcutaneous fat) or intestinal (omental fat) surgery, it was found that both of them were expressed in pre- and mature adipocytes at mRNA and protein levels (Roche et al., 2006). The treatment of 2-AG dose-dependently induced the production of intracellular cAMP in human preadipocytes, which can be attenuated by AM251 at concentration specifically antagonized the activity of CB1 (Roche et al., 2006). The treatment of palmitoylethanolamide, an inhibitor of FAAH (Di Marzo, Goparaju, et al., 2001), caused the reduction of forskolin-induced cAMP levels in human preadipocytes, which was blocked by SR144528, a specific antagonist of CB2. These observations indicated that CB1 and CB2 increase and decrease the cAMP levels in adipocytes, respectively (Roche et al., 2006).

The expression of multiple genes in adipose tissues has been analyzed in the DIO mice. HFD induces marked changes in gene expression levels in white adipose tissue (WAT) and BAT, which are mostly reversed by rimonabant treatment. Functional analysis of the expression profiles indicates that rimonabant treatment results in activation of the lipolytic and thermogenetic pathways (Jbilo et al., 2005). Moreover, the acute treatment with AM251 suppresses the expression of lipogenic genes and the tumor necrosis factor-α signaling pathway in epididymal WAT, which is considered to be a mechanism for the improved insulin sensitivity in DIO mice treated with CB1 inverse agonists (Zhao et al., 2010).

Although the results described above indicate direct effects of the ECS in adipose tissues, there are also evidences questioning this view. For example, we cannot detect CB1 immunoreactivity in WAT or BAT by IHC (Fig. 14.2). Effects of rimonabant on lipogenic gene expression in WAT appear to depend on the reduction of FI in DIO rats (Nogueiras et al., 2008). The increased thermogenesis in BAT and reduction of BW following rimonabant treatment are attenuated by sympathetic denervation, suggesting much of the effects are mediated by the CNS (Verty, Allen, & Oldfield, 2008). Furthermore, the increased lipolysis in WAT of DIO rats following acute rimonabant treatment could be mediated by noradrenaline excretion in response to rimonabant, suggesting a role of the sympathetic nervous system (Mølhøj et al., 2010).

6. NOVEL CB1 ANTAGONISTS

The classic CB1 antagonists are inverse agonists that are orally active and brain penetrating (Di Marzo & Després, 2009; Fong & Heymsfield, 2009). As shown in Table 14.1, their side effects for obesity treatment are mostly CNS associated (Després et al., 2005, 2009; Hollander et al., 2010; Nissen et al., 2008; Pi-Sunyer et al., 2006; Scheen et al., 2006; Van Gaal et al., 2008). The finding that the peripherally based CB1 receptors may significantly contribute to the antiobesity effects of rimonabant suggests an alternative strategy (Cota et al., 2003; Gómez et al., 2002). In order to avoid the CNS-mediated side effects, novel CB1 antagonists, including neutral antagonist, peripherally restricted antagonist, and allosteric inhibitors, are investigated.

LH-21, which was suggested as a neutral antagonist of CB1 (Jagerovic et al., 2004), was later shown to be a low affinity inverse agonist for the CB1, and a weak ligand for CB2 as well (Chen et al., 2008). It has been reported that LH-21 lacks brain penetration and suppresses FI and BW gain when administrated systemically in rats, implying a peripheral CB1-mediated pathway (Pavon et al., 2006). However, a measurement of rat plasma and brain concentrations of LH-21 after an intravenous injection showed the near 1:1 plasma-to-brain concentration ratio, demonstrating its brain penetrability. Only at high dose (60 mg/kg), it suppressed overnight FI and BW gain of either WT or CB1 KO mice, suggesting a CB1-independent mechanism of anorectic effect (Chen et al., 2008).

When compared with rimonabant, AM6545 (a CB1 neutral antagonist derived from the modifications of the rimonabant structure), does not antagonize cannabinoid-induced central effects. It only partially blocks the hypothermic effect of WIN55,212-2 administrated peripherally, but not centrally, a cannabinoid effect regardless of the administration route (Rawls, Cabassa, Geller, & Adler, 2002). AM6545 lacks side effects and is extruded from the brain by the ABC transporter P-glycoprotein (Tam et al., 2010). Despite of its poor brain penetration, AM6545 dose-dependently reduced FI in rats, especially for the high-carbohydrate and -fat diets, due to reduced feeding rate (Randall et al., 2010). In addition, AM6545 increases energy expenditure, improves glucose tolerance and insulin resistance, and reduces liver steatosis in *ob/ob* mice without effect on BW. Transgenic overexpression of CB1 in hepatocytes on the CB1 KO background reintroduces the insulin resistance caused by HFD, which can be corrected by treatment with AM6545, suggesting that the expression of CB1 in hepatocytes is sufficient for some of the DIO phenotypes (Tam et al., 2010). However, another study showed that AM6545 reduces FI and BW in CB1 KO, but not CB1/CB2 double KO mice, suggesting the involvement of CB2 in mediating its action (Cluny et al., 2010).

A thioamide derivative of rimonabant with improved solubility appears to be a neutral antagonist of CB1, which decreases BW gain of DIO mice (Boström et al., 2010). Another derivative of rimonabant, compound-1 ((*E*)-*N*-(3-chlorobenzyl)-3-(5-(4-chlorophenyl)-1-(2,4-dichlorophenyl)-4-methyl-1*H*-pyrazol-3-yl)acrylamide) is considered a peripherally restricted CB1 antagonist (Son et al., 2010). At dosage (10 mg/kg) which does not change the BW of DIO mice, compound-1 still causes significant reduction of hepatic SREBP-1 expression, indicating its effects on metabolism independent of BW changes (Son et al., 2010). The effects of four-substituted pyrazole-3-carboxamides on BW of DIO mice have been investigated in an attempt to identify antagonist and inverse agonist for CB1. One of the compounds was found to have the plasma/brain ratio of 13, but still displayed comparable efficacy in terms of BW reduction as rimonabant, demonstrating the potential of developing peripheral CB1 antagonist for obesity (Receveur et al., 2010).

PSNCBAM-1, an allosteric antagonist of CB1, acutely reduced FI and BW in rats (Horswill et al., 2007). Both CB1 agonists, WIN55,212-2 and CP55940, attenuate miniature inhibitory postsynaptic current (mIPSC) frequency. However, PSNCBAM-1 pretreatment abolishes CP55940-induced reductions in mIPSC frequency but not that of induced by WIN55,212-2. It also attenuates the effects of CB1 inverse agonist AM251 (Wang, Horswill, Whalley, & Stephens, 2011), which demonstrates the unique properties of this allosteric antagonist.

7. FUTURE PERSPECTIVES

7.1. Targeting endocannabinoid turnover

The BW reduction results of CB1 antagonist in the clinical trials, shown in Table 14.1, suggest that an overactive ECS may contribute to human obesity. Then, the question becomes whether endocannabinoid metabolism can be targeted to achieve comparable results without the liability of CB1 inverse agonists. As discussed previously, FAAH and MAGL KO mice have elevated endocannabinoids, but normal BW, fat mass, and food consumption (Cravatt et al., 2001; Taschler et al., 2011), indicating that the elevated levels of 2-AG or anandamide in long term may not be enough to cause obesity in mice. One explanation is that the state of the receptor, but not the alteration of the endocannabinoid levels, is critical for the manifestation of obesity. Chronic treatment of cannabinoids results in tolerance through decreased receptor binding to the brain regions and receptor expression levels (Breivogel et al., 1999; De Fonseca, Gorriti, Fernández-Ruiz, Palomo, & Ramos, 1994; Oviedo, Glowa, & Herkenham, 1993; Romero et al., 1997). It has been shown that anandamide is rapidly degraded in the brain (Willoughby, Moore, Martin, & Ellis, 1997). On the other hand, CB1 degradation after endocytosis also plays a role, which is mediated by the GPCR-associated sorting protein (Martini et al., 2007). Therefore, another way of mimicking CB1 antagonism is to accelerate its endocytosis and degradation. Whether this is achievable remains to be seen. Alternatively, the reduction of prostaglandins in the KO mice (Nomura et al., 2011) or the changes of the levels of other substrates or downstream products of FAAH and MAGL (Cravatt et al., 1996) may act against an overactive ECS.

7.2. Signal transduction of CB1

PKA plays a critical role for mediating glucose and lipid homeostasis (Cohen, 2006). CB1 is coupled to the Gi/o proteins as its activation by cannabinoids has inhibitory effects on the cAMP production in cells with activated adenylyl cyclase (Allyn, 1984; Howlett & Fleming, 1984; Martin, 1986). In human and rat brain, CB1 specific agonist WIN55,212-2 decreases forskolin-induced cAMP accumulation, while rimonabant increases basal cAMP levels (Mato, Pazos, & Valdizán, 2002). On the other hand, in CHO cells transfected with CB1, the activation of cannabinoid CB1 receptor resulted in cAMP accumulation, which caused PKA-induced gene transcription and could be blocked by rimonabant (Calandra et al., 1999), suggesting that CB1 can mediate dual intracellular signaling pathways. Then, the unanswered questions are what coupling mechanisms are utilized by CB1 at different developmental stages and in different tissues, and whether the activation of PKA is mandatory for the cellular responses. It is possible that ECS plays a role in the fine-tuning of the intracellular PKA activation in response to a variety of hormonal and nutritional stimuli. More studies are needed to determine the spaciotemporal characteristics of the CB1-mediated regulation of intracellular cAMP levels.

Besides the cAMP pathway, more recent studies suggest that the activation of CB1 is linked to phospholipases, mitogen-activated protein kinases, and β-arrestins, indicating that the signal transduction pathways elicited by CB1 are more complicated than originally thought (Turu & Hunyady, 2010). Future studies are required to assign each pathway to specific cellular and physiological functions.

7.3. Discrepancies regarding peripheral CB1 and mechanisms of actions of the novel CB1 compounds

As stated in the previous sections, the expression of CB1 and CB2 has been reported in a variety of tissues or cells. However, the results reported by different research groups are extremely inconsistent. For example, CB1 mRNA and proteins have been detected in the pancreatic exocrine and almost all the endocrine cells depending on techniques and reagents used. As we pointed out earlier, non-specific antibodies are likely accountable for the discrepancies.

Knowing the very high levels of CB1 in neural cells, it is unclear how CB1 expressed at extremely low (sometimes undetectable) levels in cells, such as hepatocytes and adipocytes, could be functional. Are the receptor coupling mechanisms more efficient in these tissues than those in neurons? Hopefully, analysis of CB1 signal transduction in single or highly purified cells using more sensitive and innovative methods would provide some answers.

Furthermore, studies of other tissue- and cell-specific CB1 KO mouse models, such as those for pancreatic ganglions, islet β-cells, and adipocytes, should help generate a more comprehensively view of the sites and actions of CB1. These models should also be used to characterize the novel peripherally restricted compounds so that their precise modes of actions can be defined.

8. CONCLUSIONS

The existence of cannabinoid compounds in plants and their specific receptors in animals is a testimony to the intriguing relationships between these two drastically different organisms, which also lend strong scientific support for traditional botanical medicine. The extensive research efforts have revealed the components of the ECS, and more importantly, the chemical matters that can specifically modulate the activities of ECS. Their potential for clinical use has been investigated, especially for CB1 antagonists for the treatment of obesity and related metabolic syndrome. Despite strong evidence for their efficacy in animal models and in the clinic, CB1 antagonists fail to establish as treatment mainly due to CNS-mediated side effects. Therefore, the peripherally restricted CB1 antagonists with limited accessibility to the brain may have the beneficial effects of rimonabant, but not the detrimental ones. However, the data obtained from the tissue-specific KO or knockdown mice indicate that the antagonism of brain CB1 appears to be essential for rimonabant's antiobesity activities (Pang et al., 2011; Quarta et al., 2010). The debate about the roles of central versus peripheral CB1 reflects the complexity and the lack of thorough understanding of the ECS. Drugs targeting ECS may regulate behavior and metabolism depending upon the cellular and tissue contexts. Conceivably, CB1 antagonists could be more tolerable in a sub-population of patients who are less likely to develop depression/anxiety, as identified using genetic and biochemical markers, and medical history (Di Marzo & Després, 2009). More studies are required to understand the endocannabinoid metabolism and CB1-mediated signaling events in different cells, which should inspire the creation of novel, effective, and safer antiobesity drugs.

ACKNOWLEDGMENTS

This work was financially supported by research grant from Allen Foundation Inc. (to G. C.), start-up fund from the University of Tennessee at Knoxville (to G. C.), and Scientist Development Grant from American Heart Association (09SDG2140003 to G. C.).

REFERENCES

Abel, E. L. (1971). Effects of marihuana on the solution of anagrams, memory and appetite. *Nature, 231*, 260–261.

Abel, E. L. (1975). Cannabis: Effects on hunger and thirst. *Behavioral Biology, 15*, 255–281.

Abood, M. E., Ditto, K. E., Noel, M. A., Showalter, V. M., & Tao, Q. (1997). Isolation and expression of a mouse CB1 cannabinoid receptor gene: Comparison of binding properties with those of native CB1 receptors in mouse brain and n18tg2 neuroblastoma cells. *Biochemical Pharmacology, 53*, 207–214.

Allyn, C. (1984). Inhibition of neuroblastoma adenylate cyclase by cannabinoid and nantradol compounds. *Life Sciences, 35*, 1803–1810.

Andersson, U., Filipsson, K., Abbott, C. R., Woods, A., Smith, K., Bloom, S. R., et al. (2004). AMP-activated protein kinase plays a role in the control of food intake. *The Journal of Biological Chemistry, 279*, 12005–12008.

Arnone, M., Maruani, J., Chaperon, F., Thiébot, M. H., Poncelet, M., Soubrié, P., et al. (1997). Selective inhibition of sucrose and ethanol intake by SR 141716, an antagonist of central cannabinoid (CB1) receptors. *Psychopharmacology, 132*, 104–106.

Bakkali-Kassemi, L., El Ouezzani, S., Magoul, R., Merroun, I., Lopez-Jurado, M., & Errami, M. (2011). Effects of cannabinoids on neuropeptide Y and β-endorphin expression in the rat hypothalamic arcuate nucleus. *The British Journal of Nutrition, 105*, 654–660.

Beal, J. E., Olson, R., Laubenstein, L., Morales, J. O., Bellman, P., Yangco, B., et al. (1995). Dronabinol as a treatment for anorexia associated with weight loss in patients with AIDS. *Journal of Pain and Symptom Management, 10*, 89–97.

Bennetzen, M. F., Wellner, N., Ahmed, S. S., Ahmed, S. M., Diep, T. A., Hansen, H. S., et al. (2011). Investigations of the human endocannabinoid system in two subcutaneous adipose tissue depots in lean subjects and in obese subjects before and after weight loss. *International Journal of Obesity, 35*, 1377–1384.

Bensaid, M., Gary-Bobo, M., Esclangon, A., Maffrand, J. P., Le Fur, G., Oury-Donat, F., et al. (2003). The cannabinoid CB1 receptor antagonist SR141716 increases Acrp30 mRNA expression in adipose tissue of obese fa/fa rats and in cultured adipocyte cells. *Molecular Pharmacology, 63*, 908–914.

Bermúdez-Silva, F., Suárez, J., Baixeras, E., Cobo, N., Bautista, D., Cuesta-Muñoz, A., et al. (2008). Presence of functional cannabinoid receptors in human endocrine pancreas. *Diabetologia, 51*, 476–487.

Berrendero, F., Sepe, N., Ramos, J. A., Di Marzo, V., & Fernàndez-Ruiz, J. J. (1999). Analysis of cannabinoid receptor binding and mRNA expression and endogenous cannabinoid contents in the developing rat brain during late gestation and early postnatal period. *Synapse, 33*, 181–191.

Berry, E. M., & Mechoulam, R. (2002). Tetrahydrocannabinol and endocannabinoids in feeding and appetite. *Pharmacology & Therapeutics, 95*, 185–190.

Bisogno, T., Howell, F., Williams, G., Minassi, A., Cascio, M. G., Ligresti, A., et al. (2003). Cloning of the first sn1-DAG lipases points to the spatial and temporal regulation of endocannabinoid signaling in the brain. *The Journal of Cell Biology, 163*, 463–468.

Blüher, M., Engeli, S., Klöting, N., Berndt, J., Fasshauer, M., Bátkai, S., et al. (2006). Dysregulation of the peripheral and adipose tissue endocannabinoid system in human abdominal obesity. *Diabetes, 55*, 3053–3060.

Boström, J., Olsson, R. I., Tholander, J., Greasley, P. J., Ryberg, E., Nordberg, H., et al. (2010). Novel thioamide derivatives as neutral CB1 receptor antagonists. *Bioorganic & Medicinal Chemistry Letters, 20*, 479–482.

Bouaboula, M., Hilairet, S., Marchand, J., Fajas, L., Fur, G. L., & Casellas, P. (2005). Anandamide induced PPARγ transcriptional activation and 3T3-L1 preadipocyte differentiation. *European Journal of Pharmacology, 517*, 174–181.

Bray, G. A., Nielsen, S. J., & Popkin, B. M. (2004). Consumption of high-fructose corn syrup in beverages may play a role in the epidemic of obesity. *The American Journal of Clinical Nutrition, 79*, 537–543.

Breivogel, C. S., Childers, S. R., Deadwyler, S. A., Hampson, R. E., Vogt, L. J., & Sim-Selley, L. J. (1999). Chronic D9-tetrahydrocannabinol treatment produces a time-dependent loss of cannabinoid receptors and cannabinoid receptor-activated G proteins in rat brain. *Journal of Neurochemistry, 73*, 2447–2459.

Breivogel, C. S., Griffin, G., Di Marzo, V., & Martin, B. R. (2001). Evidence for a new G protein-coupled cannabinoid receptor in mouse brain. *Molecular Pharmacology, 60*, 155–163.

Brown, J. E., Kassouny, M., & Cross, J. K. (1977). Kinetic studies of food intake and sucrose solution preference by rats treated with low doses of Δ9-tetrahydrocannabinol. *Behavioral Biology, 20*, 104–110.

Buckley, N. E., Hansson, S., Harta, G., & Mezey, É. (1997). Expression of the CB1 and CB2 receptor messenger RNAs during embryonic development in the rat. *Neuroscience, 82*, 1131–1149.

Buckley, N. E., McCoy, K. L., Mezey, E., Bonner, T., Zimmer, A., Felder, C. C., et al. (2000). Immunomodulation by cannabinoids is absent in mice deficient for the cannabinoid CB2 receptor. *European Journal of Pharmacology, 396*, 141–149.

Buczynski, M. W., Dumlao, D. S., & Dennis, E. A. (2009). Thematic review series: Proteomics. An integrated omics analysis of eicosanoid biology. *Journal of Lipid Research, 50*, 1015–1038.

Calandra, B., Portier, M., Kernéis, A., Delpech, M., Carillon, C., Le Fur, G., et al. (1999). Dual intracellular signaling pathways mediated by the human cannabinoid CB1 receptor. *European Journal of Pharmacology*, *374*, 445–455.

Casu, M. A., Porcella, A., Ruiu, S., Saba, P., Marchese, G., Carai, M. A. M., et al. (2003). Differential distribution of functional cannabinoid CB1 receptors in the mouse gastroenteric tract. *European Journal of Pharmacology*, *459*, 97–105.

Chen, R. Z., Frassetto, A., Lao, J. Z., Huang, R. R., Xiao, J. C., Clements, M. J., et al. (2008). Pharmacological evaluation of LH-21, a newly discovered molecule that binds to cannabinoid CB1 receptor. *European Journal of Pharmacology*, *584*, 338–342.

Cluny, N. L., Vemuri, V. K., Chambers, A. P., Limebeer, C. L., Bedard, H., Wood, J. T., et al. (2010). A novel peripherally restricted cannabinoid receptor antagonist, AM6545, reduces food intake and body weight, but does not cause malaise, in rodents. *British Journal of Pharmacology*, *161*, 629–642.

Cohen, P. (2006). The twentieth century struggle to decipher insulin signalling. *Nature Reviews. Molecular Cell Biology*, 7, 867–873.

Cohen, S., Dadi, H., Shaoul, E., Sharfe, N., & Roifman, C. M. (1999). Cloning and characterization of a lymphoid-specific, inducible human protein tyrosine phosphatase, Lyp. *Blood*, *93*, 2013–2024.

Cohen, P., & Grahame Hardie, D. (1991). The actions of cyclic AMP on biosynthetic processes are mediated indirectly by cyclic AMP-dependent protein kinase. *Biochimica et Biophysica Acta*, *1094*, 292–299.

Colombo, G., Agabio, R., Diaz, G., Lobina, C., Reali, R., & Gessa, G. L. (1998). Appetite suppression and weight loss after the cannabinoid antagonist SR 141716. *Life Sciences*, *63*, L113–L117.

Compton, D. R., Aceto, M. D., Lowe, J., & Martin, B. R. (1996). In vivo characterization of a specific cannabinoid receptor antagonist (SR141716A): Inhibition of delta 9-tetrahydrocannabinol-induced responses and apparent agonist activity. *The Journal of Pharmacology and Experimental Therapeutics*, *277*, 586–594.

Cota, D., Marsicano, G., Tschöp, M., Grübler, Y., Flachskamm, C., Schubert, M., et al. (2003). The endogenous cannabinoid system affects energy balance via central orexigenic drive and peripheral lipogenesis. *The Journal of Clinical Investigation*, *112*, 423–431.

Cottone, E., Salio, C., Conrath, M., & Franzoni, M. F. (2003). Xenopus laevis CB1 cannabinoid receptor: Molecular cloning and mRNA distribution in the central nervous system. *The Journal of Comparative Neurology*, *464*, 487–496.

Cravatt, B. F., Demarest, K., Patricelli, M. P., Bracey, M. H., Giang, D. K., Martin, B. R., et al. (2001). Supersensitivity to anandamide and enhanced endogenous cannabinoid signaling in mice lacking fatty acid amide hydrolase. *Proceedings of the National Academy of Sciences*, *98*, 9371–9376.

Cravatt, B. F., Giang, D. K., Mayfield, S. P., Boger, D. L., Lerner, R. A., & Gilula, N. B. (1996). Molecular characterization of an enzyme that degrades neuromodulatory fatty-acid amides. *Nature*, *384*, 83–87.

De Fonseca, F. R., Gorriti, M. A., Fernández-Ruiz, J. J., Palomo, T., & Ramos, J. A. (1994). Downregulation of rat brain cannabinoid binding sites after chronic D9-tetrahydrocannabinol treatment. *Pharmacology, Biochemistry, and Behavior*, *47*, 33–40.

Després, J. P., Golay, A., & Sjöström, L. (2005). Effects of rimonabant on metabolic risk factors in overweight patients with dyslipidemia. *The New England Journal of Medicine*, *353*, 2121–2134.

Després, J. P., Ross, R., Boka, G., Alméras, N., Lemieux, I., & for the ADAGIO-Lipids Investigators, (2009). Effect of rimonabant on the high-triglyceride/low-HDL-cholesterol dyslipidemia, intraabdominal adiposity, and liver fat. *Arteriosclerosis, Thrombosis, and Vascular Biology*, *29*, 416–423.

Devane, W. A., Dysarz, F. A., Johnson, M. R., Melvin, L. S., & Howlett, A. C. (1988). Determination and characterization of a cannabinoid receptor in rat brain. *Molecular Pharmacology*, *34*, 605–613.

Devane, W. A., HanuS, L., Breuer, A., Pertwee, R. G., Stevenson, L. A., Griffin, G., et al. (1992). Isolation and structure of a brain constituent that binds to the cannabinoid receptor. *Science*, *258*, 1946–1949.

Di Marzo, V. (2008a). Endocannabinoids: Synthesis and degradation. *Reviews of Physiology Biochemistry and Pharmacology*, *160*, 1–24, Springer: Berlin/Heidelberg.

Di Marzo, V. (2008b). Targeting the endocannabinoid system: To enhance or reduce? *Nature Reviews. Drug Discovery*, *7*, 438–455.

Di Marzo, V., Breivogel, C. S., Tao, Q., Bridgen, D. T., Razdan, R. K., Zimmer, A. M., et al. (2000). Levels, metabolism, and pharmacological activity of anandamide in CB1 cannabinoid receptor knockout mice. *Journal of Neurochemistry*, *75*, 2434–2444.

Di Marzo, V., Côté, M., Matias, I., Lemieux, I., Arsenault, B., Cartier, A., et al. (2009). Changes in plasma endocannabinoid levels in viscerally obese men following a 1 year lifestyle modification programme and waist circumference reduction: Associations with changes in metabolic risk factors. *Diabetologia*, *52*, 213–217.

Di Marzo, V., & Després, J. P. (2009). CB1 antagonists for obesity—What lessons have we learned from rimonabant? *Nature Reviews. Endocrinology*, *5*, 633–638.

Di Marzo, V., Fontana, A., Cadas, H., Schinelli, S., Cimino, G., Schwartz, J. C., et al. (1994). Formation and inactivation of endogenous cannabinoid anandamide in central neurons. *Nature*, *372*, 686–691.

Di Marzo, V., Goparaju, S. K., Wang, L., Liu, J., Batkai, S., Jarai, Z., et al. (2001). Leptin-regulated endocannabinoids are involved in maintaining food intake. *Nature*, *410*, 822–825.

Di Marzo, V., Melck, D., Orlando, P., Bisogno, T., Zagoory, O., Bifulco, M., et al. (2001). Palmitoylethanolamide inhibits the expression of fatty acid amide hydrolase and enhances the anti-proliferative effect of anandamide in human breast cancer cells. *The Biochemical Journal*, *358*, 249–255.

Di Marzo, V., Sepe, N., De Petrocellis, L., Berger, A., Crozier, G., Fride, E., et al. (1998). Trick or treat from food endocannabinoids? *Nature*, *396*, 636.

Dinh, T. P., Carpenter, D., Leslie, F. M., Freund, T. F., Katona, I., Sensi, S. L., et al. (2002). Brain monoglyceride lipase participating in endocannabinoid inactivation. *Proceedings of the National Academy of Sciences*, *99*, 10819–10824.

DiPatrizio, N. V., Astarita, G., Schwartz, G., Li, X., & Piomelli, D. (2011). Endocannabinoid signal in the gut controls dietary fat intake. *Proceedings of the National Academy of Sciences*, *108*, 12904–12908.

Dixon, J. B., O'Brien, P. E., Playfair, J., Chapman, L., Schachter, L. M., Skinner, S., et al. (2008). Adjustable gastric banding and conventional therapy for type 2 diabetes: A randomized controlled trial. *JAMA: The Journal of the American Medical Association*, *299*, 316–323.

Donovan, J., & Grundy, D. (2011). The differential effect of CB1 receptors on the discharge of afferent and efferent fibres supplying the rat jejunum. *Neurogastroenterology and Motility*, *23*, 567-e209.

Doyon, C., Denis, R. G., Baraboi, E. D., Samson, P., Lalonde, J., Deshaies, Y., et al. (2006). Effects of rimonabant (SR141716) on fasting-induced hypothalamic-pituitary-adrenal axis and neuronal activation in lean and obese Zucker rats. *Diabetes*, *55*, 3403–3410.

Duncan, M., Mouihate, A., Mackie, K., Keenan, C. M., Buckley, N. E., Davison, J. S., et al. (2008). Cannabinoid CB2 receptors in the enteric nervous system modulate gastrointestinal contractility in lipopolysaccharide-treated rats. *American Journal of Physiology. Gastrointestinal and Liver Physiology*, *295*, G78–G87.

Eckardt, K., Sell, H., Taube, A., Koenen, M., Platzbecker, B., Cramer, A., et al. (2009). Cannabinoid type 1 receptors in human skeletal muscle cells participate in the negative crosstalk between fat and muscle. *Diabetologia, 52*, 664–674.

Edery, H., Grunfeld, Y., Ben-Zvi, Z., & Mechoulam, R. (1971). Structural requirements for cannabinoid activity. *Annals of the New York Academy of Sciences, 191*, 40–53.

Engeli, S., Böhnke, J., Feldpausch, M., Gorzelniak, K., Janke, J., Bátkai, S., et al. (2005). Activation of the peripheral endocannabinoid system in human obesity. *Diabetes, 54*, 2838–2843.

Esposito, I., Proto, M. C., Gazzerro, P., Laezza, C., Miele, C., Alberobello, A. T., et al. (2008). The cannabinoid CB1 receptor antagonist rimonabant stimulates 2-deoxyglucose uptake in skeletal muscle cells by regulating the expression of phosphatidylinositol-3-kinase. *Molecular Pharmacology, 74*, 1678–1686.

Felder, C. C., & Glass, M. (1998). Cannabinoid receptors and their endogenous agonists. *Annual Review of Pharmacology and Toxicology, 38*, 179–200.

Floreani, A., Lazzari, R., Macchi, V., Porzionato, A., Variola, A., Colavito, D., et al. (2010). Hepatic expression of endocannabinoid receptors and their novel polymorphisms in primary biliary cirrhosis. *Journal of Gastroenterology, 45*, 68–76.

Fong, T. M., Guan, X. M., Marsh, D. J., Shen, C. P., Stribling, D. S., Rosko, K. M., et al. (2007). Antiobesity efficacy of a novel cannabinoid-1 receptor inverse agonist, N-[(1S,2S)-3-(4-chlorophenyl)-2-(3-cyanophenyl)-1-methylpropyl]-2-methyl-2-{[5-(trifluoromethyl)pyridin-2-yl]oxy}propanamide (MK-0364), in rodents. *The Journal of Pharmacology and Experimental Therapeutics, 321*, 1013–1022.

Fong, T. M., & Heymsfield, S. B. (2009). Cannabinoid-1 receptor inverse agonists: Current understanding of mechanism of action and unanswered questions. *International Journal of Obesity, 33*, 947–955.

Freund, T., Katona, I., & Piomelli, D. (2003). Role of endogenous cannabinoids in synaptic signaling. *Physiological Reviews, 83*, 1017–1066.

Fride, E., Ginzburg, Y., Breuer, A., Bisogno, T., Di Marzo, V., & Mechoulam, R. (2001). Critical role of the endogenous cannabinoid system in mouse pup suckling and growth. *European Journal of Pharmacology, 419*, 207–214.

Friedman, J. M. (2004). Modern science versus the stigma of obesity. *Nature Medicine, 10*, 563–569.

Friedman, J. M. (2009). Leptin at 14 y of age: An ongoing story. *The American Journal of Clinical Nutrition, 89*, 973S–979S.

Gadzicki, D., Müller-Vahl, K., & Stuhrmann, M. (1999). A frequent polymorphism in the coding exon of the human cannabinoid receptor (CNR1) gene. *Molecular and Cellular Probes, 13*, 321–323.

Galiègue, S., Mary, S., Marchand, J., Dussossoy, D., Carrière, D., Carayon, P., et al. (1995). Expression of central and peripheral cannabinoid receptors in human immune tissues and leukocyte subpopulations. *European Journal of Biochemistry, 232*, 54–61.

Gamber, K. M., Macarthur, H., & Westfall, T. C. (2005). Cannabinoids augment the release of neuropeptide Y in the rat hypothalamus. *Neuropharmacology, 49*, 646–652.

Gaoni, Y., & Mechoulam, R. (1964). Isolation, structure, and partial synthesis of an active constituent of hashish. *Journal of the American Chemical Society, 86*, 1646–1647.

Gary-Bobo, M., Elachouri, G., Scatton, B., Le Fur, G., Oury-Donat, F., & Bensaid, M. (2006). The cannabinoid CB1 receptor antagonist rimonabant (SR141716) inhibits cell proliferation and increases markers of adipocyte maturation in cultured mouse 3T3 F442A preadipocytes. *Molecular Pharmacology, 69*, 471–478.

Gasperi, V., Fezza, F., Pasquariello, N., Bari, M., Oddi, S., Agrò, A., et al. (2007). Endocannabinoids in adipocytes during differentiation and their role in glucose uptake. *Cellular and Molecular Life Sciences, 64*, 219–229.

Gaulton, K. J., Willer, C. J., Li, Y., Scott, L. J., Conneely, K. N., Jackson, A. U., et al. (2008). Comprehensive association study of type 2 diabetes and related quantitative traits with 222 candidate genes. *Diabetes, 57*, 3136–3144.

Gérard, C. M., Mollereau, C., Vassart, G., & Parmentier, M. (1991). Molecular cloning of a human cannabinoid receptor which is also expressed in testis. *The Biochemical Journal, 279*, 129–134.

Gluck, J. P., & Ferraro, D. P. (1974). Effects of Δ9-THC on food and water intake of deprivation experienced rats. *Behavioral Biology, 11*, 395–401.

Gómez, R., Navarro, M., Ferrer, B., Trigo, J. M., Bilbao, A., Del Arco, I., et al. (2002). A peripheral mechanism for CB1 cannabinoid receptor-dependent modulation of feeding. *The Journal of Neuroscience, 22*, 9612–9617.

Greenberg, I., Kuehnle, J., Mendelson, J. H., & Bernstein, J. G. (1976). Effects of marihuana use on body weight and caloric intake in humans. *Psychopharmacology, 49*, 79–84.

Griffin, G., Tao, Q., & Abood, M. E. (2000). Cloning and pharmacological characterization of the rat CB2 cannabinoid receptor. *The Journal of Pharmacology and Experimental Therapeutics, 292*, 886–894.

Grimsey, N. L., Goodfellow, C. E., Scotter, E. L., Dowie, M. J., Glass, M., & Graham, E. S. (2008). Specific detection of CB1 receptors; cannabinoid CB1 receptor antibodies are not all created equal!. *Journal of Neuroscience Methods, 171*, 78–86.

Grinspoon, L., & Bakalar, J. B. (1997). *Marihuana, the forbidden medicine*. New Haven: Yale University Press [updated 1997; cited].

Grunfeld, Y., & Edery, H. (1969). Psychopharmacological activity of the active constituents of hashish and some related cannabinoids. *Psychopharmacology, 14*, 200–210.

Hanus, L., Avraham, Y., Ben-Shushan, D., Zolotarev, O., Berry, E. M., & Mechoulam, R. (2003). Short-term fasting and prolonged semistarvation have opposite effects on 2-AG levels in mouse brain. *Brain Research, 983*, 144–151.

Hanuš, L., Breuer, A., Tchilibon, S., Shiloah, S., Goldenberg, D., Horowitz, M., et al. (1999). HU-308: A specific agonist for CB2, a peripheral cannabinoid receptor. *Proceedings of the National Academy of Sciences, 96*, 14228–14233.

Hao, S., Avraham, Y., Mechoulam, R., & Berry, E. M. (2000). Low dose anandamide affects food intake, cognitive function, neurotransmitter and corticosterone levels in diet-restricted mice. *European Journal of Pharmacology, 392*, 147–156.

Harris, L. S., Carchman, R. A., & Martin, B. R. (1978). Evidence for the existence of specific cannabinoid binding sites. *Life Sciences, 22*, 1131–1137.

Haslam, D. W., & James, W. P. (2005). Obesity. *The Lancet, 366*, 1197–1209.

Herkenham, M., Lynn, A. B., Johnson, M. R., Melvin, L. S., de Costa, B. R., & Rice, K. C. (1991). Characterization and localization of cannabinoid receptors in rat brain: A quantitative in vitro autoradiographic study. *The Journal of Neuroscience, 11*, 563–583.

Herling, A. W., Gossel, M., Haschke, G., Stengelin, S., Kuhlmann, J., Müller, G., et al. (2007). CB1 receptor antagonist AVE1625 affects primarily metabolic parameters independently of reduced food intake in Wistar rats. *American Journal of Physiology. Endocrinology and Metabolism, 293*, E826–E832.

Herling, A. W., Kilp, S., Elvert, R., Haschke, G., & Kramer, W. (2008). Increased energy expenditure contributes more to the body weight-reducing effect of rimonabant than reduced food intake in candy-fed Wistar rats. *Endocrinology, 149*, 2557–2566.

Hermann, H., & Lutz, B. (2005). Coexpression of the cannabinoid receptor type 1 with the corticotropin-releasing hormone receptor type 1 in distinct regions of the adult mouse forebrain. *Neuroscience Letters, 375*, 13–18.

Hiley, C. R., & Kaup, S. S. (2007). GPR55 and the vascular receptors for cannabinoids. *British Journal of Pharmacology, 152*, 559–561.

Hively, R. L., Mosher, W. A., & Hoffmann, F. W. (1966). Isolation of trans-Δ^6-tetrahydrocannabinol from Marijuana. *Journal of the American Chemical Society, 88*, 1832–1833.

Hollander, P. A., Amod, A., Litwak, L. E., Chaudhari, U., & for the ARPEGGIO Study Group, (2010). Effect of rimonabant on glycemic control in insulin-treated type 2 diabetes: The ARPEGGIO trial. *Diabetes Care, 33*, 605–607.
Hollister, L. E. (1971). Marihuana in man: Three years later. *Science, 172*, 21–29.
Horswill, J. G., Bali, U., Shaaban, S., Keily, J. F., Jeevaratnam, P., Babbs, A. J., et al. (2007). PSNCBAM-1, a novel allosteric antagonist at cannabinoid CB1 receptors with hypophagic effects in rats. *British Journal of Pharmacology, 152*, 805–814.
Howlett, A. C., & Fleming, R. M. (1984). Cannabinoid inhibition of adenylate cyclase. Pharmacology of the response in neuroblastoma cell membranes. *Molecular Pharmacology, 26*, 532–538.
Howlett, A. C., Johnson, M. R., Melvin, L. S., & Milne, G. M. (1988). Nonclassical cannabinoid analgetics inhibit adenylate cyclase: Development of a cannabinoid receptor model. *Molecular Pharmacology, 33*, 297–302.
Huang, H., Acuna-Goycolea, C., Li, Y., Cheng, H. M., Obrietan, K., & van den Pol, A. N. (2007). Cannabinoids excite hypothalamic melanin-concentrating hormone but inhibit hypocretin/orexin neurons: Implications for cannabinoid actions on food intake and cognitive arousal. *The Journal of Neuroscience, 27*, 4870–4881.
Izzo, A. A., Fezza, F., Capasso, R., Bisogno, T., Pinto, L., Iuvone, T., et al. (2001). Cannabinoid CB1-receptor mediated regulation of gastrointestinal motility in mice in a model of intestinal inflammation. *British Journal of Pharmacology, 134*, 563–570.
Jagerovic, N., Hernandez-Folgado, L., Alkorta, I., Goya, P., Navarro, M., Serrano, A., et al. (2004). Discovery of 5-(4-chlorophenyl)-1-(2,4-dichlorophenyl)-3-hexyl-1H-1,2,4-triazole, a novel in vivo cannabinoid antagonist containing a 1,2,4-triazole motif. *Journal of Medicinal Chemistry, 47*, 2939–2942.
Jamshidi, N., & Taylor, D. A. (2001). Anandamide administration into the ventromedial hypothalamus stimulates appetite in rats. *British Journal of Pharmacology, 134*, 1151–1154.
Járai, Z., Wagner, J. A., Varga, K., Lake, K. D., Compton, D. R., Martin, B. R., et al. (1999). Cannabinoid-induced mesenteric vasodilation through an endothelial site distinct from CB1 or CB2 receptors. *Proceedings of the National Academy of Sciences, 96*, 14136–14141.
Jbilo, O., Ravinet-Trillou, C., Arnone, M., Buisson, I., Bribes, E., Péleraux, A., et al. (2005). The CB1 receptor antagonist rimonabant reverses the diet-induced obesity phenotype through the regulation of lipolysis and energy balance. *The FASEB Journal, 19*, 1567–1569.
Jelsing, J., Galzin, A. M., Guillot, E., Pruniaux, M. P., Larsen, P. J., & Vrang, N. (2009). Localization and phenotypic characterization of brainstem neurons activated by rimonabant and WIN55,212-2. *Brain Research Bulletin, 78*, 202–210.
Jeong, W. I., Osei-Hyiaman, D., Park, O., Liu, J., Bátkai, S., Mukhopadhyay, P., et al. (2008). Paracrine activation of hepatic CB1 receptors by stellate cell-derived endocannabinoids mediates alcoholic fatty liver. *Cell Metabolism, 7*, 227–235.
Johnson, M. R., Melvin, L. S., Althuis, T. H., Bindra, J. S., Harbert, C. A., Milne, G. M., et al. (1981). Selective and potent analgetics derived from cannabinoids. *The Journal of Clinical Pharmacology, 21*, 271S–282S.
Jones, D. (2008). End of the line for cannabinoid receptor 1 as an anti-obesity target? *Nature Reviews. Drug Discovery, 7*, 961–962.
Jourdan, T., Djaouti, L., Demizieux, L., Gresti, J., Vergès, B., & Degrace, P. (2010). CB1 antagonism exerts specific molecular effects on visceral and subcutaneous fat and reverses liver steatosis in diet-induced obese mice. *Diabetes, 59*, 926–934.
Juan-Picó, P., Fuentes, E., Javier Bermúdez-Silva, F., Javier Díaz-Molina, F., Ripoll, C., Rodríguez de Fonseca, F., et al. (2006). Cannabinoid receptors regulate Ca^{2+} signals and insulin secretion in pancreatic β-cell. *Cell Calcium, 39*, 155–162.
Kapur, A., Zhao, P., Sharir, H., Bai, Y., Caron, M. G., Barak, L. S., et al. (2009). Atypical responsiveness of the orphan receptor GPR55 to cannabinoid ligands. *The Journal of Biological Chemistry, 284*, 29817–29827.

Karlsson, M., Contreras, J. A., Hellman, U., Tornqvist, H., & Holm, C. (1997). cDNA cloning, tissue distribution, and identification of the catalytic triad of monoglyceride lipase. *The Journal of Biological Chemistry*, *272*, 27218–27223.

Katayama, K., Ueda, N., Kurahashi, Y., Suzuki, H., Yamamoto, S., & Kato, I. (1997). Distribution of anandamide amidohydrolase in rat tissues with special reference to small intestine. *Biochimica et Biophysica Acta*, *1347*, 212–218.

Kim, W., Doyle, M. E., Liu, Z., Lao, Q., Shin, Y. K., Carlson, O. D., et al. (2011). Cannabinoids inhibit insulin receptor signaling in pancreatic β-cells. *Diabetes*, *60*, 1198–1209.

Kipnes, M. S., Hollander, P., Fujioka, K., Gantz, I., Seck, T., Erondu, N., et al. (2010). A one-year study to assess the safety and efficacy of the CB1R inverse agonist taranabant in overweight and obese patients with type 2 diabetes. *Diabetes, Obesity & Metabolism*, *12*, 517–531.

Kirkham, T. C., Williams, C. M., Fezza, F., & Marzo, V. D. (2002). Endocannabinoid levels in rat limbic forebrain and hypothalamus in relation to fasting, feeding and satiation: Stimulation of eating by 2-arachidonoyl glycerol. *British Journal of Pharmacology*, *136*, 550–557.

Kola, B., Farkas, I., Christ-Crain, M., Wittmann, G., Lolli, F., Amin, F., et al. (2008). The orexigenic effect of ghrelin is mediated through central activation of the endogenous cannabinoid system. *PLoS One*, *3*, e1797.

Kola, B., Hubina, E., Tucci, S. A., Kirkham, T. C., Garcia, E. A., Mitchell, S. E., et al. (2005). Cannabinoids and ghrelin have both central and peripheral metabolic and cardiac effects via AMP-activated protein kinase. *The Journal of Biological Chemistry*, *280*, 25196–25201.

Landi, M., Croci, T., Rinaldi-Carmona, M., Maffrand, J. P., Le Fur, G., & Manara, L. (2002). Modulation of gastric emptying and gastrointestinal transit in rats through intestinal cannabinoid CB1 receptors. *European Journal of Pharmacology*, *450*, 77–83.

Ledent, C., Valverde, O., Cossu, G., Petitet, F., Aubert, J. F., Beslot, F., et al. (1999). Unresponsiveness to cannabinoids and reduced addictive effects of opiates in CB_1 receptor knockout mice. *Science*, *283*, 401–404.

Linari, G., Agostini, S., Amadoro, G., Ciotti, M. T., Florenzano, F., Improta, G., et al. (2009). Involvement of cannabinoid CB1- and CB2-receptors in the modulation of exocrine pancreatic secretion. *Pharmacological Research*, *59*, 207–214.

Lipina, C., Stretton, C., Hastings, S., Hundal, J. S., Mackie, K., Irving, A. J., et al. (2010). Regulation of MAP kinase-directed mitogenic and protein kinase B-mediated signaling by cannabinoid receptor type 1 in skeletal muscle cells. *Diabetes*, *59*, 375–385.

Liu, Y. L., Connoley, I. P., Wilson, C. A., & Stock, M. J. (2004). Effects of the cannabinoid CB1 receptor antagonist SR141716 on oxygen consumption and soleus muscle glucose uptake in Lepob/Lepob mice. *International Journal of Obesity and Related Metabolic Disorders*, *29*, 183–187.

Liu, J., Wang, L., Harvey-White, J., Osei-Hyiaman, D., Razdan, R., Gong, Q., et al. (2006). A biosynthetic pathway for anandamide. *Proceedings of the National Academy of Sciences*, *103*, 13345–13350.

Long, J. Z., Li, W., Booker, L., Burston, J. J., Kinsey, S. G., Schlosburg, J. E., et al. (2009). Selective blockade of 2-arachidonoylglycerol hydrolysis produces cannabinoid behavioral effects. *Nature Chemical Biology*, *5*, 37–44.

Mahler, S. V., Smith, K. S., & Berridge, K. C. (2007). Endocannabinoid hedonic hotspot for sensory pleasure: Anandamide in nucleus accumbens shell enhances 'liking' of a sweet reward. *Neuropsychopharmacology*, *32*, 2267–2278.

Manning, F. J., McDonough, J. H., Jr., Elsmore, T. F., Saller, C., & Sodetz, F. J. (1971). Inhibition of normal growth by chronic administration of Δ-9-tetrahydrocannabinol. *Science*, *174*, 424–426.

Marsicano, G., & Lutz, B. (1999). Expression of the cannabinoid receptor CB1 in distinct neuronal subpopulations in the adult mouse forebrain. *The European Journal of Neuroscience, 11*, 4213–4225.

Martin, B. R. (1986). Cellular effects of cannabinoids. *Pharmacological Reviews, 38*, 45–74.

Martin, B. R., Balster, R. L., Razdan, R. K., Harris, L. S., & Dewey, W. L. (1981). Behavioral comparisons of the stereoisomers of tetrahydrocannabinols. *Life Sciences, 29*, 565–574.

Martini, L., Waldhoer, M., Pusch, M., Kharazia, V., Fong, J., Lee, J. H., et al. (2007). Ligand-induced down-regulation of the cannabinoid 1 receptor is mediated by the G-protein-coupled receptor-associated sorting protein GASP1. *The FASEB Journal, 21*, 802–811.

Massa, F., Mancini, G., Schmidt, H., Steindel, F., Mackie, K., Angioni, C., et al. (2010). Alterations in the hippocampal endocannabinoid system in diet-induced obese mice. *The Journal of Neuroscience, 30*, 6273–6281.

Matias, I., Gonthier, M. P., Orlando, P., Martiadis, V., De Petrocellis, L., Cervino, C., et al. (2006). Regulation, function, and dysregulation of endocannabinoids in models of adipose and β-pancreatic cells and in obesity and hyperglycemia. *The Journal of Clinical Endocrinology and Metabolism, 91*, 3171–3180.

Mato, S., Pazos, A., & Valdizán, E. M. (2002). Cannabinoid receptor antagonism and inverse agonism in response to SR141716A on cAMP production in human and rat brain. *European Journal of Pharmacology, 443*, 43–46.

Matsuda, L. A., Lolait, S. J., Brownstein, M. J., Young, A. C., & Bonner, T. I. (1990). Structure of a cannabinoid receptor and functional expression of the cloned cDNA. *Nature, 346*, 561–564.

Mattes, R. D., Engelman, K., Shaw, L. M., & Elsohly, M. A. (1994). Cannabinoids and appetite stimulation. *Pharmacology, Biochemistry, and Behavior, 49*, 187–195.

Mechoulam, R. (1970). Marihuana chemistry. *Science, 168*, 1159–1166.

Mechoulam, R., Ben-Shabat, S., Hanus, L., Ligumsky, M., Kaminski, N. E., Schatz, A. R., et al. (1995). Identification of an endogenous 2-monoglyceride, present in canine gut, that binds to cannabinoid receptors. *Biochemical Pharmacology, 50*, 83–90.

Mechoulam, R., Braun, P., & Gaoni, Y. (1972). Syntheses of .DELTA.1-tetrahydrocannabinol and related cannabinoids. *Journal of the American Chemical Society, 94*, 6159–6165.

Mechoulam, R., & Hanuš, L. (2000). A historical overview of chemical research on cannabinoids. *Chemistry and Physics of Lipids, 108*, 1–13.

Mechoulam, R., Varconi, H., Ben-Zvi, Z., Edery, H., & Grunfeld, Y. (1972). Synthesis and biological activity of five tetrahydrocannabinol metabolites. *Journal of the American Chemical Society, 94*, 7930–7931.

Miller, C. C., Murray, T. F., Freeman, K. G., & Edwards, G. L. (2004). Cannabinoid agonist, CP 55,940, facilitates intake of palatable foods when injected into the hindbrain. *Physiology & Behavior, 80*, 611–616.

Minokoshi, Y., Alquier, T., Furukawa, N., Kim, Y. B., Lee, A., Xue, B., et al. (2004). AMP-kinase regulates food intake by responding to hormonal and nutrient signals in the hypothalamus. *Nature, 428*, 569–574.

Mølhøj, S., Hansen, H. S., Schweiger, M., Zimmermann, R., Johansen, T., & Malmlöf, K. (2010). Effect of the cannabinoid receptor-1 antagonist rimonabant on lipolysis in rats. *European Journal of Pharmacology, 646*, 38–45.

Molina-Holgado, F., Rubio-Araiz, A., García-Ovejero, D., Williams, R. J., Moore, J. D., Arévalo-Martín, Á., et al. (2007). CB2 cannabinoid receptors promote mouse neural stem cell proliferation. *The European Journal of Neuroscience, 25*, 629–634.

Monory, K., Blaudzun, H., Massa, F., Kaiser, N., Lemberger, T., Schütz, G., et al. (2007). Genetic dissection of behavioural and autonomic effects of Δ9-tetrahydrocannabinol in mice. *PLoS Biology, 5*, e269.

Monteleone, P., Matias, I., Martiadis, V., De Petrocellis, L., Maj, M., & Di Marzo, V. (2005). Blood levels of the endocannabinoid anandamide are increased in anorexia nervosa and in binge-eating disorder, but not in bulimia nervosa. *Neuropsychopharmacology*, *30*, 1216–1221.

Motaghedi, R., & McGraw, T. E. (2008). The CB1 endocannabinoid system modulates adipocyte insulin sensitivity. *Obesity*, *16*, 1727–1734.

Muccioli, G. G., Naslain, D., Backhed, F., Reigstad, C. S., Lambert, D. M., Delzenne, N. M., et al. (2010). The endocannabinoid system links gut microbiota to adipogenesis. *Molecular Systems Biology*, *6*, 392.

Mukhopadhyay, B., Liu, J., Osei-Hyiaman, D., Godlewski, G., Mukhopadhyay, P., Wang, L., et al. (2010). Transcriptional regulation of cannabinoid receptor-1 expression in the liver by retinoic acid acting via retinoic acid receptor. *The Journal of Biological Chemistry*, *285*, 19002–19011.

Munro, S., Thomas, K. L., & Abu-Shaar, M. (1993). Molecular characterization of a peripheral receptor for cannabinoids. *Nature*, *365*, 61–65.

Nakata, M., & Yada, T. (2008). Cannabinoids inhibit insulin secretion and cytosolic Ca^{2+} oscillation in islet β-cells via CB1 receptors. *Regulatory Peptides*, *145*, 49–53.

Ndong, C., O'Donnell, D., Ahmad, S., & Groblewski, T. (2011). Cloning and pharmacological characterization of the dog cannabinoid CB2 receptor. *European Journal of Pharmacology*, *669*, 24–31.

Nissen, S. E., Nicholls, S. J., Wolski, K., Rodés-Cabau, J., Cannon, C. P., Deanfield, J. E., et al. (2008). Effect of rimonabant on progression of atherosclerosis in patients with abdominal obesity and coronary artery disease. *JAMA: The Journal of the American Medical Association*, *299*, 1547–1560.

Nogueiras, R., Veyrat-Durebex, C., Suchanek, P. M., Klein, M., Tschöp, J., Caldwell, C., et al. (2008). Peripheral, but not central, CB1 antagonism provides food intake-independent metabolic benefits in diet-induced obese rats. *Diabetes*, *57*, 2977–2991.

Nomura, D. K., Morrison, B. E., Blankman, J. L., Long, J. Z., Kinsey, S. G., Marcondes, M. C., et al. (2011). Endocannabinoid hydrolysis generates brain prostaglandins that promote neuroinflammation. *Science*, *334*, 809–813.

O'Rahilly, S., & Farooqi, I. S. (2008). Human obesity as a heritable disorder of the central control of energy balance. *International Journal of Obesity*, *32*, S55–S61.

Oka, S., Nakajima, K., Yamashita, A., Kishimoto, S., & Sugiura, T. (2007). Identification of GPR55 as a lysophosphatidylinositol receptor. *Biochemical and Biophysical Research Communications*, *362*, 928–934.

Okamoto, Y., Morishita, J., Tsuboi, K., Tonai, T., & Ueda, N. (2004). Molecular characterization of a phospholipase D generating anandamide and its congeners. *The Journal of Biological Chemistry*, *279*, 5298–5305.

Osei-Hyiaman, D., DePetrillo, M., Pacher, P., Liu, J., Radaeva, S., Bátkai, S., et al. (2005). Endocannabinoid activation at hepatic CB1 receptors stimulates fatty acid synthesis and contributes to diet-induced obesity. *The Journal of Clinical Investigation*, *115*, 1298–1305.

Osei-Hyiaman, D., Liu, J., Zhou, L., Godlewski, G., Harvey-White, J., Jeong, W. I., et al. (2008). Hepatic CB1 receptor is required for development of diet-induced steatosis, dyslipidemia, and insulin and leptin resistance in mice. *The Journal of Clinical Investigation*, *118*, 3160–3169.

Oviedo, A., Glowa, J., & Herkenham, M. (1993). Chronic cannabinoid administration alters cannabinoid receptor binding in rat brain: A quantitative autoradiographic study. *Brain Research*, *616*, 293–302.

Pacher, P., Btkai, S., & Kunos, G. (2006). The endocannabinoid system as an emerging target of pharmacotherapy. *Pharmacological Reviews*, *58*, 389–462.

Pang, Z., Wu, N. N., Zhao, W., Chain, D. C., Schaffer, E., Zhang, X., et al. (2011). The central cannabinoid CB1 receptor is required for diet-induced obesity and rimonabant's antiobesity effects in mice. *Obesity*, *19*, 1923–1934.

Patsenker, E., Stoll, M., Millonig, G., Agaimy, A., Wissniowski, T., Schneider, V., et al. (2011). Cannabinoid receptor type I modulates alcohol-induced liver fibrosis. *Molecular Medicine*, *17*, 1285–1294.

Pavon, F. J., Bilbao, A., Hernández-Folgado, L., Cippitelli, A., Jagerovic, N., Abellán, G., et al. (2006). Antiobesity effects of the novel in vivo neutral cannabinoid receptor antagonist 5-(4-chlorophenyl)-1-(2,4-dichlorophenyl)-3-hexyl-1H-1,2,4-triazole—LH 21. *Neuropharmacology*, *51*, 358–366.

Pertwee, R. G. (2001). Cannabinoids and the gastrointestinal tract. *Gut*, *48*, 859–867.

Perwitz, N., Wenzel, J., Wagner, I., Büning, J., Drenckhan, M., Zarse, K., et al. (2010). Cannabinoid type 1 receptor blockade induces transdifferentiation towards a brown fat phenotype in white adipocytes. *Diabetes, Obesity and Metabolism*, *12*, 158–166.

Petersen, K. F., & Shulman, G. I. (2002). Pathogenesis of skeletal muscle insulin resistance in type 2 diabetes mellitus. *The American Journal of Cardiology*, *90*, 11–18.

Pinto, L., Izzo, A. A., Cascio, M. G., Bisogno, T., Hospodar-Scott, K., Brown, D. R., et al. (2002). Endocannabinoids as physiological regulators of colonic propulsion in mice. *Gastroenterology*, *123*, 227–234.

Pi-Sunyer, F. X., Aronne, L. J., Heshmati, H. M., Devin, J., Rosenstock, J., & for the RIO-North America Study Group, (2006). Effect of rimonabant, a cannabinoid-1 receptor blocker, on weight and cardiometabolic risk factors in overweight or obese patients. *JAMA: The Journal of the American Medical Association*, *295*, 761–775.

Poirier, B., Bidouard, J. P., Cadrouvele, C., Marniquet, X., Staels, B., O'Connor, S. E., et al. (2005). The anti-obesity effect of rimonabant is associated with an improved serum lipid profile. *Diabetes, Obesity & Metabolism*, *7*, 65–72.

Popkin, B. M., & Gordon-Larsen, P. (2004). The nutrition transition: Worldwide obesity dynamics and their determinants. *International Journal of Obesity and Related Metabolic Disorders*, *28*, S2–S9.

Porter, A. C., Sauer, J. M., Knierman, M. D., Becker, G. W., Berna, M. J., Bao, J., et al. (2002). Characterization of a novel endocannabinoid, virodhamine, with antagonist activity at the CB1 receptor. *The Journal of Pharmacology and Experimental Therapeutics*, *301*, 1020–1024.

Proietto, J., Rissanen, A., Harp, J. B., Erondu, N., Yu, Q., Suryawanshi, S., et al. (2010). A clinical trial assessing the safety and efficacy of the CB1R inverse agonist taranabant in obese and overweight patients: Low-dose study. *International Journal of Obesity*, *34*, 1243–1254.

Puighermanal, E., Marsicano, G., Busquets-Garcia, A., Lutz, B., Maldonado, R., & Ozaita, A. (2009). Cannabinoid modulation of hippocampal long-term memory is mediated by mTOR signaling. *Nature Neuroscience*, *12*, 1152–1158.

Quarta, C., Bellocchio, L., Mancini, G., Mazza, R., Cervino, C., Braulke, L. J., et al. (2010). CB1 signaling in forebrain and sympathetic neurons is a key determinant of endocannabinoid actions on energy balance. *Cell Metabolism*, *11*, 273–285.

Randall, P. A., Vemuri, V. K., Segovia, K. N., Torres, E. F., Hosmer, S., Nunes, E. J., et al. (2010). The novel cannabinoid CB1 antagonist AM6545 suppresses food intake and food-reinforced behavior. *Pharmacology, Biochemistry, and Behavior*, *97*, 179–184.

Ravinet Trillou, C., Arnone, M., Delgorge, C., Gonalons, N., Keane, P., Maffrand, J. P., et al. (2003). Anti-obesity effect of SR141716, a CB1 receptor antagonist, in diet-induced obese mice. *American Journal of Physiology. Regulatory, Integrative and Comparative Physiology*, *284*, R345–R353.

Ravinet Trillou, C., Delgorge, C., Menet, C., Arnone, M., & Soubrié, P. (2004). CB1 cannabinoid receptor knockout in mice leads to leanness, resistance to diet-induced obesity and enhanced leptin sensitivity. *International Journal of Obesity and Related Metabolic Disorders*, *28*, 640–648.

Rawls, S. M., Cabassa, J., Geller, E. B., & Adler, M. W. (2002). CB1 receptors in the preoptic anterior hypothalamus regulate WIN 55212-2 [(4,5-dihydro-2-methyl-4

(4-morpholinylmethyl)-1-(1-naphthalenyl-carbonyl)-6H-pyrrolo[3,2,1ij]quinolin-6-one)]-induced hypothermia. *The Journal of Pharmacology and Experimental Therapeutics*, *301*, 963–968.

Reaven, G. M. (2005). The insulin resistance syndrome: Definition and dietary approaches to treatment. *Annual Review of Nutrition*, *25*, 391–406.

Receveur, J. M., Murray, A., Linget, J. M., Nørregaard, P. K., Cooper, M., Bjurling, E., et al. (2010). Conversion of 4-cyanomethyl-pyrazole-3-carboxamides into CB1 antagonists with lowered propensity to pass the blood–brain-barrier. *Bioorganic & Medicinal Chemistry Letters*, *20*, 453–457.

Riggs, P. K., Vaida, F., Rossi, S. S., Sorkin, L. S., Gouaux, B., Grant, I., et al. (2012). A pilot study of the effects of cannabis on appetite hormones in HIV-infected adult men. *Brain Research*, *1431*, 46–52.

Rinaldi-Carmona, M., Barth, F., Héaulme, M., Shire, D., Calandra, B., Congy, C., et al. (1994). SR141716A, a potent and selective antagonist of the brain cannabinoid receptor. *FEBS Letters*, *350*, 240–244.

Roche, R., Hoareau, L., Bes-Houtmann, S., Gonthier, M. P., Laborde, C., Baron, J. F., et al. (2006). Presence of the cannabinoid receptors, CB1 and CB2, in human omental and subcutaneous adipocytes. *Histochemistry and Cell Biology*, *126*, 177–187.

Rodriguez-Martin, I., Herrero-Turrion, M. J., Fdez, Marron, de Velasco, E., Gonzalez-Sarmiento, R., & Rodriguez, R. E. (2007). Characterization of two duplicate zebrafish Cb2-like cannabinoid receptors. *Gene*, *389*, 36–44.

Romero, J., Garcia-Palomero, E., Castro, J. G., Garcia-Gil, L., Ramos, J. A., & Fernandez-Ruiz, J. J. (1997). Effects of chronic exposure to D9-tetrahydrocannabinol on cannabinoid receptor binding and mRNA levels in several rat brain regions. *Molecular Brain Research*, *46*, 100–108.

Romero-Zerbo, S. Y., Rafacho, A., Díaz-Arteaga, A., Suárez, J., Quesada, I., Imbernon, M., et al. (2011). A role for the putative cannabinoid receptor GPR55 in the islets of Langerhans. *The Journal of Endocrinology*, *211*, 177–185.

Ruby, M. A., Nomura, D. K., Hudak, C. S. S., Mangravite, L. M., Chiu, S., Casida, J. E., et al. (2008). Overactive endocannabinoid signaling impairs apolipoprotein E-mediated clearance of triglyceride-rich lipoproteins. *Proceedings of the National Academy of Sciences*, *105*, 14561–14566.

Ryberg, E., Larsson, N., Sjögren, S., Hjorth, S., Hermansson, N. O., Leonova, J., et al. (2007). The orphan receptor GPR55 is a novel cannabinoid receptor. *British Journal of Pharmacology*, *152*, 1092–1101.

Ryberg, E., Vu, H. K., Larsson, N., Groblewski, T., Hjorth, S., Elebring, T., et al. (2005). Identification and characterisation of a novel splice variant of the human CB1 receptor. *FEBS Letters*, *579*, 259–264.

Scheen, A. J., Finer, N., Hollander, P., Jensen, M. D., & Van Gaal, L. F. (2006). Efficacy and tolerability of rimonabant in overweight or obese patients with type 2 diabetes: A randomised controlled study. *The Lancet*, *368*, 1660–1672.

Schulze, M. B., & Hu, F. B. (2005). Primary prevention of diabetes: What can be done and how much can be prevented? *Annual Review of Public Health*, *26*, 445–467.

Shire, D., Calandra, B., Rinaldi-Carmona, M., Oustric, D., Pessègue, B., Bonnin-Cabanne, O., et al. (1996). Molecular cloning, expression and function of the murine CB2 peripheral cannabinoid receptor. *Biochimica et Biophysica Acta*, *1307*, 132–136.

Shire, D., Carillon, C., Kaghad, M., Calandra, B., Rinaldi-Carmona, M., Fur, G. L., et al. (1995). An amino-terminal variant of the central cannabinoid receptor resulting from alternative splicing. *The Journal of Biological Chemistry*, *270*, 3726–3731.

Sinnayah, P., Jobst, E. E., Rathner, J. A., Caldera-Siu, A. D., Tonelli-Lemos, L., Eusterbrock, A. J., et al. (2008). Feeding induced by cannabinoids is mediated independently of the melanocortin system. *PLoS One*, *3*, e2202.

Sjödén, P. O., Järbe, T. U. C., & Henriksson, B. G. (1973). Influence of tetrahydrocannabinols (Δ8-THC and Δ9-THC) on body weight, food, and water intake in rats. *Pharmacology, Biochemistry, and Behavior, 1*, 395–399.

Soderstrom, K., & Johnson, F. (2001). Zebra finch CB1 cannabinoid receptor: Pharmacology and in vivo and in vitro effects of activation. *The Journal of Pharmacology and Experimental Therapeutics, 297*, 189–197.

Soderstrom, K., Leid, M., Moore, F. L., & Murray, T. F. (2000). Behavioral, pharmacological, and molecular characterization of an amphibian cannabinoid receptor. *Journal of Neurochemistry, 75*, 413–423.

Sofia, R. D., & Barry, H. (1974). Acute and chronic effects of δ9-tetrahydrocannabinol on food intake by rats. *Psychopharmacology, 39*, 213–222.

Son, M. H., Kim, H. D., Chae, Y. N., Kim, M. K., Shin, C. Y., Ahn, G. J., et al. (2010). Peripherally acting CB1-receptor antagonist: The relative importance of central and peripheral CB1 receptors in adiposity control. *International Journal of Obesity, 34*, 547–556.

Song, D., Bandsma, R., Xiao, C., Xi, L., Shao, W., Jin, T., et al. (2011). Acute cannabinoid receptor type 1 (CB1R) modulation influences insulin sensitivity by an effect outside the central nervous system in mice. *Diabetologia, 54*, 1181–1189.

Soria-Gómez, E., Matias, I., Rueda-Orozco, P. E., Cisneros, M., Petrosino, S., Navarro, L., et al. (2007). Pharmacological enhancement of the endocannabinoid system in the nucleus accumbens shell stimulates food intake and increases c-Fos expression in the hypothalamus. *British Journal of Pharmacology, 151*, 1109–1116.

Steiner, H., Bonner, T. I., Zimmer, A. M., Kitai, S. T., & Zimmer, A. (1999). Altered gene expression in striatal projection neurons in CB1 cannabinoid receptor knockout mice. *Proceedings of the National Academy of Sciences, 96*, 5786–5790.

Sugiura, T., Kondo, S., Sukagawa, A., Nakane, S., Shinoda, A., Itoh, K., et al. (1995). 2-Arachidonoylgylcerol: A possible endogenous cannabinoid receptor ligand in brain. *Biochemical and Biophysical Research Communications, 215*, 89–97.

Tam, J., Vemuri, V. K., Liu, J., Bátkai, S., Mukhopadhyay, B., Godlewski, G., et al. (2010). Peripheral CB1 cannabinoid receptor blockade improves cardiometabolic risk in mouse models of obesity. *The Journal of Clinical Investigation, 120*, 2953–2966.

Taschler, U., Radner, F. P. W., Heier, C., Schreiber, R., Schweiger, M., Schoiswohl, G., et al. (2011). Monoglyceride lipase deficiency in mice impairs lipolysis and attenuates diet-induced insulin resistance. *The Journal of Biological Chemistry, 286*, 17467–17477.

Teixeira-Clerc, F., Julien, B., Grenard, P., Van Nhieu, J. T., Deveaux, V., Li, L., et al. (2006). CB1 cannabinoid receptor antagonism: A new strategy for the treatment of liver fibrosis. *Nature Medicine, 12*, 671–676.

Tharp, W. G., Lee, Y. H., Maple, R. L., & Pratley, R. E. (2008). The cannabinoid CB1 receptor is expressed in pancreatic δ-cells. *Biochemical and Biophysical Research Communications, 372*, 595–600.

Tornqvist, H., & Belfrage, P. (1976). Purification and some properties of a monoacylglycerol-hydrolyzing enzyme of rat adipose tissue. *The Journal of Biological Chemistry, 251*, 813–819.

Trebicka, J., Racz, I., Siegmund, S. V., Cara, E., Granzow, M., Schierwagen, R., et al. (2011). Role of cannabinoid receptors in alcoholic hepatic injury: Steatosis and fibrogenesis are increased in CB2 receptor-deficient mice and decreased in CB1 receptor knockouts. *Liver International, 31*, 860–870.

Tsou, K., Brown, S., Sañudo-Peña, M. C., Mackie, K., & Walker, J. M. (1997). Immunohistochemical distribution of cannabinoid CB1 receptors in the rat central nervous system. *Neuroscience, 83*, 393–411.

Turu, G., & Hunyady, L. (2010). Signal transduction of the CB1 cannabinoid receptor. *Journal of Molecular Endocrinology, 44*, 75–85.

van der Poorten, D., Shahidi, M., Tay, E., Sesha, J., Tran, K., McLeod, D., et al. (2010). Hepatitis C virus induces the cannabinoid receptor 1. *PLoS One, 5*, e12841.

van Diepen, H., Schlicker, E., & Michel, M. (2008). Prejunctional and peripheral effects of the cannabinoid CB1 receptor inverse agonist rimonabant (SR 141716). *Naunyn-Schmiedeberg's Archives of Pharmacology*, *378*, 345–369.

Van Gaal, L. F., Scheen, A. J., Rissanen, A. M., Rössner, S., Hanotin, C., & Ziegler, O. (2008). Long-term effect of CB1 blockade with rimonabant on cardiometabolic risk factors: Two year results from the RIO-Europe Study. *European Heart Journal*, *29*, 1761–1771.

Verty, A. N. A., Allen, A. M., & Oldfield, B. J. (2008). The effects of rimonabant on brown adipose tissue in rat: Implications for energy expenditure. *Obesity*, *17*, 254–261.

Verty, A. N. A., McGregor, I. S., & Mallet, P. E. (2004). The dopamine receptor antagonist SCH 23390 attenuates feeding induced by Δ9-tetrahydrocannabinol. *Brain Research*, *1020*, 188–195.

Vickers, S. P., Webster, L. J., Wyatt, A., Dourish, C. T., & Kennett, G. A. (2003). Preferential effects of the cannabinoid CB_1 receptor antagonist, SR 141716, on food intake and body weight gain of obese (*fa/fa*) compared to lean Zucker rats. *Psychopharmacology*, *167*, 103–111.

Wadden, T. A., Fujioka, K., Toubro, S., Gantz, I., Erondu, E., Chen, M., et al. (2010). A randomized trial of lifestyle modification and taranabant for maintaining weight loss achieved with a low-calorie diet. *Obesity*, *18*, 2301–2310.

Wang, X., Horswill, J. G., Whalley, B. J., & Stephens, G. J. (2011). Effects of the allosteric antagonist 1-(4-chlorophenyl)-3-[3-(6-pyrrolidin-1-ylpyridin-2-yl)phenyl]urea (PSNCBAM-1) on CB1 receptor modulation in the cerebellum. *Molecular Pharmacology*, *79*, 758–767.

Wang, Q., Perrard, X. D., Perrard, J. L., Mansoori, A., Smith, C. W., Ballantyne, C. M., et al. (2011). Effect of the cannabinoid receptor-1 antagonist rimonabant on inflammation in mice with diet-induced obesity. *Obesity*, *19*, 505–513.

Wells, J. C. K. (2009). Thrift: A guide to thrifty genes, thrifty phenotypes and thrifty norms. *International Journal of Obesity*, *33*, 1331–1338.

Williams, C. M., & Kirkham, T. C. (1999). Anandamide induces overeating: Mediation by central cannabinoid (CB1) receptors. *Psychopharmacology*, *143*, 315–317.

Williams, C. M., & Kirkham, T. C. (2002a). Observational analysis of feeding induced by Δ9-THC and anandamide. *Physiology & Behavior*, *76*, 241–250.

Williams, C. M., & Kirkham, T. C. (2002b). Reversal of Δ9-THC hyperphagia by SR141716 and naloxone but not dexfenfluramine. *Pharmacology, Biochemistry, and Behavior*, *71*, 333–340.

Williams, C. M., Rogers, P. J., & Kirkham, T. C. (1998). Hyperphagia in pre-fed rats following oral Δ9-THC. *Physiology & Behavior*, *65*, 343–346.

Willoughby, K. A., Moore, S. F., Martin, B. R., & Ellis, E. F. (1997). The biodisposition and metabolism of anandamide in mice. *The Journal of Pharmacology and Experimental Therapeutics*, *282*, 243–247.

Wilson, R. S., May, E. L., Martin, B. R., & Dewey, W. L. (1976). 9-Nor-9-hydroxyhexahydrocannabinols. Synthesis, some behavioral and analgesic properties, and comparison with the tetrahydrocannabinols. *Journal of Medicinal Chemistry*, *19*, 1165–1167.

Wolf, G. (2009). Adipose-specific phospholipase as regulator of adiposity. *Nutrition Reviews*, *67*, 551–554.

Yamaguchi, F., Macrae, A. D., & Brenner, S. (1996). Molecular cloning of two cannabinoid type 1-like receptor genes from the puffer fish Fugu rubripes. *Genomics*, *35*, 603–605.

Yanovski, S. Z., & Yanovski, J. A. (2011). Obesity prevalence in the United States: Up, down, or sideways? *The New England Journal of Medicine*, *364*, 987–989.

Zhao, W., Fong, O., Muise, E. S., Thompson, J. R., Weingarth, D., Qian, S., et al. (2010). Genome-wide expression profiling revealed peripheral effects of cannabinoid receptor 1 inverse agonists in improving insulin sensitivity and metabolic parameters. *Molecular Pharmacology*, *78*, 350–359.

Zimmer, A., Zimmer, A. M., Hohmann, A. G., Herkenham, M., & Bonner, T. I. (1999). Increased mortality, hypoactivity, and hypoalgesia in cannabinoid CB1 receptor knockout mice. *Proceedings of the National Academy of Sciences*, *96*, 5780–5785.

Zlas, J., Stark, H., Seligman, J., Levy, R., Werker, E., Breuer, A., et al. (1993). Early medical use of cannabis. *Nature*, *363*, 215.

CHAPTER FIFTEEN

Lipocalin 13 Regulation of Glucose and Lipid Metabolism in Obesity

Yingjiang Zhou, Liangyou Rui[1]

Department of Molecular & Integrative Physiology, University of Michigan Medical School, Ann Arbor, Michigan, USA

[1]Corresponding author: e-mail address: ruily@umich.edu

Contents

Abstract

Lipocalin (LCN) family members are small secreted proteins that bind to small hydrophobic molecules via their characteristic central β-barrels. A couple of LCN family members, including major urinary protein 1, retinol-binding protein 4, LCN2, and LCN13, have been reported to regulate insulin sensitivity and nutrient metabolism. LCN13 is expressed by multiple tissues, including the liver, pancreas, epididymis, and skeletal muscle, and is secreted into the bloodstream in mice. Obesity is associated with a down-regulation of LCN13 expression and lower levels of circulating LCN13. LCN13 therapies overcome LCN13 deficiency in mice with either genetic or dietary obesity, leading to an improvement in hyperglycemia, hyperinsulinemia, insulin resistance, glucose intolerance, and hepatic steatosis. In hepatocytes, LCN13 directly suppresses hepatic gluconeogenesis and lipogenesis but increases fatty acid β oxidation. LCN13 also enhances insulin sensitivity in adipocytes. The potential mechanisms of the antidiabetes and antisteatosis actions of LCN13 are discussed.

Vitamins and Hormones, Volume 91
ISSN 0083-6729
http://dx.doi.org/10.1016/B978-0-12-407766-9.00015-8

1. INTRODUCTION

The alarmingly escalating epidemic of overweight and obesity is becoming a challenging public health problem globally. In the United States, the prevalence of obesity (body mass index (BMI) $\geq$30) and overweight (BMI 25.0–29.9) in adults was 35.7% and 33.1%, respectively, in 2009–2010 (Flegal, Carroll, Kit, & Ogden, 2012). America's obesity epidemic may not plateau until the prevalence of obesity hits approximately 42% in adults (Hill, Rand, Nowak, & Christakis, 2010). It has been well defined that obesity is closely associated with various metabolic diseases, including insulin resistance, type 2 diabetes, hypertension, dyslipidemia, nonalcoholic fatty liver disease (NAFLD), and cardiovascular disease (Despres & Lemieux, 2006; Kahn, Hull, & Utzschneider, 2006; Van Gaal, Mertens, & De Block, 2006).

Obesity develops as a result of an imbalance between food intake and energy expenditure (Morris & Rui, 2009; Spiegelman & Flier, 2001). The excess energy is stored as triacylglycerol in adipose tissues; moreover, in obesity, lipids spillover to other tissues which normally do not store a large amount of lipids, including the liver and skeletal muscle (Flier, 2004; Morris & Rui, 2009). Ectopic lipid accumulation in these tissues impairs their normal functions (called lipotoxicity), leading to insulin resistance and type 2 diabetes (Despres & Lemieux, 2006; Greenberg et al., 2011; Taube, Eckardt, & Eckel, 2009). Obesity is also associated with chronic, low-grade inflammation, particularly in adipose tissues and the liver, and proinflammatory cytokines further exacerbate insulin resistance (Hotamisligil, 2006; Shoelson & Goldfine, 2009; Tarantino, Saldalamacchia, Conca, & Arena, 2007; Taube et al., 2009). Inflammation is able to increase hepatic glucose production by augmenting the hyperglycemic response to glucagon (Chen et al., 2012; Sheng et al., 2012). Adipose tissues and the liver also secrete a variety of molecules that modulate systemic insulin sensitivity and glucose and lipid metabolism in an endocrine fashion (Ahima & Osei, 2008; Esteve, Ricart, & Fernandez-Real, 2009; Halberg, Wernstedt-Asterholm, & Scherer, 2008; Rajala & Scherer, 2003). For example, adiponectin, secreted by adipocytes, functions as an endogenous insulin sensitizer to improve glucose metabolism (Berg, Combs, Du, Brownlee, & Scherer, 2001; Yamauchi et al., 2002, 2001). FGF21, mainly secreted by the liver, promotes fatty acid oxidation as well as enhances insulin sensitivity (Badman et al., 2007; Inagaki et al., 2007). Recently, several lipocalin (LCN) family members have been reported to be involved in the regulation of insulin sensitivity and nutrient metabolism (Cho, Zhou, Sheng, & Rui, 2011; Guo et al., 2010; Hui et al., 2009; Jun, Siddall, & Rosen, 2011; Law et al., 2010; Sheng, Cho, Zhou, Shen, & Rui, 2011; Yan et al., 2007, 2005; Zhou, Jiang, & Rui, 2009). In this review, we discuss these LCNs in the setting of obesity, focusing on LCN13.

2. LCN STRUCTURE AND FUNCTION

The LCN family contains a large number of diverse small secreted proteins, typically composed of 160–180 amino acids with an N-terminal signal peptide (Ganfornina, Gutierrez, Bastiani, & Sanchez, 2000). LCN family members are found in vertebrate and invertebrate animals, plants, and even bacteria. In mammals, *LCN* genes are clustered on the same chromosomal loci and are likely to evolve from an ancestral *LCN* gene through an *in situ* tandem duplication (Grzyb, Latowski, & Strzalka, 2006; Suzuki et al., 2004). The name of "lipocalin" is derived from the Greek words "lipos" and "kalyx" (means drinking vessel), indicating the common feature of the family members: they bind small hydrophobic molecules (Grzyb et al., 2006). LCN family members have low levels of primary amino acid sequence homology (usually <30%) (Flower, North, & Sansom, 2000; Grzyb et al., 2006); however, all LCN family members contain one or more short structurally conserved regions (SCRs) (Flower, North, & Attwood, 1993). LCNs are generally divided into two subfamilies based on their SCR number. Kernel LCNs, including retinol-binding protein (RBP), β-lactoglobulin, and LCN2, contain three SCRs, while outlier LCNs, including crustacyanin, nitrophorins, and α-1-acid glycoprotein, contain one or two SCRs (Grzyb et al., 2006). Many LCNs have conserved cysteines which form disulfide bridges to stabilize their three-dimensional structures (Glasgow et al., 1998). Some LCNs are able to form dimmers or oligomers under specific conditions (e.g., a low pH and high calcium concentrations) (Grzyb et al., 2006).

Despite of very limited similarity in their amino acid sequences, all LCN family members share a highly conserved tertiary structure with a characteristic central β-barrel. This β-barrel is formed by eight antiparallel β-strands in a cylindrical manner with a "closed end" on one side and an "open end" on the opposite side (Flower, 1996; Flower et al., 1993). The "open end" provides an access into the central cavity for small hydrophobic molecules (Grzyb et al., 2006; Schlehuber & Skerra, 2005). The interior of the β-barrel consists of hydrophobic amino acid residues which define the binding capacity of individual LCNs to various hydrophobic molecules (Grzyb et al., 2006). The eight β-strands, connected by seven loops between two consecutive β-strands, coil in a right-handed manner around a central axis and interact through transversal hydrogen bonds (Grzyb et al., 2006; Schlehuber & Skerra, 2005). The first loop is a large and flexible Ω-type loop which functions as a dynamic lid for the "open end" of the β-barrel, and the other six are short hairpin-type loops (Flower et al., 1993).

The biological activity of LCN family members may largely depend on their ability to specifically bind to small hydrophobic molecules, including fatty acids, phospholipids, steroids, retinol, and pheromones (Flower et al., 2000; Grzyb et al., 2006; Schlehuber & Skerra, 2005; Zhou & Rui, 2010). LCNs control the transportation, stability, release, activation, and clearance of these bioactive hydrophobic molecules which in turn regulate diverse biological processes, including cellular metabolism, proliferation, differentiation, and death (Flower, 1996; Sharrow, Vaughn, Zidek, Novotny, & Stone, 2002; Zhou & Rui, 2010). In rodents, the LCN family members have been documented to regulate chemical communication, reproduction, immune responses, and cancer development (Bratt, 2000; Chamero et al., 2007; Logdberg & Wester, 2000; More, 2006; Oehninger, Coddington, Hodgen, & Seppala, 1995). Recently, LCNs have been reported to regulate insulin sensitivity and nutrient metabolism in obesity (Cho et al., 2011; Guo et al., 2010; Hui et al., 2009; Jun et al., 2011; Law et al., 2010; Sheng et al., 2011; Yan et al., 2007, 2005; Zhou et al., 2009).

3. LCN13 REGULATION OF GLUCOSE METABOLISM

There have only been three reports in literatures that describe LCN13 action thus far (Cho et al., 2011; Sheng et al., 2011; Suzuki et al., 2004). The *LCN13* gene was originally identified in 2004 by analyzing the epididymal cluster of the *LCN* genes on chromosome 2 in mice (Suzuki et al., 2004). This chromosomal locus also contains 12 additional *LCN* genes: *LCN2–5*, *LCN8–13*, *C8γ* (complement 8 γ), and *PGDS* (prostaglandin D2 synthase) (Suzuki et al., 2004). The *LCN13* gene is predicted to have seven exons and to encode a polypeptide containing 176 amino acids. The first 19 amino acids of its N-terminus are hydrophobic and are predicted to function as a signal peptide responsible for LCN13 secretion (Suzuki et al., 2004). LCN13 protein was verified later using anti-LCN13 antibodies and also detected in the mouse bloodstream (Cho et al., 2011). LCN13 proteins are expressed and secreted by multiple tissues in mice, including the liver, pancreas, and skeletal muscles (Cho et al., 2011). Two distinct forms of LCN13 proteins are generated from a single full-length LCN13 cDNA, suggesting that LCN13 precursors are subjected to differential proteolytic cleavages and/or posttranslational modifications (e.g., acylation and glycosylation) (Cho et al., 2011). However, the underlying mechanisms by which the LCN13 precursors are differentially processed and modified to generate the mature forms of LCN13 remain completely unknown.

The expression and secretion of LCN13 are not constant in mice and change in response to alterations of metabolic states. The levels of circulating LCN13 are lower in the fasting state than in the fed state (Cho et al., 2011), suggesting that LCN13 may be involved in nutrient sensing and metabolic regulation. Importantly, obesity is associated with downregulation of LCN13, and the levels of LCN13 in the bloodstream are extremely low in mice with either genetic (*db/db*) or high-fat diet (HFD)-induced obesity (Cho et al., 2011). The expression of LCN13 in the liver is also lower in obese mice than in lean mice (Cho et al., 2011); however, factors that regulate the expression, secretion, and clearance of LCN13 are unknown.

To study the physiological function of LCN13, we have generated three useful regents: recombinant LCN13 proteins, LCN13 adenoviruses expressing LCN13 proteins, and LCN13 transgenic (Tg) mice containing the *LCN13* transgene under the control of the constitutively active chicken β-actin/rabbit β-globin hybrid promoter. Chronic treatments with recombinant LCN13 protein, via osmotic minipumps, do not prevent HFD-induced obesity; however, the treatments improve insulin resistance and glucose intolerance in HFD-fed mice (Cho et al., 2011). To overcome LCN13 deficiency in obese mice, we infected *db/db* (lack of functional leptin receptors) or HFD-fed mice with LCN13 adenoviruses via tail vein injection, with β-gal adenoviruses as controls. LCN13 adenoviral infection results in an increase in the levels of circulating LCN13 due to the overexpression of recombinant LCN13 in the liver (Cho et al., 2011). Adenovirus-mediated overexpression of LCN13 reduces hyperglycemia, insulin resistance, and glucose intolerance in *db/db* mice as well as in HFD-fed mice (Cho et al., 2011). LCN13 Tg mice secrete higher levels of LCN13 into the circulation than WT control mice, as expected (Cho et al., 2011). Although LCN13 Tg mice are not protected from HFD-induced obesity, they resist HFD-induced insulin resistance and glucose intolerance (Cho et al., 2011). Additionally, Tg overexpression of LCN13 markedly reduces hyperglycemia, insulin resistance, and glucose intolerance in *ob/ob* mice that are deficient of leptin (Sheng et al., 2011). Moreover, neutralization of endogenous

LCN13 through tail vein injection of anti-LCN13 antibodies induces glucose intolerance in normal mice (Sheng et al., 2011). Together, these results demonstrate that obesity is associated with LCN13 deficiency, which contributes to insulin resistance, hyperglycemia, and glucose intolerance in mice. These observations raise the possibility that LCN13 and its related molecules may have therapeutic potential for treating obesity-associated type 2 diabetes.

LCN13 is able to directly regulate glucose metabolism in both adipocytes and hepatocytes. Recombinant LCN13 dose-dependently enhances insulin signaling and insulin-stimulated glucose uptake in cultured adipocytes (Cho et al., 2011). LCN13 also increases the ability of insulin to suppress glucose production in primary hepatocyte cultures (Cho et al., 2011). These observations suggest that LCN13 acts as an insulin sensitizer to improve glucose metabolism in these two cell types. The liver secretes LCN13, and neutralization of endogenous LCN13 increases glucose production in primary hepatocytes and impairs glucose tolerance in mice (Cho et al., 2011; Sheng et al., 2011). These observations suggest that hepatic LCN13 is able to suppress hepatic glucose production in an autocrine/paracrine fashion. In the absence of insulin, LCN13 still stimulates glucose uptake in adipocytes and inhibits glucose production and the expression of key gluconeogenic genes (e.g., phosphoenolpyruvate carboxykinase and glucose-6-phosphatase) in primary hepatocytes (Cho et al., 2011). These results indicate that LCN13 is able to regulate glucose metabolism by an additional insulin-independent mechanism. Since skeletal muscles, immune cells, and the brain (particularly the hypothalamus) play an important role in the regulation of systemic insulin sensitivity and glucose metabolism, it will be important to determine whether LCN13 regulates the activity of these tissues in the future.

The molecular mechanism of LCN13 action is currently unknown. Like other LCN family members, LCN13 is predicted to bind to small hydrophobic molecules via its central β-barrel. These small hydrophobic molecules may be bioactive and regulates insulin sensitivity and glucose metabolism. LCN13 may exert its biological action by controlling the transportation, stability, release, activation, and/or clearance of these molecules. Additionally, several LCN family members have been reported to bind to and activate their cognate receptors on the plasma membrane (Berry, Jin, Majumdar, & Noy, 2011; Zhou & Rui, 2010). Similarly, LCN13 may modulate insulin sensitivity and nutrient metabolism by binding to and activating its cognate receptors. In the future, it will be important to identify and characterize the putative LCN13-bound small molecules (ligands), the putative LCN13 receptors, and the signaling pathways that link LCN13 ligands and LCN13 receptors to LCN13 metabolic responses. It will also be interesting to determine whether the two forms of LCN13 have different specificity and/or affinity for LCN13 ligands, and whether these two forms differentially regulate the activation of LCN13 receptors.

4. LCN13 REGULATION OF LIPID METABOLISM

LCN13 appears not to have antiobesity capability, but it markedly decreases hepatic steatosis in obese mice (Sheng et al., 2011). LCN13 Tg mice resist HFD-induced hyperlipidemia and hepatic steatosis (Sheng et al., 2011). LCN13 Tg mice also resist hyperlipidemia and hepatic steatosis induced by leptin deficiency (Sheng et al., 2011). Similarly, adenovirus-mediated overexpression of LCN13 ameliorates hepatic steatosis in *db/db* mice (Sheng et al., 2011). LCN13 is able to suppress the hepatic lipogenic program in mice (Sheng et al., 2011). The expression of key liver lipogenic genes, including stearoyl-coenzyme A desaturase 1 (SCD1), fatty acid synthase (FAS), carbohydrate-responsive element-binding protein (ChREBP), peroxisome proliferator-activated receptor-γ (PPARγ), and sterol regulatory element-binding protein (SREBP-1c), is lower in LCN13 Tg mice as well as in mice infected with LCN13 adenoviruses (Sheng et al., 2011). In contrast, LCN13 increases the expression of carnitine palmitoyltransferase-1α (CPT-1α) in the liver (Sheng et al., 2011). CPT-1α promotes fatty acid β oxidation (Schreurs, Kuipers, & van der Leij, 2010). Therefore, LCN13 protects obesity-associated hepatic steatosis presumably by both decreasing lipogenesis and increasing fatty acid β oxidation in the liver. In addition to antidiabetes, LCN13 also has an antisteatosis effect; thus, LCN13 and its related molecules may have therapeutic potential for treating both type 2 diabetes and NAFLD.

In agreement with its antisteatosis effects, LCN13 directly suppresses lipogenesis in primary hepatocytes in a dose-dependent manner (Sheng et al., 2011). It inhibits the expression of SCD1, FAS, SREBP-1c, ChREBP, and PPARγ in hepatocytes (Sheng et al., 2011). Moreover, LCN13 directly stimulates the expression of CPT-1α, fatty acid β oxidation, and ketogenesis in primary hepatocytes (Sheng et al., 2011). Conversely, in primary hepatocytes, neutralization of endogenous LCN13 with anti-LCN13 antibody increases the expression of lipogenic genes and lipogenesis but decreases CPT-1α expression and fatty acid β oxidation (Sheng et al., 2011). These observations suggest that liver-derived LCN13 regulates lipogenesis and fatty acid β oxidation in an autocrine/paracrine fashion. Because the liver cross talks with other tissues, including adipose tissues, skeletal muscles, and the brain, it will be interesting to study whether LCN13 regulates lipid metabolism and lipid levels in the liver indirectly by regulating the cross talk among these tissues.

The molecular mechanism by which LCN13 regulates lipid metabolism is currently unknown. As discussed earlier, LCN13 is predicted to bind to bioactive hydrophobic molecules. These small hydrophobic molecules may regulate hepatocyte lipogenesis and fatty acid oxidation. LCN13 may regulate the transportation, stability, release, activation, and/or clearance of these bioactive molecules, thereby indirectly regulating lipid metabolism in the liver. Alternatively, LCN13 may bind to and activate its cognate receptors on hepatocytes, thereby directly regulating lipid metabolism in the liver. In the future, it is essential to identify and characterize these putative hydrophobic molecules and LCN13 receptors in order to advance our understanding of LCN13 action. Interestingly, LCN13 decreases hepatic steatosis but not adiposity in obese mice, suggesting that LCN13 decreases lipogenesis and increases fatty acid oxidation specifically in hepatocytes but not in adipocytes. Adipocytes do respond to LCN13 by increasing glucose uptake (Cho et al., 2011). It will be interesting to further study the differential effects of LCN13 on adipocytes and hepatocytes in order to gain insight into the molecular mechanism of LCN13 action.

5. REGULATION OF NUTRIENT METABOLISM BY OTHER LCN FAMILY MEMBERS

Three other LCN family members, major urinary protein 1 (MUP1), RBP4, and LCN2, have also been reported to be involved in the regulation of insulin sensitivity and glucose metabolism in obesity. MUP1 is abundantly expressed in mouse livers and secreted into the circulation (Zhou & Rui, 2010). *MUP* genes are highly polymorphic, and the *MUP* gene subfamily contains 21 paralogous genes clustered on chromosome 4 in mice (Hastie, Held, & Toole, 1979). The polymorphic *MUP* genes serve as a specific genetic marker of individual identity (Zhou & Rui, 2010). MUP expression is sexually dimorphic and is higher in males (Geertzen, Ouderaa, & Kassenaar, 1973; Lane & Neuhaus, 1972). Each individual adult male mouse normally secretes 8–14 different MUP isoforms, and this specific pattern of MUP isoforms serves as an individual identity signature (Zhou & Rui, 2010). MUPs bind to various volatile pheromones via their central β-barrels and function as pheromone carriers (Zhou & Rui, 2010). The MUP–pheromone complexes are excreted into the urine and mediate chemical signaling in rodents (Zhou & Rui, 2010). Interestingly, both the expression of MUP1 in the liver and the levels of circulating MUP1 are significantly lower in mice with either genetic (*db/db*) or HFD-induced obesity (Hui et al., 2009; Zhou et al., 2009). Like LCN13, recombinant MUP1 ameliorates hyperglycemia and glucose intolerance in both *db/db* and HFD-fed mice (Hui et al., 2009; Zhou et al., 2009). At cellular levels, MUP1 inhibits the expression of gluconeogenic genes and glucose production in hepatocytes in an autocrine/paracrine fashion (Zhou et al., 2009). MUP1 also improves skeletal muscle insulin sensitivity in obese mice presumably by increasing mitochondrial oxidative capacity (Hui et al., 2009).

RBP4 is secreted into the circulation from both adipose tissues and the liver (Kotnik, Fischer-Posovszky, & Wabitsch, 2011). In contrast to LCN13 and MUP1, RBP4 levels are higher in both rodents and humans with obesity (Broch et al., 2010; Graham et al., 2006; Stefan et al., 2007; Yang et al., 2005). However, clinical study results are not consistent, and several studies report no correlation between circulating RBP4 levels and the severity of obesity and insulin resistance as reviewed by Kotnik et al. (2011). Tg overexpression of RBP4 or administration of recombinant RBP4 proteins promotes insulin resistance and glucose intolerance in normal mice; conversely, genetic deletion of RBP4 improves insulin sensitivity (Yang et al., 2005). RBP4 promotes insulin resistance in skeletal muscles, adipocytes, and hepatocytes (Ost et al., 2007; Yang et al., 2005). RBP4 also stimulates cytokine secretion by macrophages, thus indirectly promoting insulin resistance (Deng et al., 2009; Norseen et al., 2012). In adipocytes, retinol-bound RBP4 activates STRA6 (stimulated by retinoic acid 6), the RBP4 receptor on the plasma membrane (Berry et al., 2011). STRA6 in turn activates the JAK2/STAT5 pathway, leading to SOCS3 expression in adipocytes (Berry et al., 2011). SOCS3 has been reported to induce insulin resistance (Rui, Yuan, Frantz, Shoelson, & White, 2002; Ueki, Kondo, & Kahn, 2004; Ueki, Kondo, Tseng, & Kahn, 2004).

Levels of circulating LCN2, also called neutrophil gelatinase-associated LCN, are higher in mice and humans with obesity (Esteve et al., 2009). The relative contribution of LCN2 to insulin resistance in obesity remains controversial, and three groups described different metabolic phenotypes of LCN2 knockout (KO) mice. Guo et al. (2010) reported that LCN2 KO mice are susceptible to HFD-induced obesity and insulin resistance. Law et al. (2010) showed that LCN2 KO mice resist HFD-induced insulin resistance and glucose intolerance. Jun et al. (2011) demonstrated that LCN2 deficiency has a very mild effect on HFD-induced insulin resistance and glucose intolerance. The reasons underlying the discrepancy between these three studies are currently unknown.

In summary, obesity is associated with an altered secretion of several LCN family members. These LCNs can be divided into two groups: A and B (Fig. 15.1). The A group contains LCN13 and MUP1 and enhances insulin sensitivity and improves glucose metabolism. Obesity is associated with deficiency of these group members. The B group contains RBP4 and LCN2 and promotes insulin resistance, and obesity is associated with an increase in the expression and secretion of these LCNs. These two groups have opposing effects on insulin sensitivity and nutrient metabolism; therefore, the ratios of A to B LCNs are likely to be an important factor to determine insulin sensitivity. Obesity is associated with a marked decrease in this A to B LCN

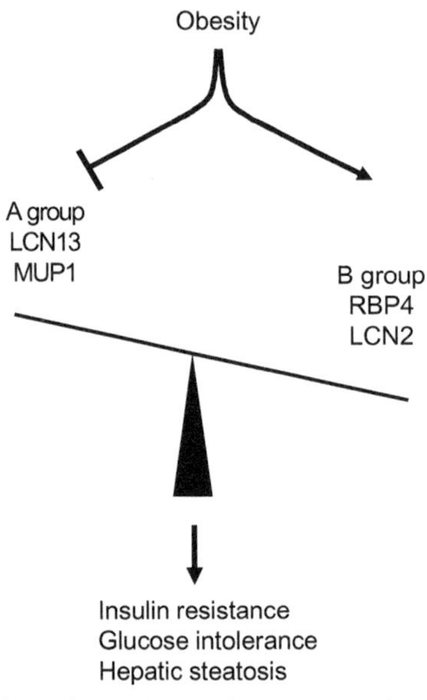

Figure 15.1 A model of lipocalin action in obesity. Lipocalins known to regulate nutrient metabolism can be divided into two groups: A and B. The A group (LCN13 and MUP1) ameliorates insulin resistance, glucose intolerance, and hepatic steatosis in obesity, whereas the B group (RBP4 and LCN2) has opposite effects. Obesity is associated with both deficiency of the A group of LCNs and increased expression/secretion of the B group of LCNs, thus disrupting balance between A and B LCNs. This A and B LCN imbalance contributes to metabolic disorders in obesity, including type 2 diabetes and NAFLD.

ratios; therefore, one goal for treatment of insulin resistance and type 2 diabetes is to restore the normal balance between A and B LCNs.

6. CONCLUSIONS AND FUTURE DIRECTIONS

LCN13 is a LCN family member that is secreted into the bloodstream by multiple tissues. It has an antidiabetes and antisteatosis effect in mice with obesity. Obesity is associated with LCN13 deficiency, which is likely to contribute to insulin resistance, hepatic steatosis, and glucose intolerance in obese mice. Aside from LCN13, other LCN family members, including MUP1, RBP4, and LCN2, are also involved in the regulation of insulin sensitivity and nutrient metabolism, particularly in the setting of obesity. In general, these LCN family members fall into two distinct categories. LCNs in the first category contain LCN13 and MUP1 and have an anti-insulin resistance and antidiabetes effect, whereas LCNs in the second category, including RBP4 and LCN2, have an opposite effect. Obesity is associated with a downregulation of the first category LCNs and an upregulation of the second category LCNs, resulting in a sharp decrease in the first to the second category LCN ratios. This imbalance between the first and the second category LCNs may contribute to metabolic disorders in obesity.

The mechanisms of LCN13 action remain largely unknown. In the future, we will determine whether LCN13 exerts its biological action directly via binding to and activating its cognate receptors or indirectly by regulating the transportation, stability, release, and/or clearance of its bound bioactive molecules (ligands). It is important to identify and characterize the putative LCN13 receptors and LCN13 ligands in order to completely understand LCN13 actions. The intracellular signaling pathways that act downstream of LCN13 receptors or LCN13 ligands and mediate LCN13's metabolic responses should be examined. Additionally, human LCN13 has not been identified yet. It is also unclear whether LCN13 has a similar antidiabetes and antisteatosis effect in humans. The relevance of LCN13's metabolic functions described in mice to human obesity and obesity-associated metabolic disorders should be vigorously examined in order to evaluate the therapeutic potentials of LCN13 and LCN13-related molecules in the treatments of type 2 diabetes, NAFLD, and other obesity-associated metabolic diseases. Furthermore, more LCNs, in addition to LCN13, MUP1, RBP4, and LCN2 as described in Fig. 15.1, are expected to be identified and characterized, and their contributions to obesity-associated metabolic disorders will be investigated.

ACKNOWLEDGMENTS

We thank Crystal W. Rui for helpful discussion. This work was supported by RO1 DK 065122 from the NIH.

REFERENCES

Ahima, R. S., & Osei, S. Y. (2008). Adipokines in obesity. *Frontiers of Hormone Research*, *36*, 182–197.

Badman, M. K., Pissios, P., Kennedy, A. R., Koukos, G., Flier, J. S., & Maratos-Flier, E. (2007). Hepatic fibroblast growth factor 21 is regulated by PPARalpha and is a key mediator of hepatic lipid metabolism in ketotic states. *Cell Metabolism*, *5*, 426–437.

Berg, A. H., Combs, T. P., Du, X., Brownlee, M., & Scherer, P. E. (2001). The adipocyte-secreted protein Acrp30 enhances hepatic insulin action. *Nature Medicine*, 7, 947–953.

Berry, D. C., Jin, H., Majumdar, A., & Noy, N. (2011). Signaling by vitamin A and retinol-binding protein regulates gene expression to inhibit insulin responses. *Proceedings of the National Academy of Sciences of the United States of America*, *108*, 4340–4345.

Bratt, T. (2000). Lipocalins and cancer. *Biochimica et Biophysica Acta*, *1482*, 318–326.

Broch, M., Gomez, J. M., Auguet, M. T., Vilarrasa, N., Pastor, R., Elio, I., et al. (2010). Association of retinol-binding protein-4 (RBP4) with lipid parameters in obese women. *Obesity Surgery*, *20*, 1258–1264.

Chamero, P., Marton, T. F., Logan, D. W., Flanagan, K., Cruz, J. R., Saghatelian, A., et al. (2007). Identification of protein pheromones that promote aggressive behaviour. *Nature*, *450*, 899–902.

Chen, Z., Sheng, L., Shen, H., Zhao, Y., Wang, S., Brink, R., et al. (2012). Hepatic TRAF2 regulates glucose metabolism through enhancing glucagon responses. *Diabetes*, *61*, 566–573.

Cho, K. W., Zhou, Y., Sheng, L., & Rui, L. (2011). Lipocalin-13 regulates glucose metabolism by both Insulin-dependent and insulin-independent mechanisms. *Molecular and Cellular Biology*, *31*, 450–457.

Deng, Z. B., Poliakov, A., Hardy, R. W., Clements, R., Liu, C., Liu, Y., et al. (2009). Adipose tissue exosome-like vesicles mediate activation of macrophage-induced insulin resistance. *Diabetes*, *58*, 2498–2505.

Despres, J. P., & Lemieux, I. (2006). Abdominal obesity and metabolic syndrome. *Nature*, *444*, 881–887.

Esteve, E., Ricart, W., & Fernandez-Real, J. M. (2009). Adipocytokines and insulin resistance: The possible role of lipocalin-2, retinol binding protein-4, and adiponectin. *Diabetes Care*, *32*(Suppl. 2), S362–S367.

Flegal, K. M., Carroll, M. D., Kit, B. K., & Ogden, C. L. (2012). Prevalence of obesity and trends in the distribution of body mass index among US adults, 1999–2010. *Journal of the American Medical Association*, *307*, 491–497.

Flier, J. S. (2004). Obesity wars: Molecular progress confronts an expanding epidemic. *Cell*, *116*, 337–350.

Flower, D. R. (1996). The lipocalin protein family: Structure and function. *The Biochemical Journal*, *318*(Pt. 1), 1–14.

Flower, D. R., North, A. C., & Attwood, T. K. (1993). Structure and sequence relationships in the lipocalins and related proteins. *Protein Science*, *2*, 753–761.

Flower, D. R., North, A. C., & Sansom, C. E. (2000). The lipocalin protein family: Structural and sequence overview. *Biochimica et Biophysica Acta*, *1482*, 9–24.

Ganfornina, M. D., Gutierrez, G., Bastiani, M., & Sanchez, D. (2000). A phylogenetic analysis of the lipocalin protein family. *Molecular Biology and Evolution*, *17*, 114–126.

Geertzen, H. G., Ouderaa, F. J. V., & Kassenaar, A. A. (1973). Isolation and metabolism of male sex-dependent urinary protein from rats. *Acta Endocrinologica (Copenhagen)*, *72*, 197–208.
Glasgow, B. J., Abduragimov, A. R., Yusifov, T. N., Gasymov, O. K., Horwitz, J., Hubbell, W. L., et al. (1998). A conserved disulfide motif in human tear lipocalins influences ligand binding. *Biochemistry*, *37*, 2215–2225.
Graham, T. E., Yang, Q., Bluher, M., Hammarstedt, A., Ciaraldi, T. P., Henry, R. R., et al. (2006). Retinol-binding protein 4 and insulin resistance in lean, obese, and diabetic subjects. *The New England Journal of Medicine*, *354*, 2552–2563.
Greenberg, A. S., Coleman, R. A., Kraemer, F. B., McManaman, J. L., Obin, M. S., Puri, V., et al. (2011). The role of lipid droplets in metabolic disease in rodents and humans. *The Journal of Clinical Investigation*, *121*, 2102–2110.
Grzyb, J., Latowski, D., & Strzalka, K. (2006). Lipocalins—A family portrait. *Journal of Plant Physiology*, *163*, 895–915.
Guo, H., Jin, D., Zhang, Y., Wright, W., Bazuine, M., Brockman, D. A., et al. (2010). Lipocalin-2 deficiency impairs thermogenesis and potentiates diet-induced insulin resistance in mice. *Diabetes*, *59*, 1376–1385.
Halberg, N., Wernstedt-Asterholm, I., & Scherer, P. E. (2008). The adipocyte as an endocrine cell. *Endocrinology and Metabolism Clinics of North America*, *37*, 753–768.
Hastie, N. D., Held, W. A., & Toole, J. J. (1979). Multiple genes coding for the androgen-regulated major urinary proteins of the mouse. *Cell*, *17*, 449–457.
Hill, A. L., Rand, D. G., Nowak, M. A., & Christakis, N. A. (2010). Infectious disease modeling of social contagion in networks. *PLoS Computational Biology*, *6*, e1000968.
Hotamisligil, G. S. (2006). Inflammation and metabolic disorders. *Nature*, *444*, 860–867.
Hui, X., Zhu, W., Wang, Y., Lam, K. S., Zhang, J., Wu, D., et al. (2009). Major urinary protein-1 increases energy expenditure and improves glucose intolerance through enhancing mitochondrial function in skeletal muscle of diabetic mice. *The Journal of Biological Chemistry*, *284*, 14050–14057.
Inagaki, T., Dutchak, P., Zhao, G., Ding, X., Gautron, L., Parameswara, V., et al. (2007). Endocrine regulation of the fasting response by PPARalpha-mediated induction of fibroblast growth factor 21. *Cell Metabolism*, *5*, 415–425.
Jun, L. S., Siddall, C. P., & Rosen, E. D. (2011). A minor role for lipocalin 2 in high-fat diet-induced glucose intolerance. *American Journal of Physiology, Endocrinology and Metabolism*, *301*, E825–E835.
Kahn, S. E., Hull, R. L., & Utzschneider, K. M. (2006). Mechanisms linking obesity to insulin resistance and type 2 diabetes. *Nature*, *444*, 840–846.
Kotnik, P., Fischer-Posovszky, P., & Wabitsch, M. (2011). RBP4: A controversial adipokine. *European Journal of Endocrinology*, *165*, 703–711.
Lane, S. E., & Neuhaus, O. W. (1972). Multiple forms of 2 u, a sex-dependent urinary protein of the adult male rat. *Biochimica et Biophysica Acta*, *263*, 433–440.
Law, I. K., Xu, A., Lam, K. S., Berger, T., Mak, T. W., Vanhoutte, P. M., et al. (2010). Lipocalin-2 deficiency attenuates insulin resistance associated with aging and obesity. *Diabetes*, *59*, 872–882.
Logdberg, L., & Wester, L. (2000). Immunocalins: A lipocalin subfamily that modulates immune and inflammatory responses. *Biochimica et Biophysica Acta*, *1482*, 284–297.
More, L. (2006). Mouse major urinary proteins trigger ovulation via the vomeronasal organ. *Chemical Senses*, *31*, 393–401.
Morris, D. L., & Rui, L. (2009). Recent advances in understanding leptin signaling and leptin resistance. *American Journal of Physiology, Endocrinology and Metabolism*, *297*, E1247–E1259.
Norseen, J., Hosooka, T., Hammarstedt, A., Yore, M. M., Kant, S., Aryal, P., et al. (2012). Retinol-binding protein 4 inhibits insulin signaling in adipocytes by inducing proinflammatory cytokines in macrophages through a c-Jun N-terminal kinase- and

toll-like receptor 4-dependent and retinol-independent mechanism. *Molecular and Cellular Biology, 32*, 2010–2019.

Oehninger, S., Coddington, C. C., Hodgen, G. D., & Seppala, M. (1995). Factors affecting fertilization: Endometrial placental protein 14 reduces the capacity of human spermatozoa to bind to the human zona pellucida. *Fertility and Sterility, 63*, 377–383.

Ost, A., Danielsson, A., Liden, M., Eriksson, U., Nystrom, F. H., & Stralfors, P. (2007). Retinol-binding protein-4 attenuates insulin-induced phosphorylation of IRS1 and ERK1/2 in primary human adipocytes. *The FASEB Journal, 21*, 3696–3704.

Rajala, M. W., & Scherer, P. E. (2003). Minireview: The adipocyte—At the crossroads of energy homeostasis, inflammation, and atherosclerosis. *Endocrinology, 144*, 3765–3773.

Rui, L., Yuan, M., Frantz, D., Shoelson, S., & White, M. F. (2002). SOCS-1 and SOCS-3 block insulin signaling by ubiquitin-mediated degradation of IRS1 and IRS2. *The Journal of Biological Chemistry, 277*, 42394–42398.

Schlehuber, S., & Skerra, A. (2005). Lipocalins in drug discovery: From natural ligand-binding proteins to "anticalins" *Drug Discovery Today, 10*, 23–33.

Schreurs, M., Kuipers, F., & van der Leij, F. R. (2010). Regulatory enzymes of mitochondrial beta-oxidation as targets for treatment of the metabolic syndrome. *Obesity Reviews, 11*, 380–388.

Sharrow, S. D., Vaughn, J. L., Zidek, L., Novotny, M. V., & Stone, M. J. (2002). Pheromone binding by polymorphic mouse major urinary proteins. *Protein Science, 11*, 2247–2256.

Sheng, L., Cho, K. W., Zhou, Y., Shen, H., & Rui, L. (2011). Lipocalin 13 protein protects against hepatic steatosis by both inhibiting lipogenesis and stimulating fatty acid beta-oxidation. *The Journal of Biological Chemistry, 286*, 38128–38135.

Sheng, L., Zhou, Y., Chen, Z., Ren, D., Cho, K. W., Jiang, L., et al. (2012). NF-kappaB-inducing kinase (NIK) promotes hyperglycemia and glucose intolerance in obesity by augmenting glucagon action. *Nature Medicine, 18*(6), 943–949.

Shoelson, S. E., & Goldfine, A. B. (2009). Getting away from glucose: Fanning the flames of obesity-induced inflammation. *Nature Medicine, 15*, 373–374.

Spiegelman, B. M., & Flier, J. S. (2001). Obesity and the regulation of energy balance. *Cell, 104*, 531–543.

Stefan, N., Hennige, A. M., Staiger, H., Machann, J., Schick, F., Schleicher, E., et al. (2007). High circulating retinol-binding protein 4 is associated with elevated liver fat but not with total, subcutaneous, visceral, or intramyocellular fat in humans. *Diabetes Care, 30*, 1173–1178.

Suzuki, K., Lareyre, J. J., Sanchez, D., Gutierrez, G., Araki, Y., Matusik, R. J., et al. (2004). Molecular evolution of epididymal lipocalin genes localized on mouse chromosome 2. *Gene, 339*, 49–59.

Tarantino, G., Saldalamacchia, G., Conca, P., & Arena, A. (2007). Non-alcoholic fatty liver disease: Further expression of the metabolic syndrome. *Journal of Gastroenterology and Hepatology, 22*, 293–303.

Taube, A., Eckardt, K., & Eckel, J. (2009). Role of lipid-derived mediators in skeletal muscle insulin resistance. *American Journal of Physiology, Endocrinology and Metabolism, 297*(5), E1004–E1012.

Ueki, K., Kondo, T., & Kahn, C. R. (2004). Suppressor of cytokine signaling 1 (SOCS-1) and SOCS-3 cause insulin resistance through inhibition of tyrosine phosphorylation of insulin receptor substrate proteins by discrete mechanisms. *Molecular and Cellular Biology, 24*, 5434–5446.

Ueki, K., Kondo, T., Tseng, Y. H., & Kahn, C. R. (2004). Central role of suppressors of cytokine signaling proteins in hepatic steatosis, insulin resistance, and the metabolic syndrome in the mouse. *Proceedings of the National Academy of Sciences of the United States of America, 101*, 10422–10427.

Van Gaal, L. F., Mertens, I. L., & De Block, C. E. (2006). Mechanisms linking obesity with cardiovascular disease. *Nature*, *444*, 875–880.

Yamauchi, T., Kamon, J., Minokoshi, Y., Ito, Y., Waki, H., Uchida, S., et al. (2002). Adiponectin stimulates glucose utilization and fatty-acid oxidation by activating AMP-activated protein kinase. *Nature Medicine*, *8*, 1288–1295.

Yamauchi, T., Kamon, J., Waki, H., Terauchi, Y., Kubota, N., Hara, K., et al. (2001). The fat-derived hormone adiponectin reverses insulin resistance associated with both lipoatrophy and obesity. *Nature Medicine*, *7*, 941–946.

Yan, Q. W., Yang, Q., Mody, N., Graham, T. E., Hsu, C. H., Xu, Z., et al. (2007). The adipokine lipocalin 2 is regulated by obesity and promotes insulin resistance. *Diabetes*, *56*, 2533–2540.

Yang, Q., Graham, T. E., Mody, N., Preitner, F., Peroni, O. D., Zabolotny, J. M., et al. (2005). Serum retinol binding protein 4 contributes to insulin resistance in obesity and type 2 diabetes. *Nature*, *436*, 356–362.

Zhou, Y., Jiang, L., & Rui, L. (2009). Identification of MUP1 as a regulator for glucose and lipid metabolism in mice. *The Journal of Biological Chemistry*, *284*, 11152–11159.

Zhou, Y., & Rui, L. (2010). Major urinary protein regulation of chemical communication and nutrient metabolism. *Vitamins and Hormones*, *83*, 151–163.

CHAPTER SIXTEEN

Sirtuin 1 Deacetylase: A Key Regulator of Hepatic Lipid Metabolism

Jongsook Kim Kemper[1], Sung-E. Choi, Dong Hyun Kim
Department of Molecular and Integrative Physiology, University of Illinois at Urbana, Urbana, Illinois, USA
[1]Corresponding author: e-mail address: jongsook@illinois.edu

Contents

Abstract

Obesity is a serious medical problem worldwide and disruption of metabolic/energy homeostasis plays a pivotal role in this global epidemic. In obese people, fatty liver (steatosis) develops, which increases the risk for diabetes, cardiovascular disease, and even, liver cancer. Sirtuin 1 (SIRT1) is a NAD^+-dependent deacetylase that functions as a key metabolic/energy sensor and mediates homeostatic responses to nutrient availability. Accumulating evidence indicates that SIRT1 is a master regulator of the transcriptional networks that control hepatic lipid metabolism. During energy-deprived conditions, SIRT1 deacetylates and alters the expression and activities of key transcriptional regulators involved in hepatic lipogenesis, fatty acid β-oxidation, and cholesterol/bile acid metabolism. This review will discuss the latest advances in this field, focusing on beneficial roles of SIRT1 in hepatic lipid metabolism including its potential as a therapeutic target for treatment of steatosis and other obesity-related metabolic diseases.

Vitamins and Hormones, Volume 91
ISSN 0083-6729
http://dx.doi.org/10.1016/B978-0-12-407766-9.00016-X

1. INTRODUCTION

Obesity is a serious medical problem worldwide and dysregulation of metabolic pathways, caused by imbalances in energy metabolism, plays a pivotal role in this global epidemic (Grundy, Brewer, Cleeman, Smith, & Lenfant, 2004; Russell & Proctor, 2006). In nearly 80% of obese individuals, fatty liver (hepatic steatosis) develops, which substantially increases the risk for type 2 diabetes, cardiovascular disease, and even, liver cancer (Cohen, Horton, & Hobbs, 2011; Stefan, Kantartzis, & Haring, 2008). Despite its clinical importance, what causes hepatic steatosis and possible therapeutic options to treat steatosis remain poorly understood.

The liver is the chief organ that is responsible for maintaining lipid and glucose levels in the body. In response to nutritional and hormonal signals, the liver controls lipid and carbohydrate metabolic pathways to maintain homeostasis. A growing body of evidence indicates that SIRT1 is a master regulator of hepatic lipid (fatty acid, triglyceride, cholesterol, bile acid) metabolism. SIRT1 is a NAD^+-dependent deacetylase that links cellular energy and metabolic status to transcriptional outcomes to meet energy deficits under energy-deprived conditions, such as fasting or exercise (Canto & Auwerx, 2009; Chalkiadaki & Guarente, 2012; Guarente, 2007; Houtkooper, Pirinen, & Auwerx, 2012). SIRT1 regulates transcriptional programs of hepatic metabolism by deacetylating transcriptional regulatory proteins, as well as, histones at its metabolic target genes (Chalkiadaki & Guarente, 2012; Haigis & Sinclair, 2010; Houtkooper et al., 2012). In this review, we provide an update of the latest advances in the SIRT1 field, with a focus on the role of SIRT1 in hepatic lipid metabolism. We discuss how the expression and activity of hepatic SIRT1 are regulated under normal physiological conditions and how they are dysregulated in fatty liver in obesity. We also discuss how SIRT1 modulates transcriptional regulators involved in hepatic lipid metabolism and further discuss the underlying mechanisms by which SIRT1 regulates their expression levels and activities by protein deacetylation. Finally, we discuss the therapeutic potential of SIRT1 as a novel target for treating hepatic steatosis and other obesity-related disease, such as type 2 diabetes, cardiovascular disease, and certain cancers.

2. SIRT1 AS A MASTER REGULATOR OF METABOLISM

2.1. SIRT1 as a link between metabolic/energy status and transcriptional outcomes

In mammals, there are seven sirtuin proteins (SIRT1–7) (Guarente, 2007; Haigis & Sinclair, 2010; Yamamoto, Schoonjans, & Auwerx, 2007). They are not only ubiquitously expressed and share a common catalytic core domain but also have distinct features in their subcellular localization, enzymatic activities, and target proteins. Of the sirtuins, SIRT1 is the best-characterized protein. SIRT1 is a mammalian ortholog of the yeast silent information regulator 2 (sir2) (Klar, Fogel, & Macleod, 1979; Sinclair & Guarente, 1997). Yeast sir2 was originally discovered as one of several genes that encode transcriptional regulators that silence gene expression and was shown to be a longevity protein that increased life span (Gottlieb & Esposito, 1989; Sinclair & Guarente, 1997). SIRT1 was then shown to be a NAD^+-dependent enzyme that catalyzes deacetylation of histones (Imai, Armstrong, Kaeberlein, & Guarente, 2000). Intensive studies during the past decade have demonstrated that SIRT1 also deacetylates a wide range of nonhistone regulatory proteins including transcriptional nuclear receptors involved in diverse biological pathways, such as metabolism, stress response, and possibly the aging process (Canto & Auwerx, 2009; Finkel, 2003; Sinclair, Lin, & Guarente, 2006). Well-known transcriptional regulators involved in the cellular stress response and metabolism that are modulated by SIRT1-mediated deacetylation include p53, FOXOs, PGC-1α, NF-κB, CREB-regulated transcriptional coactivator 2 (CRTC2), liver X receptors (LXRs), farnesoid X receptor (FXR), and sterol response element binding proteins (SREBPs).

NAD^+-dependent SIRT1 activity is particularly important in the regulation of metabolic/energy homeostasis because it can sense nutrient availability and connect cellular energy/metabolic status to transcriptional outcomes. A large body of evidence indicates that cellular NAD^+ levels in mammalian tissues are increased in response to metabolic stress conditions, such as fasting, calorie restriction, and vigorous exercise, which leads to increased SIRT1 deacetylase activity (Chalkiadaki & Guarente, 2012; Guarente, 2007; Houtkooper et al., 2012). Further, recent studies demonstrated that NAD^+ levels also fluctuate in response to circadian rhythm and that the clock genes are regulated by SIRT1. In addition to SIRT1, AMP-activated kinase (AMPK) is another major cellular energy sensor that maintains metabolic/energy homeostasis in the body (Kahn, Alquier, Carling, & Hardie, 2005). AMPK activator agonists like AICAR were shown to increase NAD^+ levels and activate the SIRT1 activity (Canto et al., 2009). All these studies demonstrate that SIRT1 senses changes in cellular NAD^+ levels that reflect intracellular energy levels and adjusts energy outputs by protein deacetylation, which modulates levels and activities of transcriptional metabolic regulators and alters usually, but not always, transcriptional programs of cellular metabolic pathways.

2.2. Regulation of SIRT1 in physiology and disease

Fluctuating nutrient and hormonal signals during fasting and feeding cycles dynamically regulate the expression of SIRT1 (Fig. 16.1A). Recent studies have shown that cAMP-responsive element binding protein (CREB) positively activates *SIRT1* gene transcription during fasting, while carbohydrate response element binding protein (ChREBP) negatively regulates SIRT1 expression after feeding (Chalkiadaki & Guarente, 2011; Noriega et al., 2011). After short-term fasting (5–6 h in mice), increased glucagon levels lead to the dephosphorylation of CRTC2, which results in transport of CRTC2 into the nucleus (Liu et al., 2008). The CREB/CRTC2 transcriptional complex then activates transcription of the *SIRT1* gene as well as gluconeogenic genes.

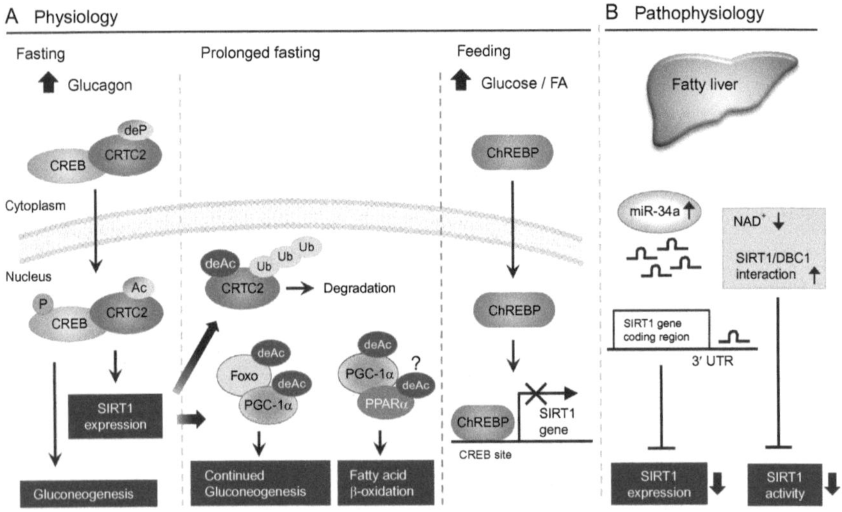

Figure 16.1 Regulation of hepatic SIRT1 in physiology and disease. (A) In response to glucagon signaling during short-term fasting, CREB is phosphorylated and CRTC2 is dephosphorylated, which results in the translocation of these proteins into the nucleus. The CREB/CRTC2 transcriptional complex drives expression of gluconeogenesis genes and the *SIRT1* gene. During prolonged fasting, SIRT1 deacetylates CRTC2, resulting in its ubiquitination and proteasomal degradation. SIRT1 also deacetylates PGC-1α and Foxo-1, resulting in continued gluconeogenesis in the absence of CREB, and stimulation of fatty acid (FA) β-oxidation. It remains unknown whether PPARα is deacetylated by SIRT1. Upon feeding, glucose and FA levels rise which causes the translocation of ChREBP into the nucleus. ChREBP binds to a site in the *SIRT1* promoter that overlaps the CREB site, resulting in inhibition of SIRT1 gene transcription. (B) In pathological conditions, such as fatty liver in obesity, aberrantly elevated miR-34a binds to the 3′UTR of the SIRT1 mRNA and downregulates SIRT1. Further, in fatty liver, cellular NAD^+ levels are reduced and the interaction between SIRT1 and DBC1 is increased, which leads to decreased SIRT1 deacetylation activity. The model is adapted from Chalkiadaki and Guarente, 2011.

Under more prolonged fasting conditions (at least 12–18 h in mice), SIRT1 deacetylates CRTC2, which results in its ubiquitination and proteasomal degradation (Noriega et al., 2011). Furthermore after prolonged fasting, SIRT1 deacetylates key regulators of gluconeogenesis and fatty acid β-oxidation, PGC-1α and Foxo-1, thus increasing their transcriptional activities (Brunet et al., 2004; Rodgers et al., 2005). These coordinated transcriptional events lead to fatty acid β-oxidation and continued gluconeogenesis to meet energy demands during prolonged fasting. Conversely, under feeding conditions, the expression of SIRT1 is transiently decreased by ChREBP, a transcriptional factor that couples hepatic glucose utilization and lipid synthesis (Uyeda & Repa, 2006). ChREBP is transported into the nucleus and binds to a site in the *SIRT1* promoter that overlaps the CREB binding site, resulting in repression of SIRT1 gene transcription (Noriega et al., 2011).

Recent studies have suggested that the expression levels and deacetylase activity of SIRT1 are constitutively decreased under pathological conditions, such as in fatty livers of obese mice (Fig. 16.1B) and that small noncoding microRNAs (miRs) play a key role in this abnormal regulation (Coste et al., 2008; Lee et al., 2010). MiRs have been recently demonstrated as important cellular regulators that function as negative gene regulators by directly binding to the 3′untranslated region (3′UTR) of target genes (Bartel, 2004; Neilson & Sharp, 2008). Consistent with their critical biological functions, miRs are aberrantly expressed in human diseases (Cermelli, Ruggieri, Marrero, Ioannou, & Beretta, 2011; Cheung et al., 2008; Esau et al., 2006). MiR-34a is the most highly elevated hepatic miR in fatty liver disease patients and obese mice (Cermelli et al., 2011; Cheung et al., 2008; Lee et al., 2010; Trajkovski et al., 2011). In recent studies, our group has shown that SIRT1 expression is posttranscriptionally inhibited by abnormally elevated levels of miR-34a in fatty livers of dietary or genetic obese mice (Lee et al., 2010). SIRT1 deacetylase activity was also shown be inhibited by decreased levels of intracellular NAD^+ levels in fatty liver of obese mice (Yoshino, Mills, Yoon, & Imai, 2011). In obesity, SIRT1 activity is also inhibited by an unusual protein interaction between SIRT1 and a negative regulator of SIRT1, DBC1. Metabolic stress induced by a high-fat diet increases the interaction between SIRT1 and DBC1, resulting in decreased SIRT1 deacetylase activity (Escande et al., 2010). Unlike normal physiological regulation of SIRT1, transcriptional regulation of SIRT1 under pathological conditions has not been well studied. Further, it remains largely unknown whether aberrant posttranslational modifications of SIRT1 negatively regulate the expression levels and activities of SIRT1 under pathological conditions.

2.3. Metabolic functions of SIRT1

The role of SIRT1 in cellular metabolism has been demonstrated in a number of studies using small molecule SIRT1 activators or transgenic mice that are deficient for SIRT1 expression or that overexpress SIRT1. The small molecule compounds activating SIRT1 include resveratrol, a polyphenolic compound abundant in red-grape skin, and synthetic compounds, such as SRT1720 (Baur et al., 2006; Feige et al., 2008; Lagouge et al., 2006). *In vivo* animal studies using these natural or synthetic compounds have revealed that activation of SIRT1 ameliorates insulin resistance, increases mitochondrial content, improves metabolic profiles, and increases survival in mice challenged with a high-fat diet (Baur et al., 2006; Feige et al., 2008; Lagouge et al., 2006). Remarkably, a recent human study has shown that calorie restriction-like beneficial effects on energy metabolism and metabolic profiles were observed in obese humans when treated with a resveratrol-supplemented diet for 30 days (Timmers et al., 2011).

In transgenic studies, SIRT1-overexpressing mice were resistant to body weight gain and also, insulin resistance and glucose intolerance was ameliorated when these mice were fed a high-fat diet compared to wild-type control mice (Banks et al., 2008; Pfluger, Herranz, Velasco-Miguel, Serrano, & Tschop, 2008; Rodgers & Puigserver, 2007). Consistent with these SIRT1 gain-of-function studies, detrimental metabolic effects were observed in loss-of-function studies. In SIRT1 liver-specific knockout (SIRT1 LKO) mice fed a high-fat diet, fatty acid metabolism was altered and the fat (triglyceride, TG) was accumulated abnormally in the liver, resulting in hepatic steatosis and elevated inflammatory responses (Purushotham et al., 2009). Further, in SIRT1 LKO mice generated by a different group, mTorc2/Akt signaling was impaired, which led to hyperglycemia, oxidative stress, and insulin resistance (Wang et al., 2011).

All told, these recent studies demonstrate that SIRT1 is a master regulator of cellular metabolism and mediates beneficial metabolic effects in various metabolic tissues. In the following sections, we discuss beneficial metabolic functions of SIRT1, with a focus on hepatic lipid metabolism.

3. REGULATION OF HEPATIC FAT (TG) METABOLISM BY SIRT1

3.1. Regulation of hepatic TG levels

The liver is the central organ in the body that ensures metabolic homeostasis. The liver controls nearly all aspects of lipid (fatty acid and TG, cholesterol/

bile acids) and carbohydrate (glucose, glycogen) metabolic pathways. In these sections, we focus on the regulation of hepatic fatty acid and TG metabolic pathways in response to metabolic signals. Hepatic fat (TG) metabolism largely involves the uptake, synthesis, storage, and utilization of fatty acids and TG in response to nutrient and hormonal metabolic signals under fasting and feeding conditions (Cohen et al., 2011). During fasting, the liver converts TG into usable energy through the process of fatty acid β-oxidation to meet energy deficits. Conversely, under feeding conditions, lipid metabolic programs in the liver are reversed in order to store extra energy in the form of lipid TG droplets. Under normal physiological conditions, these processes are tightly controlled, but in disease states, such as nonalcoholic fatty liver disease (NAFLD), regulation of these processes is impaired, resulting in an imbalance between hepatic storage and removal of TG (fat) and abnormal accumulations of fat in the liver. In NAFLD, hepatic insulin resistance, ER stress, and chronic inflammation are increased (Baker, Hayden, & Ghosh, 2011; Meshkani & Adeli, 2009), which substantially increases the risk for type 2 diabetes, cardiovascular disease, and hepatobiliary liver diseases, such as gallstones, steatohepatitis, cirrhosis, and liver cancer.

As shown in Fig. 16.2, there are four major pathways controlling hepatic TG levels: (1) TG import through uptake of dietary TG, (2) TG synthesis (lipogenesis), (3) utilization of TG through fatty acid β-oxidation, and (4)

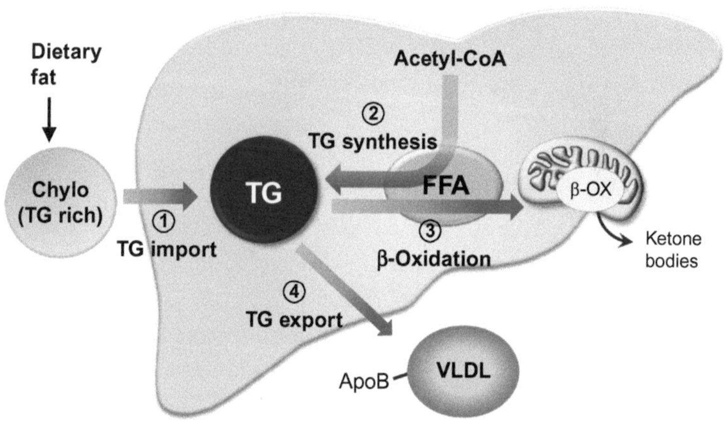

Figure 16.2 Four major pathways controlling hepatic fat (TG) levels. (1) Hepatic TG levels are increased by TG import through the uptake of TG-rich chylomicrons. (2) Hepatic TG levels are also increased by synthesis of free fatty acid (FFA) from acetyl CoA and subsequently TG synthesis. (3) Hepatic TG levels are decreased by the release of FFAs followed by their β-oxidation in mitochondria. (4) Hepatic TG levels are decreased by TG export into the blood as VLDLs that contain ApoB lipoprotein. The model is adapted from Cohen, Horton, and Hobbs, 2011.

TG export through very-low-density lipoprotein (VLDL) lipoproteins. First, after a meal, dietary lipids taken up in the small intestine are packaged as TG-rich chylomicrons and released into the systemic circulation. About 20% of chylomicrons are taken up by the liver to provide a source of lipid for hepatic TG (Cohen et al., 2011). Second, hepatic TG levels are also increased by lipogenesis by conversion of fatty acids into TG. Two major transcription factors, sterol response element binding protein-1c (SREBP-1c) and ChREBP, are largely responsible for the activation of the hepatic lipogenic program of fatty acid synthesis and subsequent TG production (Brown & Goldstein, 1997; Uyeda & Repa, 2006). Third, during prolonged fasting, TG levels are decreased by release of fatty acids from TG followed by fatty acid β-oxidation in mitochondria that provides energy fuel sources for the body, especially ketone bodies for brain. In addition, fatty acids produced by β-oxidation can also be re-esterified and converted back to TG in the liver, which partially restores hepatic TG levels. Fourth, hepatic TG levels are decreased by export into the blood as the VLDL lipoprotein particle. Human subjects with a heterozygous mutation in the gene encoding apolipoprotein B (ApoB), a structural protein of VLDL, showed threefold increase in hepatic TG levels compared to normal individuals (Tanoli, Yue, Yablonskiy, & Schonfeld, 2004).

Recent studies have shown that SIRT1 decreases hepatic TG levels by inhibiting lipogenesis and stimulating fatty acid β-oxidation (Ponugoti et al., 2010; Purushotham et al., 2009; Walker et al., 2010). SIRT1 inhibits hepatic fatty acid and TG synthesis by targeting SREBP-1c and stimulates β-oxidation by targeting PPARα and its coactivator, PGC-1α, which will be discussed further in the following sections.

3.2. Inhibition of hepatic lipogenesis by SIRT1

Recent studies from our group and A. Narr and his colleagues have shown that SIRT1 inhibits hepatic lipogenesis by deacetylation of SREBP-1c, a key lipogenic activator, during fasting (Ponugoti et al., 2010; Walker et al., 2010). SREBPs are DNA binding transcription factors that are critically involved in the regulation of cellular lipid and cholesterol levels (Brown & Goldstein, 1997). There are three isoforms of SREBPs (SREBP-1c, -1a, and -2) encoded by two genes (Horton, 2002). SREBP-1a and -1c are synthesized from the same gene and SREBP-2 from a separate gene. SREBP-1c is more abundantly expressed in adult liver, whereas SREBP-1a is the predominant isoform in cultured hepatic cells. SREBP-1a and -1c regulate fatty acid and TG synthesis and SREBP-2 is a critical regulator of cholesterol synthesis

(Horton, Bashmakov, Shimomura, & Shimano, 1998; Shimano et al., 1997). In response to increased insulin levels in the fed state, SREBP-1c binds to and activates its lipogenic target genes, such as fatty acid synthase, acetyl CoA carboxylase, stearyl CoA desaturase, and also its own gene, *SREBP-1c*, which results in increased fatty acid synthesis (Chen, Liang, Ou, Goldstein, & Brown, 2004; Repa et al., 2000). In contrast, SREBP-2 is not transcriptionally regulated upon feeding (Horton et al., 1998; Shimano et al., 1997).

Recent studies independently carried out by two groups, including us, support the conclusion that SIRT1 deacetylates and thus, downregulates hepatic SREBP-1c lipogenic activity during fasting (Fig. 16.3) (Ponugoti et al., 2010; Walker et al., 2010). Biochemical proteomic analysis along with mechanistic studies have revealed that SREBP-1c is acetylated by p300 and deacetylated by SIRT1 at Lys-289 and Lys-309 and that deacetylation of SREBP-1c by SIRT1 downregulates SREBP-1c transcriptional activity by

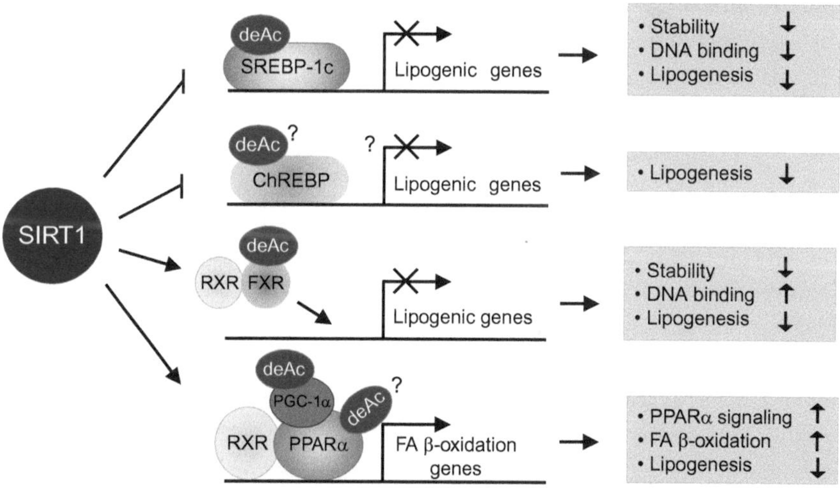

Figure 16.3 Regulation of hepatic fat (TG) metabolism by SIRT1. During fasting, SIRT1 deacetylates and modulates activities of metabolic regulators involved in transcriptional programs of hepatic lipogenesis and β-oxidation. First, SIRT1 deacetylates SREBP-1c and thus, decreases stability and DNA binding to lipogenic genes, which leads to inhibition of hepatic lipogenesis. SIRT1 may inhibit hepatic lipogenesis by deacetylating ChREBP but this has not been reported. Second, SIRT1 deacetylates and increases transactivation potential of FXR which is known to indirectly repress hepatic FA and TG synthesis. Deacetylation of FXR leads to increased heterodimerization with RXRα and DNA binding, although it decreases FXR stability. Third, SIRT1 activates PPARα signaling, resulting in increased expression of β-oxidation genes. SIRT1 also enhances the activity of PGC-1α by deacetylation. However, it is not known whether SIRT1 increases PPARα activity by deacetylation.

promoting ubiquitination and possibly, proteasomal degradation (Ponugoti et al., 2010; Walker et al., 2010). Consistently, treatment with small molecule activators of SIRT1 decreased nuclear SREBP-1c levels, whereas SIRT1 inhibitors increased the levels (Walker et al., 2010). Further, deacetylation by SIRT1 inhibited binding of SREBP-1c to its target lipogenic genes. Remarkably, while SREBP-1c acetylation levels were dynamically regulated in response to fasting and feeding under normal physiological conditions, SREBP-1c acetylation levels were constitutively elevated in fatty livers of obese mice (Ponugoti et al., 2010). Consistent with the role of SIRT1 in downregulating SREBP-1c, adenoviral-mediated hepatic overexpression of SIRT1 or treatment with resveratrol daily for 1 week decreased SREBP-1c acetylation levels in these mice with hepatic gene expression profiles consistent with beneficial metabolic effects (Ponugoti et al., 2010).

In addition to SREBP-1c, ChREBP is also an important lipogenic activator (Fig. 16.3). ChREBP is a transcription factor that couples hepatic glucose utilization and fatty acid and TG synthesis and is translocated into the nucleus when glucose and fatty acid levels are elevated during feeding (Uyeda & Repa, 2006). Increased ChREBP expression was observed in SIRT1 LKO mice and further, hepatic steatosis was evident even under normal feeding conditions (Wang, Li, & Deng, 2010). However, it is not known whether the expression of ChREBP-1c is a direct SIRT1 target for deacetylation or is indirectly regulated by SIRT1.

The nuclear bile acid receptor, FXR, is also a direct target of SIRT1 in the regulation of hepatic lipid metabolism (Kemper, 2011; Kemper et al., 2009). FXR is the primary biosensor for bile acids and plays a central role in maintaining bile acid homeostasis (Lee, Lee, Hubbert, Edwards, & Zhang, 2006; Lefebvre, Cariou, Lien, Kuipers, & Staels, 2009). In addition to controlling bile acid levels, FXR also indirectly regulates transcription of genes involved in TG synthesis and clearance, leading to decreased hepatic TG levels (Watanabe et al., 2004; Zhang, Castellani, Sinal, Gonzalez, & Edwards, 2004). In recent studies, our group has shown that FXR acetylation levels were dynamically regulated during fasting and feeding in normal mice, but constitutively elevated in fatty liver of obese mice (Fig. 16.3) (Kemper et al., 2009). SIRT1 deacetylates FXR during fasting and enhances FXR transactivation potential by increasing occupancy of FXR/RXRα heterodimers at their target genes. Notably, downregulation of hepatic SIRT1 increases FXR acetylation levels, resulting in gene expression patterns expected to be detrimental, including increased expression of the lipogenic *SREBP-1c* and *FAS* genes. Further, treatment of obese mice with resveratrol daily for 1 week substantially decreased FXR acetylation levels and beneficial metabolic effects (Kemper et al., 2009). These findings indicate an intriguing functional link between elevated FXR acetylation levels and decreased hepatic SIRT1 activity with deleterious effects, further providing evidence that SIRT1 is a potential therapeutic target for treating metabolic disease.

3.3. Stimulation of hepatic fatty acid β-oxidation by SIRT1

Fatty acid β-oxidation is a major means of TG utilization in liver under energy-limited conditions, and the transcriptional regulators, PPARα and PGC-1α, play key roles in this process (Evans, Barish, & Wang, 2004; Rodgers, Lerin, Gerhart-Hines, & Puigserver, 2008). While SIRT1 increases transcriptional activity of the coactivator PGC-1α by deacetylation, it is not known whether PPARα is also a direct target of SIRT1 deacetylation during fasting (Fig. 16.3). Studies utilizing liver-specific SIRT1 LKO mice or adenoviral-mediated liver-specific acute downregulation of SIRT1 in mice have shown that SIRT1 stimulates β-oxidation and decreases hepatic TG levels in response to fasting (Purushotham et al., 2009). Acute downregulation of hepatic SIRT1 by adenoviral shRNA resulted in reduced expression of β-oxidation genes in livers of fasted mice. Consistent with these findings, a recent SIRT1 LKO mouse study showed that hepatic deletion of SIRT1 results in impaired PPARα signaling and decreased expression of genes involved in β-oxidation, whereas overexpression of SIRT1 leads to increased PPARα transcriptional signaling. Further, SIRT1 LKO mice were more susceptible to high-fat diet-induced gain weight and hepatic steatosis (Purushotham et al., 2009).

However, contradictory results were observed in another study using similarly generated SIRT1 LKO mice. In these SIRT1 LKO mice, a reduction in weight gain and reduced liver TG accumulation were observed when the mice were fed a high-fat diet (Chen et al., 2008). Because these mice are the same mouse strain CV57BL and similarly generated, the differences in dietary composition (different carbohydrate and cholesterol levels in the high-fat diet) and/or the ages of the mice may account for the discrepancies in these two studies.

4. REGULATION OF HEPATIC CHOLESTEROL AND BILE ACID METABOLISM BY SIRT1

The liver is the chief metabolic organ that maintains whole body cholesterol and bile acid homeostasis. Cholesterol and bile acids are steroid molecules that play essential roles in biological functions. Cholesterol serves as a component of the cell membrane and is also a precursor for biosynthesis of steroid hormones, lipid-soluble vitamins, and bile acids (Russell, 2003; Russell & Setchell, 1992). Bile acids are detergent-like amphipathic molecules that facilitate digestion of lipid-soluble nutrients. In addition to this traditional role, bile acids have been recently recognized as key signaling molecules that control integrative metabolism and energy balance in the body (Chiang, 2009; Hylemon et al., 2009). Consistent with their critical functions in the body, excesses of cholesterol and bile acids lead to human diseases, such as cardiovascular disease, hepatobiliary disease, and liver/colon cancer. Therefore, the levels of cholesterol and bile acids are tightly regulated under normal physiological conditions. Liver plays the key role in accomplishing these important functions because first, cholesterol conversion to bile acids exclusively occurs in the liver and second, this process is the major route for elimination of cholesterol from the body because a minor fraction (about 5%) of cholesterol (as bile acids) is excreted daily through feces (Chiang, 2009; Russell, 1999). In this section, we discuss how SIRT1 mediates beneficial metabolic effects to maintain cholesterol and bile acid homeostasis by modulating key regulators of hepatic cholesterol and bile acid metabolism.

4.1. The role of SIRT1 in hepatic cholesterol metabolism

SIRT1 plays a beneficial metabolic role in cholesterol metabolism by deacetylating and thus, increasing the activity of the oxysterol receptors, LXRs. LXRs (LXRα and LXRβ) are nuclear receptors that function as sterol sensors and are key regulators of body lipid homeostasis (Kalaany & Mangelsdorf, 2006; Tontonoz & Mangelsdorf, 2003). LXRs promote reverse hepatic cholesterol transport by promoting efflux of HDL cholesterol-containing lipoproteins from peripheral tissues and uptake in the liver, where lipoproteins can be degraded and the released cholesterol can then be removed from the body through its conversion into bile acids (Tontonoz & Mangelsdorf, 2003). Therefore, this LXR-mediated reverse cholesterol transport prevents the formation of foam cells in arterial walls, which reduces atherosclerosis and the incidence of cardiovascular diseases.

Recently, Li et al. have shown that LXRs are acetylated at Lys-432 in LXRα and Lys-433 in LXRβ adjacent to the ligand-regulated activation function 2 domain of the receptor (Li et al., 2007). Upon activation by ligands, LXRs interact with SIRT1, which promotes deacetylation and consequently ubiquitination/degradation of LXRs. Mutation of Lys-432 eliminates transcriptional activation of LXRs mediated by the SIRT1, demonstrating that deacetylation of LXRs by SIRT1 is important for their activity. Further, in SIRT1 KO mice, decreased expression of LXR downstream target genes involved in cholesterol metabolism including the ATP binding cassette transporter, impaired cholesterol homeostasis, and hepatic cholesterol accumulation were observed. The role of SIRT1 in cholesterol metabolism was also shown by other studies in which adenoviral-mediated acute hepatic downregulation of SIRT1 reduced expression of CYP7A1, the rate-limiting hepatic enzyme for bile acid biosynthesis from cholesterol (Rodgers & Puigserver, 2007). These results are consistent with the abnormal cholesterol accumulation in the liver observed in the SIRT1 LKO mice describe above (Li et al., 2007).

There is a major puzzling issue in the activation of LXRs by SIRT1 deacetylation. In addition to reversed cholesterol transport, LXRs also promote hepatic lipogenesis by activating the transcriptional program of fatty acid synthesis, especially by increasing transcription of SREBP-1c, a key lipogenic activator (Repa et al., 2000). SIRT1-mediated deacetylation of LXRs, therefore, can lead to both beneficial (promoting reverse cholesterol transport) and deleterious (promoting hepatic lipogenesis) effects in hepatic lipid metabolism. However, SIRT1 deacetylates not only LXRs but also SREBP-1c (Walker et al., 2010). Under fasting, SIRT1 deacetylates SREBP-1c and thus, reduces its transcriptional activity. It is known that both LXRs and SREBP-1c contribute to transcriptional activation of the *SREBP-1c* gene (Chen et al., 2004; Yoshikawa et al., 2001), and therefore, abrogation of SREBP-1c activity by SIRT1-mediated deacetylation may allow SIRT1 to show beneficial effects on hepatic cholesterol metabolism. It will be important to elucidate precise molecular mechanisms by which deacetylation of LXRs by SIRT1 selectively regulates hepatic cholesterol and fatty acid metabolism in gene- and metabolic pathway-specific manners.

4.2. The role of SIRT1 in hepatic bile acid metabolism

SIRT1 also plays beneficial roles in maintaining cholesterol and bile acid homeostasis by modulating hepatic FXR signaling. FXR belongs to the nuclear receptor superfamily and functions as the primary biosensor for endogenous bile acids (Kalaany & Mangelsdorf, 2006; Mangelsdorf & Evans, 1995). Upon activation by elevated hepatic bile acid levels, FXR mediates transcriptional programs for every aspect of hepatic bile acid metabolism, including biosynthesis from cholesterol, transport (import and export), and bile acid metabolism (Kemper, 2011; Lee et al., 2006;

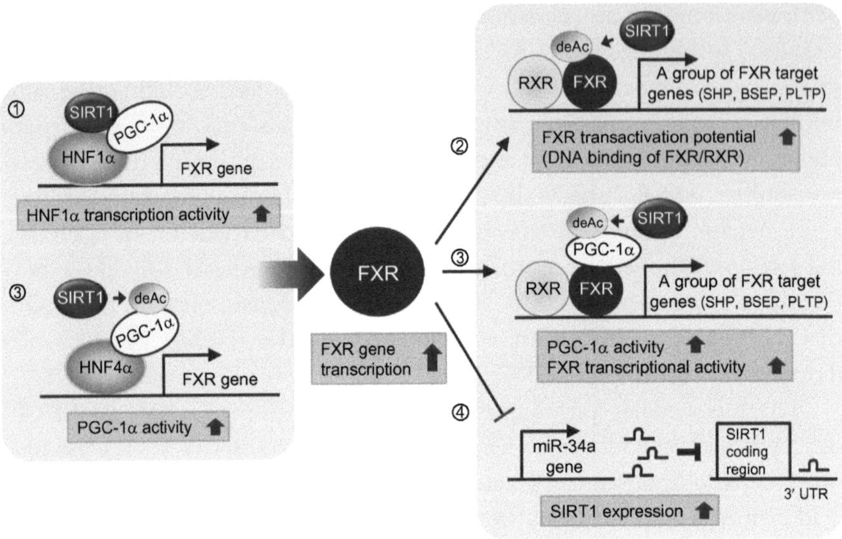

Figure 16.4 Regulation of hepatic bile acid metabolism by SIRT1. SIRT1 and FXR comprise interactive regulatory loops in the liver during fasting. (1) SIRT1 increases the interaction between HNF-1α/PGC-1α and increases FXR gene transcription. (2) SIRT1 deacetylates FXR and increases its DNA binding and transactivation potential of regulating bile acid-responsive FXR target genes. (3) SIRT1 deacetylates PGC-1α, which increases both FXR gene transcription and FXR transcriptional activity. (4) FXR, in turn, upregulates hepatic SIRT1 expression by inhibiting gene transcription of miR-34a, a negative regulator of SIRT1, by directly binding to the 3′UTR of the SIRT1 mRNA.

Lefebvre et al., 2009). Since bile acids are detergent-like molecules and toxic to hepatocytes, dysregulation, resulting in excess bile acids can cause hepatobiliary disease and liver cancer (Lefebvre et al., 2009; Russell & Setchell, 1992). FXR protects the liver from elevated toxic bile acid levels by mediating these processes and SIRT1 directly targets hepatic FXR signaling in hepatocytes.

Recent studies from several laboratories including ours have shown that SIRT1 regulates hepatic transcriptional programs of FXR signaling at multiple levels (Fig. 16.4). SIRT1 and FXR can form interactive regulatory loops in the liver during fasting. First, SIRT1 increases *FXR* gene transcription by upregulating the transcriptional complex, HNF-1α and PGC-1α (Purushotham et al., 2012), although the underlying mechanisms remain unclear. Occupancy of HNF-1α at the *FXR* gene was decreased in SIRT1 LKO mice, suggesting that SIRT1 may increase HNF-1α DNA binding activity. Notably, SIRT1 LKO mice had deranged bile acid metabolism including impaired transport of biliary bile acids and phospholipids, which predisposed the mice to the development of cholesterol gallstones when fed a lithogenic diet containing high fat, cholesterol, and bile acid (Purushotham et al., 2012). Second, SIRT1 deacetylates FXR and enhances its transactivation potential by increasing heterodimerization with RXRα and binding to FXR target genes (Kemper et al., 2009). Well-known direct FXR target genes involved in hepatic lipid metabolism include small heterodimer partner, a key negative regulator of bile acid synthesis, and hepatic bile acid export transporters, like BSEP, as well as PLTP and Apo CII, stimulators of TG clearance (Kemper, 2011; Lee et al., 2006; Lefebvre et al., 2009). Notably, FXR acetylation levels are constitutively elevated in fatty liver of obese mice, which results in abnormal transcriptional programs with deleterious metabolic effects (Kemper et al., 2009). Third, SIRT1 further activates transcription of FXR target genes by deacetylating and increasing PGC-1α activity. It was shown that PGC-1α enhances hepatic FXR activity during fasting by both increasing transcription of FXR and also by acting as a coactivator of FXR (Zhang et al., 2004). Fourth, FXR, in turn, reciprocally increase the expression of SIRT1 by inhibiting miR-34a, which directly binds to the 3′UTR of the SIRT1 mRNA (Lee & Kemper, 2010; Lee et al., 2010; Yamakuchi, Ferlito, & Lowenstein, 2008). These interactive regulatory loops are critical for maintaining bile acid homeostasis and dysregulation of these networks can lead to metabolic abnormalities, resulting in hepatobiliary disease, such as gallstones, cholestasis, steatosis, and hepatocellular carcinoma (liver cancer).

Combined, SIRT1 plays beneficial roles in maintaining cholesterol and bile acid levels by assisting the key bile acid regulator FXR.

5. CONCLUSION AND FUTURE PERSPECTIVES

SIRT1 was originally discovered as a longevity protein in yeast and soon after, also identified in multicellular organisms like *Caenorhabditis elegans* and drosophila (Kaeberlein, McVey, & Guarente, 1999; Tissenbaum & Guarente, 2001). Over a decade ago, SIRT1 was further identified as a NAD^+-dependent deacetylase that mediates mammalian metabolic responses to nutrient availability (Imai et al., 2000). Since then, SIRT1 has been intensively studied because of its great potential for human health benefits. As discussed in this review, SIRT1 plays beneficial roles in hepatic lipid metabolism by inhibiting hepatic lipogenesis, stimulating fatty acid β-oxidation, and maintaining cholesterol and bile acid levels. Whether or not SIRT1 is a life span extension protein in human has not yet been established, but it is evident that

SIRT1 mediates beneficial effects on mammalian physiology and, therefore, activation of SIRT1 should enhance the quality of life by affecting human health in many ways. Not surprisingly, there have been a number of studies done to understand how SIRT1 levels and activity are regulated at multiple levels: transcriptionally, posttranscriptionally through miRs, and posttranslationally through its covalent modifications (Chalkiadaki & Guarente, 2011; Houtkooper et al., 2012; Kwon & Ott, 2008). Also, numerous studies have been done to understand how the SIRT1 activity can be enhanced by small molecule compounds, SIRT1-interacting regulatory proteins and by increasing cellular NAD^+ levels (Canto & Auwerx, 2012). All these SIRT1-based studies should be extremely valuable for the development of therapeutic drugs for the treatment of age-related human diseases, such as obesity-related steatosis discussed in this review, diabetes, cardiovascular disease, neurodegenerative disease, and cancer.

ACKNOWLEDGMENTS

We apologize to those authors whose original works were not discussed in this review due to space limitations. We thank Byron Kemper for critical reading of the manuscript. Research discussed from the authors' laboratory was supported by grants, NIH DK062777, NIH DK095842, and a Basic Science Award from the American Diabetes Association.

REFERENCES

Baker, R. G., Hayden, M. S., & Ghosh, S. (2011). NF-kappaB, inflammation, and metabolic disease. *Cell Metabolism, 13*, 11–22.

Banks, A. S., Kon, N., Knight, C., Matsumoto, M., Gutierrez-Juarez, R., Rossetti, L., et al. (2008). SirT1 gain of function increases energy efficiency and prevents diabetes in mice. *Cell Metabolism, 8*, 333–341.

Bartel, D. P. (2004). MicroRNAs: Genomics, biogenesis, mechanism, and function. *Cell, 116*, 281–297.

Baur, J. A., Pearson, K. J., Price, N. L., Jamieson, H. A., Lerin, C., Kalra, A., et al. (2006). Resveratrol improves health and survival of mice on a high-calorie diet. *Nature, 444*, 337–342.

Brown, M. S., & Goldstein, J. L. (1997). The SREBP pathway: Regulation of cholesterol metabolism by proteolysis of a membrane-bound transcription factor. *Cell, 89*, 331–340.

Brunet, A., Sweeney, L. B., Sturgill, J. F., Chua, K. F., Greer, P. L., Lin, Y., et al. (2004). Stress-dependent regulation of FOXO transcription factors by the SIRT1 deacetylase. *Science, 303*, 2011–2015.

Canto, C., & Auwerx, J. (2009). Caloric restriction, SIRT1 and longevity. *Trends in Endocrinology and Metabolism, 20*, 325–331.

Canto, C., & Auwerx, J. (2012). Targeting sirtuin 1 to improve metabolism: All you need is NAD(+)? *Pharmacological Reviews, 64*, 166–187.

Canto, C., Gerhart-Hines, Z., Feige, J. N., Lagouge, M., Noriega, L., Milne, J. C., et al. (2009). AMPK regulates energy expenditure by modulating NAD+ metabolism and SIRT1 activity. *Nature, 458*, 1056–1060.

Cermelli, S., Ruggieri, A., Marrero, J. A., Ioannou, G. N., & Beretta, L. (2011). Circulating microRNAs in patients with chronic hepatitis C and non-alcoholic fatty liver disease. *PLoS One, 6*, e23937.

Chalkiadaki, A., & Guarente, L. (2011). Metabolic signals regulate SIRT1 expression. *EMBO Reports, 12*, 985–986.

Chalkiadaki, A., & Guarente, L. (2012). Sirtuins mediate mammalian metabolic responses to nutrient availability. *Nature Reviews. Endocrinology, 8*, 287–296.

Chen, D., Bruno, J., Easlon, E., Lin, S. J., Cheng, H. L., Alt, F. W., et al. (2008). Tissue-specific regulation of SIRT1 by calorie restriction. *Genes & Development, 22*, 1753–1757.

Chen, G., Liang, G., Ou, J., Goldstein, J. L., & Brown, M. S. (2004). Central role for liver X receptor in insulin-mediated activation of Srebp-1c transcription and stimulation of fatty acid synthesis in liver. *Proceedings of the National Academy of Sciences of the United States of America, 101*, 11245–11250.

Cheung, O., Puri, P., Eicken, C., Contos, M. J., Mirshahi, F., Maher, J. W., et al. (2008). Nonalcoholic steatohepatitis is associated with altered hepatic microRNA expression. *Hepatology, 48*, 1810–1820.

Chiang, J. Y. (2009). Bile acids: Regulation of synthesis. *Journal of Lipid Research, 50*, 1955–1966.

Cohen, J. C., Horton, J. D., & Hobbs, H. H. (2011). Human fatty liver disease: Old questions and new insights. *Science, 332*, 1519–1523.

Coste, A., Louet, J. F., Lagouge, M., Lerin, C., Antal, M. C., Meziane, H., et al. (2008). The genetic ablation of SRC-3 protects against obesity and improves insulin sensitivity by reducing the acetylation of PGC-1{Alpha}. *Proceedings of the National Academy of Sciences of the United States of America, 105*, 17187–17192.

Esau, C., Davis, S., Murray, S. F., Yu, X. X., Pandey, S. K., Pear, M., et al. (2006). MiR-122 regulation of lipid metabolism revealed by in vivo antisense targeting. *Cell Metabolism, 3*, 87–98.

Escande, C., Chini, C. C., Nin, V., Dykhouse, K. M., Novak, C. M., Levine, J., et al. (2010). Deleted in breast cancer-1 regulates SIRT1 activity and contributes to high-fat diet-induced liver steatosis in mice. *The Journal of Clinical Investigation, 120*, 545–558.

Evans, R. M., Barish, G. D., & Wang, Y. X. (2004). PPARs and the complex journey to obesity. *Nature Medicine, 10*, 355–361.

Feige, J. N., Lagouge, M., Canto, C., Strehle, A., Houten, S. M., Milne, J. C., et al. (2008). Specific SIRT1 activation mimics low energy levels and protects against diet-induced metabolic disorders by enhancing fat oxidation. *Cell Metabolism, 8*, 347–358.

Finkel, T. (2003). Ageing: A toast to long life. *Nature, 425*, 132–133.

Gottlieb, S., & Esposito, R. E. (1989). A new role for a yeast transcriptional silencer gene, SIR2, in regulation of recombination in ribosomal DNA. *Cell, 56*, 771–776.

Grundy, S. M., Brewer, H. B., Jr., Cleeman, J. I., Smith, S. C., Jr., & Lenfant, C. (2004). Definition of metabolic syndrome: Report of the National Heart, Lung, and Blood Institute/American Heart Association conference on scientific issues related to definition. *Circulation, 109*, 433–438.

Guarente, L. (2007). Sirtuins in aging and disease. *Cold Spring Harbor Symposia on Quantitative Biology, 72*, 483–488.

Haigis, M. C., & Sinclair, D. A. (2010). Mammalian Sirtuins: Biological insights and disease relevance. *Annual Review of Pathology, 5*, 253–295.

Horton, J. D. (2002). Sterol regulatory element-binding proteins: Transcriptional activators of lipid synthesis. *Biochemical Society Transactions, 30*, 1091–1095.

Horton, J. D., Bashmakov, Y., Shimomura, I., & Shimano, H. (1998). Regulation of sterol regulatory element binding proteins in livers of fasted and refed mice. *Proceedings of the National Academy of Sciences of the United States of America, 95*, 5987–5992.

Houtkooper, R. H., Pirinen, E., & Auwerx, J. (2012). Sirtuins as regulators of metabolism and healthspan. *Nature Reviews. Molecular Cell Biology, 13*, 225–238.

Hylemon, P. B., Zhou, H., Pandak, W. M., Ren, S., Gil, G., & Dent, P. (2009). Bile acids as regulatory molecules. *Journal of Lipid Research, 50*, 1509–1520.

Imai, S., Armstrong, C. M., Kaeberlein, M., & Guarente, L. (2000). Transcriptional silencing and longevity protein Sir2 is an NAD-dependent histone deacetylase. *Nature, 403*, 795–800.

Kaeberlein, M., McVey, M., & Guarente, L. (1999). The SIR2/3/4 complex and SIR2 alone promote longevity in Saccharomyces cerevisiae by two different mechanisms. *Genes & Development, 13*, 2570–2580.

Kahn, B. B., Alquier, T., Carling, D., & Hardie, D. G. (2005). AMP-activated protein kinase: Ancient energy gauge provides clues to modern understanding of metabolism. *Cell Metabolism, 1*, 15–25.

Kalaany, N. Y., & Mangelsdorf, D. J. (2006). LXRS and FXR: The yin and yang of cholesterol and fat metabolism. *Annual Review of Physiology, 68*, 159–191.

Kemper, J. K. (2011). Regulation of FXR transcriptional activity in health and disease: Emerging roles of FXR cofactors and post-translational modifications. *Biochimica et Biophysica Acta, 1812*, 842–850.

Kemper, J. K., Xiao, Z., Ponugoti, B., Miao, J., Fang, S., Kanamaluru, D., et al. (2009). FXR acetylation is normally dynamically regulated by p300 and SIRT1 but constitutively elevated in metabolic disease states. *Cell Metabolism, 10*, 392–404.

Klar, A. J., Fogel, S., & Macleod, K. (1979). MAR1—A regulator of the HMa and HMalpha loci in Saccharomyces cerevisiae. *Genetics, 93*, 37–50.

Kwon, H. S., & Ott, M. (2008). The ups and downs of SIRT1. *Trends in Biochemical Sciences, 33*, 517–525.

Lagouge, M., Argmann, C., Gerhart-Hines, Z., Meziane, H., Lerin, C., Daussin, F., et al. (2006). Resveratrol improves mitochondrial function and protects against metabolic disease by activating SIRT1 and PGC-1alpha. *Cell, 127*, 1109–1122.

Lee, J., & Kemper, J. K. (2010). Controlling SIRT1 expression by microRNAs in health and metabolic disease. *Aging (Albany NY), 2*, 527–534.

Lee, F. Y., Lee, H., Hubbert, M. L., Edwards, P. A., & Zhang, Y. (2006). FXR, a multipurpose nuclear receptor. *Trends in Biochemical Sciences, 31*, 572–580.

Lee, J., Padhye, A., Sharma, A., Song, G., Miao, J., Mo, Y. Y., et al. (2010). A pathway involving farnesoid X receptor and small heterodimer partner positively regulates hepatic Sirtuin 1 levels via microRNA-34a inhibition. *The Journal of Biological Chemistry, 285*, 12604–12611.

Lefebvre, P., Cariou, B., Lien, F., Kuipers, F., & Staels, B. (2009). Role of bile acids and bile acid receptors in metabolic regulation. *Physiological Reviews, 89*, 147–191.

Li, X., Zhang, S., Blander, G., Tse, J. G., Krieger, M., & Guarente, L. (2007). SIRT1 deacetylates and positively regulates the nuclear receptor LXR. *Molecular Cell, 28*, 91–106.

Liu, Y., Dentin, R., Chen, D., Hedrick, S., Ravnskjaer, K., Schenk, S., et al. (2008). A fasting inducible switch modulates gluconeogenesis via activator/coactivator exchange. *Nature, 456*, 269–273.

Mangelsdorf, D. J., & Evans, R. M. (1995). The RXR heterodimers and orphan receptors. *Cell, 83*, 841–850.

Meshkani, R., & Adeli, K. (2009). Hepatic insulin resistance, metabolic syndrome and cardiovascular disease. *Clinical Biochemistry, 42*, 1331–1346.

Neilson, J. R., & Sharp, P. A. (2008). Small RNA regulators of gene expression. *Cell, 134*, 899–902.

Noriega, L. G., Feige, J. N., Canto, C., Yamamoto, H., Yu, J., Herman, M. A., et al. (2011). CREB and ChREBP oppositely regulate SIRT1 expression in response to energy availability. *EMBO Reports, 12*, 1069–1076.

Pfluger, P. T., Herranz, D., Velasco-Miguel, S., Serrano, M., & Tschop, M. H. (2008). Sirt1 protects against high-fat diet-induced metabolic damage. *Proceedings of the National Academy of Sciences of the United States of America, 105*, 9793–9798.

Ponugoti, B., Kim, D. H., Xiao, Z., Smith, Z., Miao, J., Zang, M., et al. (2010). SIRT1 deacetylates and inhibits SREBP-1C activity in regulation of hepatic lipid metabolism. *The Journal of Biological Chemistry, 285*, 33959–33970.

Purushotham, A., Schug, T. T., Xu, Q., Surapureddi, S., Guo, X., & Li, X. (2009). Hepatocyte-specific deletion of SIRT1 alters fatty acid metabolism and results in hepatic steatosis and inflammation. *Cell Metabolism, 9*, 327–338.

Purushotham, A., Xu, Q., Lu, J., Foley, J. F., Yan, X., Kim, D. H., et al. (2012). Hepatic deletion of SIRT1 decreases hepatocyte nuclear factor 1alpha/farnesoid X receptor signaling and induces formation of cholesterol gallstones in mice. *Molecular and Cellular Biology, 32*, 1226–1236.

Repa, J. J., Liang, G., Ou, J., Bashmakov, Y., Lobaccaro, J. M., Shimomura, I., et al. (2000). Regulation of mouse sterol regulatory element-binding protein-1c Gene (SREBP-1c) by oxysterol receptors, LXRalpha and LXRbeta. *Genes & Development, 14*, 2819–2830.

Rodgers, J. T., Lerin, C., Gerhart-Hines, Z., & Puigserver, P. (2008). Metabolic adaptations through the PGC-1 alpha and SIRT1 pathways. *FEBS Letters, 582*, 46–53.

Rodgers, J. T., Lerin, C., Haas, W., Gygi, S. P., Spiegelman, B. M., & Puigserver, P. (2005). Nutrient control of glucose homeostasis through a complex of PGC-1alpha and SIRT1. *Nature, 434*, 113–118.

Rodgers, J. T., & Puigserver, P. (2007). Fasting-dependent glucose and lipid metabolic response through hepatic sirtuin 1. *Proceedings of the National Academy of Sciences of the United States of America, 104*, 12861–12866.

Russell, D. W. (1999). Nuclear orphan receptors control cholesterol catabolism. *Cell, 97*, 539–542.

Russell, D. W. (2003). The enzymes, regulation, and genetics of bile acid synthesis. *Annual Review of Biochemistry, 72*, 137–174.

Russell, J. C., & Proctor, S. D. (2006). Small animal models of cardiovascular disease: Tools for the study of the roles of metabolic syndrome, dyslipidemia, and atherosclerosis. *Cardiovascular Pathology, 15*, 318–330.

Russell, D. W., & Setchell, K. D. R. (1992). Bile acid biosynthesis. *Biochemistry (NY), 31*, 4737–4749.

Shimano, H., Horton, J. D., Shimomura, I., Hammer, R. E., Brown, M. S., & Goldstein, J. L. (1997). Isoform 1c of sterol regulatory element binding protein is less active than isoform 1a in livers of transgenic mice and in cultured cells. *The Journal of Clinical Investigation, 99*, 846–854.

Sinclair, D. A., & Guarente, L. (1997). Extrachromosomal rDNA circles—A cause of aging in yeast. *Cell, 91*, 1033–1042.

Sinclair, D. A., Lin, S. J., & Guarente, L. (2006). Life-span extension in yeast. *Science, 312*, 195–197 author reply 195–197.

Stefan, N., Kantartzis, K., & Haring, H. U. (2008). Causes and metabolic consequences of fatty liver. *Endocrine Reviews, 29*, 939–960.

Tanoli, T., Yue, P., Yablonskiy, D., & Schonfeld, G. (2004). Fatty liver in familial hypobetalipoproteinemia: Roles of the APOB defects, intra-abdominal adipose tissue, and insulin sensitivity. *Journal of Lipid Research, 45*, 941–947.

Timmers, S., Konings, E., Bilet, L., Houtkooper, R. H., van de Weijer, T., Goossens, G. H., et al. (2011). Calorie restriction-like effects of 30 days of resveratrol supplementation on energy metabolism and metabolic profile in obese humans. *Cell Metabolism, 14*, 612–622.

Tissenbaum, H. A., & Guarente, L. (2001). Increased dosage of a Sir-2 gene extends lifespan in Caenorhabditis elegans. *Nature, 410*, 227–230.

Tontonoz, P., & Mangelsdorf, D. J. (2003). Liver X receptor signaling pathways in cardiovascular disease. *Molecular Endocrinology, 17*, 985–993.

Trajkovski, M., Hausser, J., Soutschek, J., Bhat, B., Akin, A., Zavolan, M., et al. (2011). MicroRNAs 103 and 107 regulate insulin sensitivity. *Nature, 474*, 649–653.

Uyeda, K., & Repa, J. J. (2006). Carbohydrate response element binding protein, ChREBP, a transcription factor coupling hepatic glucose utilization and lipid synthesis. *Cell Metabolism, 4*, 107–110.

Walker, A. K., Yang, F., Jiang, K., Ji, J. Y., Watts, J. L., Purushotham, A., et al. (2010). Conserved role of SIRT1 orthologs in fasting-dependent inhibition of the lipid/cholesterol regulator SREBP. *Genes & Development, 24*, 1403–1417.

Wang, R. H., Kim, H. S., Xiao, C., Xu, X., Gavrilova, O., & Deng, C. X. (2011). Hepatic Sirt1 deficiency in mice impairs mTorc2/Akt signaling and results in hyperglycemia, oxidative damage, and insulin resistance. *The Journal of Clinical Investigation, 121*, 4477–4490.

Wang, R. H., Li, C., & Deng, C. X. (2010). Liver steatosis and increased ChREBP expression in mice carrying a liver specific SIRT1 null mutation under a normal feeding condition. *International Journal of Biological Sciences, 6*, 682–690.

Watanabe, M., Houten, S. M., Wang, L., Moschetta, A., Mangelsdorf, D. J., Heyman, R. A., et al. (2004). Bile acids lower triglyceride levels via a pathway involving FXR, SHP, and SREBP-1c. *The Journal of Clinical Investigation, 113*, 1408–1418.

Yamakuchi, M., Ferlito, M., & Lowenstein, C. J. (2008). MiR-34a repression of SIRT1 regulates apoptosis. *Proceedings of the National Academy of Sciences of the United States of America, 105*, 13421–13426.

Yamamoto, H., Schoonjans, K., & Auwerx, J. (2007). Sirtuin functions in health and disease. *Molecular Endocrinology, 21*, 1745–1755.

Yoshikawa, T., Shimano, H., Amemiya-Kudo, M., Yahagi, N., Hasty, A. H., Matsuzaka, T., et al. (2001). Identification of liver X receptor-retinoid X receptor as an activator of the sterol regulatory element-binding protein 1c gene promoter. *Molecular and Cellular Biology, 21*, 2991–3000.

Yoshino, J., Mills, K. F., Yoon, M. J., & Imai, S. (2011). Nicotinamide mononucleotide, a key NAD(+) intermediate, treats the pathophysiology of diet- and age-induced diabetes in mice. *Cell Metabolism, 14*, 528–536.

Zhang, Y., Castellani, L. W., Sinal, C. J., Gonzalez, F. J., & Edwards, P. A. (2004). Peroxisome proliferator-activated receptor-gamma coactivator 1alpha (PGC-1alpha) regulates triglyceride metabolism by activation of the nuclear receptor FXR. *Genes & Development, 18*, 157–169.

CHAPTER SEVENTEEN

Protein Tyrosine Phosphatase 1B (PTP1B) and Obesity

Hyeongjin Cho[1]
Department of Chemistry, Inha University, Incheon, Korea
[1]Corresponding author: e-mail address: hcho@inha.ac.kr

Contents

Abstract

Protein tyrosine phosphatase 1B (PTP1B) is a negative regulator of the leptin and insulin signaling pathways. The important roles of PTP1B related to obesity and diabetes were confirmed by a deletion of PTP1B gene in mice. Mice with the whole body deletion of PTP1B were protected against the development of obesity and diabetes. When PTP1B gene was deleted selectively in the brain of mice, the major effects on weight and glucose control were consistent with the whole body deletion of PTP1B. This is in contrast to the muscle-, liver-, and adipocyte-specific deletion, which had no beneficial effects on obesity. While these results indicate the importance of neuronal PTP1B in maintaining energy homeostasis, the peripheral PTP1B is also being investigated for their potential roles in the control of energy balance. Validation of PTP1B as a therapeutic target for obesity and diabetes prompted efforts to develop potent and selective inhibitors of PTP1B. Among the small molecule inhibitors investigated, trodusquemine, which acts

Vitamins and Hormones, Volume 91
ISSN 0083-6729
http://dx.doi.org/10.1016/B978-0-12-407766-9.00017-1

both centrally and peripherally, is currently in phase 2 clinical trials. An approach using PTP1B-directed antisense oligonucleotides is also in phase 2 clinical trials.

1. INTRODUCTION

Protein tyrosine phosphatases (PTPs) are a family of enzymes that catalyze the dephosphorylation of tyrosine (Tyr)-phosphorylated proteins. Compared to the protein tyrosine kinases (PTKs), which have been known for more than half a century, the cellular presence of PTP was recognized much later in 1988 by the purification of PTP1B from human placenta (Tonks, Diltz, et al., 1988). Biological studies have shown that PTP1B is expressed in multiple tissues and is involved in multiple signal transduction pathways (Kenner, Anyanwu, et al., 1996). Among those, PTP1B is a key enzyme in the dephosphorylation of the insulin receptor (IR) and its downstream signaling components (Elchebly, Payette, et al., 1999). PTP1B is also involved in the negative regulation of the leptin signaling pathway by dephosphorylating the upstream signaling molecules (Zabolotny, Bence-Hanulec, et al., 2002). The important roles of PTP1B related to diabetes and obesity were further confirmed by a genetic deletion of PTP1B in mice (Elchebly, Payette, et al., 1999; Klaman, Boss, et al., 2000). PTP1B-deficient mice were resistant to weight gain and remained insulin sensitive on a high-fat diet (HFD). This is in contrast to the wild-type mice, which became obese and insulin resistant on the same diet. Together with the global prevalence of obesity and diabetes, the establishment of PTP1B as a therapeutic target for these metabolic syndromes prompted studies of small molecule inhibitors of PTP1B by pharmaceutical companies and academic laboratories.

2. PTP1B

2.1. Classification of PTPs

An analysis of the human genome sequence revealed 107 PTP genes, among which 38 are Tyr-specific. These can be classified further into two classes as transmembrane and nonreceptor PTPs with 21 and 17 members, respectively (Alonso, Sasin, et al., 2004). Transmembrane PTPs generally have an extracellular domain, a transmembrane domain, and one or, more often, two cytoplasmic catalytic domains. Nonreceptor PTPs, exemplified by PTP1B, contain a single catalytic domain with N- or C-terminal extensions (Fischer, Charbonneau, et al., 1991).

2.2. Structure and subcelluar localization of PTP1B

PTP1B has a C-terminal side extension containing a proline-rich region and a hydrophobic region. The proline-rich region interacts with the Src homology 3 domains to recruit substrates (Liu, Hill, et al., 1996). The hydrophobic sequence is inserted into the endoplasmic reticulum (ER) membrane and tethers PTP1B to the cytoplasmic side of the ER (Frangioni, Beahm, et al., 1992). The substrate selectivity of PTP1B is controlled partly by the modular protein interaction and the spatial restriction imposed by subcellular compartmentalization (Tonks, 2003).

3. BIOLOGICAL ROLES OF PTP1B

3.1. PTP1B in insulin signal transduction

PTP1B is expressed in multiple tissues including the skeletal muscle, liver, adipose tissue, and brain. PTP1B is involved in multiple signal transduction pathways, as anticipated by the presence of PTP1B in a range of tissues (Kenner et al., 1996). Among these, deregulation of the insulin and leptin signalings are important processes of PTP1B related to diabetes and obesity (Asante-Appiah, Kennedy, et al., 2003; Zabolotny et al., 2002). When insulin binds to its receptor, the cytosolic tyrosine kinase is activated by the autophosphorylation of three clustered Tyr residues within the kinase domain, which in turn phosphorylates the target proteins, such as IR substrate (IRS) proteins. IRS phosphorylation induces the sequential activation of phosphatidylinositol 3-kinase (PI3K) and AKT, which leads to further downstream events, such as translocation of the glucose transporter to the cell surface and stimulation of glycogen synthesis (Asante-Appiah et al., 2003). The removal of phosphates from the phosphotyrosine (pTyr) residues of IR is an important mechanism for the negative regulation of the IR. PTP1B was established as a key enzyme in the dephosphorylation of IR, as well as its downstream signaling component, IRS-1 (Elchebly et al., 1999).

3.2. PTP1B in leptin signal transduction

The growing interest in obesity control has brought into focus the role of PTP1B as a deregulator of leptin signaling (Morris & Rui, 2009). Leptin is synthesized and secreted mainly from the adipocytes of white adipose tissue (Zhang, Proenca, et al., 1994). White adipose tissue (WAT), along with brown adipose tissue (BAT), is a major site to store fat as triglycerides and cholesteryl ester. Instead of being limited to the role of fat stores, WAT also secretes leptin and other hormones known as adipokines, including adiponectin, IL-6, and TNF-α (Wellen & Hotamisligil, 2005). The leptin levels in the blood are proportional to the body fat mass and are normally elevated in human obesity (Considine, Sinha, et al., 1996). Leptin acts on the receptors in the hypothalamus of the brain to regulate body weight and energy expenditure. Although the main site of leptin action is the hypothalamus, the leptin receptor (LepR) is also observed in peripheral tissues (Tartaglia, Dembski, et al., 1995). Multiple isoforms of LepRs are produced as a result of the alternate splicing of the obese (*db*) gene. Among those, the long isoform (Ob-Rb) exists at high levels in the hypothalamus and plays a critical role in regulating the metabolic rate (Houseknecht, Portocarrero, et al., 1998).

The LepR does not exhibit intrinsic kinase activity and mediates the signal by Janus kinase 2 (JAK2), a PTK associated with the cytoplasmic region of the receptor (Ghilardi & Skoda, 1997). Leptin binding to its receptor causes the autophosphorylation and activation of JAK2, as well as the subsequent phosphorylation of Tyr residues on the cytoplasmic region of LepR. pTyr residues on LepR act as docking sites for downstream signaling proteins, such as STAT3, STAT5, and SHP-2, to branch the leptin signaling pathway. STAT3 and STAT5, when phosphorylated by JAK2, dimerize and translocate into the nucleus to induce the transcription of the target genes (Sahu, 2011). Phosphorylation of LepR-bound SHP-2 by JAK2 causes activation of the ERK pathway (Banks, Davis, et al., 2000). Autophosphorylated Tyr residue on JAK2 also acts as a docking site for an adaptor protein SH2B1 which, in turn, associates with IRS-1 or IRS-2 (Morris & Rui, 2009). This complex formation leads to the activation of PI3K, suggesting a potential cross talk between insulin and leptin signaling (Niswender, Morton, et al., 2001). Branches downstream of PI3K include AKT–FOXO, PDE3B-cAMP, and mTOR (Cota, Proulx, et al., 2006; Kim, Pak, et al., 2006; Kitamura, Feng, et al., 2006; Zhao, Huan, et al., 2002). Leptin also causes inhibition of α2AMPK activity in some regions of the hypothalamus, and this regulates food intake and body weight (Minokoshi, Alquier, et al., 2004). Even though the role of leptin is not limited in brain and leptin exerts its effects via multiple pathways, the JAK2–STAT3 pathway in the hypothalamus is crucial in maintaining energy homeostasis, evidenced by the morbid obesity of mice lacking STAT3 in brain (Gao, Wolfgang, et al., 2004).

Dephosphorylation of JAK2 is an important mechanism for terminating leptin signal transduction (Zabolotny et al., 2002). JAK2 was indicated as a direct substrate of PTP1B based on the observation of coimmunoprecipitation of JAK2 and catalytically inactive PTP1B. In another study, PTP1B overexpression in a mouse hypothalamic cell line caused a decrease in the Tyr phosphorylation of JAK2 and STAT3. Consistent with this observation, PTP1B overexpression resulted in a decrease in the mRNA levels of leptin-inducible genes and an increase in the mRNA levels of genes downregulated by leptin (Kaszubska, Falls, et al., 2002).

3.3. Other functions of PTP1B

Recent studies identified a role of PTP1B as a negative modulator of brown fat adipogenesis. PTP1B-deficient mice exhibited a trend for enhanced differentiation of BAT. However, the mice expressing proteolysis-resistant PTP1B failed to differentiate. Given that active BAT is present in healthy adult humans and plays a significant role in metabolism, stimulation of BAT differentiation might help increasing energy expenditure, thus providing additional rationale for the inhibition of PTP1B to control obesity (Matsuo, Bettaieb, et al., 2011; Nedergaard, Bengtsson, et al., 2007).

4. GENETIC DELETION OF PTP1B IN MICE

4.1. Deletion of PTP1B in the whole body of mice

The biological role of PTP1B as a regulator of body weight and glucose homeostasis was examined by the genetic ablation of PTP1B in all tissues of mice. The mouse homolog of the PTP1B gene was disrupted completely (PTP1B$^{-/-}$) or in one of the two haploid chromosomes (PTP1B$^{+/-}$). In the heterozygotic (PTP1B$^{+/-}$) mice, the PTP1B expression level was approximately half that of the wild-type mice. The whole body PTP1B-deficient mice were born and grew healthy, decreasing the concern regarding the possible toxic effect of PTP1B inhibition. The PTP1B-deficient mice were resistant to weight gain, and enhanced insulin sensitivity was observed in both PTP1B$^{-/-}$ and PTP1B$^{+/-}$ mice (Elchebly, Payette, et al., 1999; Klaman, Boss, et al., 2000). The antiobesity effects of the whole body PTP1B knockouts were accompanied by a decrease in the adipocyte cell mass without a change in cell number. In female PTP1B$^{+/-}$ mice, some discrepancies in the effects on weight gain were observed between two laboratories, but they are possibly due to differences in experimental conditions including the genetic background of the mice and dietary fat content.

The excess energy obtained by PTP1B-deficient mice was apparently dissipated as heat, rather than being stored as fat. The lower metabolic efficiency was not due to an increase in uncoupling protein (UCP) expression, which is the most common mechanism for heat production. In PTP1B-deficient mice, the leptin levels in serum were maintained low, which is consistent with their decreased adiposity. Interestingly, PTP1B$^{-/-}$ mice on a HFD exhibited much lower leptin levels than those of the wild-type control on a chow diet. The mechanism underlying the increase in metabolic rate needs to be determined. However, Klaman et al. raised the possibility of PTP1B regulating distinct signaling pathways as well as unidentified roles of insulin, based on their observation of no remarkable increase in BAT mass and UCP expression in BAT and skeletal muscle in PTP1B$^{-/-}$ mice (Klaman, Boss, et al., 2000).

4.2. Deletion of PTP1B in the specific tissues of mice

To identify the tissues responsible for the phenotype of the whole body PTP1B-knockouts, transgenic mice were generated with the PTP1B gene deleted selectively in the skeletal muscle, brain, adipocyte, and liver. Table 17.1 summarizes the selected metabolic responses of the whole body and tissue-specific mutant mice to HFD feeding. Muscle- and liver-specific PTP1B deletion in mice had no beneficial effects on the body weight or adipocity, except an improvement of the lipid profiles in a liver-specific deletion but not in a muscle-specific deletion. Improved glucose tolerance was observed in both of the muscle- and liver-specific deletion mutants with the enhancement of insulin signaling in the respective tissues. Compared to the glucose tolerance, the insulin tolerance was improved only in a muscle-specific PTP1B deletion but not in a liver-specific deletion (Delibegovic et al., 2007).

The absence of PTP1B in the liver also attenuated the HFD-induced ER stress. ER stress refers to the conditions impeding the function of ER and can have many causes including obesity. ER stress upregulates PTP1B expression. Recent studies suggested an association of ER stress in the pathogenesis of diabetes. Based on these, the attenuation of ER stress was suggested to be one of the underlying mechanisms for the improvement of insulin sensitivity and lipid profiles in a liver-specific deletion of PTP1B (Delibegovic et al., 2009). The attenuation of ER stress by PTP1B depletion in other tissues might also be a possible mechanism for the antiobesity effects of PTP1B deficiency in mice. ER stress was further suggested to be a possible link between obesity and insulin resistance (Eizirik, Cardozo, et al., 2008).

Table 17.1 Effects of a deletion of the PTP1B gene in the whole body or specific tissues in mice

		Antiobesity effect		Insulin sensitivity	
PTP1B deleted tissue	**Genotype**	**M**	**F**	**M**	**F**
Whole body[a]	$PTP1B^{-/-}$	Yes	Yes	↑	
	$PTP1B^{+/-}$	Yes	Yes	↑	
Whole body[b]	$PTP1B^{-/-}$	Yes	Yes	↑	↑
	$PTP1B^{+/-}$	Yes	No		
Muscle[c]	$PTP1B^{-/-}$	No		↑	
	$PTP1B^{+/-}$	No		↑	
Muscle[d]	$PTP1B^{-/-}$	No		↑[f]	
Neuron[d]	$PTP1B^{-/-}$	Yes	Yes	↑	↑
	$PTP1B^{+/-}$	Yes	Yes		
Adipocyte[d]	$PTP1B^{-/-}$	No[g]			
Liver[e]	$PTP1B^{-/-}$	No[d]		↑[h]	
	$PTP1B^{+/-}$			↑[h]	
POMC[f]	$PTP1B^{-/-}$	Yes[i]	Yes	↑	
	$PTP1B^{+/-}$	No	No[j]		

[a]Elchebly et al., 1999.
[b]Klaman et al. (2000).
[c]Delibegovic, Bence, et al. (2007).
[d]Bence, Delibegovic, et al. (2006) and Owen, Czopek, et al. (2012).
[e]Delibegovic, Zimmer, et al. (2009).
[f]Banno, Zimmer, et al. (2010).
[g]More weight gain compared to the wild-type.
[h]Glucose tolerance enhanced but insulin tolerance remained unchanged.
[i]Leptin sensitivity increased.
[j]Tendency to suppress the diet-induced weight gain, but the suppression was not statistically significant.

An adipocyte-specific deletion of PTP1B in mice enhanced the body weight gain compared to wild-type mice (Bence et al., 2006; Owen et al., 2012). Interestingly, an adipocyte-specific deletion of the IR in mice resulted in resistance to weight gain with a reduced fat mass, as well as resistance to obesity-related glucose intolerance (Blüher, Dodson Michael, et al., 2002).

A neuronal-specific PTP1B deletion in mice reduced their weight and adipocity, increased their activity and energy expenditure, and improved glucose homeostasis (Bence et al., 2006). These effects are consistent with the effects of a whole body deletion of PTP1B, suggesting that the inhibition of neuronal PTP1B is sufficient for appreciable weight reduction and is essential for optimal control of glucose homeostasis. Both neuronal and whole body PTP1B$^{-/-}$ mice also showed enhanced leptin sensitivity. A neuronal PTP1B deletion increased leptin signaling in the hypothalamus and corresponding biological responses, which was consistent with a whole body deletion. Upon leptin injection, the neuronal PTP1B$^{-/-}$ mice showed an enhanced suppression of body weight gain and food intake, which was consistent with the whole body knockouts. As a crucial regulator for the systemic effects of leptin, neuronal PTP1B acts directly on the leptin signaling pathway (Bence et al., 2006).

Mice with pro-opiomelanocortin (POMC) neuron-specific deletion of the gene encoding PTP1B were generated to examine the function of PTP1B in specific neurons that control the energy balance. On a HFD, both male and female POMC–PTP1B$^{-/-}$ mice showed significantly lower body weight gain and fat mass due to the increased energy expenditure and enhanced leptin sensitivity (Banno et al., 2010; De Jonghe, Hayes, et al., 2011). These results suggest that POMC neurons are important sites of PTP1B action, even though it is unclear if they were the results of enhanced neuronal leptin signaling or other signaling pathways, such as insulin signaling. The central action of insulin regulates both energy and glucose homeostasis (Plum, Belgardt, et al., 2006).

5. INHIBITORS OF PTP1B

5.1. pTyr mimetics in PTP inhibitors

pTyr residue is the key element for the substrate recognition by PTPs. A range of scaffolds were developed as pTyr analogs that bind to the active site of PTPs without being hydrolyzed (Fig. 17.1). As an initial approach, replacement of the phosphate group with methylene phosphonate moiety, as in **2**, prevented hydrolysis of the phosphate group from pTyr (Burke, Kole, et al., 1994). The introduction of two F atoms on the methylene carbon, as in **3**, improved the inhibitory potency by compensating for the decrease in electronegativity from a change of O to C. The difluoromethylene phosphate (DFMP) -containing PTP inhibitors are currently being assessed by Merck Frosst. Thiophosphoryl or sulfotyrosylated peptides were PTP inhibitors prepared according to the same paradigm (Hiriyanna, Baedke, et al., 1994; Liotta, Kole, et al., 1994). A highly basic environment in the active site of PTPs favors inhibitors containing phosphonates or modified phosphates.

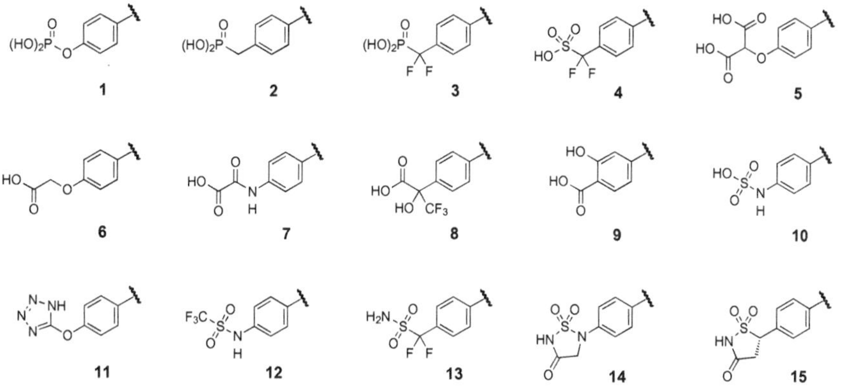

Figure 17.1 pTyr mimetics in PTP inhibitors.

However, the highly anionic nature of the phosphonates might be an obstacle for inhibitors to move across the cellular plasma membrane. Prodrug approaches to improve intracellular delivery are described below. A difluoromethylenesulfonic acid group (**4**) was more than 1000-fold less effective in mimicking pTyr than the DFMP moiety (Hussain, Ahmed, et al., 2008).

Carboxylate was also introduced as pTyr mimetics. Malonate derivative (**5**) included two carboxylate moieties, which might be beneficial for binding to the active site of PTPs but not favorable for cell penetration (Akamatsu, Roller, et al., 1997). Monocarboxylates (**6–9**) achieved improved cell permeability at the cost of sacrificing the potency of inhibition (Adams, Abraham, et al., 2007; Akamatsu et al., 1997; Andersen, Iversen, et al., 2000; Burke, Yao, et al., 1998, Iversen, Andersen, et al., 2000; Shrestha, Bhattarai, et al., 2007a, 2007b). Sulfamic acid (**10**) was reported to be a phosphate mimic, but these compounds were strong inhibitors of PTPs only when additional sulfamic acid or carboxylate functionality is present within the same molecule (Klopfenstein, Evdokimov, et al., 2006).

Noncharged phosphate mimics were also developed to improve the cell permeability of the inhibitors. Tetrazole (**11**) and trifluoromethylsulfonamide (**12**) are the earliest of those. (Aminosulfonyl) difluoromethyl moiety (**13**) was reported recently to replace carboxylate (Huang, Ramphal, et al., 2003; Liljebris, Larsen, et al., 2002; Patel, Jain, et al., 2012). These phosphate or carboxylate mimetics were moderately successful in realizing both cell permeability and inhibitory potency. 1,2,5-Thiazolidin-3-one 1,1-dioxide (TZD, **14**) and (S)-isomer of isothiazolidinone ((S)-IZD, **15**) are improved versions of pTyr mimetics that achieved modest cell permeability without sacrificing the inhibitory activity against PTPs. Several pharmaceutical companies are actively investigating TZD and ITZ containing compounds as PTP1B-targeting drug candidates. These compounds are discussed below.

5.2. Small molecule inhibitors of PTP1B

An inspection of PTPs revealed that they share high homology in both the amino acid sequence and three-dimensional structure. Given the cellular expression of multiple PTPs, the development of PTP1B-selective inhibitors has been a challenge for medicinal chemists. Most challenging is achieving selectivity against TC-PTP, which is the most homologous to PTP1B with >70% sequence identity in their catalytic domains and 94% identity for the active site residues (Andersen, Mortensen, et al., 2001; Barr, 2010; Iversen, Møller, et al., 2002). One of the distinguishing structural features of PTP1B is the presence of additional noncatalytic phosphate-binding sites near the active site (Puius, Zhao, et al., 1997). These are often called secondary and third aryl phosphate-binding sites, or more briefly as B and C sites with the active site being called the A site. Compared to the deep cleft of the active site, the B site is shallower and the C site is less clearly defined by the presence of positively charged amino acid residues, Arg47 and Lys41. The B and C sites are not present in most PTPs except in TC-PTP (Puius et al., 1997). These sites have been exploited to differentiate PTP1B from most PTPs. Although the B and C sites are also present in TC-PTP, small differences in the amino acid residues in these sites of PTP1B and TC-PTP suggest the possibility for differentiation of the two closely related PTPs (Iversen et al., 2002; Lau, Bayly, et al., 2004).

Compound **A** uses two DFMP moieties as phosphate mimics to achieve inhibitory potency against PTP1B (Ki=2.4 nM) and showed 11-fold selectivity over TC-PTP. The DFMP moieties were shown to occupy the A and C sites of PTP1B in the **A**-PTP1B crystal structure (Shen, Keng, et al., 2001). The highly anionic nature of the phosphonates achieved nanomolar inhibitory potency but at the same time prevented the inhibitor from moving across the cellular membrane. To circumvent the low cell permeability of DFMP compounds, the phosphate moiety was modified with protecting groups that could provide inhibitors with sufficient lipophilicity for cell penetration. Moreover, once inside the cytoplasm, the moiety is removed via cellular enzymes to regenerate the original phosphonate. This prodrug approach was successful in a recent study demonstrating the intracellular delivery and *in vivo* efficacy of a DFMP-based PTP1B inhibitor (Boutselis, Yu, et al., 2007).

Merck Frosst reported compound **B** as a mixture of four diastereomers, all of which are potent inhibitors of PTP1B. The most potent among the isomers exhibited an IC_{50} value of 5 nM against PTP1B and sevenfold selectivity over TC-PTP. The two pTyr mimetic groups were embedded in the A and B sites of PTP1B, as demonstrated by an X-ray crystallographic study (Lau et al., 2004).

Compound **C**, which was developed by Abbott lab, contained carboxylate moieties as a replacement for the phosphate moiety. Reduced potency was observed (Ki = 40 nM against PTP1B) with 6-fold selectivity over TC-PTP. The methoxy salicylate moiety was bound in the B site of the crystal structure (Liu, Xin, et al., 2003).

Compound **D** from Wyeth contained thiophene dicarboxylate as a pTyr surrogate. A high inhibitory potency against PTP1B (Ki = 4 nM) demonstrated the effective interaction of the thiophene dicarboxylate with the active site of PTP1B. The oxygen atoms of the sulfonamide moiety interacted with the positively charged residues (Arg221 and Lys120) in the B site (Wilson, Wan, et al., 2007). No discrimination of TC-PTP was observed. The monoalkyl esters of compound **D** were examined as prodrugs, but the obesity or diabetes related effects of these esters were below the statistically significant level (Erbe, Klaman, et al., 2009).

Compound **E** was reported by Abbott lab. With a single carboxylate group, reduced inhibitory potency was observed for PTP1B (Ki = 2.1 μM). The methyl salicylate moiety occupied the B site, as in compound **C**, and exhibited >15-fold selectivity over TC-PTP. The cellular effects of PTP1B inhibition were observed for compound **E** but not for a more potent derivative (Ki = 0.92 μM) with a —NH_2 group replaced by —CH_2OH (Zhao, Liu, et al., 2004).

Potent PTP1B inhibitors possessing TZD (**14**) or (S)-IZD (**15**) were reported independently by several companies including AstraZeneca, Norvatis, Vertex, Wyeth, and Incyte. Although they are negatively charged in physiological pH, the delocalized nature of the charge was expected to ameliorate the burden of the charged molecule to diffuse across the cell membrane.

AstraZeneca used compound **F** to confirm the validity of the TZD moiety as a phosphate mimic (Black, Breed, et al., 2005). Norvatis reported compound **G** with an improved inhibitory potency (IC_{50} = 126 nM against PTP1B) utilizing the TZD moiety without selectivity or bioavailability data (Barnes and Bebernitz, 2007).

Wyeth used TZD–thiophene couple and reported that compound **H** had some cell permeability (Ki=4.3 μM) (Wan, Follows, et al., 2007). The benzylsulfonyl amide moiety of compound **H** is likely to bind to the B site, based on the drug-PTP1B crystal structure of its analogues, but no selectivity data were reported.

Compounds **I** and **J** were produced by Incyte. Compound **I** was a highly potent inhibitor of PTP1B (IC_{50}=23 nM) with an interaction with the A and B sites of the enzyme (Douty, Wayland, et al., 2008). This compound, however, showed no selectivity over TC-PTP, and no biological data were reported. Compound **J** was less potent (IC_{50}=480 nM) with low Caco-2 cell permeability (Sparks, Polam, et al., 2007).

Compound **K** from Kyorin is expected to bind to the A and B sites based on a computational docking study. Although the cell permeability data were not reported, the IC_{50} of 3.2 μM in the cell-based assay was comparable to that (IC_{50}=3.0 μM) observed in a cell-free PTP1B assay, indicating compound **K** to have good cell permeability (Adams et al., 2007).

Compounds **L–U** (Fig. 17.3) were largely from more recent publications with *in vivo* data. Most of them exhibited antiobesity and/or antidiabetic

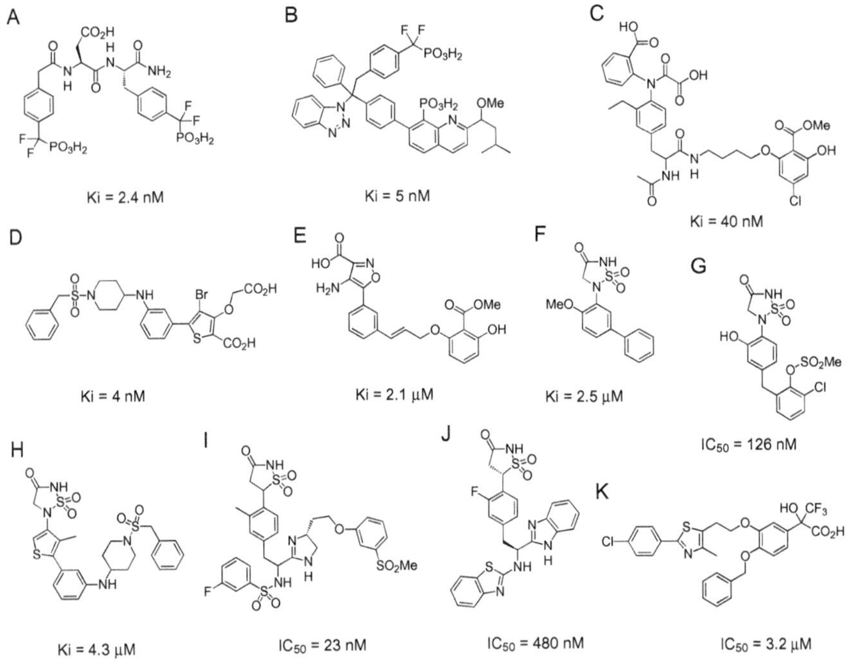

Figure 17.2 Inhibitors of PTP1B in literature 1.

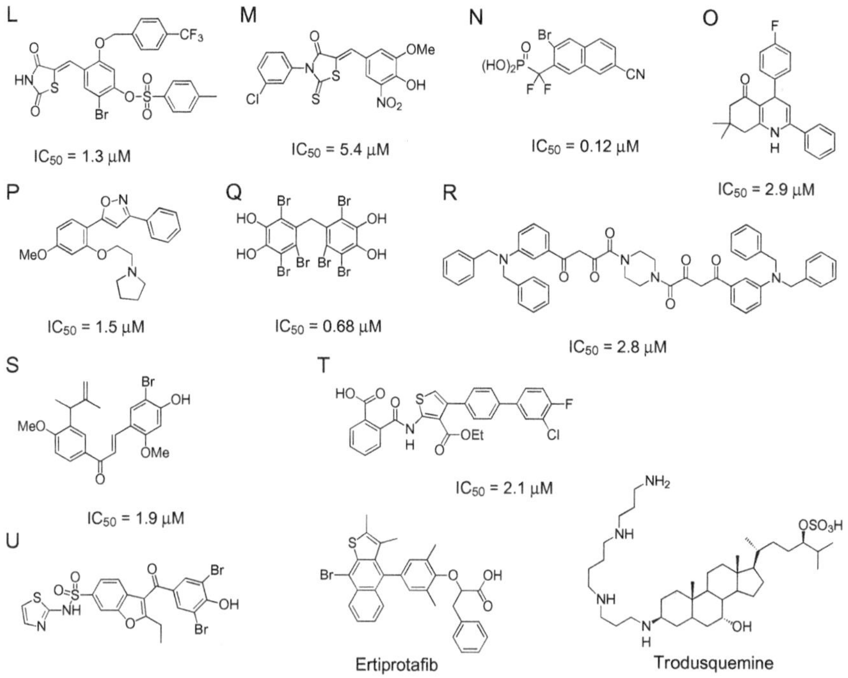

Figure 17.3 Inhibitors of PTP1B in literature 2.

effects in animal studies or cell-based studies, suggesting the cell permeable nature of the compounds. Thiazolidinedione derivatives, which are exemplified by compound **L**, were identified as novel class PTP1B inhibitors. Among these, **L** exhibited the lowest IC_{50} of 1.3 μM toward PTP1B and 3-fold selectivity over TC-PTP. In a mouse model system, compound **L** suppressed weight gain and improved lipid-related blood parameters but had no significant effect on controlling the blood glucose level (Bhattarai, Kafle, et al., 2010). However, another derivative in this series effectively improved glucose tolerance. Similar to the glitazone series of drugs, compound **L** activated PPAR-γ (Bhattarai, Kafle, et al., 2009). The mechanisms underlying the effects of thiazolidinedione derivatives might be complex and require additional study. Given that the glitazone series of drugs have tendency to induce weight gain, it is interesting to note that another thiazolidinedione derivative (not shown in Fig. 17.3) analogous to compound **L** improved the glucose tolerance while suppressing weight gain.

Compound **M** exhibited broad inhibitory activity toward cytosolic PTPs with medium potency (IC_{50}=5.4 μM). In a cell-based assay, treatment with compound **M** increased the Tyr phosphorylation level of AKT, suggesting that compound **M** could be a lead compound for the development of cell-permeable PTP1B inhibitors (Stuible, Zhao, et al., 2007).

The DFMP derivative, compound **N**, in the pipeline of Merck Frosst, exhibited an IC_{50} value of 0.12 μM against PTP1B. In a diabetic mouse model, compound **N** showed good oral bioavailability and other pharmacokinetic profiles. Furthermore, an oral dose of compound **N** resulted in an improvement in the glucose tolerance of mice (Han, Belley, et al., 2008).

Compounds **O** and **P** were reported from the same laboratory and showed IC_{50} values of 2.9 and 1.5 μM, respectively, against PTP1B. No TC-PTP selectivity data were provided. The oral administration of compounds **O** or **P** in rats showed improved lipid profiles and glucose tolerance (Kumar, Maurya, et al., 2009; Kumar, Sharma, et al., 2010).

The multiple brominated compound **Q** exhibited >70-fold PTP1B selectivity over TC-PTP. The basis of the discrimination awaits structural explanation. In a mouse model system, the oral administration of compound **Q** for a 6-week period decreased the plasma glucose levels to an extent comparable to those of rosiglitazone based on equal doses (Shi, Li, et al., 2012).

Compound **R** is in an amide-linked dimeric form of aryl diketoacid, which itself is a PTP1B inhibitor with an IC_{50} value of 20 μM. Dimeric **R** was more potent than the monomeric diketoacid and bound to the active site as the monomeric form but in a different conformation. As a neutral dimer, **R** was cell permeable and enhanced insulin signaling in hepatoma cells (Liu, Zeng, et al., 2008).

Compound **S** is one of the derivatives of natural product licochalcone, showing an IC_{50} value of 1.9 μM (Liu, Lee, et al., 2011). Thiophene derivative **T** was membrane permeable and activated PI3K/AKT pathway in a cell-based assay (Ye, Zhang, et al., 2010).

Compound **U** is an allosteric inhibitor of medium potency (IC_{50} = 8 μM) produced by Sunesis Pharmaceuticals. A crystallographic study showed that compound **U** binds 20 Å away from the active site and inhibits PTP1B by preventing the conformational changes in the catalytic loop to a catalytically active form. Compound **U** is cell permeable and enhanced IR phosphorylation in a cell-based assay. Differences in amino acids were also recognized in the allosteric site of PTP1B and the corresponding region of TC-PTP. These differences might be valuable for designing PTP1B selective inhibitors (Wiesmann, Barr, et al., 2004).

Ertiprotafib was developed by Wyeth as the first PTP1B inhibitor that progressed to clinical trials before being discontinued in the second phase.

In animal studies, ertiprotafib reduced the fasting glucose and insulin levels and improved the glucose tolerance. Ertiprotafib also lowered triglycerides and free fatty acid levels significantly. Ertiprotafib inhibited PTP1B *in vitro* at low nanomolar concentrations and exhibited inhibitory effects against PTP1B in cell-based studies. However, some inconsistencies were observed between the PTP1B-deficient and ertiprotafib-treated mice groups. For example, an ertiprotafib treatment resulted in a significant decrease in the serum-free fatty acid levels, which is in contrast to the tendency of an increase due to a PTP1B deficiency. An explanation of the inconsistencies required additional mechanisms for the action of ertiprotafib in addition to PTP1B inhibition. Ertiprotafib was found to act as a PPARα/PPARγ dual agonist and an IκB kinase β (IKK-β) inhibitor (Erbe, Wang, et al., 2005; Shrestha et al., 2007a, 2007b). Interestingly, a heterozygous deletion of the IKK-β gene in mice offered protection against the development of insulin resistance in response to obesity (Arkan, Hevener, et al., 2005; Yuan, Konstantopoulos, et al., 2001).

Trodusquemine is a PTP1B inhibitor currently in phase 2 clinical trials by Genaera as a potential treatment for obesity and diabetes (Lantz, Emeigh Hart, et al., 2010). Being identified by its antimicrobial activity from the dog fish shark, the PTP1B inhibitory activity of trodusquemine was recognized later with its allosteric mode of inhibition (Lantz, Emeigh Hart, et al., 2010; Rao, Shinnar, et al., 2000). By binding to PTP1B on a site away from the active site, trodusquemine inhibits PTP1B with 200-fold selectivity over TC-PTP. In addition to its potential to resolve selectivity issues, trodusquemine crosses the blood–brain barrier to act both peripherally and centrally to inhibit PTP1B. Trodusquemine enhanced the insulin-stimulated Tyr phosphorylation of IR in a cell-based study. In *in vivo* experiments, trodusquemine, being capable of crossing the blood–brain barrier, increased the insulin-stimulated Tyr phosphorylation of IR and STAT3 significantly in the hypothalamic tissues of mice. Trodusquemine also normalized the plasma insulin and leptin levels of DIO mice to those of the low-fat diet-fed control mice.

Trodusquemine suppressed food intake and reduced the body weight in DIO mice. When a HFD-fed obese mouse group was injected with trodusquemine, the drug-treated group consumed ~50% of the diet compared to the saline-treated control group. The decrease in body weight continued for the 3-week drug treatment, resulting in an approximate 40% decrease in body weight. In a parallel experiment, a DIO mice group was provided with the restricted amount of diet consumed daily by the trodusquemine-treated DIO mice. In the pair-fed mice group, the weight loss stopped after 2 weeks in contrast to the continuous loss during the 3 week experiment period in the trodusquemine-treated group. This suggests that the body weight loss by trodusquemine is not only the result of reduced food intake. In animals, a large weight reduction in a short time period provokes recovery mechanisms to restore the weight loss. The sustained weight reduction by trodusquemine indicated additional effects and suggests that the drug might override the feedback mechanisms to regain the body weight. The overall weight loss by trodusquemine was attributed to the selective reduction of the body fat content without a significant decrease in lean body mass.

5.3. Antisense oligonucleotides

Another clinical approach to inhibiting PTP1B is to use Antisense oligonucleotides (ASOs) with improved selectivity toward PTP1B compared to the small molecule inhibitors discussed earlier. PTP1B-directed ASO was developed by ISIS Pharmaceuticals and is currently in phase 2 clinical trials.

ACKNOWLEDGEMENTS

This work was supported by the Basic Science Research Program through the National Research Foundation of Korea (NRF) funded by the Ministry of Education, Science and Technology (2010–0022853).

REFERENCES

Adams, D. R., Abraham, A., et al. (2007). 2-Aryl-3,3,3-trifluoro-2-hydroxypropionic acids: A new class of protein tyrosine phosphatase 1B inhibitors. *Bioorganic & Medicinal Chemistry Letters, 17*(23), 6579–6583.

Akamatsu, M., Roller, P. P., et al. (1997). Potent inhibition of protein-tyrosine phosphatase by phosphotyrosine-mimic containing cyclic peptides. *Bioorganic & Medicinal Chemistry, 5* (1), 157–163.

Alonso, A., Sasin, J., et al. (2004). Protein tyrosine phosphatases in the human genome. *Cell, 117*(6), 699–711.

Andersen, H. S., Iversen, L. F., et al. (2000). 2-(Oxalylamino) benzoic acid is a general, competitive inhibitor of protein tyrosine phosphatases. *Journal of Biological Chemistry, 275*(10), 7101–7108.

Andersen, J. N., Mortensen, O. H., et al. (2001). Structural and evolutionary relationships among protein tyrosine phosphatase domains. *Molecular and Cellular Biology, 21*(21), 7117–7136.

Arkan, M. C., Hevener, A. L., et al. (2005). IKK-β links inflammation to obesity-induced insulin resistance. *Nature Medicine, 11*(2), 191–198.

Asante-Appiah, E., Kennedy, B. P., et al. (2003). Protein tyrosine phosphatases: The quest for negative regulators of insulin action. *American Journal of Physiology, Endocrinology and Metabolism, 284*(4), E663–E670.

Banks, A. S., Davis, S. M., et al. (2000). Activation of downstream signals by the long form of the leptin receptor. *Journal of Biological Chemistry, 275*(19), 14563–14572.

Banno, R., Zimmer, D., et al. (2010). PTP1B and SHP2 in POMC neurons reciprocally regulate energy balance in mice. *The Journal of Clinical Investigation, 120*(3), 720–734.
Barnes, D., Bebernitz, G. R., et al. (2007). 1,2,5-Thiadiazolidine derivatives useful for treating conditions mediated by protein tyrosine phosphatases (PTPase). PCT Int. WO2007067613, 38 pp (Novartis A.-G., Switzerland; Novartis Pharma G.m.b.H.). Novatis Institute for Biomedical Research Institute Inc., USA.
Barr, A. J. (2010). Protein tyrosine phosphatases as drug targets: Strategies and challenges of inhibitor development. *Future Medicinal Chemistry, 2*(10), 1563–1576.
Bence, K. K., Delibegovic, M., et al. (2006). Neuronal PTP1B regulates body weight, adiposity and leptin action. *Nature Medicine, 12*(8), 917–924.
Bhattarai, B. R., Kafle, B., et al. (2009). Thiazolidinedione derivatives as PTP1B inhibitors with antihyperglycemic and antiobesity effects. *Bioorganic & Medicinal Chemistry Letters, 19*(21), 6161–6165.
Bhattarai, B. R., Kafle, B., et al. (2010). Novel thiazolidinedione derivatives with anti-obesity effects: Dual action as PTP1B inhibitors and PPAR-γ activators. *Bioorganic & Medicinal Chemistry Letters, 20*(22), 6758–6763.
Black, E., Breed, J., et al. (2005). Structure-based design of protein tyrosine phosphatase-1B inhibitors. *Bioorganic & Medicinal Chemistry Letters, 15*(10), 2503–2507.
Blüher, M., Dodson Michael, M., et al. (2002). Adipose tissue selective insulin receptor knockout protects against obesity and obesity-related glucose intolerance. *Developmental Cell, 3*(1), 25–38.
Boutselis, I. G., Yu, X., et al. (2007). Synthesis and cell-based activity of a potent and selective protein tyrosine phosphatase 1B inhibitor prodrug. *Journal of Medicinal Chemistry, 50*(4), 856–864.
Burke, T. R., Jr., Kole, H. K., et al. (1994). Potent inhibition of insulin receptor dephosphorylation by a hexamer peptide containing the phosphotyrosyl mimetic F2Pmp. *Biochemical and Biophysical Research Communications, 204*(1), 129–134.
Burke, T. R., Jr., Yao, Z.-J., et al. (1998). Enantioselective synthesis of nonphosphorous-containing phosphotyrosyl mimetics and their use in the preparation of tyrosine phophatase inhibitory peptides. *Tetrahedron, 54*(34), 9981–9994.
Considine, R. V., Sinha, M. K., et al. (1996). Serum immunoreactive-leptin concentrations in normal-weight and obese humans. *The New England Journal of Medicine, 334*(5), 292–295.
Cota, D., Proulx, K., et al. (2006). Hypothalamic mTOR signaling regulates food intake. *Science, 312*(5775), 927–930.
De Jonghe, B. C., Hayes, M. R., et al. (2011). Deficiency of PTP1B in POMC neurons leads to alterations in energy balance and homeostatic response to cold exposure. *American Journal of Physiology, Endocrinology and Metabolism, 300*(6), E1002–E1011.
Delibegovic, M., Bence, K. K., et al. (2007). Improved glucose homeostasis in mice with muscle-specific deletion of protein-tyrosine phosphatase 1B. *Molecular and Cellular Biology, 27*(21), 7727–7734.
Delibegovic, M., Zimmer, D., et al. (2009). Liver-specific deletion of protein-tyrosine phosphatase 1B (PTP1B) improves metabolic syndrome and attenuates diet-induced endoplasmic reticulum stress. *Diabetes, 58*(3), 590–699.
Douty, B., Wayland, B., et al. (2008). Isothiazolidinone inhibitors of PTP1B containing imidazoles and imidazolines. *Bioorganic & Medicinal Chemistry Letters, 18*(1), 66–71.
Eizirik, D. L., Cardozo, A. K., et al. (2008). The role for endoplasmic reticulum stress in diabetes mellitus. *Endocrine Reviews, 29*(1), 42–61.
Elchebly, M., Payette, P., et al. (1999). Increased insulin sensitivity and obesity resistance in mice lacking the protein tyrosine phosphatase-1B gene. *Science, 283*(5407), 1544–1548.
Erbe, D. V., Klaman, L. D., et al. (2009). Prodrug delivery of novel PTP1B inhibitors to enhance insulin signaling. *Diabetes, Obesity & Metabolism, 11*(6), 579–588.

Erbe, D. V., Wang, S., et al. (2005). Ertiprotafib improves glycemic control and lowers lipids via multiple mechanisms. *Molecular Pharmacology*, *67*(1), 69–77.

Fischer, E. H., Charbonneau, H., et al. (1991). Protein tyrosine phosphatases: A diverse family of intracellular and transmembrane enzymes. *Science*, *253*(5018), 401–406.

Frangioni, J. V., Beahm, P. H., et al. (1992). The nontransmembrane tyrosine phosphatase PTP-1B localizes to the endoplasmic reticulum via its 35 amino acid C-terminal sequence. *Cell*, *68*(3), 545–560.

Gao, Q., Wolfgang, M. J., et al. (2004). Disruption of neural signal transducer and activator of transcription 3 causes obesity, diabetes, infertility, and thermal dysregulation. *Proceedings of the National Academy of Sciences*, *101*(13), 4661–4666.

Ghilardi, N., & Skoda, R. C. (1997). The leptin receptor activates Janus Kinase 2 and signals for proliferation in a factor-dependent cell line. *Molecular Endocrinology*, *11* (4), 393–399.

Han, Y., Belley, M., et al. (2008). Discovery of [(3-bromo-7-cyano-2-naphthyl) (difluoro)-methyl]phosphonic acid, a potent and orally active small molecule PTP1B inhibitor. *Bioorganic & Medicinal Chemistry Letters*, *18*(11), 3200–3205.

Hiriyanna, K. T., Baedke, D., et al. (1994). Thiophosphorylated substrate analogs are potent active site-directed inhibitors of protein-tyrosine phosphatases. *Analytical Biochemistry*, *223*(1), 51–58.

Houseknecht, K. L., Portocarrero, C. P., et al. (1998). Leptin and its receptors: Regulators of whole-body energy homeostasis. *Domestic Animal Endocrinology*, *25*(6), 457–475.

Huang, P., Ramphal, J., et al. (2003). Structure-based design and discovery of novel inhibitors of protein tyrosine phosphatases. *Bioorganic & Medicinal Chemistry*, *11*(8), 1835–1849.

Hussain, M., Ahmed, V., et al. (2008). A re-examination of the difluoromethylenesulfonic acid group as a phosphotyrosine mimic for PTP1B inhibition. *Bioorganic & Medicinal Chemistry*, *16*(14), 6764–6777.

Iversen, L. F., Andersen, H. S., et al. (2000). Structure based design of a low molecular weight, non-phosphorous, non-peptide, and highly selective inhibitor of protein-tyrosine phosphatase 1B. *Journal of Biological Chemistry*, *275*(14), 10300–10307.

Iversen, L. F., Møller, K. B., et al. (2002). Structure determination of T cell protein-tyrosine phosphatase. *Journal of Biological Chemistry*, *277*(22), 19982–19990.

Kaszubska, W., Falls, H. D., et al. (2002). Protein tyrosine phosphatase 1B negatively regulates leptin signaling in a hypothalamic cell line. *Molecular and Cellular Endocrinology*, *195*(1–2), 109–118.

Kenner, K. A., Anyanwu, E., et al. (1996). Protein-tyrosine phosphatase 1B is a negative regulator of insulin- and insulin-like growth factor-I-stimulated signaling. *Journal of Biological Chemistry*, *271*(33), 19810–19816.

Kim, M. S., Pak, Y. K., et al. (2006). Role of hypothalamic Foxo1 in the regulation of food intake and energy homeostasis. *Nature Neuroscience*, *9*(7), 901–906.

Kitamura, T., Feng, Y., et al. (2006). Forkhead protein FoxO1 mediates Agrp-dependent effects of leptin on food intake. *Nature Medicine*, *12*(5), 534–540.

Klaman, L. D., Boss, O., et al. (2000). Increased energy expenditure, decreased adiposity, and tissue-specific insulin sensitivity in protein-tyrosine phosphatase 1B-deficient mice. *Molecular and Cellular Biology*, *20*(15), 5479–5489.

Klopfenstein, S. R., Evdokimov, A. G., et al. (2006). 1,2,3,4-Tetrahydroisoquinolinyl sulfamic acids as phosphatase PTP1B inhibitors. *Bioorganic & Medicinal Chemistry Letters*, *16* (6), 1574–1578.

Kumar, A., Maurya, R. A., et al. (2009). Design and synthesis of 3,5-diarylisoxazole derivatives as novel class of anti-hyperglycemic and lipid lowering agents. *Bioorganic & Medicinal Chemistry*, *17*(14), 5285–5292.

Kumar, A., Sharma, S., et al. (2010). Design and synthesis of 2,4-disubstituted polyhydroqunolines as prospective anti-hyperglycemic and lipid modulating agents. *Bioorganic & Medicinal Chemistry*, *18*(11), 4138–4148.

Lantz, K. A., Emeigh Hart, S. G., et al. (2010). Inhibition of PTP1B by trodusquemine (MSI-1436) causes fat-specific weight loss in diet-induced obese mice. *Obesity*, *18*(8), 1516–1523.

Lau, C. K., Bayly, C. I., et al. (2004). Structure based design of a series of potent and selective non peptidic PTP-1B inhibitors. *Bioorganic & Medicinal Chemistry Letters*, *14*(4), 1043–1048.

Liljebris, C., Larsen, S. D., et al. (2002). Investigation of potential bioisosteric replacements for the carboxyl groups of peptidomimetic inhibitors of protein tyrosine phosphatase 1B: Identification of a tetrazole-containing inhibitor with cellular activity. *Journal of Medicinal Chemistry*, *45*(9), 1785–1798.

Liotta, A. S., Kole, H. K., et al. (1994). A synthetic tris-sulfotyrosyl dodecapeptide analogue of the insulin receptor 1146-kinase domain inhibits tyrosine dephosphorylation of the insulin receptor in situ. *Journal of Biological Chemistry*, *269*(37), 22996–23001.

Liu, F., Hill, D. E., et al. (1996). Direct binding of the proline-rich region of protein tyrosine phosphatase 1B to the Src homology 3 domain of p130(Cas). *Journal of Biological Chemistry*, *271*(49), 31290–31295.

Liu, Z., Lee, W., et al. (2011). Design, synthesis, and evaluation of bromo-retrochalcone derivatives as protein tyrosine phosphatase 1B inhibitors. *Bioorganic & Medicinal Chemistry Letters*, *21*(12), 3755–3758.

Liu, G., Xin, Z., et al. (2003). Selective protein tyrosine phosphatase 1B inhibitors: Targeting the second phosphotyrosine binding site with non-carboxylic acid-containing ligands. *Journal of Medicinal Chemistry*, *46*(16), 3437–3440.

Liu, S., Zeng, L.-F., et al. (2008). Targeting inactive enzyme conformation: Aryl diketoacid derivatives as a new class of PTP1B inhibitors. *Journal of the American Chemical Society*, *130*(50), 17075–17084.

Matsuo, K., Bettaieb, A., et al. (2011). Regulation of brown fat adipogenesis by protein tyrosine phosphatase 1B. *PLoS One*, *6*(1), e16446.

Minokoshi, Y., Alquier, T., et al. (2004). AMP-kinase regulates food intake by responding to hormonal and nutrient signals in the hypothalamus. *Nature*, *428*(6982), 569–574.

Morris, D. L., & Rui, L. (2009). Recent advances in understanding leptin signaling and leptin resistance. *American Journal of Physiology, Endocrinology and Metabolism*, *297*(6), E1247–E1259.

Nedergaard, J., Bengtsson, T., et al. (2007). Unexpected evidence for active brown adipose tissue in adult humans. *American Journal of Physiology, Endocrinology and Metabolism*, *293*(2), E444–E452.

Niswender, K. D., Morton, G. J., et al. (2001). Intracellular signalling: Key enzyme in leptin-induced anorexia. *Nature*, *413*(6858), 794–795.

Owen, C., Czopek, A., et al. (2012). Adipocyte-specific protein tyrosine phosphatase 1B deletion increases lipogenesis, adipocyte cell size and in a minor regulator of glucose homeostasis. *PLoS One*, 7(2), e32700.

Patel, D., Jain, M., et al. (2012). Discovery of potent, selective and orally bioavailable triarylsulfonamide based PTP1B inhibitors. *Bioorganic & Medicinal Chemistry Letters*, *22*(2), 1111–1117.

Plum, L., Belgardt, B. F., et al. (2006). Central insulin action in energy and glucose homeostasis. *The Journal of Clinical Investigation*, *116*(7), 1761–1766.

Puius, Y. A., Zhao, Y., et al. (1997). Identification of a second aryl phosphate-binding site in protein-tyrosine phosphatase 1B: A paradigm for inhibitor design. *Proceedings of the National Academy of Sciences*, *94*(25), 13420–13425.

Rao, M. N., Shinnar, A. E., et al. (2000). Aminosterols from the Dogfish Shark *Squalus acanthias*. *Journal of Natural Products*, *63*(5), 631–635.

Sahu, A. (2011). Intracellular leptin-signaling pathways in hypothalamic neurons: The emerging role of phosphatidylinositol-3 kinase-phosphodiesterase-3B-cAMP pathway. *Neuroendocrinology*, *93*(4), 201–210.

Shen, K., Keng, Y.-F., et al. (2001). Acquisition of a specific and potent PTP1B inhibitor from a novel combinatorial library and screening procedure. *Journal of Biological Chemistry*, *276*(50), 47311–47319.

Shi, D., Li, J., et al. (2012). Bromophenols as inhibitors of protein tyrosine phosphatase 1B with antidiabetic properties. *Bioorganic & Medicinal Chemistry Letters*, *22*(8), 2827–2832.

Shrestha, S., Bhattarai, B. R., et al. (2007a). PTP1B inhibitor Ertiprotafib is also a potent inhibitor of IκB kinase β (IKK-β). *Bioorganic & Medicinal Chemistry Letters*, *17*(10), 2728–2730.

Shrestha, S., Bhattarai, B. R., et al. (2007b). Mono- and disalicylic acid derivatives: PTP1B inhibitors as potential anti-obesity drugs. *Bioorganic & Medicinal Chemistry*, *15*(20), 6535–6548.

Sparks, R. B., Polam, P., et al. (2007). Benzothiazole benzimidazole (*S*)-isothiazolidinone derivatives as protein tyrosine phosphatase-1B inhibitors. *Bioorganic & Medicinal Chemistry Letters*, *17*(3), 736–740.

Stuible, M., Zhao, L., et al. (2007). Cellular inhibition of protein tyrosine phosphatase 1B by uncharged thioxothiazolidinone derivatives. *ChemBioChem*, *8*(2), 179–186.

Tartaglia, L. A., Dembski, M., et al. (1995). Identification and expression cloning of a leptin receptor, OB-R. *Cell*, *83*(7), 1263–1271.

Tonks, N. K. (2003). PTP1B: From the sidelines to the front lines. *FEBS Letters*, *546*(1), 140–148.

Tonks, N. K., Diltz, C. D., et al. (1988). Purification of the major protein-tyrosine-phosphatases of human placenta. *Journal of Biological Chemistry*, *263*(14), 6722–6730.

Wan, Z.-K., Follows, B., et al. (2007). Probing acid replacements of thiophene PTP1B inhibitors. *Bioorganic & Medicinal Chemistry Letters*, *17*(10), 2913–2920.

Wellen, K. E., & Hotamisligil, G. S. (2005). Inflammation, stress, and diabetes. *The Journal of Clinical Investigation*, *115*(5), 1111–1119.

Wiesmann, C., Barr, K. J., et al. (2004). Allosteric inhibition of protein tyrosine phosphatase 1B. *Nature Structural and Molecular Biology*, *11*(8), 730–737.

Wilson, D. P., Wan, Z.-K., et al. (2007). Structure-based optimization of protein tyrosine phosphatase 1B inhibitors: From the active site to the second phosphotyrosine binding site. *Journal of Medicinal Chemistry*, *50*(19), 4681–4698.

Ye, D., Zhang, Y., et al. (2010). Novel thiophene derivatives as PTP1B inhibitors with selectivity and cellular activity. *Bioorganic & Medicinal Chemistry*, *18*(5), 1773–1782.

Yuan, M., Konstantopoulos, N., et al. (2001). Reversal of obesity- and diet-induced insulin resistance with salicylates or targeted disruption of *Ikkβ*. *Molecular Sciences*, *293*(5535), 1673–1677.

Zabolotny, J. M., Bence-Hanulec, K. K., et al. (2002). PTP1B regulates leptin signal transduction in vivo. *Developmental Cell*, *2*(4), 489–495.

Zhang, Y., Proenca, R., et al. (1994). Positional cloning of the mouse *obese* gene and its human homologue. *Nature*, *372*(6505), 425–432.

Zhao, A. Z., Huan, J.-N., et al. (2002). A phosphatidylinositol 3-kinase phosphodiesterase 3B-cyclic AMP pathway in hypothalamic action of leptin on feeding. *Nature Neuroscience*, *5*(8), 727–728.

Zhao, H., Liu, G., et al. (2004). Isoxazole carboxylic acids as protein tyrosine phosphatase 1B (PTP1B) inhibitors. *Bioorganic & Medicinal Chemistry Letters*, *14*(22), 5543–5546.

CHAPTER EIGHTEEN

Nomilin as an Anti-Obesity and Anti-Hyperglycemic Agent

Ryuichiro Sato[1]
Department of Applied Biological Chemistry, The University of Tokyo, Bunkyo, Tokyo, Japan
[1]Corresponding author: e-mail address: aroysato@mail.ecc.u-tokyo.ac.jp

Contents

Abstract

Recent scientific findings support the notion that bile acids, which are cholesterol catabolites, are bioactive signaling molecules that function as ligands for the farnesoid X receptor or a G-protein-coupled receptor, TGR5. Through these receptors, bile acids can maintain not only bile acid homeostasis but also lipid and carbohydrate homeostasis. An intriguing finding regarding the role of TGR5 in energy metabolism and glucose homeostasis suggests a potential approach to combat obesity and insulin resistance by targeting this receptor to increase thermogenesis and incretin secretion. In this review, I have summarized the latest findings related to TGR5 agonists, in particular, a citrus limonoid, nomilin, and the roles of these agonists in energy metabolism and glucose homeostasis.

1. INTRODUCTION

The prevalence of obesity has been increasing worldwide, which has also led to a dramatic increase in the risk of cardiovascular diseases, type 2 diabetes, and metabolic syndrome. Excess visceral abdominal fat accumulation

Vitamins and Hormones, Volume 91
ISSN 0083-6729
http://dx.doi.org/10.1016/B978-0-12-407766-9.00018-3

is a major feature of abdominal obesity that is responsible for development of metabolic syndrome. Abdominal visceral fat secretes several types of adipokines that reduce insulin receptor sensitivity in various tissues, thereby leading to insulin resistance. Because insulin resistance almost always precedes the onset of type 2 diabetes and serves as a basis for development of metabolic syndrome, early intervention through not only pharmacological management but also improved dietary life is highly required to delay or prevent its progression. Given the potential for preventing insulin resistance, improvement of glucose tolerance through increased insulin sensitivity in several tissues and lessening of marked hyperglycemia is highly recommended. For this purpose, the function of the incretin hormone glucagon-like peptide-1 (GLP-1) has been receiving significant attention (McIntosh, Widenmaier, & Kim, 2010). GLP-1 promotes insulin biosynthesis and secretion prior to an increase in blood glucose levels, reduces appetite, impairs gastric emptying, suppresses glucagon secretion, and markedly improves insulin sensitivity in various tissues. Inhibitors of DPP-IV, an enzyme responsible for rapid degradation of serum GLP-1, have been recently introduced for treatment of type 2 diabetes (Moller, 2012). Among the several ways to prevent obesity, an increase in energy expenditure through enhanced heat production in skeletal muscles and brown adipose tissues (BATs) is one of the promising approaches. Unlike the white adipose tissue (WAT), BAT is a major site of adaptive thermogenesis in which heat is generated by the actions of the uncoupling protein-1 (UCP-1; Azzu, Jastroch, Divakaruni, & Brand, 2010). Skeletal muscle, the most important thermogenic site in humans, also generates heat through UCP-1 and UCP-3 and thereby increasing energy expenditure (Parker, Affourtit, Vidal-Puig, & Brand, 2008).

A recent study demonstrated that administration of bile acids, that is, catabolic products of cholesterol, to mice increases energy expenditure through the bile acid receptor TGR5, a G-protein-coupled receptor with seven transmembrane domains, thereby preventing obesity and insulin resistance (Watanabe et al., 2006). Enteroendocrine L cells are present throughout the small and large intestines, where TGR5 associating with bile acids promotes GLP-1 release. After being absorbed, bile acids interact with TGR5 present in BAT in rodents and skeletal muscle in humans, thereby increasing the intracellular cAMP level and eventually increasing energy expenditure. Given that TGR5 is thought to be a crucial mediator for energy and glucose homeostasis, this receptor is an attractive target for the potential treatment of metabolic disorders. Simultaneously, adequate intake of TGR5 ligands contained in foods should be highly recommended to increase insulin sensitivity and lower serum glucose levels. For this purpose, TGR5 agonists have been searched among a variety of food constituents and plant extracts, and several types of triterpenoids have been identified as antiobesity TGR5 agents (Genet et al., 2010; Ono, Inoue, Hashidume, Shimizu, & Sato, 2011; Sato et al., 2007). This review attempts to provide an up-to-date summary of current knowledge related to the functions of the bile acid receptor TGR5 and its agonists in energy and glucose homeostasis.

2. BILE ACID BIOSYNTHESIS

Cholesterol is an essential component of animal cell membranes as well as a precursor in biosynthesis of steroid hormones, vitamin D, and bile acids. In addition to dietary cholesterol uptake, biosynthesis in the liver is a major source of cholesterol in our body. Through more than 20 enzymatic reactions, approximately 1 g of cholesterol is synthesized daily from acetyl-coenzyme A (Goldstein, DeBose-Boyd, & Brown, 2006). The fact that the amount of cholesterol synthesized in the liver is much higher than that of dietary-derived cholesterol is implicated as a reason why statins, inhibitors of a rate-limiting enzyme (HMG CoA reductase) in cholesterol biosynthesis, are now used as first-line therapy worldwide. Hepatic cholesterol that is synthesized and taken up is secreted together with triglycerides and phospholipids as a form of lipoprotein, very low-density lipoprotein (VLDL). Plasma VLDL is eventually metabolized into LDL in the bloodstream. These LDL particles bind to LDL receptors that are localized on cell surfaces and are then engulfed by the cells. Conversely, excess cholesterol in all tissues is excreted through an ABC transporter, ABCA1, as a form of high-density lipoprotein (HDL) and is finally transported back to the liver (Nagao, Tomioka, & Ueda, 2011). Although cholesterol circulates in our body, this steroid, having a complex structure involving 27 carbon atoms, is incapable of being degraded into smaller molecules. The liver, the only organ that can dispose of excess cholesterol, converts it to bile acids, that is, molecules containing 24 carbon atoms, through several enzymatic processes and excretes them in the bile (Chiang, 2009).

Bile acid synthesis in the liver is the most important mechanism for eliminating and degrading cholesterol. The first step in bile acid synthesis is the conversion of cholesterol to 7-α-hydroxycholesterol that is catalyzed by CYP7a1, a cytochrome P_{450} and the rate-limiting enzyme in this biosynthetic pathway. Bile acids synthesized through several more reactions, namely cholic and deoxycholic acids, are converted to bile salts by conjugation with glycine or taurine to increase their water solubility. After bile salts are secreted into the bile ducts, they are stored in the gallbladder until they are secreted into the upper small intestinal tract. After each meal, the gallbladder contracts to expel the bile into the intestine, where bile acids serve as solubilizers to facilitate absorption of lipids and fat-soluble vitamins. Most bile salts (>90%) are reabsorbed in the lower small intestine through a bile acid transporter, ileal bile acid transporter (IBAT), and are transported back to the liver. It has been estimated that bile acids are recycled more than 10 times before they are finally eliminated.

3. BIOLOGICAL FUNCTIONS OF BILE ACIDS AS FARNESOID X RECEPTOR LIGANDS

Bile acids have been considered to be simple cholesterol catabolites and lipid solubilizers. The finding that bile acids can activate the nuclear receptor farnesoid X receptor (FXR) and exert a direct influence on gene expression in the small intestine and liver has dramatically changed our concept of these cholesterol catabolites (Makishima et al., 1999; Parks et al., 1999; Wang, Chen, Hollister, Sowers, & Forman, 1999). Therefore, bile acids are now considered to be bioactive molecules that modulate lipid and carbohydrate metabolism (Sinal et al., 2000). The ligand-bound FXR forms a heterodimer with the retinoid X receptor, and this dimer is capable of binding to a specific DNA sequence, an inverted repeat separated by one base pair (5′-AGGGTCAxTGACCT-3′). In the small intestine, bile acids activate FXR, thereby stimulating gene expression of the ileal bile acid-binding protein to facilitate increases in bile acid uptake by eliminating the toxicity of these detergents (Grober et al., 1999; Nakahara et al., 2005). Simultaneously, FXR augments gene expression of the *fibroblast growth factor 15/19 (FGF15/19)* in the intestinal epithelial cells (Itoh, 2010). The human *FGF* gene family comprises 22 members, and three of these, including FGF19, 21, and 23, are hormone-like (Kliewer & Mangelsdorf, 2010). *FGF15* is the mouse ortholog of human *FGF19*. Plasma FGF15/19 secreted from the small intestine in response to FXR activation by bile acids interacts with the FGF15/19 receptor in the liver and thereafter suppresses expression of CYP7a1. This is thought to be one of the negative feedback regulatory systems for bile acid biosynthesis.

In the liver, bile acids returned from the small intestine activate FXR and promote gene expression of another nuclear receptor, *small heterodimer partner (SHP)*. SHP is a unique nuclear receptor that lacks the DNA-binding domain and suppresses transcriptional activities of several nuclear receptors by direct interaction with them (Hirokane, Nakahara, Tachibana, Shimizu, & Sato, 2004). Gene expression of *CYP7a1*, the rate-limiting enzyme of bile acid synthesis, is highly regulated by two nuclear receptors, liver receptor homolog-1 (LRH-1) and liver X receptor (LXR). SHP represses gene expression of *CYP7a1* through inactivation of these two receptors (Goodwin et al., 2000; Shin & Osborne, 2008). This FXR–SHP–(LRH-1 and LXR) pathway is another major negative feedback system for bile acid synthesis. Inactivation of LXR by SHP reduces gene expression of the *sterol regulatory element-binding protein-1c* that orchestrates fatty acid and triglyceride metabolism (Watanabe et al., 2004). Simultaneously, SHP suppresses the transcription of several gluconeogenic genes through inactivation of a couple of nuclear receptors (Ma, Saha, Chan, & Moore, 2006). Overall, FXR activation by bile acids exerts a substantial impact on bile acid, cholesterol, and carbohydrate metabolism in the small intestine and liver (Duran-Sandoval et al., 2005).

4. TGR5, A BILE ACID RECEPTOR

Bile acids absorbed through IBAT in the ileum are secreted into the portal vein and eventually taken up by the liver. However, some of them circulate throughout the body at a concentration of 5–15 μM. Because expression of FXR is limited in several organs, including the liver, intestine, kidney, and adrenal, the physiological function of plasma bile acids had remained unclear. In 2002, a G-protein-coupled receptor, TGR5, comprising 330 amino acid residues, was found to be responsive to bile acids as a cell-surface receptor (Maruyama et al., 2002). In humans, TGR5 is expressed in various tissues such as heart, spleen, skeletal muscle, kidney, liver, small intestine, placenta, lung, and peripheral blood leukocytes, and it binds to not only free forms of bile acids but also taurine- and glycine-conjugated forms (Kawamata et al., 2003). Ligand binding to TGR5 triggers an increase in intracellular cAMP concentration by transduction through a G_s protein and receptor internalization. *In vitro* experiments show that lithocholic acid, deoxycholic acid, chenodeoxycholic acid, and cholic acid increase intracellular cAMP, in this order, but that ursodeoxycholic acid, an isomer of chenodeoxycholic acid, is much less active than the aforementioned acids. Interestingly, TGR5 is highly expressed in the gallbladder, the volume of which is regulated by TGR5 activation causing smooth muscle relaxation (Li et al., 2011).

TGR5-defecient mice challenged with a high-fat diet for 2 months show significant fat accumulation with body weight gain, suggesting that this receptor is a potential new player in energy homeostasis (Maruyama et al., 2006). Indeed, energy homeostasis is maintained through TGR5 activation by bile acids (Watanabe et al., 2006). Administration of cholic acid to diet-induced obese mice normalizes body weight and increases energy expenditure in BAT (Fig. 18.1). This effect of bile acids is dependent on induction of the cAMP-dependent thyroid hormone-activating enzyme type 2 iodothyronine deionase (D2). D2 converts inactive thyroxine (T4) into the active thyroid hormone 3,2,3′-tri-iodothyronine (T3). Three types of iodothyronine deionases (D1, D2, and D3) are capable of regulating thyroid hormone activity by removal of specific iodine moieties from T4. D2 is

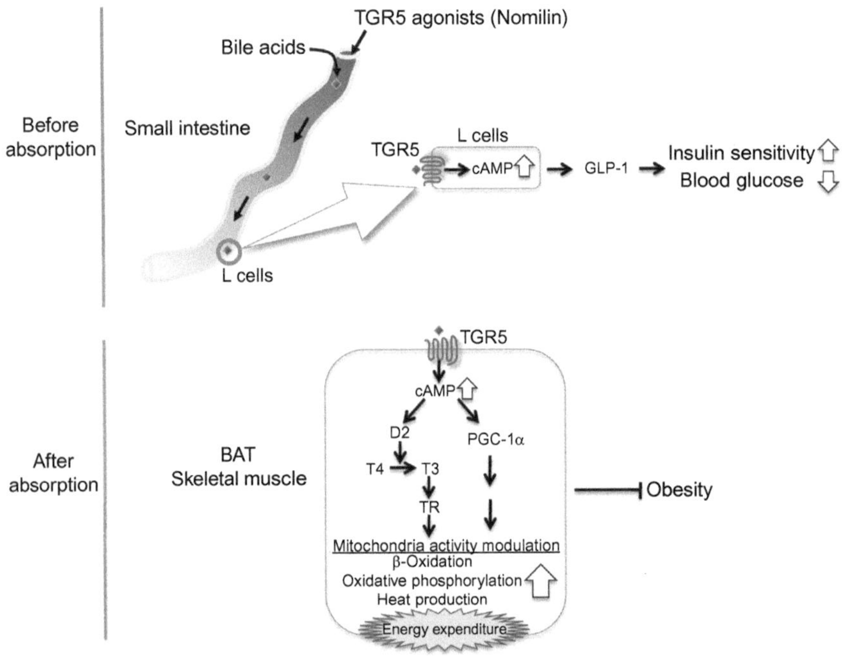

Figure 18.1 Proposed pathways by which TGR5 agonists exert their actions on the small intestine, BAT, and skeletal muscle. In response to TGR5 activation by its agonists, GLP-1 is released from L cells in the intestinal mucosa. GLP-1 can stimulate insulin secretion, lower blood glucose levels, and improve insulin sensitivity. After absorption, TGR5 agonists activate TGR5 in BAT and skeletal muscles, thereby raising the intracellular cAMP level. This signal modulates mitochondrial activity with accompanying increased fatty acid β-oxidation, oxidative phosphorylation, and heat production. Rising energy expenditure attenuates diet-induced obesity. (See Color Insert.)

predominantly responsible for deionation of T4, thereby generating active T3 and controlling the thermogenesis (Bianco & Kim, 2006). A critical aspect of D2-mediated T3 generation is providing the higher concentration of intracellular active T3 necessary to occupy the number of thyroid hormone receptors (TRs) in selected tissues (de Jesus et al., 2001). Increased intracellular cAMP levels mediated by TGR5 activation also induces gene expression of *peroxisome proliferator-activated receptor γ-coactivator 1α (PGC-1α)* that is originally described as a cold-inducible coactivator, which regulates adaptive thermogenesis in BAT and skeletal muscle by enhancing mitochondrial biogenesis and oxidative metabolism. PGC-1α regulates energy expenditure by interacting with a selected subset of nuclear receptors, including the peroxisome proliferator-activated receptors (PPARs) and TRs. Three distinct uncoupling protein isoforms, UCP-1, -2, and -3, have been identified and implicated as thermogenesis mediators. PPARγ or PPARα activation regulates the expression of all the three UCP isoforms (Kelly et al., 1998). Transcriptional regulation of genes related to β-oxidation of fatty acids is also under the control of a combination of PPARα and PGC-1α (Vega, Huss, & Kelly, 2000). Consequently, PGC-1α activation stimulates energy expenditure through modulation of mitochondrial activity.

Another potential function for TGR5 is highlighted in intestinal cells (Fig. 18.1). Bile acids induce GLP-1 secretion from intestinal enteroendocrine secretin tumor cells by a mechanism that is dependent on TGR5 activation and the subsequent elevation of intracellular cAMP levels (Katsuma, Hirasawa, & Tsujimoto, 2005). TGR5 transgenic mice are associated with robust GLP-1 secretion within 10 min after an oral glucose load, suggesting that bile acids secreted in response to ingested glucose provide a sufficient signal to induce GLP-1 release in the context of high expression of TGR5 (Thomas et al., 2009). Further studies are required to validate the contribution of TGR5 to intestinal GLP-1 production *in vivo*.

5. TGR5 AGONISTS

With the emerging roles of bile acids in regulating glucose and energy metabolism through TGR5, the discovery and development of potent small molecule agonists of TGR5 is undoubtedly expected strongly (Thomas, Auwerx, & Schoonjans, 2008; Tiwari & Maiti, 2009). Similarly, screening of various naturally occurring compounds mimicking bile acid functions has been tried. Taking advantage of an increase in intracellular cAMP in response to activation of TGR5 by agonists, a luciferase assay system can be established using the cAMP response element (CRE)-driven reporter gene, which is enhanced by endogenous activated CRE-binding protein. This assay system manages to identify several TGR5 agonists, including betulic, oleanolic, and ursolic acid, which are triterpenoids (Genet et al., 2010). Oleanolic acid lowers serum glucose and insulin levels in mice fed with a high-fat diet and enhances glucose tolerance (Sato et al., 2007). Mice fed with a high-fat diet and treated with betulic acid also show decreases body weights; abdominal fat accumulation; and serum glucose, serum triglycerides, and serum total cholesterol levels (de Melo et al., 2009).

We have established the same type of assay system using human HEK293 cells that overexpress human TGR5 together with a CRE-containing (four repeats) reporter gene (Fig. 18.2A; Ono et al., 2011). Approximately 160 commercially available purified food constituents and their derivatives, which are categorized as flavonoids (~45%), terpenoids (~20%), and steroids (~15%), have been evaluated as potential TGR5 agonists. During the search for such agonists, three forms of flavonoids and an unrelated structural compound have been identified as weak TGR5 agonists; however, a citrus limonoid, nomilin, has been found to be the most potent agonist. Nomilin is classed as a tetranorterpenoid, and it is converted into limonin in citrus fruits through several reactions via obacunone. Strangely, nomilin is a more potent agonist than obacunone, whereas the final conversion form limonin hardly exerts any TGR5 ligand activity (Fig. 18.2C). This result indicates that the structure shared between nomilin and obacunone, including the ring designated as A (Fig. 18.2B) that is altered in limonin must be essential for a TGR5 agonist.

6. A CITRUS LIMONOID, NOMILIN

Limonoids occur naturally only in plant species of the Rutaceae and Meliaceae plant families (Manners, 2007). Limonoids are highly oxygenated modified triterpenes that are biosynthesized from the acetate–mevalonate pathway in citrus fruits (Roy & Saraf, 2006). They occur in significant amounts as aglycones and glycosides in seeds and fruit tissues. Citrus fruits contain approximately 36 limonoid aglycones and 17 limonoid glucosides (Hasegawa & Miyake, 1996). Major citrus species accumulate limonin, nomilin, obacunone, and deacetylnomilin. Limonoid aglycones that cause bitterness in numerous citrus fruits are converted into tasteless limonoid glucosides during fruit maturation (Endo et al., 2002). Soluble limonin

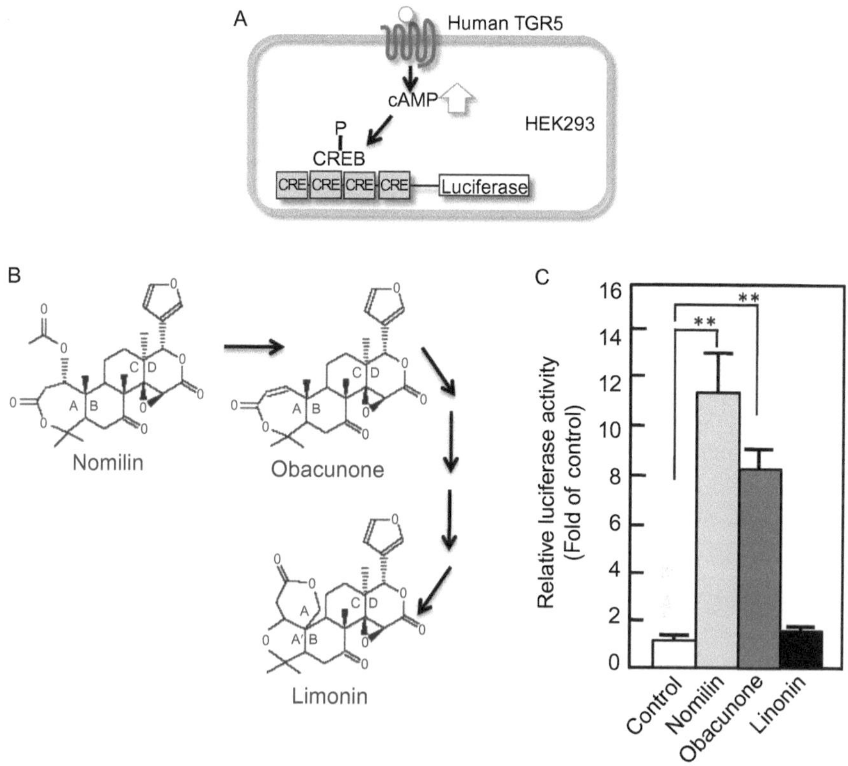

Figure 18.2 A citrus limonoid, nomilin, as a TGR5 agonist. (A) In an effort to isolate novel TGR5 agonists, HEK293 cells expressing human TGR5 and the CRE-driven reporter were cultured in the presence of various food constituents (Ono et al., 2011). (B) The structures of nomilin, obacunone, and limonin with the number of rings in each illustrated. Nomilin is metabolized to obacunone and finally to limonin through several steps in citrus fruits. (C) The reporter assay reveals that nomilin and obacunone but not limonin are potent TGR5 agonists. (See Color Insert.)

glucoside is the predominant limonoid glucoside present in citrus juices, and it occurs at an amount twice that of the other limonoid glucosides combined (Herman, Fong Chi, Ou, & Hasegawa, 1990). In contrast, citrus seeds constitute the only natural repository of limonoid aglycones (Vikram, Jayaprakasha, & Patil, 2007). The seeds contain limonoid aglycones in amounts up to 1% of their fresh weight (Hasegawa, Bennett, & Verdon, 1980). In citrus seeds, nomilin glucoside occurs at a higher concentration than that of limonin glucoside (Ozaki et al., 1995).

Several studies have established citrus limonoids, including limonin, nomilin, and their glucosides, as having significant biological activity in mammalian systems. Limonoids have been shown to inhibit the growth of estrogen receptor-positive and -negative human breast cancer cells in culture (Guthrie et al., 2000). Nomilin glucoside is found to be the most effective limonoid glucoside, and nomilin is the most effective limonoid aglycone. Nomilin has also been confirmed to significantly inhibit the growth of MCF-7 human breast cancer cells (Tian, Miller, Ahmad, Tang, & Patil, 2001). It is believed that enhancement of the elimination of toxic electrophiles by phase-II enzymes, including glutathione-*S*-transferase (GST) and quinine reductase, is one way of cellular anticarcinogenic protection. Among the limonoids and flavonoids tested, nomilin is the most potent inducer of GST in mice when administered over a 1-week period (Ahmad et al., 2006). This GST-inducing effect is commonly observed in compounds with a furan structure, for example, nomilin.

Limonoids exert significant antibacterial and antifungal activities (Abdelgaleil, Hashinaga, & Nakatani, 2005, Abdelgaleil, Iwagawa, Doe, & Nakatani, 2004). Nomilin has also been shown to inhibit HIV-1 replication by inhibiting p-24 antigen activity and protease activity in cell systems (Battinelli et al., 2003). Furthermore, limonoids possess antifeedant activity against insects (Poulose, Harris, & Patil, 2006).

Evidence of the contribution of citrus fruits to human health is obtained in a study of the effect of orange and grapefruit juice consumption on cholesterol metabolism in rabbits (Kurowska, Borradaile, Spence, & Carroll, 2000). This study showed that replacing drinking water with either orange juice or grapefruit juice reduced serum LDL cholesterol levels by 43% and 32%, respectively. A human study revealed that administration of a 750-mL dose of orange juice daily to 15 human subjects with elevated serum cholesterol levels increased HDL cholesterol levels by 21% and decreased the LDL/HDL cholesterol ratio by 16% (Elzbieta et al., 2000). These results suggest that secondary metabolites, including flavonoids and limonoids, could function endogenously to influence cholesterol metabolism.

7. ANTI-OBESITY AND ANTI-HYPERGLYCEMIC EFFECTS OF TGR5 AGONISTS

Olive tree leaves are known to be a traditional antidiabetic and antihypertensive herbal drug. In the course of isolating new agonists for TGR5 among plant extracts, an extract obtained from olive tree leaves has been found to activate TGR5 using the abovementioned reporter assay with the CRE-driven reporter gene (Sato et al., 2007). Further isolation of the active components in the hydrophobic fraction that contains most of the activity identified oleanolic acid as a novel TGR5 agonist with action comparable with that of the potent natural TGR5 ligand lithocholic acid. Oleanolic acid, a triterpenoid, does not activate FXR and is therefore a selective TGR5 agonist unlike bile acids. Obese mice fed a high-fat diet for 10 weeks and further fed a high-fat diet supplemented with oleanolic acid (100 mg/kg) for 7 days exhibited a slight tendency to lower body weight gain, but showed a significant decrease of the epididymal fat pad mass compared with that or mice fed with an oleanolic acid-free high-fat diet. Furthermore, oleanolic acid treatment clearly decreases serum glucose and insulin levels. An intraperitoneal glucose tolerance test revealed improved glucose tolerance in mice treated with oleanolic acid. These results clearly show that oleanolic acid, a triterpenoid from olive leaves, is a highly specific TGR5 agonist with potent antihyperglycemic activity.

In an effort to isolate new TGR5 agonists among the triterpenoid family members, betulic acid was found to be the most potent agonist (Genet et al., 2010). An independent study has demonstrated that betulic acid prevents abdominal fat accumulation in mice fed a high-fat diet (de Melo et al., 2009). Mice treated with betulic acid and fed a high-fat diet show decreased body weight gain; abdominal fat accumulation; and serum glucose, serum triglycerides, and total cholesterol levels compared to their respective controls fed no betulic acid. Taken together, it is likely that betulic acid exerts its antiobesity and antihyperglycemic effects through actions as a TGR5 agonist.

A novel TGR5 agonist, the limonoid nomilin, exerts antiobesity and antihyperglycemic effects in mice fed a high-fat diet (Ono et al., 2011). Similar to oleanolic acid, nomilin does not activate FXR and is therefore a selective TGR5 ligand. Mice fed a nomilin-containing high-fat diet, the nomilin content of which was gradually increased from 0.02% to 0.2%, for 160 days showed decreased weight gain and increased *D2* and *PGC-1α* gene expression in BAT, suggesting that dietary nomilin enhances energy expenditure in BAT through TGR5 activation. Obese mice fed a 0.2% nomilin-containing high-fat diet for 77 days showed notably lower body weight gain; serum glucose, insulin, and nonesterified fatty acid levels; and epididymal WAT weight compared to those fed a nomilin-free high-fat diet. Importantly, this long-term nomilin treatment did not affect serum hepatic enzyme activities (alanine amonitransferase and aspartate aminotransferase), suggesting the absence of toxicity. An oral glucose tolerance test using these mice revealed that nomilin significantly enhances glucose tolerance in mice fed a high-fat diet. It is possible that dietary nomilin promotes GLP-1 secretion from enteroendocrine L cells in the small intestine and therefore improves insulin sensitivity and glucose metabolism in nomilin-fed mice. An increase in blood GLP-1 in response to nomilin treatment has not yet been ascertained simply because of the low sensitivity of the GLP-1 detection assay. However, given the dramatic alterations in glucose metabolism in nomilin-fed mice, it seems likely that dietary nomilin has a potential to improve glucose homeostasis in the gut prior to absorption and to enhance energy expenditure in BAT after absorption.

8. THE FUTURE

As discussed in this review, TGR5 exhibits a remarkable spectrum of actions, many of which are related to development of metabolic syndrome. A recent report has also demonstrated that TGR5 activation inhibits atherosclerosis by reducing macrophage inflammation and lipid loading (Pols et al., 2011). The key involvement of TGR5 in mediating glucose and energy homeostasis makes it an attractive target in our fight against obesity and the metabolic syndrome. The citrus limonoid, nomilin, that has been found to be a novel potent TGR5 agonist among a number of food constituents appears to be a safe and effective agent for promoting weight loss and insulin sensitivity and could be used as a nutraceutical or as a healthful fortifier in functional foods.

ACKNOWLEDGMENTS

Work in the author's laboratory was supported by research grants from the Ministry of Education, Culture, Sports, Science, and Technology of Japan.

REFERENCES

Abdelgaleil, S. A. M., Hashinaga, F., & Nakatani, M. (2005). Antifungal activity of limonoids from *Khaya ivorensis*. *Pest Management Science*, *61*, 186–190.

Abdelgaleil, S. A. M., Iwagawa, T., Doe, M., & Nakatani, M. (2004). Antifungal limonoids from the fruits of *Khaya senegalensis*. *Fitoterapia*, *75*, 566–572.

Ahmad, H., Li, J., Polson, M., Mackie, K., Quiroga, W., & Patil, B. S. (2006). Citrus limonoids and flavanoids: Enhancement of phase II detoxification enzymes and their potential in chemoprevention. In B. S. Patil, N. D. Turner, E. D. Miller & J. S. Brodbelt (Eds.), *Potential health benefits of citrus*, Vol. 936, (pp. 130–143). Washington, DC: American Chemical Society.

Azzu, V., Jastroch, M., Divakaruni, A. S., & Brand, M. D. (2010). The regulation and turnover of mitochondrial uncoupling proteins. *Biochimica et Biophysica Acta*, *1797*, 785–791.

Battinelli, L., Mengoni, F., Lichtner, M., Mazzanti, G., Saija, A., Mastroianni, C. M., et al. (2003). Effect of limonin and nomilin on HIV-1 replication on infected human mononuclear cells. *Planta Medica*, *69*, 910–913.

Bianco, A. C., & Kim, B. W. (2006). Deiodinases: Implications of the local control of thyroid hormone action. *Journal of Clinical Investigation, 116*, 2571–2579.

Chiang, J. Y. (2009). Bile acids: Regulation of synthesis. *Journal of Lipid Research, 50*, 1955–1966.

de Jesus, L. A., Carvalho, S. D., Ribeiro, M. O., Schneider, M., Kim, S. W., Harney, J. W., et al. (2001). The type 2 iodothyronine deiodinase is essential for adaptive thermogenesis in brown adipose tissue. *Journal of Clinical Investigation, 108*, 1379–1385.

de Melo, C. L., Queiroz, M. G., Filho, A. A. C., Rodrigues, A. M., de Sousa, D. F., Almeida, J. G., et al. (2009). Betulinic acid, a natural pentacyclic triterpenoid, prevents abdominal fat accumulation in mice fed a high-fat diet. *Journal of Agricultural and Food Chemistry, 57*, 8776–8781.

Duran-Sandoval, D., Cariou, B., Percevault, F., Hennuyer, N., Grefhorst, A., van Dijk, T. H., et al. (2005). The farnesoid X receptor modulates hepatic carbohydrate metabolism during the fasting-refeeding transition. *Journal of Biological Chemistry, 280*, 29971–29979.

Elzbieta, M., Kurowska, J., Spence, D., Jordan, J., Wetmore, S., Freeman, D. J., et al. (2000). HDL-cholesterol-raising effect of orange juice in subjects with hypercholesterolemia. *The American Journal of Clinical Nutrition, 72*, 1095–1100.

Endo, T., Kita, M., Shimada, T., Moriguchi, T., Hidaka, T., Matsumoto, R., et al. (2002). Modification of limonoid metabolism in suspension cell culture of citrus. *Plant Biotechnology, 19*, 397–403.

Genet, C., Strehle, A., Schmidt, C., Boudjelal, G., Lobstein, A., Schoonjans, K., et al. (2010). Structure-activity relationship study of betulinic acid, a novel and selective TGR5 agonist, and its synthetic derivatives: Potential impact in diabetes. *Journal of Medicinal Chemistry, 53*, 178–190.

Goldstein, J. L., DeBose-Boyd, R. A., & Brown, M. S. (2006). Protein sensors for membrane sterols. *Cell, 124*, 35–46.

Goodwin, B., Jones, S. A., Price, R. R., Watson, M. A., McKee, D. D., Moore, L. B., et al. (2000). A regulatory cascade of the nuclear receptors FXR, SHP-1, and LRH-1 represses bile acid biosynthesis. *Molecular Cell, 6*, 517–526.

Grober, J., Zaghini, I., Fujii, H., Jones, S. A., Kliewer, S. A., Willson, T. M., et al. (1999). Identification of a bile acid-responsive element in the human ileal bile acid-binding protein gene. Involvement of the farnesoid X receptor/9-cis-retinoic acid receptor heterodimer. *Journal of Biological Chemistry, 274*, 29749–54297.

Guthrie, N., Hasegawa, S., Manners, G. D., Carroll, K. K., Berhow, M. A., Hasegawa, S., et al. (2000). Inhibition of human breast cancer cells by citrus limonoids. In M. A. Berhow, S. Hasegawa & G. D. Manners (Eds.), *Citrus limonoids: Functional chemicals in agriculture and foods*, Vol. 758, (pp. 164–174). Washington, DC: American Chemical Society.

Hasegawa, S., Bennett, R. D., & Verdon, C. P. (1980). Limonoids in citrus seeds: Origin and relative concentration. *Journal of Agricultural and Food Chemistry, 28*, 922–925.

Hasegawa, S., & Miyake, M. (1996). Biochemistry and biological functions of citrus limonoids. *Food Reviews International, 12*, 413–435.

Herman, Z., Fong Chi, H., Ou, P., & Hasegawa, S. (1990). Limonoid glucosides in orange juice by HPLC. *Journal of Agricultural and Food Chemistry, 38*, 1860–1861.

Hirokane, H., Nakahara, M., Tachibana, S., Shimizu, M., & Sato, R. (2004). Bile acid reduces the secretion of very low density lipoprotein by repressing microsomal triglyceride transfer protein gene expression mediated by hepatocyte nuclear factor-4. *Journal of Biological Chemistry, 279*, 45685–45692.

Itoh, N. (2010). Hormone-like (endocrine) Fgfs: Evolutionary history and roles in development, metabolism, and disease. *Cell and Tissue Research, 342*, 1–11.

Katsuma, S., Hirasawa, A., & Tsujimoto, G. (2005). Bile acids promote glucagon-like peptide-1 secretion through TGR5 in a murine enteroendocrine cell line STC-1. *Biochemical and Biophysical Research Communications, 329*, 386–390.

Kawamata, Y., Fujii, R., Hosoya, M., Harada, M., Yoshida, H., Miwa, M., et al. (2003). A G protein-coupled receptor responsive to bile acids. *Journal of Biological Chemistry, 278*, 9435–9440.

Kelly, L. J., Vicario, P. P., Thampson, G. M., Candelore, M. R., Doebber, T. W., Ventre, J., et al. (1998). Peroxisome proliferators-activated receptors and mediate in vivo regulation of uncoupling protein (UCP-1, UCP-2, UCP-3) gene expression. *Endocrinology, 139*, 4920–4927.

Kliewer, S. A., & Mangelsdorf, D. J. (2010). Fibroblast growth factor 21: From pharmacology to physiology. *The American Journal of Clinical Nutrition, 91*, 254S–257S.

Kurowska, E. M., Borradaile, N. M., Spence, J. D., & Carroll, K. K. (2000). Hypocholesterolemic effects of dietary citrus juices in rabbits. *Nutrition Research, 20*, 121–129.

Li, T., Holmstrom, S. R., Kir, S., Umetani, M., Schmidt, D. R., Kliewer, S. A., et al. (2011). The G Protein-coupled bile acid receptor, TGR5, stimulates gallbladder filling. *Molecular Endocrinology, 25*, 1066–1071.

Ma, K., Saha, P. K., Chan, L., & Moore, D. D. (2006). Farnesoid X receptor is essential for normal glucose homeostasis. *Journal of Clinical Investigation, 116*, 1102–1109.

Makishima, M., Okamoto, A. Y., Repa, J. J., Tu, H., Learned, R. M., Luk, A., et al. (1999). Identification of a nuclear receptor for bile acids. *Science, 284*, 1362–1365.

Manners, G. D. (2007). Citrus limonoids: Analysis, bioactivity, and biomedical prospects. *Journal of Agricultural and Food Chemistry, 55*, 8285–8294.

Maruyama, T., Miyamoto, Y., Nakamura, T., Tamai, Y., et al. (2002). Identification of membrane-type receptor for bile acids (M-BAR). *Biochemical and Biophysical Research Communications, 298*, 714–719.

Maruyama, T., Tanaka, K., Suzuki, J., Miyoshi, H., Harada, N., Nakamura, T., et al. (2006). Targeted disruption of G protein-coupled bile acid receptor 1(Gpbar1/M-Bar) in mice. *Journal of Endocrinology, 191*, 197–205.

McIntosh, C. H. S., Widenmaier, S., & Kim, S.-J. (2010). Pleiotropic actions of the incretin hormones. *Vitamins and Hormones, 84*, 21–79.

Moller, D. E. (2012). Metabolic disease drug discovery-"hitting the target" is easier said than done. *Cell Metabolism, 15*, 19–24.

Nagao, K., Tomioka, M., & Ueda, K. (2011). Function and regulation of ABCA1—Membrane meso-domain organization and reorganization. *FEBS Journal, 278*, 3190–3203.

Nakahara, M., Furuya, N., Takagaki, K., Sugaya, T., Hirota, K., Fukamizu, A., et al. (2005). Ileal bile acid-binding protein, functionally associated with the farnesoid X receptor or the ileal bile acid transporter, regulates bile acid activity in the small intestine. *Journal of Biological Chemistry, 280*, 42283–42289.

Ono, E., Inoue, J., Hashidume, T., Shimizu, M., & Sato, R. (2011). Anti-obesity and anti-hyperglycemic effects of the dietary citrus limonoid nomilin in mice fed a high-fat diet. *Biochemical and Biophysical Research Communications, 410*, 677–681.

Ozaki, Y., Ayano, S., Inaba, N., Miyake, M., Berhow, M. A., & Hasegawa, S. (1995). Limonoid glucosides in fruit, juice and processing by-products of Satsuma Mandarin (*Citrus unshiu* Marcov.). *Journal of Food Science, 60*, 186–189.

Parker, N., Affourtit, C., Vidal-Puig, A., & Brand, M. D. (2008). Energization-dependent endogenous activation of proton conductance in skeletal muscle mitochondria. *Biochemical Journal, 412*, 131–139.

Parks, D. J., Blanchard, S. G., Bledsoe, R. K., Chandra, G., Consler, T. G., Kliewer, S. A., et al. (1999). Bile acids: Natural ligands for an orphan nuclear receptor. *Science, 284*, 1365–1368.

Pols, T. W. H., Nomura, M., Harach, T., Sasso, L. J., Oosterveer, M. H., Thomas, C., et al. (2011). TGR5 activation inhibits atherosclerosis by reducing macrophage inflammation and lipid loading. *Cell Metabolism, 14*, 747–757.

Poulose, S. M., Harris, E. D., & Patil, B. S. (2006). Antiproliferative effects of citrus limonoids against human neuroblastoma and colonic adenocarcinoma cells. *Nutrition and cancer, 56*, 103–112.

Roy, A., & Saraf, S. (2006). Limonoids: Overview of significant bioactive triterpenes distributed in plants kingdom. *Biological and Pharmaceutical Bulletin, 29*, 191–201.

Sato, H., Genet, C., Strehle, A., Thomas, C., Lobstein, A., Wagner, A., et al. (2007). Anti-hyperglycemic activity of a TGR5 agonist isolated from *Olea europaea. Biochemical and Biophysical Research Communications, 362*, 793–798.

Shin, D.-J., & Osborne, T. F. (2008). Peroxisome proliferator-activated receptor-coactivator-1 activation of CYP7A1 during food restriction and diabetes is still inhibited by small heterodimer partner. *Journal of Biological Chemistry, 283*, 15089–15096.

Sinal, C. J., Tohkin, M., Miyata, M., Ward, J. M., Lambert, G., & Gonzalez, F. J. (2000). Targeted disruption of the nuclear receptor FXR/BAR impairs bile acid and lipid homeostasis. *Cell, 102*, 731–744.

Thomas, C., Auwerx, J., & Schoonjans, K. (2008). Bile acids and the membrane bile acid receptor TGR5-connecting nutrition and metabolism. *Thyroid, 18*, 167–174.

Thomas, C., Gioiello, A., Noriega, L., Strehle, A., Oury, J., Rizzo, G., et al. (2009). TGR5-mediated bile acid sensing controls glucose homeostasis. *Cell Metabolism, 10*, 167–177.

Tian, Q., Miller, E. G., Ahmad, H., Tang, L., & Patil, B. S. (2001). Differential inhibition of human cancer cell proliferation by citrus limonoids. *Nutrition and Cancer, 40*, 180–184.

Tiwari, A., & Maiti, P. (2009). TGR5: An emerging bile acid G-protein coupled receptor target for the potential treatment of metabolic disorders. *Drug Discovery Today, 14*, 523–530.

Vega, R. B., Huss, J. M., & Kelly, D. P. (2000). The coactivator PGC-1 cooperates with peroxisome proliferator-activated receptor alpha in transcriptional control of nuclear genes encoding mitochondrial fatty acid oxidation enzymes. *Molecular and Cellular Biology, 20*, 1868–1876.

Vikram, A., Jayaprakasha, G. K., & Patil, B. S. (2007). Simultaneous determination of citrus limonoid aglycones and glucosides by high performance liquid chromatography. *Analytica Chimica Acta, 590*, 180–186.

Wang, H., Chen, J., Hollister, K., Sowers, L. C., & Forman, B. M. (1999). Endogenous bile acids are ligands for the nuclear receptor FXR/BAR. *Molecular Cell, 3*, 543–553.

Watanabe, M., Houten, S. M., Mataki, C., Christoffolete, M. A., Kim, B. W., Sato, H., et al. (2006). Bile acids induce energy expenditure by promoting intracellular thyroid hormone activation. *Nature, 439*, 484–489.

Watanabe, M., Houten, S. M., Wang, L., Moschetta, A., Mangelsdorf, D. J., Heyman, R. A., et al. (2004). Bile acids lower triglyceride levels via a pathway involving FXR, SHP, and SREBP-1c. *Journal of Clinical Investigation, 113*, 1408–1418.

INDEX

Note: Page numbers followed by "*f*" indicate figures, and "*t*" indicate tables.

D

E

H

P

Q

R

T

U

V

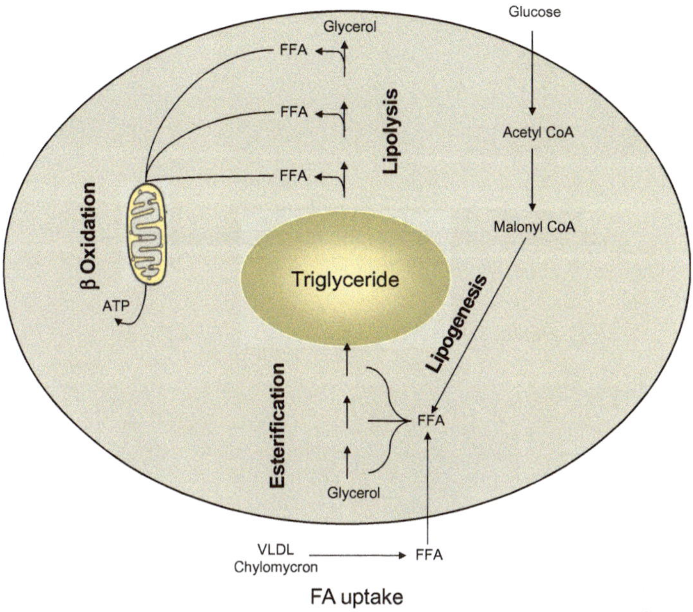

Laura L. Gathercole *et al.*, Figure 1.1 Pathways of lipid metabolism. VLDL, very low density lipoproteins; FFA, free fatty acids; ATP, adenosine triphosphate.

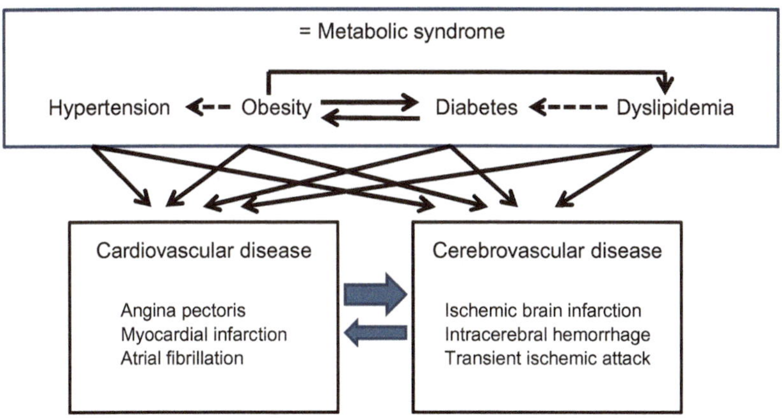

Yaroslav Winter *et al.*, Figure 2.2 Interrelations between obesity, other risk factors, and vascular diseases.

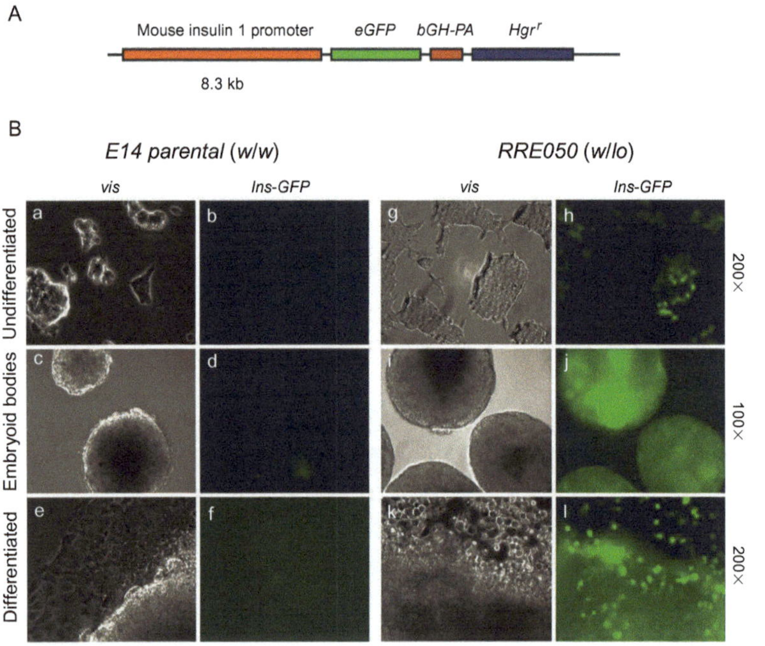

Fangnian Wang *et al.*, Figure 3.1 Reduced expression of Brd2 potentiates insulin transcription in embryonic stem cells. (A) Schematic of mouse insulin 1 promoter that drives expression of eGFP (*eGFP*). Promoter elements are also shown: bovine growth hormone poly A sequence (*bGH-PA*) and hygromycin resistance (*Hgr^r*). (B) Visual (*vis*) and fluorescence (*Ins-GFP*) micrographs of ES cells transfected with the construct in (A) and permitted to undergo differentiation. Undifferentiated cells were cultured on gelatin-coated tissue culture plastic in the presence of leukemia inhibitory factor (LIF) (undifferentiated), then LIF was withdrawn to permit the formation of embryoid bodies for 2 weeks (embryoid bodies), and culture was continued for an additional 2 weeks to permit additional differentiation (differentiated). Magnification is shown at the right-hand edge of the figure. The parental ES cells (*E14 parental*; *w/w*) were compared to Brd2 KO ES cells (*RRE050*; *w/lo*) as described in Wang et al. (2009), at the same stage of differentiation.

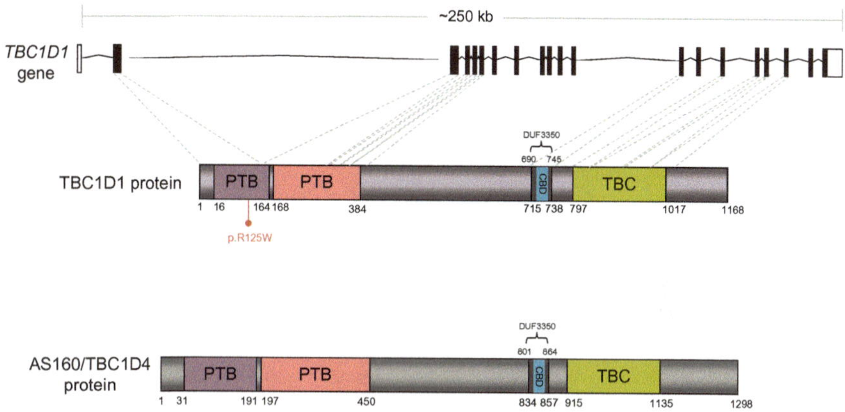

Luca Fontanesi and Francesca Bertolini, Figure 4.1 Structure of the human *TBC1D1* gene and functional domains in the TBC1D1 and TBC1D4 proteins. Domains have been identified using Simple Modular Architecture Research Tool—SMART (http://smart.embl-heidelberg.de/) analysis on Q86TI0 (TBC1D1) and O60343 (TBC1D4) or by BLAST and PFAM analysis (CBD). The protein domains are PTB: phosphotyrosine-binding domains; TBC: domains in Tre-2, BUB2p, and Cdc16p—Rab-GAP (GTPase-activating protein) domains; CBD: calmodulin-binding domains. DUF3350 regions (domain of unknown function) are annotated in the NP_055988 (TBC1D1) and NP_055647 (TBC1D4) protein sequences. To identify the putative phosphorylation sites, please see these entries in the NCBI database: http://www.ncbi.nlm.nih.gov/guide/proteins/. Vertical bars in the schematic representation of the gene structure representing exons are not on scale.

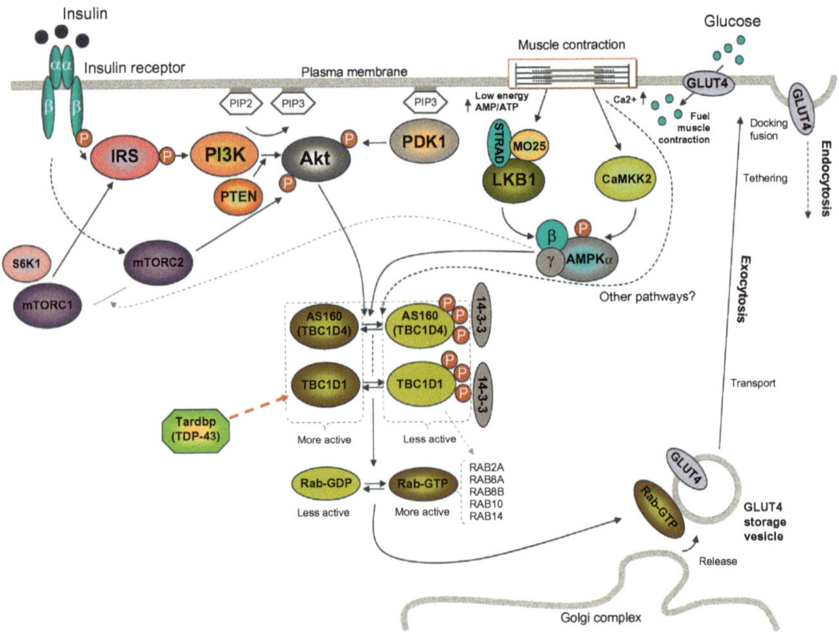

Luca Fontanesi and Francesca Bertolini, Figure 4.2 Schematic representation of the insulin and muscle contraction-dependent pathways regulating GLUT4 translocation in skeletal muscle showing the role of TBC1D1 and TBC1D4. Explanation of the different steps is given in the text. 14-3-3: 14-3-3 protein; AKT: protein kinase B (PKB); AMP: adenosine monophosphate; AMPK: 5′ adenosine monophosphate-activated protein kinase; AS160: Akt substrate of 160 kDa; ATP: adenosine-5′-triphosphate; CAMKK2: calcium/calmodulin-dependent protein kinase kinase 2; GLUT4: glucose transporter type 4; GMP: guanosine monophosphate; GTP: guanosine-5′-triphosphate; IRS: insulin receptor substrate protein; LKB1: liver kinase B1; MO25: mouse protein-25; mTORc1: mammalian target of rapamycin, in multiprotein complex 1; mTORc2: mammalian target of rapamycin, in multiprotein complex 2; PDK1: pyruvate dehydrogenase lipoamide kinase isozyme 1; PI3K: phosphatidylinositol 3-kinase; PIP2: phosphatidylinositol (3,4,5)-biphosphate; PIP3: phosphatidylinositol (3,4,5)-triphosphate; PTEN: phosphatase and tensin homolog; Rab: rab protein (member of the Ras superfamily of monomeric G proteins); STRAD: STe20-related adaptor; Tardbp: Tat activating regulatory DNA-binding protein; TBC1D1: TBC1 (tre-2/USP6, BUB2, cdc16) domain family, member 1; TBC1D4: TBC1 (tre-2/USP6, BUB2, cdc16) domain family, member 4; TDP-43: TAR DNA-binding protein 43.

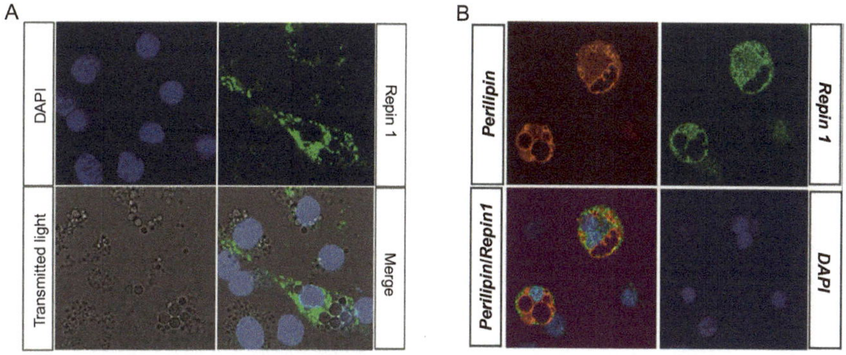

John T. Heiker and Nora Klöting, Figure 5.1 Cellular localization of Repin1. Antibody staining of mature 3T3-L1 cells against Repin1 (green). Nuclei were stained with DAPI (blue). (A) Repin1 is localized in the cytoplasm with contact to lipid droplets (see transmission picture). (B) Repin1 is in part colocalized with perilipin in 3T3-L1 cells.

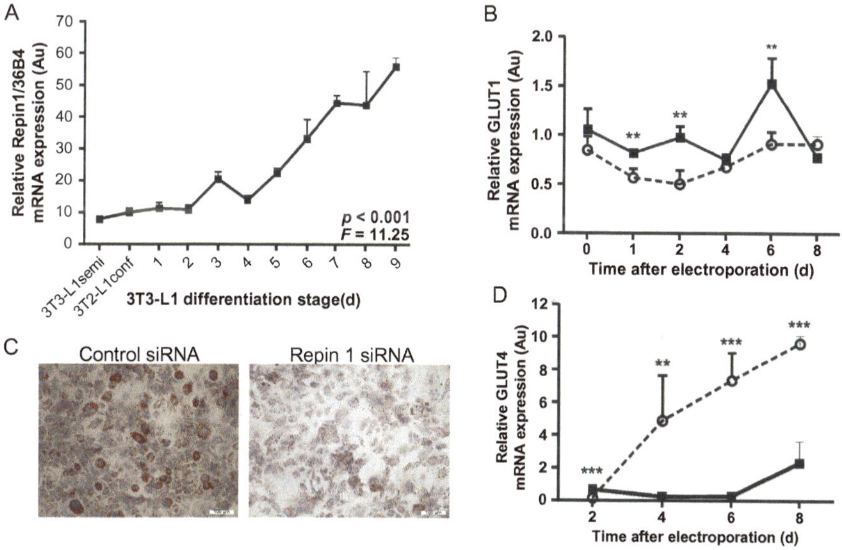

John T. Heiker and Nora Klöting, Figure 5.2 Repin1 during adipogenesis. (A) Expression of *Repin1* in 3T3-L1 adipocytes during differentiation. At the indicated time points (semiconfluent preadipocytes, confluent preadipocytes, day 1–9 after induction), total RNA was extracted and subjected to quantitative real-time PCR. *Repin1* is significantly upregulated during adipogenesis ($p < 0.001$) (magnification 630×). (B) Oil-Red-O staining of mature 3T3-L1 cells of controls and Repin1 siRNA-treated cells. (C) *GLUT1* and (D) *GLUT4* expression in 3T3-L1 cells during differentiation in controls and *Repin1* siRNA-treated cells.

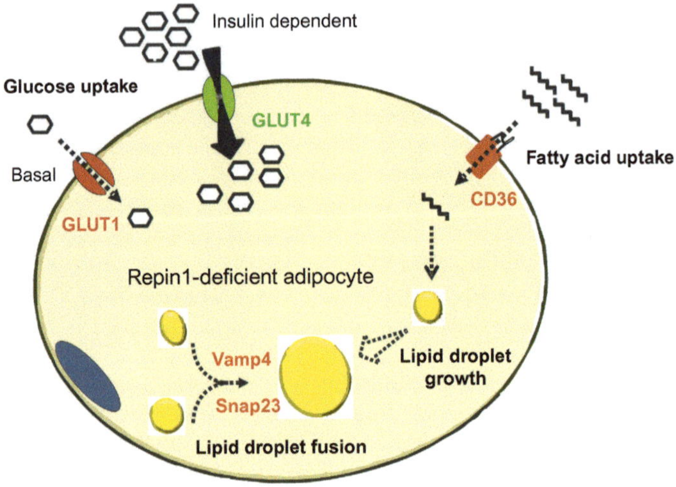

John T. Heiker and Nora Klöting, Figure 5.4 Schematic overview of Repin1 action in adipose tissue. Repin1 deficiency leads to downregulation (red) and upregulation (green) of proteins resulting in altered adipose tissue function.

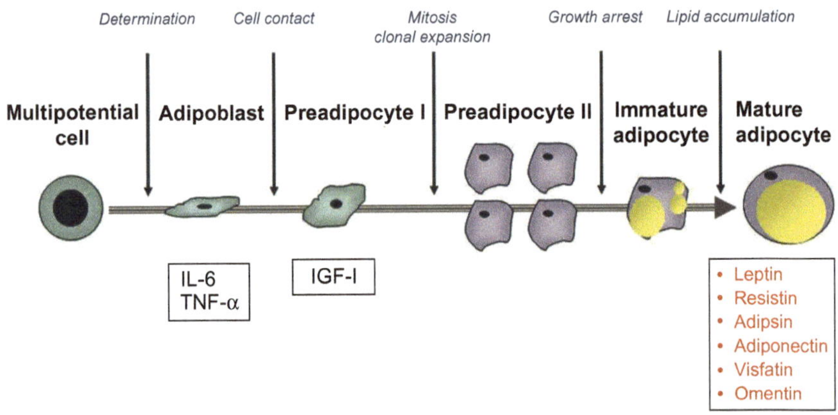

Gabriel Ángel Martos-Moreno *et al.*, Figure 6.1 Adipogenic differentiation of multipotential mesenchymal cells and cell markers (in boxes). Of note, adipokine expression is restricted to the mature adipocyte, the end stage of this process. *Adapted from Feve (2005).*

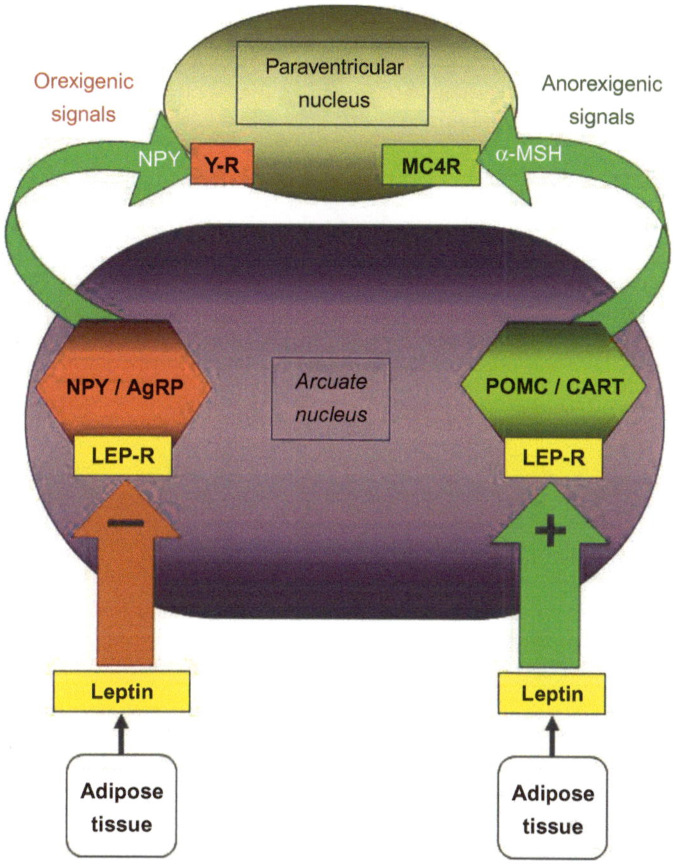

Gabriel Ángel Martos-Moreno *et al.*, Figure 6.2 Schematic representation of the excitatory (+) and inhibitory (−) effects of leptin in NPY/AgRP and POMC/CART producing neurons in the hypothalamus. These neuronal groups reciprocally inhibit each other, through direct signaling, as well self-limiting their own activity to prevent over-activation. AgRP, agouti-related protein; CART, cocaine- and amphetamine-regulated transcript; LEP-R, leptin receptor; MC4R, melanocortin receptor number 4; NPY, neuropeptide Y; POMC, proopiomelanocortin; Y-R, Y receptor; α-MSH, alpha melanocyte-stimulating hormone.

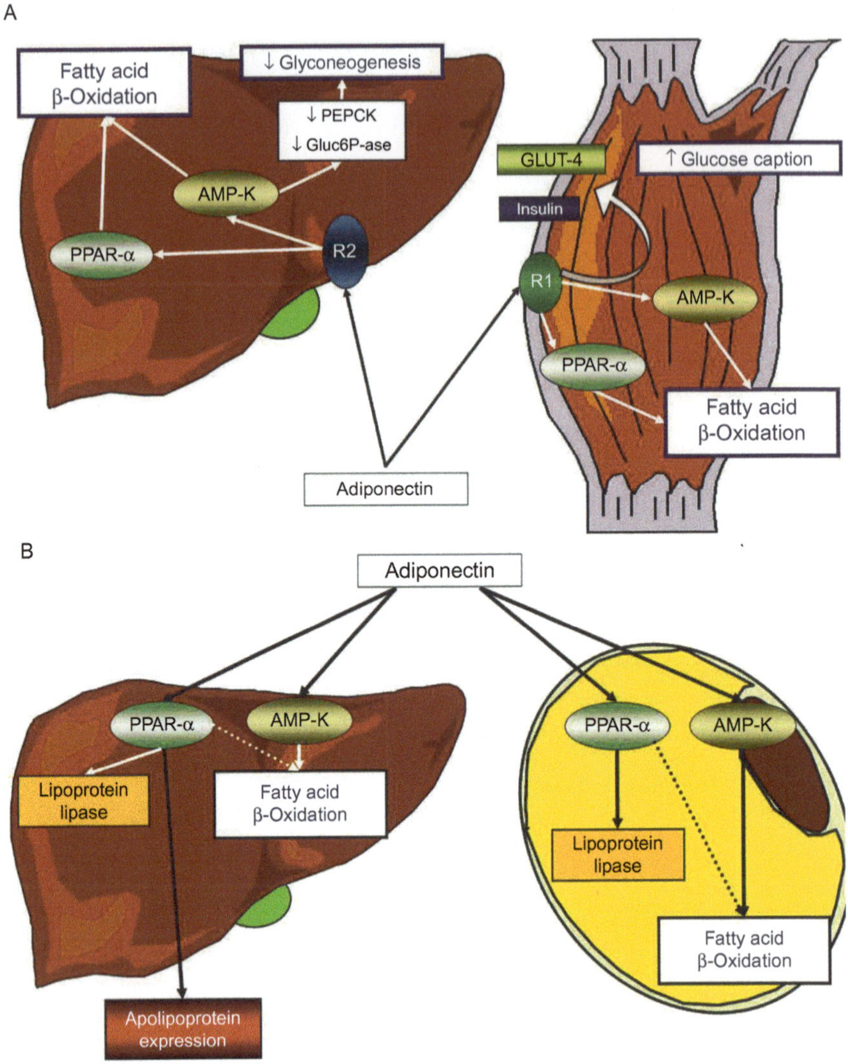

Gabriel Ángel Martos-Moreno *et al.*, Figure 6.3 Schematic representation of the main metabolic effects of adiponectin on glucose metabolism in the liver and muscle (A) and on lipid metabolism in the liver and adipocytes (B). The stimulation of muscle adipoR1 determines the activation of AMP-dependent kinase (AMP-K) and induces the expression of the peroxisome proliferator activated receptor alpha (PPAR-α) and, as a consequence, the expression of enzymes involved in fatty acid catabolism and glucose uptake. Additionally, adiponectin enhances the expression and externalization of glucose transporter 4 (Glut-4) in myocytes and directly modulates insulin receptor activity. The activation of adipoR2 and liver AMP-K inhibits glyconeogenesis by modulating glucose-6-phosphatase (Gluc6P-ase) and phosphoenolpyruvatecarboxykinase (PEPCK) activity, enhancing lipoprotein lipase activity, and modulating apolipoprotein gene expression. Finally, adiponectin favors insulin-induced glucose uptake, fatty acid oxidation, and lipoprotein lipase activity in WAT adipocytes (Chinetti-Gbaguidi et al., 2005; Kadowaki & Yamauchi, 2005; Kershaw & Flier, 2004; Meier & Gressner, 2004; Yamauchi et al., 2003; Yu & Ginsberg, 2005).

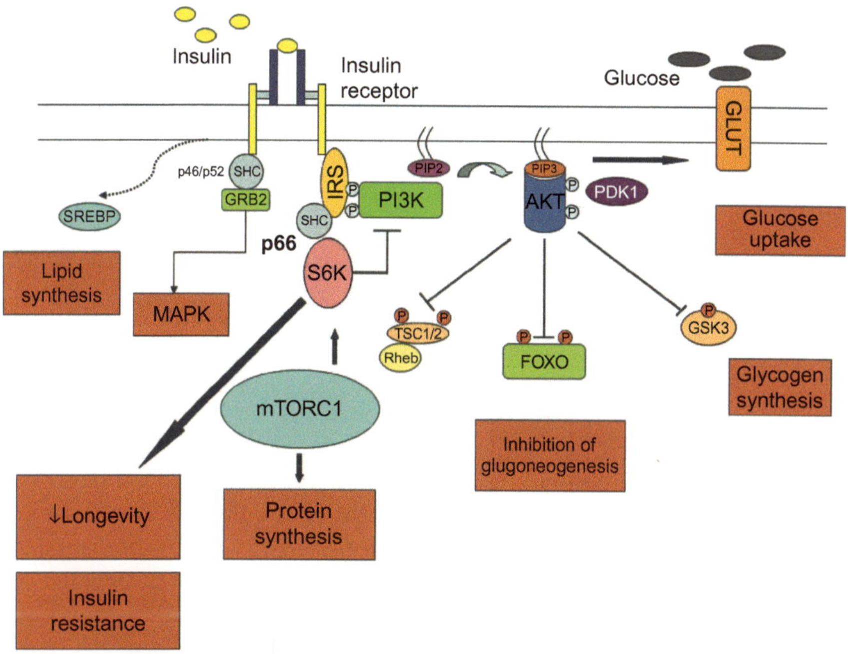

Sofia Chiatamone Ranieri *et al.*, Figure 9.1 Insulin signaling pathways. Activation of the insulin receptor induces transcription of SREBP (Sterol regulatory element-binding protein) and phosphorylation of members of the IRS and SHC families resulting in the activation of a variety of signaling pathways, including MAPK, PI3K, and consequently Akt. These pathways act in a coordinated manner to regulate glucose, lipid, and protein metabolism. Insulin inhibits hepatic gluconeogenesis, promotes glucose uptake and storage into cells by inducing translocation of the glucose transporter and increasing glycogen synthesis, and in addition, stimulates amino acid and fatty acid uptake and protein and lipid synthesis. Chronic abnormal activation of these cascades alters insulin sensibility and accelerates physiological aging of the organism.

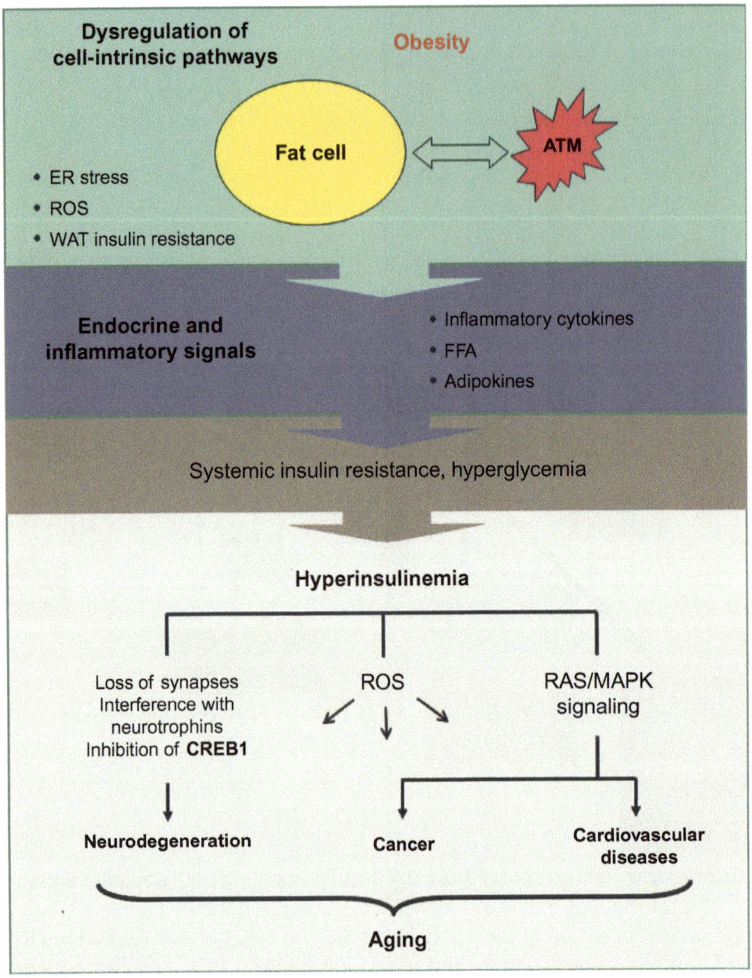

Sofia Chiatamone Ranieri *et al.*, Figure 9.2 Mechanisms linking obesity-associated insulin resistance and aging. Increased adipose mass predisposes to systemic insulin resistance through interplay of cell-extrinsic and cell-intrinsic mechanisms. Adipocytes and adipose tissue macrophages (ATMs) produce adipokines and inflammatory cytokines interfering with insulin signaling in all tissues (in particular, liver, muscle, and adipose tissue); excess of nutrients and fatty acids (FAs) has a negative impact on insulin signaling and increases reactive oxygen species (ROS) generation by mitochondrial β oxidation, leading to mitochondrial dysfunction and induction of cellular endoplasmic reticulum (ER) stress. Hyperinsulinemia plays a central role in obesity-associated pathology, by acting through canonical and noncanonical signaling branches; the latter include hyperstimulation of the ERK cascade in endothelial cells (atherosclerosis) and epithelial cells (cancer), as well as interference with neurotrophin signaling in the brain (neurodegeneration, Alzheimer's Disease).

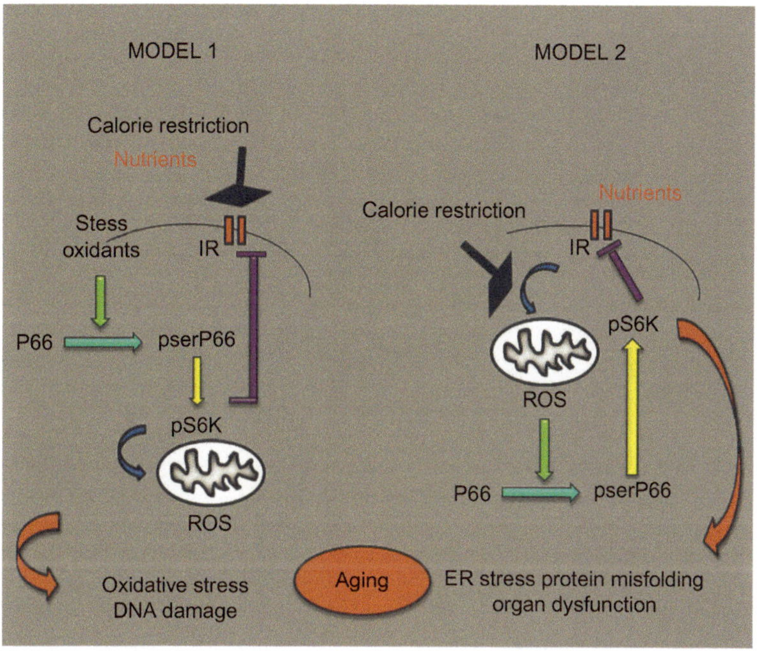

Sofia Chiatamone Ranieri *et al.*, Figure 9.3 P66 Shc:TOR and ROS. Hypothetical redox circuits linking p66Shc and mTOR/S6K to mitochondrial nutrient signaling, calorie restriction, and aging: the two different models are not necessarily alternative. Model 1: p66Shc may rise intracellular ROS by activating mTOR/S6K and promoting the stimulatory effect of this cascade on mitochondrial metabolism (Scheme 1: p66 → S6K → ROS). Model 2: p66Shc as a redox sensor activates S6K in response to nutrient catabolism favoring feedback inibitory effects on insulin responsiveness (Scheme 2: ROS → p66 → S6K).

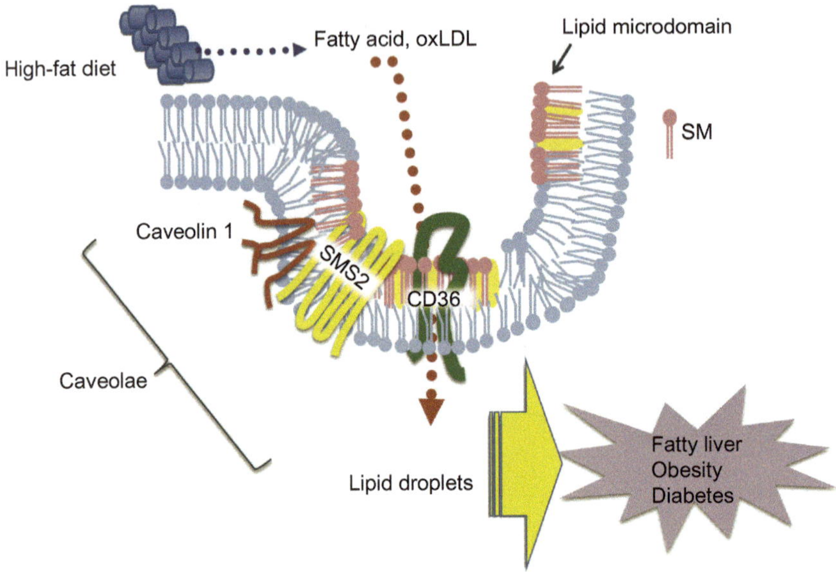

Susumu Mitsutake and Yasuyuki Igarashi, Figure 12.4 SMS2 could be a novel regulator of lipid microdomain structure and function. After a high-fat diet, increased fatty acid or oxidized low density lipoprotein (oxLDL) would be incorporated into the cells via CD36/FAT, resulting in triglyceride (TG) accumulation in lipid droplets. SMS2 is localized in lipid microdomains, where it associates with CD36/FAT and caveolin-1. Thus, SMS2 could be a novel regulator of lipid microdomain structure and function; such regulation would be crucial for obesity and diabetes. *Part of this figure was originally published in Mitsutake et al. (2011) © 2011. The American Society for Biochemistry and Molecular Biology.*

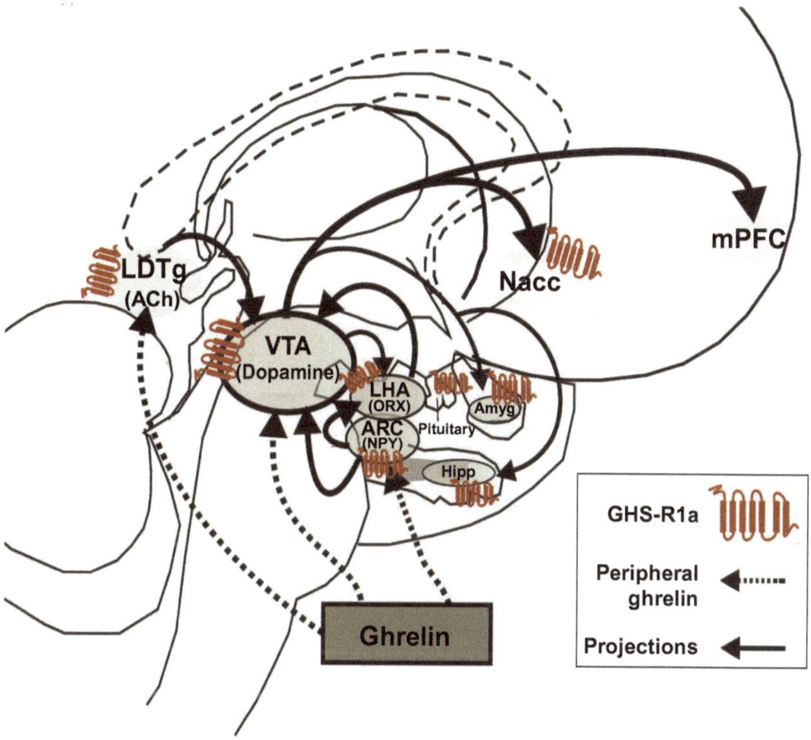

Harriët Schellekens *et al.*, Figure 13.3 Ghrelin signaling in the mesolimbic reward circuitry in the human brain. Schematic representation of ghrelin-mediated regulation of homeostatic food intake and food reward in the human brain. Key pathways and regions implicated in the regulation of food intake and reward-seeking behavior are depicted. Peripheral ghrelin increases food intake via homeostatic circuits in the ARC including direct activation of NPY-expressing neurons. Moreover, ghrelin directly activates VTA dopaminergic neurons which project to the NAcc. The involvement of ghrelin in signaling in projections from VTA to other brain regions, including the LHA, amygdala, hippocampus, and mPFC, has also been suggested. A cholinergic-dopaminergic reward link has been suggested via GHS-R1a signaling in the LDTg. Expression of the GHS-R1a is indicated. Amyg, amygdala; ARC, arcuate nucleus; LDTg, laterodorsal tegmental area; LHA, lateral hypothalamic area; Hipp, hippocampus; mPFC, medial prefrontal cortex; NAcc, nucleus accumbens; NPY, neuropeptide Y; ORX, orexin; VTA, ventral tegmental area.

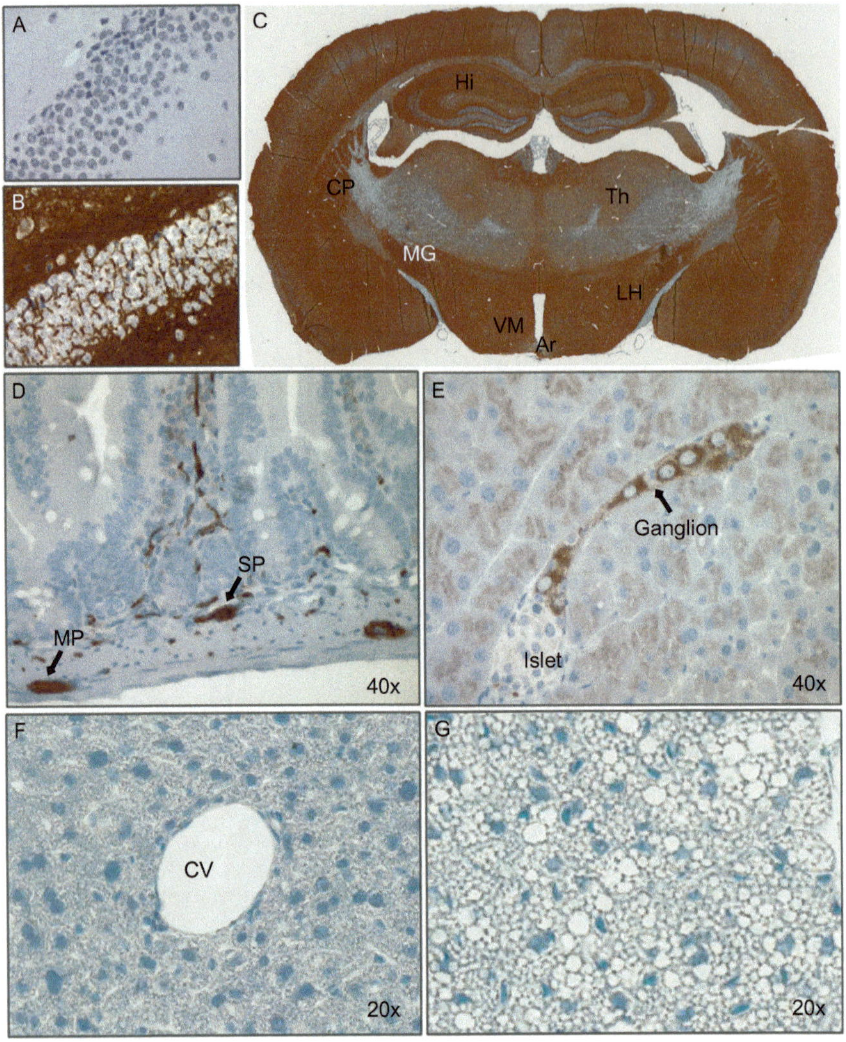

Guoxun Chen and Zhen Pang, Figure 14.2 Detection of CB1 in the brain and peripheral tissues by immunohistochemistry. Fixed and paraffin-embedded tissues are sectioned and stained using an anti-CB1 polyclonal antibody, as described previously (Pang et al., 2011). Brown color indicates CB1 immunoreactivity. Specificity of the antibody is demonstrated by the complete absence of staining in CB1 knockout tissues, for example, the hippocampal dentate gyrus (A), compared to the wild-type mouse (B). CB1 is detected mainly in neuronal processes in various brain regions (C), as well as in neuronal cells residing in the small intestines (D) and the pancreas (E). CB1 immunoreactivity is undetectable in the liver (F), WAT (not shown) and BAT (G). Hi, hippocampus; CP, caudate putamen; MG, medial globus pallidus; Th, thalamus; Ar, arcuate nucleus; LH, lateral hypothalamus area; VM, ventromedial hypothalamic nucleus; MP, myenteric (Auerbach's) plexus; SP, submucosal (Meissner's) plexus; CV, central vein.

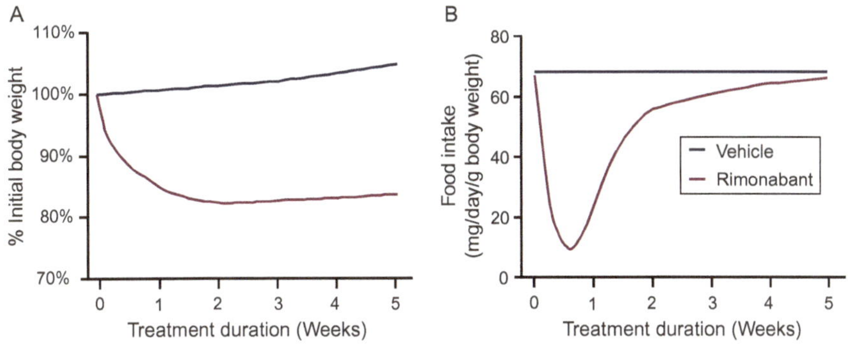

Guoxun Chen and Zhen Pang, Figure 14.3 A schematic view of the changes of body weight (A) and food intake (B) upon the administration of rimonabant. Treatment with rimonabant results in transient reduction of food intake, but sustained decrease in body weight in mice fed high-fat diet.

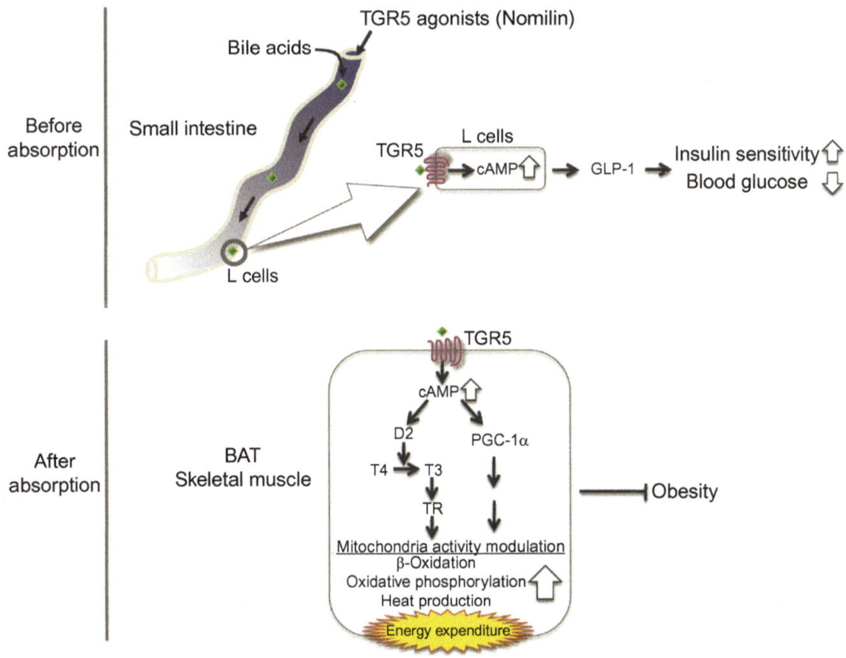

Ryuichiro Sato, Figure 18.1 Proposed pathways by which TGR5 agonists exert their actions on the small intestine, BAT, and skeletal muscle. In response to TGR5 activation by its agonists, GLP-1 is released from L cells in the intestinal mucosa. GLP-1 can stimulate insulin secretion, lower blood glucose levels, and improve insulin sensitivity. After absorption, TGR5 agonists activate TGR5 in BAT and skeletal muscles, thereby raising the intracellular cAMP level. This signal modulates mitochondrial activity with accompanying increased fatty acid β-oxidation, oxidative phosphorylation, and heat production. Rising energy expenditure attenuates diet-induced obesity.

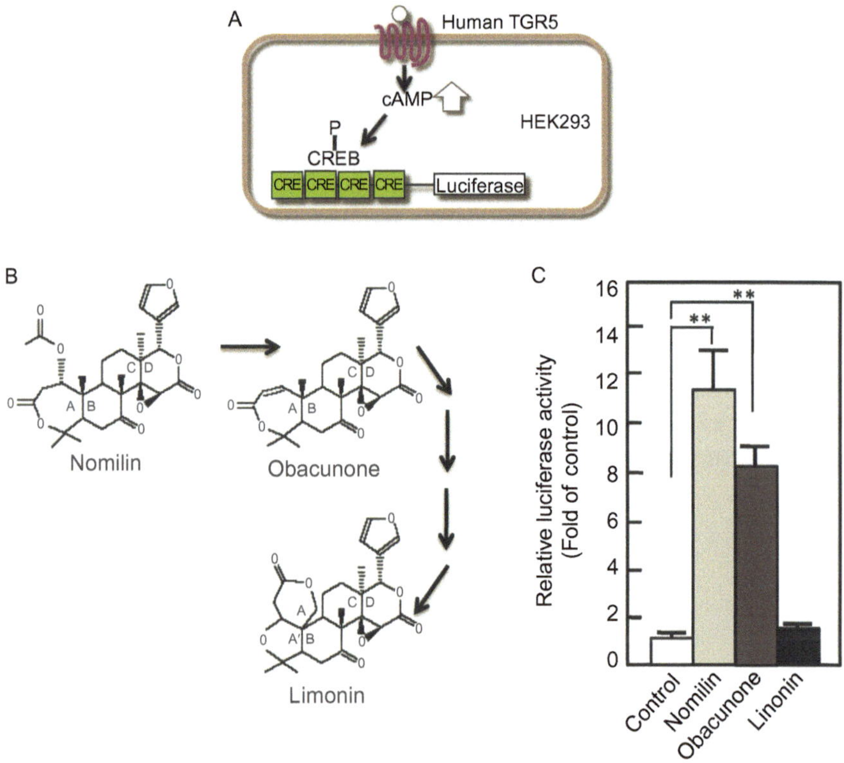

Ryuichiro Sato, Figure 18.2 A citrus limonoid, nomilin, as a TGR5 agonist. (A) In an effort to isolate novel TGR5 agonists, HEK293 cells expressing human TGR5 and the CRE-driven reporter were cultured in the presence of various food constituents (Ono et al., 2011). (B) The structures of nomilin, obacunone, and limonin with the number of rings in each illustrated. Nomilin is metabolized to obacunone and finally to limonin through several steps in citrus fruits. (C) The reporter assay reveals that nomilin and obacunone but not limonin are potent TGR5 agonists.

FSC
www.fsc.org
MIX
Papier | Fördert gute Waldnutzung
FSC® C083411

Kolibri 360 GmbH
Ferdinand-Jühlke-Straße 7
99095 Erfurt, Deutschland
produktsicherheit@kolibri360.de